Atmosphere, weather and climate

SEVENTH EDITION

ROGER G. BARRY AND RICHARD J. CHORLEY

ROUTLEDGE

London and New York

First published in 1968 by
Methuen & Co. Ltd
Second edition 1971
Third edition 1976
Fourth edition 1982
Fifth edition 1987
Reprinted by Routledge 1989, 1990
Sixth edition 1992
Reprinted 1995
Seventh edition 1998 by Routledge
11 New Fetter Lane, London EC4P 4EE

Simultaneously published in the USA and Canada
by Routledge
29 West 35th Street, New York, NY 10001

© 1968, 1971, 1976, 1982, 1987, 1992
and 1998
Roger G. Barry and Richard J. Chorley

Typeset in Sabon by Florencetype
Printed and bound in Great Britain by
Clays Ltd, St Ives Plc

British Library Cataloguing in Publication Data
A catalogue record for this book is available from
the British Library

*Library of Congress Cataloguing in Publication
Data*
Barry, Roger Graham.
    Atmosphere, weather, and climate / Roger G.
Barry and Richard J. Chorley. --- 7th ed.
       p. cm.
    Includes bibliographical references and index.
    1. Meteorology. 2. Atmospheric physics.
    3. Climatology.
    I. Chorley, Richard J. II. Title.
QC861.2.B36   1998
551.5--dc21   97-22866

ISBN 0–415–16019–7 (hbk)
ISBN 0–415–16020–0 (pbk)

Contents

Preface to the seventh edition

When the first edition of this book appeared in 1968, it was greeted as being 'remarkably up to date' (*Meteorological Magazine*). Since that time, several new editions have extended and sharpened its description and analysis of atmospheric processes and global climates. Indeed, succeeding prefaces provide a virtual commentary on recent advances in meteorology and climatology of relevance to students in these fields and to scholars in related disciplines.

In 1971, major additions were made regarding the energy budget of the earth, the spatial pattern of the heat budget components, atmospheric stability, orographic precipitation, oceanic circulation and associated climatic effects, vorticity, mesoscale systems in middle latitudes, rainfall variability, and aspects of the climate of the sub-Arctic, the Mediterranean and Eastern Asia.

The third edition of 1976 involved the substantial recasting of Chapters 2, 4 and 7 and the addition of a new appendix on synoptic weather maps. New material was introduced on atmospheric composition and its variations with time, the radiation budget, adiabatic temperature changes, the effects of topographic barriers on winds, land and sea breezes, the southern hemisphere circulation and air masses, depression structure and the spatial distribution of precipitation, long-range forecasting, the climate of the Mediterranean, the intertropical confluence, tropical disturbances and subsynoptic systems, and urban climates.

In 1982, Chapters 7 and 8 were substantially rewritten and the units were standardized throughout. Major additions and revisions involved the material on solar radiation, thunderstorm mechanisms, drought, mesoscale rainfall systems, tornado structure, disturbances within subtropical high-pressure belts, the energy balance of vegetated surfaces, urban climatology, atmospheric pollution, and the nature and causes of climatic change. This fourth edition contained over 100 new or revised figures and many new plates, compared with the first edition.

The fifth edition of 1987 included new sections on modelling the atmospheric circulation, forecasting, the African monsoon and ENSO events, together with a new appendix on data sources. Additional material was provided on greenhouse gases and volcanic dust, sunspot activity, albedo, satellite measurements of the planetary energy budget, atmospheric stability, condensation nuclei and cloud processes, mechanisms of orographic precipitation, convergence and divergence, subtropical high-pressure behaviour, hemispheric circulation patterns, southern hemisphere air masses, frontal structures, mesoscale convective systems, super-cell thunderstorms, European blocking conditions, regional winds of Iberia, rainstorms of the south-west USA, the Intertropical Convergence Zone, tropical weather systems, cloud clusters, the Asian monsoon circulation, climate of the Sahara, climatic change in the Sahel Zone and Europe, the climatic effect of global vegetation changes, and the use of mathematical models for climatic forecasting. Some seventy new or revised figures and plates were introduced.

The sixth edition of 1992 responded to the recognition of the reality and possible effects of human activities on the environment, as well as to other current scientific advances, by the most extensive revision to date. New sections were added on acid precipitation, the Walker circulation, modelling the atmosphere–earth–ocean system, the climates of Amazonia and southern Africa, tropical urban climates, recent climatic changes, forcing and feedback mechanisms of the atmosphere–earth–ocean system, anthropogenic changes in atmospheric composition and in the greenhouse effect, as well as on model predictions of climate and associated environmental changes which might be expected during the twenty-first century. Additional material was provided on the greenhouse effect, aerosols, ozone, the carbon cycle, global cloud cover, the thermal role of the oceans, the exosphere and the magnetosphere, orographic precipitation, aerodynamic roughness and airflow over topographic obstructions, medium- and long-range forecasting, wind activity over Western Europe, hurricanes, Asian monsoon mechanisms, monsoon depressions, the timing intensity and phases of the Asian summer monsoon, the West African monsoon, tropical forest climatology, urban pollution, urban climatology, and Pleistocene climatic changes. SI units of energy flux density were used throughout and some ninety new or redrawn figures and tables were included.

Since the previous edition appeared in 1992, interest in the world's climate at a wide range of scales, accompanied by the increasing tempo of scientific research, has necessitated that the new (1998) edition should undergo equally rigorous modification and updating. While the existing well-tried format has been broadly retained, the original three basic chapters have each been subdivided into two so that this seventh edition is comprised of eleven main chapters, replacing the pre-existing eight. Chapters 1 and 2 contain additional material on methane production, volcanic aerosols and sunspots. Chapters 3 and 4 contain a new section on the global hydrological cycle, together with further treatment of evaporation, forms of precipitation, rainfall intensity, duration and frequency, desertification, cloud forcing, and severe thunderstorms. Chapter 5 gives added emphasis to the planetary boundary layer and Chapter 6 devotes considerable new space to modern views of atmosphere–ocean interaction, the climatic role of the oceans, oceanic structure and circulation, and to the significance of the deep ocean circulation. Chapter 7 updates the treatment of mid-latitude synoptic systems, and Chapter 8 has been radically extended by additional regional treatments of the climates of Australia, New Zealand, the southern

westerlies, the sub-arctic and polar regions, and the former USSR. Additional material is included on Lamb's airflow types, European thunderstorms, and the location of heavy rainstorms in North America. Chapter 9 introduces a new section on teleconnections, as well as an enlarged treatment of ENSO events. The structure and movement of the ITCZ, hurricane forecasting, depression and typhoon paths and frontal locations over Eastern Asia, the north Australian monsoon, and forecasting tropical weather are also highlighted. Chapter 10 sees emphasis placed on the boundary layer dominance of local climates. New material is included on the energy balances of oceanic, water-covered and frozen surfaces and on the climates of boreal forests; urban climates in tropical environments are further emphasized. Chapter 11 has been completely rewritten and extended to stress the structure and operation of the atmosphere–earth–ocean system and the causes of its climate changes. A discussion of the various modelling strategies adopted for the prediction of climate change is undertaken, in particular relating to the IPCC 1990–95 models. A consideration of other environmental impacts of climate change is also included. Appendix 1 has been modified to reflect new applications of climate classifications and recent studies of wind chill and heat stress. A selection of WWW addresses for climatic data are included in Appendix 4.

Throughout the book, more than 100 new or redrawn figures, revised tables and new plates are presented.

Wherever possible, the criticisms and suggestions of colleagues and reviewers have been taken into account in preparing this latest edition.

R. G. BARRY
Cooperative Institute for Research in Environmental Sciences and the Department of Geography, University of Colorado, Boulder

R. J. CHORLEY
Department of Geography and Sidney Sussex College, University of Cambridge

Acknowledgements

The authors are very much indebted to Mr A. J. Dunn for his considerable contribution to the first edition; also to Dr F. Kenneth Hare of Birkbeck College, London, now at the University of Toronto, Ontario, for his thorough and authoritative criticism of the preliminary text and his valuable suggestions for its improvement; also to Mr Alan Johnson of Barton Peveril School, Eastleigh, Hampshire, for helpful comments on Chapters 1–3; and to Dr C. Desmond Walshaw, formerly of the Cavendish Laboratory, Cambridge, and Mr R. H. A. Stewart of the Nautical College, Pangbourne, for offering valuable criticisms and suggestions at an early stage in the preparation of the original manuscript. Gratitude is also expressed to the following persons for their helpful comments with respect to the fourth edition: Dr Brian Knapp of Leighton Park School, Reading; Dr L. F. Musk of the University of Manchester; Dr A. H. Perry of University College, Swansea; Dr R. Reynolds of the University of Reading; and Dr P. Smithson of the University of Sheffield. Dr C. Ramage, of the University of Hawaii, made numerous helpful suggestions on the revision of Chapter 6 for the fifth edition. Dr Z. Toth and Dr D. Gilman of the National Meteorological Center, Washington, DC, kindly helped in the updating of Chapter 4, I and Dr M. Tolbert of the University of Colorado assisted with the environmental chemistry in the seventh edition. Thanks are also due to Dr N. Cox of Durham University for his many textual suggestions which contributed significantly to the improvement of the seventh edition. The authors accept complete responsibility for any remaining textual errors.

The figures were prepared by the cartographic and photographic staffs in the Geography Departments at Cambridge University (Mr I. Agnew, Mr R. Blackmore, Mr R. Coe, Mr I. Gulley, Mrs S. Gutteridge, Miss L. Judge, Miss R. King, Mr C. Lewis, Mrs P. Lucas, Miss G. Seymour, Mr A. Shelley and Miss J. Wyatt and, especially, Mr M. Young); at Southampton University (Mr A. C. Clarke, Miss B. Manning and Mr R. Smith); and at the University of Colorado, Boulder (Mr T. Wiselogel). Every edition of this book, including the present seventh, has been graced by the illustrative imagination and cartographic expertise of Mr M. Young of the Department of Geography, Cambridge University, to whom the authors owe a considerable debt of gratitude.

Thanks are also due to Carol Pedigo, NSIDC, for word processing support for the seventh edition.

Our grateful thanks go to out families for their constant encouragement and forbearance.

The authors would like to thank the following learned societies, editors, publishers, organizations and individuals for permission to reproduce figures, tables and plates.

Learned societies

American Association for the Advancement of Science for Figure 6.35 from *Science*.

American Geographical Society for Figure 1.16 from the *Geographical Review*.

American Geophysical Union for Figures 2.13, 7.14, 7.15 and 11.3 from the *Review of Geophysics and Space Physics*; for Figures 1.3, 1.9, 2.13, 2.23 and 2.27 from the *Journal of Geophysical Research*; for Figure 10.6 from the *Transactions* and for Figure 8.43 from *Arctic Meteorology and Climatology* by D. H. Bromwich and C. R. Stearns (eds).

American Meteorological Society for Figures 1.2, 2.23A, 2.27C, 4.11 and 8.38 from the *Bulletin*; for Figures 6.16, 7.8, 8.42, 10.1 and 11.11 from the *Journal of Applied Meteorology*; for Figures 7.2 and 7.4 from the *Meteorological Monographs*; for Figures 4.28, 6.39 and 9.56 from the *Journal of*

Atmospheric Sciences; and for Figures 3.12, 5.11, 6.8, 6.10, 6.26, 6.27, 6.37, 7.6, 7.11, 7.26, 9.2, 9.11, 9.12, 9.33, 9.50 and 9.59 from the *Monthly Weather Review*; for Figure 6.30 from the *Journal of Physical Oceanography* and for Figures 7.9, 7.16 and 7.18 from *Extratropical Cyclones* by C. W. Newton and E. D. Holspainen (eds).

American Planning Association for Figure 10.39 from the *Journal*.

Association of American Geographers for Figure 3.21 from the *Annals*; and for Figure 7.32 from *Resource Paper 11*.

Chinese Meteorological Society for Figures 9.24 and 9.34.

European Space Agency, Darmstadt, for Plates 1 and 28.

Geographical Association for Figure 8.4 from *Geography*.

Geographical Society of China for Figure 9.37.

Indian National Science Academy, New Delhi, for Figure 9.28.

Institute of British Geographers for Figures 3.15, 3.16, 10.26 and 10.40 from the *Transactions*; and for Figures 3.22 and 11.12 from the *Atlas of Drought in Britain 1975–76* by J. C. Doornkamp and K. J. Gregory (eds).

Institution of Civil Engineers for Figure 3.16A from the *Proceedings*.

International Glaciological Society for Figure 10.9.

Marine Technology Society, Washington, DC, for Figure 9.52 from the *Journal*.

National Geographic Society for Plate 20 from the *National Geographic Picture Atlas of Our Fifty States*.

Royal Geographical Society for Figure 6.20 from the *Journal*.

Royal Meteorological Society for Figures 7.13, 8.7, 8.8, 9.3 and 10.19 from the *Quarterly Journal*; for Figure 11.6 from *World Climate 8000–0 BC*; for Figures 4.17, 8.9, 9.43 and 11.13 from the *Journal of Climatology*; and for Figures 2.10, 3.8, 3.9, 4.9, 4.14, 4.15, 4.23, 4.24, 8.5, 8.12, 8.20, 9.57, 10.27 and 11.7, and for Plates 11, 17 and 18 from *Weather*.

Royal Society of Canada for Figure 2.15 from *Special Publication 5*.

Royal Society of London for Figure 4.21 from the *Proceedings, Section A*.

US National Academy of Sciences for Figure 11.5 and for Figure 11.4 from *Natural Climate Variability on Decade-to-Century Time Scales* by P. Grootes.

Editors

Advances in Space Research for Figures 2.8 and 4.13.

American Scientist for Figure 9.51.

Climate Monitor for Figure 11.14.

Climatic Change for Figures 4.24 and 11.21.

Endeavour for Figure 4.19.

Erdkunde for Figures 9.21, 10.41, A1.1B and A1.2.

Geografia Fisica e Dinamica Quaternaria for Figure 11.25.

Geographical Magazine for Figure 9.40.

Geographical Reports of Tokyo Metropolitan University for Figure 9.36.

International Journal of Climatology (John Wiley & Sons, Chichester) for Figures 3.17, 8.19, 8.35, 8.37, 9.3 and A1.2.

Japanese Progress in Climatology for Figure 10.36.

Meteorological Magazine for Figures 6.23, 7.10, 7.12 and 8.6, and Plates 24, D and E.

Meteorological Monographs for Figures 7.2 and 7.4.

Meteorologische Rundschau for Figure 10.12.

Meteorologiya Gidrologiya (Moscow) for Figure 9.17.

Nature for Figure 1.11.

New Scientist for Figures 4.12, 4.22, 4.26 and 7.27.

Progress in Physical Geography for Figures 10.31, 10.32 and 10.34.

Science for Figures 6.35 and 10.30.

Tellus for Figures 8.10, 8.11 and 9.25.

Zeitschrift für Geomorphologie for Figure 10.5 from *Supplement 21*.

Publishers

Academic Press, New York, for Figures 6.38, 7.14, 7.15, 4.25, 9.10, 9.11 and 9.12 from *Advances in Geophysics*; for Figure 9.15 from *Monsoon Meteorology* by C. S. Ramage; and for Figure 9.40 from *Quaternary Research*.

Academic Press, Orlando, Fla., for Figure 3.13 from *Water at the Surface of the Earth* by D. H. Miller.

Allen and Unwin, London, for Figures 2.14 and 2.16B from *Oceanography for Meteorologists* by H. V. Sverdrup.

Belhaven Press, London, for Figure 9.55 from *Recent Climatic Change* by S. Gregory (ed.).

Butterworth-Heinemann for Figure 6.29 from *Ocean Circulation* by G. Beerman.

Cambridge University Press for Figure 4.8 from *Clouds, Rain and Rainmaking* by B. J. Mason;

for Figure 6.7 from *World Weather and Climate* by D. Riley and L. Spalton; for Figure 8.31 from *The Warm Desert Environment* by A. Goudie and J. Wilkinson; for Figure 10.24 from *The Tropical Rain Forest* by P. W. Richards; for Figure 10.28 from *Air: Composition and Chemistry* by P. Brimblecombe (ed.); for Figures 7.24 and 9.9 from *Weather Systems* by L. F. Musk; for Figures 1.4, 1.8, 11.19 and 11.18 from *Climate Change: The IPCC Scientific Assessment 1992*; for Figure 6.40 from *Climate System Modelling* by K. E. Trenberth; and for Figure 9.54 from *Teleconnections Linking Worldwide Climate Anomalies* by M. H. Glantz *et al.* (eds).

Chapman and Hall, New York, for Figure 6.33 from *Elements of Dynamic Oceanography* by D. Tolmazin; and for Figure 8.44 from *Encyclopedia of Climatology* by J. Oliver and R. W. Fairbridge (eds).

Cleaver-Hulme Press, London, for Figure 5.15 from *Realms of Water* by Ph. H. Kuenen.

The Controller, Her Majesty's Stationery Office (Crown Copyright Reserved) for Figure 3.3 from *Geophysical Memoir 102* by J. K. Bannon and L. P. Steele; for Figure 9.31 from *Geophysical Memoir 115* by J. Findlater; for Figures 6.23, 7.10 and 8.6, and for Plates 21 D and E from the *Meteorological Magazine*; for Figures 7.12 and 7.17 from *A Course in Elementary Meteorology* by D. E. Pedgley; for Figures 8.27 and 8.28 from *Weather in the Mediterranean 1*, 2nd edn (1962); for the tephigram base of Figure 4.1 from *RAF Form 2810*; and for Figure 6.36 from *Global Ocean Surface Temperature Atlas* by M. Bottomley *et al.*

Elsevier, Amsterdam, for Figure 9.38 from *Palaeogeography, Palaeoclimatology, Palaeoecology*; and for Figures 9.48 and 9.49 from *Climates of Central and South America* by W. Schwerdtfeger (ed.); for Figures 6.13 and 8.30 from *Climates of the World* by D. Martyn; and for Figure 8.41 from *Climates of the Soviet Union* by P. E. Lydolph.

Generalstabens Litografiska Anstalt, Stockholm, for Figure 7.19 from *Klimatologi* by G. H. Liljequist.

Harvard University Press, Cambridge, Mass., for Figures 2.20, 10.16, 10.17 and 10.18 from *The Climate near the Ground* (2nd edn) by R. Geiger.

Hutchinson, London, for Figures 10.27 and 10.35 from the *Climate of London* by T. J. Chandler; and for Figures 9.41 and 9.42 from *The Climatology of West Africa* by D. F. Hayward and J. S. Oguntoyinbo.

Kluwer Academic Publishers, Dordrecht, Holland for Figure 1.1 from *Air–Sea Exchange of Gases and Particles* by P. S. Liss and W. G. N. Slinn (eds); and for Figure 11.21 from *Climate Change* (Vol. 16) by G. A. Meehl and W. M. Washington.

Longman, London, for Figure 6.18 from *Contemporary Climatology* by A. Henderson-Sellers and P. J. Robinson; and for Figures 3.5, 3.6A and 3.18 from *Variations in the Global Water Budget* by A. Street-Perrott *et al.* (eds).

McGraw-Hill Book Company. New York, for Figure 6.25 from *Dynamical and Physical Meteorology* by G. J. Haltiner and F. L. Martin; for Figures 10.17 and 10.18 from *Forest Influences* by J. Kittredge; for Figures 3.10, 4.18 and 6.9 from *Introduction to Meteorology* by S. Petterssen; for Figures 9.5 and 9.6 from *Tropical Meteorology* by H. Riehl; and for Figure 9.21 from *The Earth's Problem Climates* by G. T. Trewartha.

Methuen, London, for Figures 2.19, 3.20 and 9.44 from *Mountain Weather and Climate* by R. G. Barry; for Figures 3.1, 6.19 and 6.22 from *Models in Geography* by R. J. Chorley and P. Haggett (eds); and for Figures 10.6, 10.20, 10.25, 10.29, 10.30, 10.33 and 10.37 from *Boundary Layer Climates* by T. R. Oke.

National Academy Press, Washington, DC, for Figure 11.5.

North-Holland Publishing Company, Amsterdam, for Figure 3.19 from the *Journal of Hydrology*.

Oliver and Boyd, Edinburgh, for Figure 10.15 from *Fundamentals of Forest Biogeocoenology* by V. Sukachev and N. Dylis.

Oxford University Press, Cape Town, for Figures 6.24, 9.45, 9.46 and 9.47 from *Climatic Change and Variability in Southern Africa* by P. D. Tyson.

Plenum Publishing Corp., New York, for Figure 8.39 from *The Geophysics of Sea Ice* by N. Untersteiner (ed.).

Princeton University Press for Figure A1.5 from *Design with Climate* by V. Olgyay.

D. Reidel, Dordrecht, for Figures 6.12 and 11.9 from *The Climate of Europe: Past, Present and Future* by H. Flohn and R. Fantechi (eds); for Figure 8.32 from *Climatic Change*; for Figure 10.34 from *Interactions of Energy and Climate* by W. Bach, J. Pankrath and J. Williams (eds).

Routledge, London, for Figures 9.41 and 9.42 from *The Climatology of West Africa* by D. F.

Hayward and J. S. Oguntoyinbo; for Figure 9.53 from *Climate Since AD 1500* by R. S. Bradley and P. D. Jones (eds); and for Figures 10.8, 10.10 and 10.14 from *Boundary Layer Climates* (2nd edn) by T. R. Oke.

Scientific American Inc., New York, for Figure 2.26 by R. E. Newell; for Figure 1.12B by M. R. Rapino and S. Self; for Figure 4.26 by J. Snow; and for Figure 2.2 by P. V. Foukal.

Springer-Verlag, Heidelberg, for Figures 9.22, 9.24, 9.34 and 9.37.

Springer-Verlag, Vienna and New York, for Figure 2.21 from the *Meteorologische Rundschau*; and for Figures 4.15 and 5.10 from *Archiv für Meteorologie, Geophysik und Bioklimatologie*.

Time-Life Inc., Amsterdam, for Plate 4 from *The Grand Canyon* by R. Wallace.

University of California Press, Berkeley, for Figure 9.7 and Plate 27 from *Cloud Structure and Distributions over the Tropical Pacific Ocean* by J. S. Malkus and H. Riehl.

University of Chicago Press for Figures 2.1, 2.5, 2.20, 10.7, 10.11 and 10.13 from *Physical Climatology* by W. D. Sellers.

University of Wisconsin Press, Madison, for Figure 9.36 from *The Earth's Problem Climates* by G. T. Trewartha.

Van Nostrand Reinhold Company, New York, for Figure 9.58 from *The Encyclopedia of Atmospheric Sciences and Astrogeology* by R. W. Fairbridge (ed.).

Walter De Gruyter, Berlin, for Figure 8.2 from *Allgemeine Klimageographie* by J. Blüthgen.

Weidenfeld and Nicolson, London, for Figure 7.23 from *Climate and Weather* by H. Flohn.

Westview Press, Boulder, for Figure 1.5 from *Climate Change and Society* by W. W. Kellogg and R. Schware.

John Wiley, Chichester, for Figures 8.9, 9.43 and 11.13 from the *Journal of Climatology*; for Figures 1.7 and 1.10 from *The Greenhouse Effect, Climatic Change, and Ecosystems* by G. Bolin *et al.*; and for Figure 10.38 from *Human Activity and Environmental Processes* by K. J. Gregory and D. E. Walling (eds).

John Wiley, New York, for Figure 2.17 from *Physical Geography* (2nd edn) by A. N. Strahler; for Figures A1.3, A1.4 and Table A1.1 from *Physical Geography* (3rd edn) by A. N. Strahler; for Figures 2.3C, 2.4 and 4.10 from *Introduction to Physical Geography* by A. N. Strahler; for Figure 2.6 from *Meteorology, Theoretical and Applied* by E. W. Hewson and R. W. Longley; for Figure 4.20 from *Weather and Climate Modification* by W. N. Hess (ed.); for Figure 7.29 from *Paleoclimate Analysis and Modeling* by A. D. Hecht; for Figures 9.16, 9.29, 9.32 and 9.34 from *Monsoons* by J. S. Fein and P. L. Stephens (eds); for Figure 6.32 from *Modern Physical Geography* (4th edn) by A. N. Strahler and A. H. Strahler; and for Figure 6.34 from *Ocean Science* by K. Stowe.

Organizations

Center for Climatic Research, University of Delaware, Newark, Delaware, for Figure 8.25.

Department of Electronics (NERC Satellite Station), University of Dundee, for Plates 21 and 24.

Deutscher Wetterdienst, Zentralamt, Offenbach am Main, for Figure 9.27.

Directorate-General Science, Research and Development, European Commission, Brussels, for Figure 8.26.

Environmental Science Service Administration (ESSA) for Plates 8 and 18.

Geographical Branch, Department of Energy, Mines and Resources, Ottawa, for Figure 8.15 from *Geographical Bulletin*.

IGU Study Group on Recent Climate Change.

Intellicast Corp. and WSI Corp. for Plate 26.

Intergovernmental Panel on Climate Change for Figures 11.8, 11.10, 11.15, 11.17, 11.20 (B and C), 11.22, 11.23 and 11.26 from *Climate Change 1995: The Science of Climate Change* by J. T. Houghton *et al.* (eds).

Laboratory of Climatology, Centerton, New Jersey, for Figure 8.24.

National Academy of Sciences, Washington, DC, for Figure 11.2.

National Aeronautics and Space Administration (NASA) for Figures 1.15 and 6.28 and for Plates 5, 14, 16, 25, 30 and C.

National Geophysical Data Center, Boulder, for Figure 2.2.

National Hurricane Center, Miami, for Plate 34.

National Meteorological Center, Washington, DC, for Figure 7.28.

National Oceanic and Atmospheric Administration (NOAA), United States Department of Commerce, Washington, DC, for Figures 7.30, 7.31 and 7.33, and for Plates 2 and 3 from *Technical Memo. NESS 95*; and for Plates 22 and 35.

National Snow and Ice Data Center, Boulder, for Plates 9, 23, 27 and A.

National Space Development Agency of Japan for Plate 33.

Natural Environmental Research Council for Figure 1.6 from *Our Future World* and for Figure 3.4A from *NERC News, July 1993* by K. A. Browning.

Naval Weather Service Command, Washington, DC, for Figures 6.3 and 6.11.

New Zealand Alpine Club for Figure 4.16.

New Zealand Meteorological Service, Wellington, New Zealand, for Figures 9.26 and 9.59 from the *Proceedings of the Symposium on Tropical Meteorology* by J. W. Hutchings (ed.).

Nigerian Meteorological Service for Figure 9.39 from *Technical Note 5*.

Press Association–Reuters Ltd, London, for Plate 6.

Quartermaster Research and Engineering Command, Natick, Mass., for Figure 8.18 by J. N. Rayner.

Risø National Laboratory, Roskilde, Denmark, for Table 5.1 and Figures 5.12 and 8.1 from *European Wind Atlas* by I. Troen and E. L. Petersen.

Smithsonian Institution, Washington, DC, for Figure 1.12A.

United Nations Food and Agriculture Organization, Rome, for Figure 10.23 from *Forest Influences*.

United States Department of Agriculture, Washington, DC, for Figures 10.21 and 10.22 from *Climate and Man*.

United States Department of Commerce for Figure 8.13.

United States Department of Energy, Washington, DC, for Figure 2.12.

United States Environmental Data Service for Figure 3.11.

United States Geological Survey, Washington, DC, for Figure 8.33 from *Professional Paper 1052* and for Figure 8.23 mostly from *Circular 1120-A*.

United States National Air Pollution Administration, Washington, DC, for Figures 10.25 and 10.29 from *Public Health Service Publication No. AP-63*.

United States Naval Oceanographic Office for Figure 6.31.

United States Weather Bureau for Figures 3.12, 3.14, 5.11, 6.10, 6.26, 6.27, 6.37, 7.6, 7.26, 9.2, 9.11 and 9.59 from the *Monthly Weather Review*; and for Figure 7.22 from *Research Paper 40*.

University of Tokyo for Figure 9.35 from *Bulletin of the Department of Geography*.

World Meteorological Organization for Plates 15 and 29 from *Technical Note 124*; for Figure 9.52 from *The Global Climate System 1982–84*; for Figure 2.25 from *GARP Publications Series, Rept No. 16*; and for Figure 11.1 from WMO Publication No. 537 by F. K. Hare.

Individuals

Arizona State Climatologist, Phoenix, Arizona, for Figure 8.34.

Dr C. F. Armstrong and Dr C. K. Stidd, of the Desert Research Institute, University of Nevada, for Figure 3.19.

Dr B. W. Atkinson, of Queen Mary College, London, for Figure 10.38.

Dr August H. Auer, Jr., of the New Zealand Meteorological Service, for Plate 36.

Mr P. E. Baylis, of the University of Dundee, and Dr R. Reynolds, of the University of Reading, for Plate 19.

Dr R. P. Beckinsale, of Oxford University, for suggested modification to Figure 7.7.

Dr H. N. Bhalme, of Pune, India, for Figure 9.33.

Dr B. Bolin, of the University of Stockholm, for Figures 1.7 and 1.10.

Dr P. Brimblecombe, of the University of East Anglia, for Figure 10.28.

Dr Otis B. Brown, of the University of Miami, for Plate B.

Mr R. Bumpas, of the National Center for Atmospheric Research, Boulder, for Plate 7.

Dr T. J. Chinn of the Institute of Geological Sciences, Dunedin, for Figure 4.16.

Dr G. C. Evans, of the University of Cambridge, for Figure 10.24.

Dr H. Flohn, of the University of Bonn, for Figures 6.15 and 9.14.

Dr S. Gregory, of the University of Sheffield, for Figures 9.13 and 9.55.

Mr Ernst Haas for Plate 4.

Dr S. L. Hastenrath, of the University of Wisconsin, for Figures 2.21 and 3.19.

Dr L. H. Horn and Dr R. A. Bryson, of the University of Wisconsin, for Figure 8.16.

Dr J. Houghton, of the Meteorological Office, Bracknell, for Figures 1.8 and 11.18 from *Climate Change 1992;* and 11.8, 11.15, 11.17, 11.20, 11.22, 11.23 and 11.26 from *Climate Change 1995: The Science of Climate Change*, IPCC.

Dr R. A. Houze, Jr., of the University of Washington, for Figures 4.25, 7.14, 7.17, 9.11 and 9.12.

Mr S. Jones for the description of Plate 33.

Dr V. E. Kousky, of São Paulo, for Figure 9.50.

Dr Y. Kurihara, of Princeton University, for Figure 9.10.

Mr E. Lantz for Plate 12.

Dr F. H. Ludlam, of Imperial College, London, for Plates 11 and 17.

Dr Kiuo Maejima, of Tokyo Metropolitan University, for Figure 9.36.

Dr J. Maley, of the Université des Sciences et des Techniques du Languedoc, for Figure 9.40.

Dr Brooks Martner, of the University of Wyoming, for Plate 10.

Dr J. R. Mather, of the University of Delaware, for Figure 8.25.

Dr D. H. Miller, of the University of Wisconsin, Madison, for Figure 3.13.

Dr Yale Mintz, of the University of California, for Figure 6.18.

Dr L. F. Musk, of the University of Manchester, for Figures 7.24 and 9.9.

Dr T. R. Oke, of the University of British Columbia, for Figures 5.14, 10.3, 10.6, 10.8, 10.10. 10.14. 10.20, 10.25, 10.29, 10.30, 10.31, 10.32 and 10.37.

Dr W. Palz for Figure 8.26.

Dr L. R. Ratisbona, of the Servicio Meteorologico Nacional, Rio de Janeiro, for Figures 9.48 and 9.49.

Mr D. A. Richter, of Analysis and Forecast Division, National Meteorological Center, Washington, DC, for Figure 7.26.

Dr J. C. Sadler, of the University of Hawaii, for Figure 9.19.

Dr B. Saltzman, of Yale University, for Figure 6.38.

Dr R. S. Scorer, of Imperial College, London, and Mrs Robert F. Symons, for Plate 13.

Dr Glenn E. Shaw, of the University of Alaska, for Figure 1.1A.

Dr K. P. Shine, of the University of Reading, for Figure 11.19.

Dr Tao Shi-yan of the Chinese Meteorological Society for Figures 9.24 and 9.34.

Dr W. G. N. Slinn, for Figure 1.1B.

Dr K. Stowe, of the California State Polytechnic College, for Figure 6.34.

Dr A. N. Strahler, of Santa Barbara, California, for Figures 2.3C, 2.4, 2.17, 4.10, 6.32, A1.3 and A1.4; and for Table A1.1.

Dr R. T. Watson, of NASA, Houston, for Figure 1.4.

Introduction

In this book we aim to provide a non-technical account of how the atmosphere works, thereby developing an understanding of weather phenomena and of global climates. The atmosphere, which is vital to terrestrial life, is a shallow envelope equivalent in thickness to less than 1 per cent of the earth's radius. Most weather systems form and decay within its lowest 10 km. It is thought that the earth's atmosphere evolved to its present form and composition at least 400 million years ago, when extensive vegetation developed on land. Its presence provides an indispensable shield from harmful radiation from the sun and its gaseous contents sustain the plant and animal biosphere on which human life depends.

Over much of the globe, the state of the atmosphere is far from constant in response to varying *weather* processes. Weather extremes – gales, blizzards, tornadoes, floods – drastically affect human activities and frequently result in loss of life, even when anticipated. Hence, by seeking to understand atmospheric phenomena, we can hope to forecast their vagaries and in some instances control or modify them in a beneficial way. This broad endeavour constitutes the field of the atmospheric sciences. *Meteorology* is specifically concerned with the physics of weather processes. *Weather systems* – which produce the variety of instantaneous states of the atmosphere – differ in their size and lifespan. Four scales are commonly recognized: *mesoscale* systems, such as thunderstorms, extend some 10 km horizontally with a lifetime of a few hours; *synoptic-scale* systems, like mid-latitude cyclones and tropical storms, have a diameter of a few thousand km and a lifetime of about five days; *planetary-scale* waves in the atmospheric circulation span 5,000–10,000 km and usually persist for several weeks. In addition, small-scale wind eddies near the earth's surface and processes operating within vegetation canopies are the concern of *micro-meteorology*.

Climate introduces the longer time scales operating in the atmosphere. It is sometimes loosely regarded as 'average weather', but it is more meaningful to define climate as the long-term state of the atmosphere encompassing the aggregate effect of weather phenomena – the extremes as well as the mean values. It is also usual to distinguish regional and global *macroclimate*, on the one hand, from *local* or *topo-climates* related to terrain features (valleys, hill slopes), on the other hand.

A recent view of climate, which has emerged over the past twenty years, envisages a *climate system* involving interactions between the atmosphere, oceans, land surface, snow and ice cover, and biosphere. Couplings between these components operating on short time scales (hours to months), which may involve two-way interactions, are considered to represent processes *internal* to the system. Changes, usually over years to millennia, in the earth's surface, biosphere, atmospheric composition, and incoming solar radiation that may affect the whole climate system without being affected by it are considered to be *external* processes. This view of the climate system has resulted in a widening interest in climate processes and their effects by biologists, chemists, oceanographers, glaciologists, solar physicists and scientists in many other disciplines. This interest in climate processes has been the major theme of this book since its inception some thirty years ago. In this new edition, the concept of the atmosphere–earth–ocean system is a central theme.

The structure of the book represents this viewpoint. We will look first at the composition and structure of the atmosphere and their role in the global exchange of energy, the moisture cycle and wind systems. Then weather and climate in middle and high latitudes and in the tropics and small-scale climates are discussed. The final chapter examines the mechanisms and characteristics of changes in the earth's climate system and projections of conditions in the twenty-first century. The key to atmospheric processes is the radiant energy that the earth and its atmosphere receive from the sun. In order to study the receipt of this energy we need to begin by considering the nature of the atmosphere – its composition and basic properties.

1

Atmospheric composition, mass and structure

A COMPOSITION OF THE ATMOSPHERE

1 Total atmosphere

Air is a mechanical mixture of gases, not a chemical compound. Table 1.1, illustrating the average composition of dry air, shows that three gases – nitrogen, oxygen and argon – account for 99.9 per cent of the air by volume. Moreover, rocket observations show that these gases are mixed in remarkably constant proportions up to about 80 km (50 miles).

Of especial significance, despite their relative scarcity, are the so-called *greenhouse gases*, which play an important role in the thermodynamics of the atmosphere by trapping long-wave terrestrial reradiation, producing the *greenhouse effect* (see Chapter 2C). The concentrations of these gases are particularly susceptible to human (i.e. anthropogenic) activities:

1 Carbon dioxide (CO_2) is involved in a complex global cycle (see A.4, this chapter). It is released from the interior of the earth and produced by respiration of biota, soil processes, combustion and oceanic evaporation. Conversely, it is dissolved in the oceans and consumed by the process of plant photosynthesis.

2 Methane (CH_4) is produced primarily through anaerobic (i.e. oxygen-deficient) processes by natural wetlands and rice paddies (together about 40 per cent of the total), as well as by enteric fermentation in animals, by termites, through coal and oil extraction, biomass burning, and from landfills.

$$CO_2 + 4H_2 \rightarrow CH_4 + 2H_2O$$

Almost two-thirds of the total is related to anthropogenic activity.

Table 1.1 Average composition of the dry atmosphere below 25 km.

Component	Symbol	Volume % (dry air)	Molecular weight
Nitrogen	N_2	78.08	28.02
Oxygen	O_2	20.95	32.00
*‡Argon	Ar	0.93	39.88
Carbon dioxide	CO_2	0.035	44.00
‡Neon	Ne	0.0018	20.18
*‡Helium	He	0.0005	4.00
†Ozone	O_3	0.00006	48.00
Hydrogen	H	0.00005	2.02
‡Krypton	Kr	0.0011	
‡Xenon	Xe	0.00009	
§Methane	CH_4	0.00017	

Notes: *Decay products of potassium and uranium.
　　　　†Recombination of oxygen.
　　　　‡Inert gases.
　　　　§At surface.

Methane is oxidized to CO_2 and H_2O by a complex photochemical reaction system.

$$CH_4 + O_2 + 2x \rightarrow CO_2 + 2x\ H_2$$

where x denotes any specific methane destroying species (such as H, OH, NO, Cl or Br).

3 Nitrous oxide (N_2O) is produced by biological mechanisms in the oceans and soils, by industrial combustion, automobiles, aircraft, biomass burning, and as a result of the use of chemical fertilizers. It is destroyed by photochemical reactions in the stratosphere involving the production of nitrogen oxides (NO_x).

4 Ozone (O_3) is produced by the high-level breakup of oxygen molecules by solar ultraviolet radiation and destroyed by reactions involving nitrogen oxides and chlorine (Cl) (the latter generated by CFCs, volcanic eruptions and vegetation burning) in the middle and upper stratosphere.

5 Chlorofluorocarbons (CFCs: chiefly $CFCl_3$(F-11) and CF_2Cl_2(F-12)) are entirely anthropogenically produced by aerosol propellants, refrigerator coolants (e.g. 'freon'), cleansers and air conditioners, and were not present in the atmosphere until the 1930s. CFC molecules rise slowly into the stratosphere and then move poleward, being decomposed by photochemical processes into chlorine after an estimated average lifetime of some 65–130 years.

6 Hydrogenated halocarbons (HFCs and HCFCs) are also entirely anthropogenic gases. They have increased sharply in the atmosphere over the last few decades, following their use as substitutes for CFCs. Trichloroethane ($C_2H_3Cl_3$), for example, which is used in dry-cleaning and degreasing agents, increased fourfold in the 1980s and has a seven-year residence time in the atmosphere. They generally have lifetimes of a few years, but still have substantial greenhouse effects.

Water vapour (H_2O), the primary greenhouse gas, is a vital atmospheric constituent. It averages about 1 per cent by volume but is very variable both in space and time, being involved in a complex global hydrological cycle (see Chapter 3).

In addition to the greenhouse gases, important *reactive gas species* are produced by the cycles of sulphur, nitrogen and chlorine. These play key roles in acid precipitation and in ozone destruction. Sources of these species are as follows:

Nitrogen species. The reactive species of nitrogen are nitric oxide (NO) and nitrogen dioxide (NO_2).

NO_x refers to these and other odd nitrogen species with oxygen. Fossil fuel combustion (approximately two-thirds for heating, one-third for cars and other transport) is the primary source of NO_x (mainly NO) accounting for $15–25 \times 10^9$ kg N/year. Biomass burning and lightning activity are other important sources. NO_x emissions increased by some 200 per cent between 1940 and 1980. The total source of NO_x is about 40×10^9 kg N/year. About 25 per cent of this goes into the stratosphere, where it undergoes photochemical dissociation. It is also removed as nitric acid (HNO_3) in snowfall. Odd nitrogen is also released as NH_x by ammonia oxidation in fertilizers and by domestic animals ($6–10 \times 10^9$ kg N/year).

Sulphur species. Reactive species are sulphur dioxide (SO_2) and reduced sulphur (H_2S, DMS). Atmospheric sulphur is almost entirely anthropogenic in origin: 90 per cent from coal and oil combustion, and much of the remainder from copper smelting. The major sources are sulphur dioxide ($80–100 \times 10^9$ kg S/year), hydrogen sulphide (H_2S) ($20–40 \times 10^9$ g S/year) and dimethyl sulphide (DMS) ($35–55 \times 10^9$ kg S/year). DMS is primarily produced by biological productivity near the ocean surface. SO_2 emissions increased by about 50 per cent between 1940 and 1980. Volcanic activity releases approximately 10^9 kg S/year as sulphur dioxide. Because the lifetime of SO_2 and H_2S in the atmosphere is only about one day, atmospheric sulphur occurs largely as carbonyl sulphur (COS), which has a lifetime of about one year. The conversion of H_2S gas to sulphur particles is an important source of atmospheric aerosols.

Despite its short lifetime, sulphur dioxide is readily transported over long distances. It is removed from the atmosphere when condensation nuclei of SO_2 are precipitated as acid rain containing sulphuric acid (H_2SO_4). The acidity of fog deposition can be more serious because up to 90 per cent of the fog droplets may be deposited. In Californian coastal fogs, pH values of only 2.0–2.5 are not uncommon. Peak pH readings in the eastern United States and Europe are ≤4.3 (pH = 7 is neutral, see Chapter 3F). In these areas and central southern China, rainfall deposits > 1 g m^{-2} of SO_2 annually.

The role of *halogens* of carbon (CFCs and HCFCs) in the destruction of ozone in the stratosphere is described in the next section.

There are also significant quantities of *aerosols* in the atmosphere. These are suspended particles of sea salt, mineral dust (particularly silicates), organic

Table 1.2 Aerosol production estimates, less than 5 μm radius (10⁹ kg/year) and typical concentrations near the surface (μg m⁻³).

	Production	Concentration Remote	Urban
Natural			
Primary production:			
Sea salt	1300	5–10	
Mineral particles	1500	0.5–5*	
Volcanic	33		
Forest fires and biological debris	50		
Secondary production (gas → particle):			
Sulphates from H_2S	100	1–2	
Nitrates from NO_x	22		
Converted plant hydrocarbons	75		
Total Natural	3060		
Anthropogenic			
Primary production:			
Industrial dust	100		
Combustion (soot)	8		100–500†
Biomass burning (soot)	5		
Secondary production (gas → particle):			
Sulphate from SO_2	140	0.5–1.5	10–20
Nitrates from NO_x	30	0.2	0.5
Biomass combustion			
(organics)	80		
Total Anthropogenic	370		

Sources: Schimel *et al.* 1996, Bridgman 1990.
Notes: *10–60 μg m⁻³ during dust episodes from the Sahara over the Atlantic.
              †Total suspended particles.

matter and smoke. Aerosols enter the atmosphere from a variety of natural and anthropogenic sources (Table 1.2). Some originate as particles – soil grains and mineral dust from dry surfaces, carbon soot from coal fires and biomass burning, and volcanic dust. Others are converted into particles from inorganic gases (sulphur from anthropogenic SO_2 and natural H_2S; ammonium salts from NH_3; nitrogen from NO_x). Sulphate aerosols, two-thirds of which come from coal-fired power station emissions, are now playing an important role in countering global warming effects by reflecting incoming solar radiation (see Chapter 11). Other aerosol sources are sea salts and organic matter (plant hydrocarbons and anthropogenically derived) (Figure 1.1). Natural sources are about eight times larger than anthropogenic ones on a global scale, but the estimates are wide-ranging. Moreover, there is considerable spatial variability. For example, some 1,500 Tg (10¹² g) of crustal material is picked up by the air annually, about one-half from the Sahara and the Arabian Peninsula. Most of this is deposited downwind over the Atlantic. Large particles originate from mineral dust, sea salt spray, fires and plant spores (Figure 1.1 A); these sink rapidly back to the surface or are washed out (scavenged) by rain after a few days. Small (Aitken) particles form by the condensation of gas-phase reaction products and from organic molecules and polymers, including natural and synthetic fibres, plastics, rubber, and vinyl. Fine particles from volcanic eruptions may reside in the stratosphere above the level of weather processes for 1–3 years. Intermediate-sized particles originate from natural sources, such as soil surfaces, from combustion, or they accumulate by random coagulation and by repeated cycles of condensation and evaporation (Figure 1.1). Particles with diameters of 0.1–1.0 μm are highly effective in scattering solar radiation (Chapter 2B.2), and those of about 0.1 μm diameter are important in cloud condensation.

Having made these generalizations about the atmosphere, we now examine the variations that occur in composition with height, latitude and time.

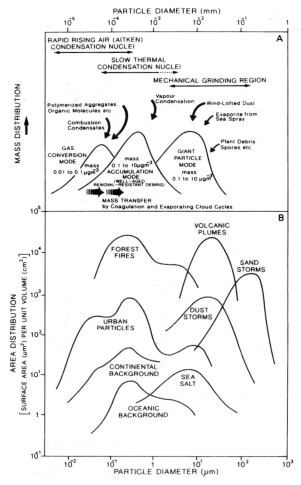

Figure 1.1 Atmospheric particles. A: Mass distribution, together with a depiction of the surface–atmosphere processes that create and modify atmospheric aerosols, illustrating the three size modes. Aitken nuclei are solid and liquid particles that act as condensation nuclei and capture ions, thus playing a role in cloud electrification. B: Distribution of surface area per unit volume.

Sources: A: After Glenn E. Shaw, University of Alaska, Geophysics Institute. B: After Slinn (1983).

2 Variations with height

The light gases (hydrogen and helium especially) might be expected to become more abundant in the upper atmosphere, but large-scale turbulent mixing of the atmosphere prevents such diffusive separation even at heights of many tens of kilometres above the surface. The height variations that do occur are related to the source-locations of the two major non-permanent gases – water vapour and ozone. Since both absorb some solar and terrestrial radiation, the heat budget and vertical temperature structure of the atmosphere are considerably affected by the distribution of these two gases.

Water vapour comprises up to 4 per cent of the atmosphere by volume (about 3 per cent by weight) near the surface, but only 3–6 ppmv (parts per million by volume) above 10 to 12 km. It is supplied to the atmosphere by evaporation from surface water or by transpiration from plants and is transferred upwards by atmospheric turbulence. Turbulence is most effective below about 10 km and as the maximum possible water vapour density of cold air is anyway very low (see B.2, this chapter), there is little water vapour in the upper layers of the atmosphere.

Ozone (O_3) is concentrated mainly between 15 and 35 km. The upper layers of the atmosphere are irradiated by ultraviolet radiation from the sun (see C.1, this chapter), which causes the break-up of oxygen molecules at altitudes above 30 km (i.e. $O_2 \rightarrow O + O$). These separated atoms ($O + O$) may then combine individually with other oxygen molecules to create ozone, as illustrated by the simple photochemical scheme:

$$O_2 + O + M \rightarrow O_3 + M$$

where M represents the energy and momentum balance provided by collision with a third atom or molecule; this Chapman cycle is shown schematically in Figure 1.2A. Such three-body collisions are rare at 80 to 100 km because of the very low density of the atmosphere, while below about 35 km most of the incoming ultraviolet radiation has already been absorbed at higher levels. Therefore ozone is mainly formed between 30 and 60 km, where collisions between O and O_2 are more likely. Ozone itself is unstable; its abundance is determined by three distinctly different photochemical interactions. Above 40 km odd oxygen is destroyed primarily by a cycle involving molecular oxygen; between 20 and 40 km NO_x cycles are dominant; while below 20 km a hydrogen–oxygen radical (HO_2) is responsible. Additional important cycles involve chlorine (ClO) and bromine (BrO) chains at various altitudes. Collisions with monatomic oxygen may recreate oxygen (see Figure 1.2B), but ozone is mainly destroyed through cycles involving catalytic reactions, some of which are photochemical associated with longer wavelength ultraviolet radiation

A Ozone formation
(The Chapman cycle)

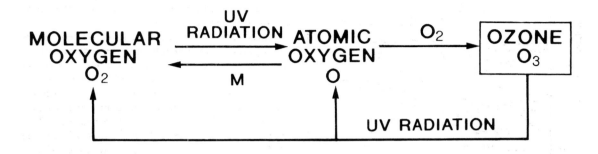

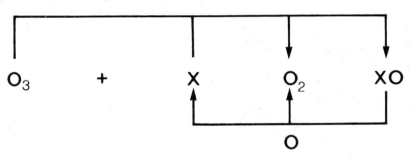

B Ozone destruction

Figure 1.2 Schematic illustrations of (A) the Chapman cycle of ozone formation and (B) ozone destruction. X is any ozone-destroying species (e.g. H, OH, NO, CR, Br).

Source: After Hales 1996. From *Bulletin of the American Meteorological Society*, by permission of the American Meteorological Society.

(2.3–2.9 μm). The destruction of ozone involves a recombination with atomic oxygen, causing a net loss of the odd oxygen. This takes place through the catalytic effect of a radical such as OH (hydroxyl):

$$\left. \begin{array}{l} H + O_2 \rightarrow HO_2 \\ HO_2 + O \rightarrow OH + O_2 \\ OH + O \rightarrow H + O_2 \end{array} \right\} \text{ net: } 2O \rightarrow O_2$$

The odd hydrogen atoms and OH result from the dissociation of water vapour, molecular hydrogen and methane (CH_4).

Stratospheric ozone is similarly destroyed in the presence of nitrogen oxides (NO_x, i.e. NO_2 and NO) and chlorine radicals (Cl, ClO). The source gas of the NO_x is nitrous oxide (N_2O), which is produced by combustion and fertilizer use, while chlorofluoro-carbons (CFCs), manufactured for 'freon', give rise to the chlorines. These source gases are transported into the stratosphere from the surface and are converted by oxidation into NO_x, and by UV photodecomposition into chlorine radicals, respectively.

The chlorine chain involves:

$$2 \, (Cl + O_3 \rightarrow ClO + O_2)$$
$$ClO + ClO \rightarrow Cl_2O_2$$

and

$$Cl + O_3 \rightarrow ClO + O_2$$
$$OH + O_3 \rightarrow HO_3 + 2O_2$$

Both reactions result in a conversion of O_3 to O_2 and the removal of all odd oxygens. Another cycle may involve an interaction of the oxides of chlorine and bromine. It appears that the increases of CR and Br species during the decades 1970–90 are sufficient to explain the observed decrease of stratospheric ozone over Antarctica (see pp. 10–11). A mechanism that may enhance the catalytic process involves polar stratospheric clouds. These can form readily during the austral spring (October), when temperatures decrease to 185–195 K, permitting the formation of particles of nitric acid (HNO_3) ice and water ice. It is apparent, however, that anthropogenic sources of the trace gases are a primary factor in the ozone decline. Conditions in the Arctic are somewhat different as the stratosphere is warmer and there is more mixing of air from lower latitudes.

The constant metamorphosis of oxygen to ozone and from ozone back to oxygen involves a very complex set of photochemical processes, which tend to maintain an approximate equilibrium above about 40 km. However, the ozone mixing ratio is at its maximum at about 35 km, whereas maximum ozone concentration (see Note 1) occurs lower down, between 20 and 25 km in low latitudes and between 10 and 20 km in high latitudes. This is the result of some circulation mechanism transporting ozone downwards to levels where its destruction is less likely, allowing an accumulation of the gas to occur. Despite the importance of the ozone layer, it is essential to realize that if the atmosphere were compressed to sea level (at normal sea-level temperature and pressure) ozone would contribute only about 3 mm to the total atmospheric thickness of 8 km (Figure 1.3).

3 Variations with latitude and season

Variations of atmospheric composition with latitude and season are particularly important in the case of water vapour and ozone.

Ozone content is low over the equator and high in subpolar latitudes in spring (see Figure 1.3). If the distribution were solely the result of photochemical processes, the maximum would occur in June near the equator, and the anomalous pattern must be due to a poleward transport of ozone. The movement is apparently from higher levels (30–40 km) in low latitudes towards lower levels (20–25 km) in high latitudes during winter months. Here the ozone is stored during the *polar night*, giving rise to an ozone-rich layer in early spring under natural conditions. It is this springtime feature that has been disrupted by the Antarctic ozone 'hole' (see pp. 10–11). The type of circulation responsible for this

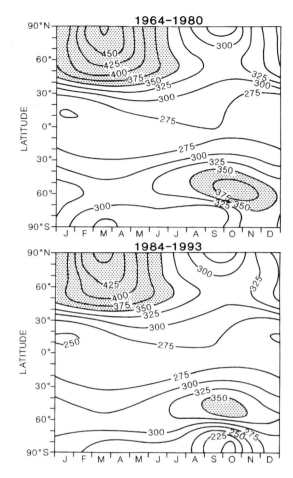

Figure 1.3 Variation of total ozone with latitude and season in Dobson units (milliatmosphere centimeters) for two time intervals, (top) 1964–1980 and (bottom) 1984–1993. Values over 350 units are stippled.
Source: From Bojkov and Fioletov 1995. From *Journal of Geophysical Research* **100**(D), Fig. 15, p. 16, 548. Courtesy American Geophysical Union.

transfer is not yet known with certainty, although it does not seem to be a simple direct one.

The water vapour content of the atmosphere is closely related to air temperature (see B.2, this chapter, and Chapter 3B and C) and is therefore greatest in summer and in low latitudes. There are, however, obvious exceptions to this generalization, such as the tropical desert areas of the world.

The carbon dioxide content of the air (averaging almost 360 parts per million (ppm)) has a large seasonal range in higher latitudes in the northern hemisphere associated with photosynthesis and decay of the biosphere. At 50 °N, the concentration ranges from 351 ppm in late summer to 363 ppm in spring. The low summer values are related to the assimilation of CO_2 by the cold polar seas. Over the year, a small net transfer of CO_2 from low to high altitudes takes place to maintain an equilibrium content in the air.

4 Variations with time

The quantities of carbon dioxide, other greenhouse gases and particles in the atmosphere may be subject to long-term variations and these are of special significance because of their possible effect on the radiation budget.

Measurements of atmospheric trace gases show increases in nearly all of them since the Industrial Revolution began (Table 1.3). The burning of fossil fuels is the primary source of these increasing trace gas concentrations. Heating, transportation and industrial activities generate almost 5×10^{20} J/year of energy. Oil and natural gas consumption account for 60 per cent of global energy and coal about 25 per cent. Natural gas is almost 90 per cent methane (CH_4), whereas the burning of coal and oil releases not only CO_2 but also odd nitrogen (NO_x), sulphur and carbon monoxide (CO). Other factors relating to agricultural practices (land clearance, farming and cattle raising) also contribute to modifying the atmospheric composition. The concentrations and sources of the most important greenhouse gases are considered in turn.

Carbon dioxide (CO_2). The major reservoirs of carbon are in limestone sediments and fossil fuels on land and in the world's oceans. The atmosphere contains about 750×10^{12} kg of carbon (C), corresponding to a CO_2 concentration of 358 ppm (Figure 1.4). The major fluxes of atmospheric carbon dioxide are a result of solution/dissolution in the ocean and photosynthesis/respiration and decomposition by biota. The average time for a CO_2 molecule to be dissolved in the ocean or taken up by plants is about four years. Photosynthetic activity leading to primary production on land involves 50×10^{12} kg of carbon annually, representing 7 per cent of atmospheric carbon; this accounts for the 14 ppm annual oscillation in CO_2 observed in the northern hemisphere due to its extensive land biosphere.

The oceans play a key role in the global carbon cycle. Photosynthesis by phytoplankton generates organic compounds of aqueous carbon dioxide. Eventually, some of the biogenic matter sinks into deeper water, where it undergoes decomposition and oxidation back into carbon dioxide. This process transfers carbon dioxide from the surface water and sequesters it in the ocean deep water. As a consequence, atmospheric concentrations of CO_2 can be maintained at a lower level than otherwise. This mechanism is known as a 'biologic pump'; long-term changes in its operation may have caused the rise in atmospheric CO_2 at the end of the last glaciation. Ocean biomass productivity is limited by the availability of nutrients and by light. Hence, unlike the land biosphere, increasing CO_2 levels will not necessarily affect ocean productivity; inputs of fertilizers in river runoff may be a more significant factor. In the oceans, the carbon dioxide ultimately

Table 1.3 Anthropogenically induced changes in concentration of atmospheric trace gases.

Gas	Concentration		Annual Increase (%)	Sources
	1850*	1995	1980s	
Carbon dioxide	280 ppm	358 ppm	0.4	Fossil fuels
Methane	800 ppbv	1720ppbv	0.6	Rice paddies, cows, wetlands
Nitrous oxide	280 ppbv	312 ppbv	0.25	Microbiological activity, fertilizer, fossil fuel
CFC-11	0	0.27 ppbv	~0	Freon[†]
HCFC-22	0	0.11 ppbv	5	CFC substitute
Ozone (troposphere)	?	10–50 ppbv	~0	Photochemical reactions

Source: After Schimel *et al.* 1996, in Houghton *et al.* 1996.
Notes: *Pre-industrial levels are primarily derived from measurements in ice cores where air bubbles are trapped as snow accumulates on polar ice sheets.
†Production began in the 1930s.

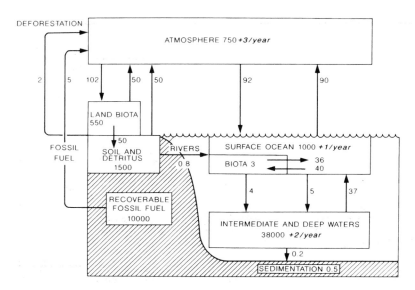

Figure 1.4 Global carbon reservoirs (gigatonnes of carbon (GtC): where 1 Gt = 10^9 metric tons = 10^{12} kg) and gross annual fluxes (GtC yr^{-1}). Numbers italicized in the reservoirs suggest the net annual accumulation due to anthropogenic causes.

Source: After Sundquist, Trabalka, Bolin and Siegenthaler; from IPCC 1990.

goes to produce carbonate of lime, partly in the form of shells and the skeletons of marine creatures. On land, the dead matter becomes humus, which may subsequently form a fossil fuel. These transfers within the oceans and lithosphere involve very long time scales compared with exchanges involving the atmosphere. As Figure 1.4 shows, the exchanges between the atmosphere and the other reservoirs are more or less balanced.

Yet this balance is not an absolute one; between AD 1750 and 1995 the concentration of atmospheric CO_2 is estimated to have increased by 28 per cent, from 280 to 358 ppm (Figure 1.5). Half of this increase has taken place since the mid-1960s; currently, atmospheric CO_2 levels are increasing by 1.5 ppmv per year. The primary net source is fossil fuel combustion, accounting for 5×10^{12} kg C/year. Tropical deforestation and fires may contribute a further 2×10^{12} kg C/year; the figure is still uncertain. Fires destroy only above-ground biomass, and a large fraction of the carbon is stored as charcoal in the soil. The consumption of fossil fuels should actually have produced an increase almost twice as great as is observed. The difference is primarily accounted for by uptake and dissolution in the oceans and the terrestrial biosphere.

Carbon dioxide has a significant impact on global temperature by its absorption and re-emission of radiation from the earth and atmosphere (see Figure 2.21 and Chapter 2C). Calculations suggest that the increase to 370 ppm (expected by AD 2002) could raise the mean air temperature near the surface by 0.5 °C compared with the 1960s (in the absence of other factors).

Research on deep ice cores taken from Antarctica has enabled changes in past atmospheric composition to be studied by extracting air bubbles trapped in the old ice. This shows large natural variations in CO_2 concentration over the last ice age cycle (Figure 1.6). These variations of up to 100 ppm were contemporaneous with temperature changes estimated to be up to 10 °C. These long-term variations in carbon dioxide and climate are discussed further in Chapter 11.

Methane (CH_4) concentrations (1,750 ppbv) are more than double the pre-industrial level (800 ppbv) and are increasing by about 10 ppbv annually (Figure 1.7). Methane has an atmospheric lifetime of 12 years and is responsible for some 15 per cent of the greenhouse effect. Cattle populations have increased by 5 per cent/year over 30 years and paddy rice area by 7 per cent/year, although it is uncertain whether these account quantitatively for the annual increase of 120 ppbv in methane over the last decade. Table 1.4, showing the mean annual release and consumption, indicates the uncertainties in our knowledge of its sources and sinks.

Nitrous oxide (N_2O), which is relatively inert, originates primarily from microbial activity (nitrification) in soils and in the oceans (4 to 8×10^9 kg N/year), with a further 0.2×10^9 kg N/year from fuel combustion. Other anthropogenic sources are

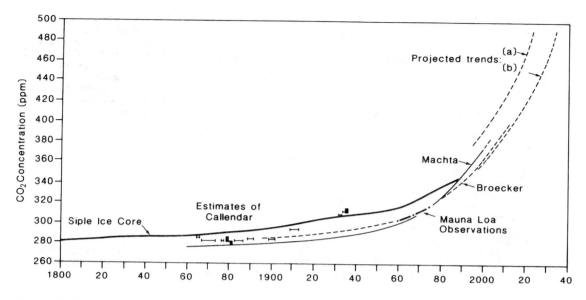

Figure 1.5 Observations of atmospheric CO_2 increase at Mauna Loa, Hawaii (1957–75), estimates from 1860–1960 based on early measurements, and projected trends into the twenty-first century.

Source: After Keeling, Callendar, Machta, Broecker and others.

Note: (a) and (b) indicate different scenarios of global fossil fuel use (from Kellogg and Schware 1981). The thick line gives readings from air bubbles in the Siple Station ice core since 1800.

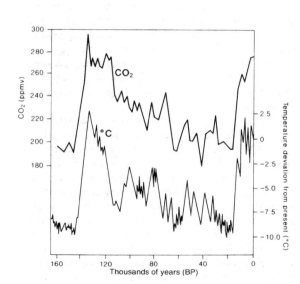

Figure 1.6 Changes in atmospheric CO_2 (ppmv: parts per million by volume) and estimates of the resulting global temperature deviations from the present value obtained from air trapped in ice bubbles in cores at Vostok, Antarctica.

Source: *Our Future World*, Natural Environment Research Council (NERC) 1989.

nitrogen fertilizers and biomass burning. The concentration of N_2O has increased from a pre-industrial level of about 285 ppbv to 310 ppbv (in clean air). Its increase began around 1940 and is now about 0.8 ppbv/year (Figure 1.8A). The major sink of N_2O is in the stratosphere, where it is oxidized into NO_x.

Table 1.4 Mean annual release and consumption of CH_4 (Tg).

	Mean	Range
A *Release*		
Natural wetlands	115	100–200
Rice paddies	110	25–170
Enteric fermentation (mammals)	80	65–110
Gas drilling	45	25–50
Biomass burning	40	20–80
Termites	40	10–100
Landfills	40	20–70
TOTAL	c. 530	
B *Consumption*		
Soils	30	15–30
Reaction with OH	500	400–600
TOTAL	c. 530	

Source: Tetlow-Smith 1995.

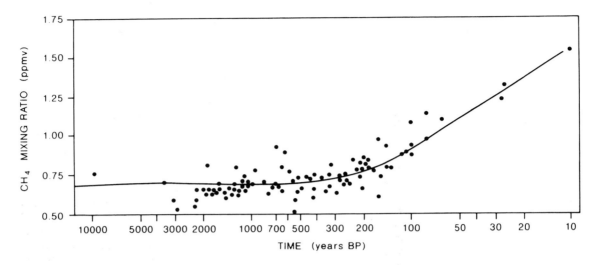

Figure 1.7 Methane concentration (parts per million by volume) in air bubbles trapped in ice dating back to 10,000 years BP obtained from ice cores in Greenland and Antarctica.

Source: Data from Rasmussen and Khalil, Craig and Chou, and Robbins; from Bolin et al. (eds) The Greenhouse Effect, Climatic Change, and Ecosystems (SCOPE 29). Copyright ©1986. Reprinted by permission of John Wiley & Sons, Inc.

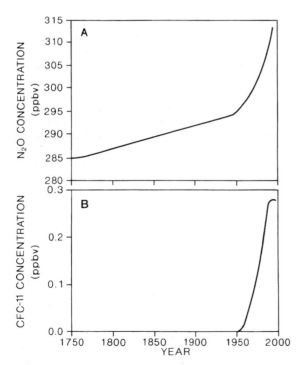

Figure 1.8 Concentrations of (A) nitrous oxide, which have increased since the mid-eighteenth century and especially since 1950; and of (B) CFC-11 since 1950. Both in parts per billion by volume.

Source: IPCC 1990.

Chlorofluorocarbons (CF_2Cl_2 and $CFCl_3$), better known as 'freons' F-11 and F-12, respectively, were first produced in the 1930s and now have a total atmospheric burden of 10^{10} kg. They increased at 4–5 per cent per year up to 1990, but are now static or slowly declining as a result of the Montreal Protocol agreements to curtail production and use substitutes (see Figure 1.8B). Although their concentration is <1 ppbv, they are a significant contributor to the greenhouse effect. They have a residence time of 55–116 years in the atmosphere. However, while the replacement of CFCs by hydrohalocarbons (HCFCs) can significantly reduce the depletion of stratospheric ozone, HCFCs still have a large greenhouse potential.

Ozone (O_3) is distributed very unevenly with height and latitude (see Figure 1.3) as a result of the complex photochemistry involved in its production (A.2, this chapter). Since the late 1970s, dramatic declines in springtime total ozone have been detected over high southern latitudes. The normal increase in stratospheric ozone associated with increasing solar radiation in spring apparently failed to develop. Observations in Antarctica show a decrease in total ozone in September–October from 320 Dobson units (10^{-3} cm at standard atmospheric temperature and pressure) in the 1960s to around 100 in the 1990s. The results from one specific location shown in Figure 1.9 illustrate the

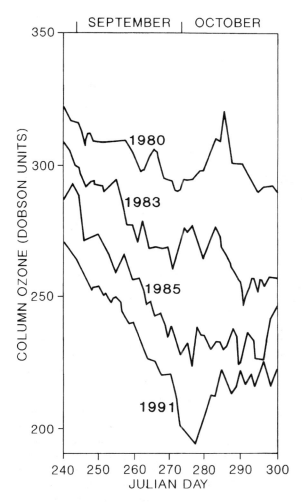

Figure 1.9 Total ozone measurements in spring over Halley Bay, Antarctica, for 1980, 1983, 1985 and 1991, showing deepening of the Antarctic ozone hole.

Sources: From Jiang *et al.* 1996. From *Journal of Geophysical Research* **101**(D4), Plate 2, p. 8,990. Courtesy American Geophysical Union.

trends (see Chapter 11). The global mean column total decreased from 306 Dobson units for 1964–80 to 297 for 1984–93 (see Figure 1.3). The decline over the last 25 years has exceeded 7 per cent in middle and high latitudes.

The effects of reduced stratospheric ozone are particularly important for their potential biological damage to living cells and human skin. It is estimated that a 1 per cent reduction in total ozone will increase ultraviolet-B radiation by 2 per cent, for example, and ultraviolet radiation at 0.30 μm is a thousand times more damaging for the skin than at 0.33 μm. The ozone decrease would also be greater in higher latitudes. However, it should be noted that the mean latitudinal and altitudinal gradients of UV-B radiation imply that the effects of a 2 per cent UV-B increase in mid-latitudes could be offset by moving poleward 20 km or 100 m lower in altitude! Recent polar observations suggest more dramatic changes. Stratospheric ozone totals in 1990 over Palmer Station, Antarctica (65 °S), were found to have maintained low levels from September until early December, instead of recovering in November. Hence, the altitude of the sun was higher and the incoming radiation much greater than in previous years, especially at wavelengths ≤ 0.30 μm. It remains, however, to determine the specific effects of increased UV radiation on marine biota, for instance.

Changes in atmospheric particle concentration derived from volcanic dust are extremely irregular (see Figure 1.11), but individual volcanic emissions are rapidly diffused geographically (see Figure 1.12). Volcanic aerosols are measured in terms of the dust veil index, which takes account of:

1 Maximum percentage depletion of monthly average direct incoming radiation measured in the middle latitudes of the hemisphere concerned.
2 Maximum spatial extent of the dust veil.
3 Persistence of the dust veil.

As shown in Figure 1.12, despite the fact that a strong westerly wind circulation carried the El Chichón dust cloud at an average velocity of 20 m s⁻¹ (45 mph) so that it encircled the globe in less than three weeks, the spread of the Krakatoa dust in 1883 was more rapid and extensive due to the greater amounts of fine dust that were blasted into high levels of the atmosphere. Only about twelve eruptions have produced measurable dust veils in the last 120 years. They occurred mainly between 1882 and 1912, and 1982 and 1992. In contrast,

presence of an 'ozone hole' over the south polar region. Similar reductions are also evident in the Arctic and at lower latitudes (Figure 1.10). Between 1979 and 1986, there was a 30 per cent decrease in ozone at 30–40 km altitude between latitudes 20 and 50 °N and S; along with this there has been an increase in ozone in the lowest 10 km as a result of anthropogenic activities. These changes in the vertical distribution of ozone concentration are likely to lead to changes in atmospheric heating (Chapter 2C), with implications for future climate

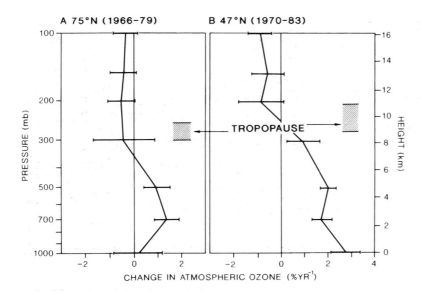

Figure 1.10 Change in atmospheric ozone with elevation (% per year) during 13-year periods at (A) Resolute, northern Canada and (B) Hohenpeissenberg, northern Germany. The horizontal bars indicate the 90 per cent confidence interval.

Source: Data from Logan; from Bolin *et al.* (eds) *The Greenhouse Effect, Climatic Change, and Ecosystems* (SCOPE 29). Copyright © 1986. Reprinted by permission of John Wiley & Sons, Inc.

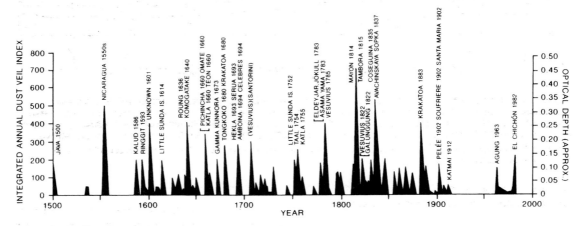

Figure 1.11 Integrated annual dust veil index for the northern hemisphere (AD 1500–1982) (after Lamb), together with approximate optical depths, indicating names and dates of major volcanic eruptions.

Source: LaMarche and Hirschboeck 1984; Crowley and North 1991; Rampino and Self 1984. Reprinted by permission from *Nature* **307**, Fig. 3, p. 123. Copyright ©1984 Macmillan Magazines Ltd.

the contribution of man-made particles (particularly of sulphates and soil) has been progressively increasing, at present accounting for about 30 per cent of the total; this figure could double by AD 2000.

The overall effect of aerosols on the lower atmosphere is uncertain; urban pollutants generally warm the atmosphere through absorption and reduce solar radiation reaching the surface (see Chapter 2C). Aerosols may lower the planetary albedo above a high-albedo desert or snow surface but increase it over an ocean surface. Thus, the global role of tropospheric aerosols is difficult to evaluate, although many authorities now consider it to be one of cooling. Volcanic eruptions, which inject dust and sulphur dioxide high into the stratosphere, are known to cause a small deficit in surface heating with a global effect of −0.1° to −0.2 °C, but the effect is short-lived, lasting only a year or so after the event. Also, unless the eruption is in low

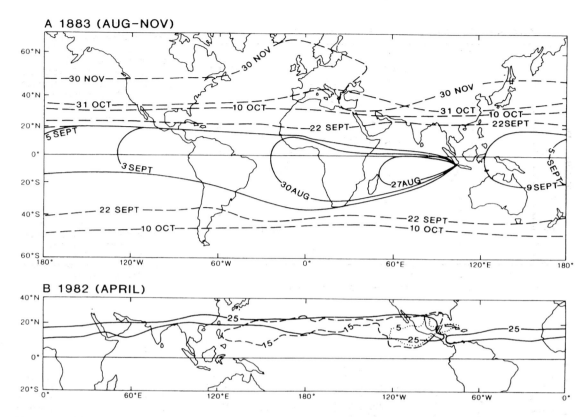

A 1883 (AUG–NOV)

B 1982 (APRIL)

Figure 1.12 The spread of volcanic material in the atmosphere following major eruptions. (A) Approximate distributions of observed optical sky phenomena associated with the spread of Krakatoa volcanic dust between the eruption of 26 August and 30 November 1883. (B) The spread of the volcanic dust cloud following the main eruption of the El Chichón volcano in Mexico on 3 April 1982. Distributions on 5, 15 and 25 April are shown.

Sources: Russell and Archibald 1888, Simkin and Fiske 1983, Rampino and Self 1984, Robock and Matson 1983. (A: by permission Smithsonian Institution.)

latitudes, the dust and sulphate aerosols remain in one hemisphere and do not cross the equator.

B MASS OF THE ATMOSPHERE

Atmospheric gases obey a few simple laws in response to changes in pressure and temperature. The first, Boyle's Law, states that, at a constant temperature, the volume (V) of a mass of gas varies inversely as its pressure (P), i.e.

$$P = \frac{k_1}{V}$$

(k_1 is a constant); and the second, Charles's Law, that, at a constant pressure, volume varies directly with absolute temperature (T) measured in degrees Kelvin (see Note 2):

$$V = k_2 T$$

These laws imply that the three qualities of pressure, temperature and volume are completely interdependent, such that any change in one of them will cause a compensating change to occur in one, or both, of the remainder. The gas laws may be combined to give the following relationship:

$$PV = RmT$$

where m = mass of air, and R = a gas constant for dry air (287 J kg^{-1} K^{-1}) (see Note 3).

If m and T are held fixed, we obtain Boyle's Law; if m and P are held fixed, we obtain Charles's Law. Since it is convenient to use density, ρ (= mass/volume), rather than volume when studying the atmosphere, we can rewrite the equation in the form known as the equation of state:

$$P = R\rho T$$

Thus, at any given pressure, an increase in temperature causes a decrease in density, and vice versa.

1 Total pressure

Air is highly compressible, such that its lower layers are much more dense than those above. Fifty per cent of the total mass of air is found below 5 km (see Figure 1.13), and the average density decreases from about 1.2 kg m^{-3} at the surface to 0.7 kg m^{-3} at 5,000 m (approximately 16,000 ft), close to the extreme limit of human habitation.

Pressure is measured as a force per unit area. The units used by meteorologists are called millibars (mb), 1 mb being equal to a force of 100 newtons acting on 1 m^2 (see Appendix 2). Pressure readings are made with a mercury barometer, which in effect measures the height of the column of mercury that the atmosphere is able to support in a vertical glass tube. The closed upper end of the tube has a vacuum space and its open lower end is immersed in a cistern of mercury. By exerting pressure downwards on the surface of mercury in the cistern, the atmosphere is able to support a mercury column in the tube of about 760 mm (29.9 in or approximately 1,013 mb). The weight of air on a surface at sea level is about 10,000 kg per square metre.

Pressures are standardized in three ways. The readings from a mercury barometer are adjusted to correspond to those for a standard temperature of 0 °C (to allow for the thermal expansion of mercury); they are referred to a standard gravity value of 9.81 ms^{-2} at 45° latitude (to allow for the slight latitudinal variation in g from 9.78 ms^{-2} at the equator to 9.83 ms^{-2} at the poles); and they are calculated for mean sea level to eliminate the effect of station elevation. This third correction is the most significant, because near sea level pressure decreases with height by about 1 mb per 8 m. A fictitious temperature between the station and sea level has to be assumed and in mountain areas this commonly causes bias in the calculated mean sea-level pressure. (The method is summarized in Note 4.)

The mean sea level pressure (p_0) can be estimated by taking account of the total mass of the atmosphere (M), the mean acceleration due to gravity (g_0) and the mean earth radius (R):

$$p_0 = g_0 \, (M/4 \, \pi \, R_E{}^2)$$

where the denominator is the surface area of a spherical earth. Substituting appropriate values into

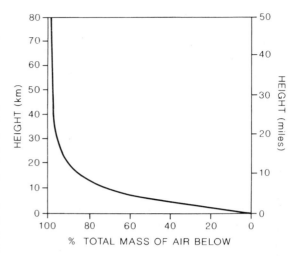

Figure 1.13 The percentage of the total mass of the atmosphere lying below elevations up to 80 km (50 miles). This illustrates the shallow character of the earth's atmosphere.

this expression: $M = 5.14 \times 10^{18}$ kg, $g_0 = 9.8$ ms^{-2}, $R_E = 6.36 \times 10^6$ m, we find $p_0 \approx 10^5$ kg ms^{-2} = 10^5 Nm^{-2}, or 10^5 pascals (Pa is the SI unit of pressure). Meteorologists still use the millibar (mb) unit; 1 millibar = 10^2 Pa (or hPa; h = hecto).

Hence the mean sea level pressure is approximately 10^5 Pa or 1,000 mb. The global mean value is 1,013.25 mb. On average, nitrogen contributes about 760 mb, oxygen 240 mb and water vapour 10 mb. In other words, each gas exerts a partial pressure independent of the others.

Atmospheric pressure, depending as it does on the weight of the overlying atmosphere, decreases logarithmically with height. This relationship is expressed by the *hydrostatic equation*:

$$\frac{\partial p}{\partial z} = -g\rho$$

i.e. the rate of change of pressure (p) with height (z) is dependent on gravity (g) multiplied by the air density (ρ). With increasing height, the drop in air density causes a decrease in this rate of pressure decrease. The temperature of the air also affects this rate, which is greater for cold dense air (see Chapter 6A.1), although the relationship between pressure and height is so significant that meteorologists often express elevations in millibars: 1,000 mb represents sea level, 500 mb about 5,500 m and 300 mb about 9,000 m. A conversion nomogram for an idealized (standard) atmosphere is given in Appendix 2.

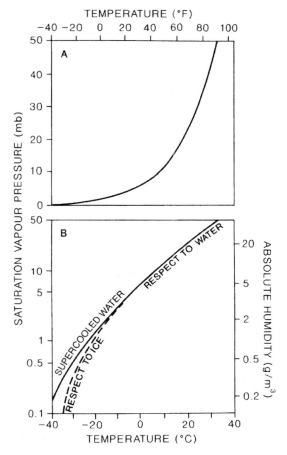

Figure 1.14 Plots of saturation vapour pressure as a function of temperature (i.e. the dew-point curve). (A) The semi-logarithmic plot. (B) shows that below 0 °C the atmospheric saturation vapour pressure is less with respect to an ice surface than with respect to a water drop. Thus, condensation may take place on an ice crystal at lower air humidity than is necessary for the growth of water drops.

2 Vapour pressure

At any given temperature, there is a limit to the density of water vapour in the air, with a consequent upper limit to the vapour pressure. This is termed the *saturation vapour pressure* (e_s) and Figure 1.14A illustrates how it increases with temperature (the Clausius–Clapeyron relationship), reaching a maximum of 1,013 mb (1 atmosphere) at boiling point. Attempts to introduce more vapour into the air when the vapour pressure is at saturation produce condensation of an equivalent amount of vapour. Figure

1.14B shows that whereas the saturation vapour pressure has a single value at any temperature above freezing point, below 0 °C the saturation vapour pressure above an ice surface is lower than that above a supercooled water surface. The significance of this will be discussed in Chapter 4D.1.

Vapour pressure (e) varies with latitude and season from about 0.2 mb over northern Siberia in January to over 30 mb in the tropics in July, but this is not reflected in the pattern of surface pressure. Pressure decreases at the surface when some of the overlying air is displaced horizontally, and in fact the air in high-pressure areas is generally dry owing to dynamic factors, particularly vertical air motion (see Chapter 6A.1), whereas air in low-pressure areas is usually moist.

C THE LAYERING OF THE ATMOSPHERE

The atmosphere can be divided conveniently into a number of rather well-marked horizontal layers, mainly on the basis of temperature. The evidence for this structure comes from regular rawinsonde (radar wind-sounding) balloons, radio-wave investigations, and, more recently, from rocket flights and satellite sounding systems. Broadly, the pattern (Figure 1.15) consists of three relatively warm layers (near the surface; between 50 and 60 km; and above about 120 km) separated by two relatively cold layers (between 10 and 30 km; and 80–100 km). Mean January and July temperature sections illustrate the considerable latitudinal variations and seasonal trends that complicate the scheme (see Figure 1.16).

1 Troposphere

The lowest layer of the atmosphere is called the troposphere. It is the zone where weather phenomena and atmospheric turbulence are most marked, and it contains 75 per cent of the total molecular or gaseous mass of the atmosphere and virtually all the water vapour and aerosols. Throughout this layer, there is a general decrease of temperature with height at a mean rate of about 6.5 °C/km. The decrease occurs because air is compressible and its density decreases with height, allowing rising air to expand and thereby cool. Additionally, the atmosphere is heated mainly by turbulent heat transfer from the surface, not by absorption of radiation. The troposphere is capped in most places by a temperature inversion level (i.e.

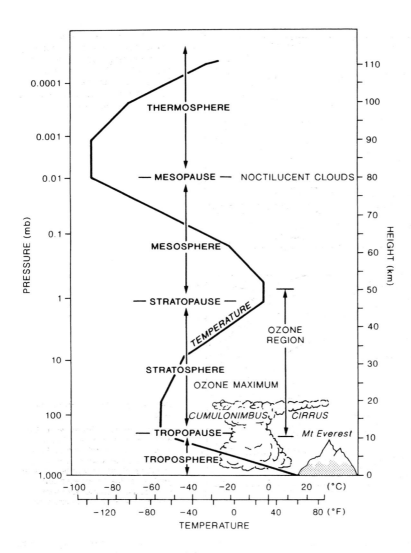

Figure 1.15 The generalized vertical distribution of temperature and pressure up to about 110 km. Note particularly the tropopause and the zone of maximum ozone concentration with the warm layer above.

Source: NASA (n.d.). Courtesy NASA.

a layer of relatively warm air above a colder one) and in others by a zone that is isothermal with height. The troposphere thus remains to a large extent self-contained, because the inversion acts as a 'lid' that effectively limits convection (see Chapter 3E). This inversion level or weather ceiling is called the *tropopause* (see Note 5). Its height is not constant in either space or time. It seems that the height of the tropopause at any point is correlated with sea level temperature and pressure, which are in turn related to the factors of latitude, season and daily changes in surface pressure. There are marked variations in the altitude of the tropopause with latitude (Figure 1.16), from about 16 km at the

equator, where there is great heating and vertical convective turbulence, to only 8 km at the poles.

The meridional temperature gradients in the troposphere in summer and winter are roughly parallel, as are the tropopauses (see Figure 1.16), and the strong lower mid-latitude temperature gradient in the troposphere is reflected in the tropopause breaks (see also Figure 6.8). In these zones, important interchanges can occur between the troposphere and stratosphere, and vice versa. Traces of water vapour probably penetrate into the stratosphere by this means, while dry, ozone-rich stratospheric air may be brought down into the mid-latitude troposphere. For example, above-average

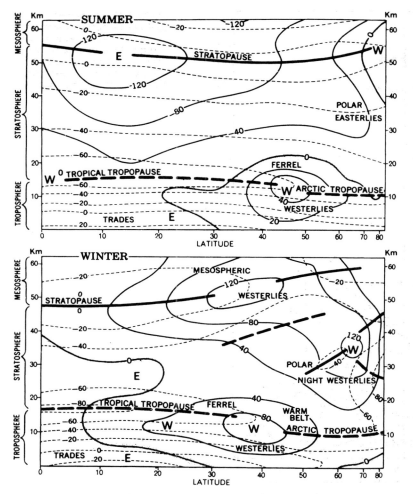

Figure 1.16 Mean zonal (westerly) winds (solid isolines, in knots; negative values from the east) and temperatures (in °C, dashed isolines), showing the broken tropopause near the mean Ferrel jet stream.

Source: After Boville (from Hare 1962).

Notes: The term 'Ferrel Westerlies' was proposed by F. K. Hare in honour of W. Ferrel (see p. 127). The heavy black lines denote reversals of the vertical temperature gradient of the tropopause and stratosphere. Summer and winter refer to the northern hemisphere.

concentrations of ozone are observed in the rear of mid-latitude low-pressure systems, where the tropopause elevation tends to be low. Both facts are probably the result of stratospheric subsidence, which warms the lower stratosphere and causes downward transfer of the ozone.

2 Stratosphere

The second major atmospheric layer is the stratosphere, which extends upwards from the tropopause to about 50 km. Although the stratosphere contains much of the total atmospheric ozone (it reaches a peak density at approximately 22 km), the maximum temperatures associated with the absorption of the sun's ultraviolet radiation by ozone occur at the *stratopause*, where temperature may exceed 0 °C (see Figure 1.16). The air density is much less here, so even limited absorption produces a large temperature increase. Temperatures increase fairly generally with height in summer, with the coldest air at the equatorial tropopause. In winter, the structure is more complex with very low temperatures, averaging −80 °C, at the equatorial tropopause, which is highest at this season. Similar low temperatures are found in the middle stratosphere at high latitudes, whereas over 50–60 °N there is a marked warm region with nearly isothermal conditions at about −45 to −50 °C. Marked seasonal changes of temperature affect the stratosphere. The cold 'polar night' winter stratosphere often undergoes dramatic *sudden warmings* associated with subsidence due to circulation changes in late winter or early spring, when temperatures at about 25 km may jump from −80 to −40 °C over a two-day period. The autumn cooling is a more gradual process. In the tropical stratosphere, there is a quasi-biennial (26-month) wind regime, with easterlies in the layer 18 to 30 km for 12 to 13 months, followed by westerlies for a similar period. The reversal begins first at high levels and takes approximately 12 months to descend from 30 to 18 km (10 to 60 mb).

How far these events in the stratosphere are linked with temperature and circulation changes in the troposphere remains a topic of meteorological research. Any interactions that do exist, however, are likely to be complex, otherwise they would already have become evident.

3 Mesosphere

Above the stratopause, average temperatures decrease to a minimum of about −133 °C (140 K) or around 90 km. This layer is commonly termed the mesosphere, although it must be noted that as yet there is no universal acceptance of terminology for the upper atmospheric layers. The layers between the tropopause and the lower thermosphere are now commonly referred to as the *middle atmosphere*, with the upper atmosphere designating the regions above about 100 km altitude. Above 80 km, temperatures again begin rising with height and this inversion is referred to as the 'mesopause'. Molecular oxygen and ozone absorption bands contribute to heating around 85 km altitude. It is in this region that 'noctilucent clouds' are observed over high latitudes in summer. Their presence appears to be due to meteoric dust particles, which act as ice crystal nuclei when traces of water vapour are carried upwards by high-level convection caused by the vertical decrease of temperature in the mesosphere. However, their formation is also thought to be related to the production of water vapour through the oxidation of atmospheric methane, since apparently they were not observed prior to the Industrial Revolution.

Pressure is very low in the mesosphere, decreasing from about 1 mb at 50 km to 0.01 mb at 90 km.

4 Thermosphere

Above the mesopause, atmospheric densities are extremely low, although the tenuous atmosphere still effects drag on space vehicles above 250 km. The lower portion of the thermosphere is composed mainly of nitrogen (N_2) and oxygen in molecular (O_2) and atomic (O) forms, whereas above 200 km atomic oxygen predominates over nitrogen (N_2 and N). Temperatures rise with height, owing to the absorption of extreme ultraviolet radiation (0.125–0.205 μm) by molecular and atomic oxygen, probably approaching 800–1,200 K at 350 km, but these temperatures are essentially theoretical. For example, artificial satellites do not acquire such temperatures because of the rarefied air. 'Temperatures' in the upper thermosphere and exosphere undergo wide diurnal and seasonal variations. They are higher by day and are also higher during a sunspot maximum, although the changes are only represented in varying velocities of the sparse air molecules.

Above 100 km, the atmosphere is increasingly affected by cosmic radiation, solar X-rays and ultraviolet radiation, which cause *ionization*, or electrical charging, by separating negatively charged electrons from neutral oxygen atoms and nitrogen molecules, leaving the atom or molecule with a net positive charge (an *ion*). The Aurora Borealis and Aurora Australis are produced by the penetration of ionizing particles through the atmosphere from about 300 km to 80 km, particularly in zones about 10–20° latitude from the earth's magnetic poles. On occasion, however, the aurorae may appear at heights up to 1,000 km, demonstrating the immense extension of a rarefied atmosphere. The term *ionosphere* is commonly applied to the layers above 80 km.

5 Exosphere and magnetosphere

The base of the exosphere is between about 500 km and 750 km. Here atoms of oxygen, hydrogen and helium (about 1 per cent of which are ionized) form the tenuous atmosphere, and the gas laws (see B, this chapter) cease to be valid. Neutral helium and hydrogen atoms, which have low atomic weights, can escape into space since the chance of molecular collisions deflecting them downwards becomes less with increasing height. Hydrogen is replaced by the breakdown of water vapour and methane (CH_4) near the mesopause, while helium is produced by the action of cosmic radiation on nitrogen and from the slow but steady breakdown of radioactive elements in the earth's crust.

Ionized particles increase in frequency through the exosphere and, beyond about 200 km, in the magnetosphere there are only electrons (negative) and protons (positive) derived from the solar wind – a plasma of electrically conducting gas.

SUMMARY

The atmosphere is a mixture of gases with constant proportions up to 80 km or more. The exceptions are ozone, which is concentrated in the lower stratosphere, and water vapour in the lower troposphere. Carbon dioxide, methane and other trace gases are increasing in this century due to the burning of fossil fuels and other anthropogenic effects, but large natural fluctuations have also occurred in the past. Air is highly compressible, so that half of its mass occurs in the lowest 5 km, and pressure decreases logarithmically with height from an average sea level value of 1,013 mb.

The vertical structure of the atmosphere comprises three relatively warm layers – the lower troposphere, the stratopause and the upper thermosphere – separated by a cold layer above the tropopause (in the lower stratosphere), and the mesopause. The temperature profile is determined by atmospheric absorption of solar radiation, and the decrease of density with height.

2

Solar radiation and the global energy budget

A SOLAR RADIATION

The prime source of the energy injected into our atmosphere is the sun, which is continually shedding part of its mass by radiating waves of electromagnetic energy and high-energy particles into space. This constant emission is important because it represents in the long run almost all the energy available to the earth (except for a small amount emanating from the radioactive decay of earth minerals). The amount of energy received by the earth, assuming for the moment that there is no interference from the atmosphere, is affected by four factors: solar output, the sun–earth distance, the altitude of the sun, and day length.

1 Solar output

Solar energy, which originates from nuclear reactions within the sun's hot core (16×10^6 K), is transmitted to the sun's surface by radiation and hydrogen convection. Visible solar radiation (light) comes from a 'cool' ($\sim$6,000 K) outer surface layer called the *photosphere*. Temperatures rise again in the outer chromosphere (10,000 K) and corona (10^6 K), which is continually expanding into space. The outflowing hot gases (plasma) from the sun, referred to as the *solar wind* (with a speed of 1.5×10^6 km hr^{-1}), interact with the earth's magnetic field and upper atmosphere. The earth intercepts both the normal electromagnetic radiation and energetic particles emitted by the sun during solar flares.

The sun behaves virtually as a *black body*, meaning that it both absorbs all energy received and in turn radiates energy at the maximum rate possible for a given temperature. The energy emitted at a particular wavelength by a perfect radiator of given

temperature is described by a relationship due to Max Planck. The black-body curves in Figure 2.1 illustrate this relationship. The total energy emitted by a black body is given by the area under each curve; its value is found by integration of Planck's equation, known as Stefan's Law:

$$F = \sigma T^4$$

where $\sigma = 5.67 \times 10^{-8}$ W m^{-2} K^{-4} (the Stefan–Boltzmann constant), i.e. the energy emitted (F) is proportional to the fourth power of the absolute temperature of the body (T).

The total solar output to space, assuming a temperature of 5,760 K for the sun, is 3.84×10^{26} W, but only a tiny fraction of this is intercepted by the earth, because the energy received is inversely proportional to the square of the solar distance (150 million km) (see Note 1).

The energy received at the top of the atmosphere on a surface perpendicular to the solar beam for mean solar distance is termed the *solar constant* (see Note 1). The most recent satellite measurements indicate a value of about 1,368 Wm^{-2}. Figure 2.1 shows the wavelength range of solar (short-wave) radiation and the long-wave (infrared) radiation emitted by the earth and atmosphere. For solar radiation, 8 per cent is ultraviolet and shorter wavelength emission, 39 per cent visible light (0.4–0.7 μm) and 53 per cent near-infrared (>0.7μm). The figure illustrates the black-body radiation curves for 6,000 K at the top of the atmosphere (which slightly exceeds the observed extraterrestrial radiation) for 300 K, and for 263 K. The mean temperature of the earth's surface is about 288 K (15 °C) and of the atmosphere about 250 K (–23 °C). Gases do not behave as black bodies, and Figure 2.1 shows the absorption bands in the atmosphere, which cause its

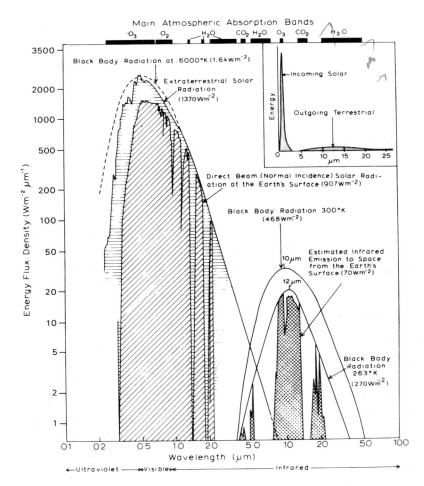

Source: Mostly after Sellers 1965.

Figure 2.1 Spectral distribution of solar and terrestrial radiation, plotted logarithmically, together with the main atmospheric absorption bands. The cross-hatched areas in the infrared spectrum indicate the 'atmospheric windows' where radiation escapes to space. The black-body radiation at 6,000 K is that proportion of the flux which would be incident on the top of the atmosphere. The inset shows the same curves for incoming and outgoing radiation with the wavelength plotted arithmetically on an arbitrary vertical scale.

emission to be much less than that from an equivalent black body. The wavelength of maximum emission (λ_{max}) varies inversely with the absolute temperature of the radiating body:

$$\lambda_{max} = \frac{2,897}{T} \times 10^{-6} \text{ m (Wien's Law)}$$

Thus solar radiation is very intense and is mainly short-wave between about 0.2 and 4.0 μm, with a maximum (per unit wavelength) at 0.5 μm, whereas the much weaker terrestrial radiation has a peak intensity at about 10 μm and a range of about 4 to 100 μm (1 μm = 1 micrometre = 10^{-6} m).

Satellite data show that the solar constant undergoes small periodic variations of about 0.1 per cent, related to sunspot activity. *Sunspots* are dark (i.e. cooler) areas visible on the sun's surface. Their

number and positions change in a regular manner, known as the sunspot cycles. These cycles have wavelengths averaging 11 years (varying in length between 8 and 13 years), the 22-year (Hale) magnetic cycle, much less importantly 37.2 years (18.6 years – the luni–solar oscillation) and possibly 80–90 years. Figure 2.2 shows the variation of sunspot activity since 1610. Between the thirteenth and eighteenth centuries, sunspot activity was generally low, except for the periods AD 1350–1400, and 1600–1645. Output within the ultraviolet part of the spectrum shows considerable variability, with up to twenty times more ultraviolet radiation emitted at certain wavelengths during a sunspot maximum than during a sunspot minimum. The relation between sunspot activity and terrestrial temperatures is a matter of some dispute. However,

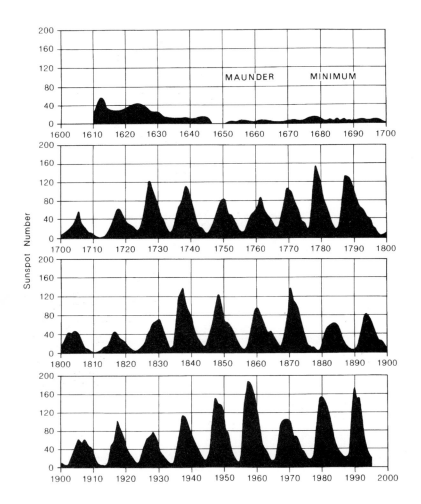

Figure 2.2 Yearly sunspot numbers for the sun's visible surface for the period 1700–1995.

Sources: Reproduced by courtesy of the National Geophysical Data Center, NOAA, Boulder, Colorado. Data before AD 1700 courtesy of Foukal (1990) and *Scientific American*.

some authorities believe that prolonged time-spans of sunspot minima (e.g. AD 1645–1705, the Maunder Minimum) and maxima (e.g. 1895–1940 and post 1970) can produce significant global cooling and warming, respectively.

Shorter-term relationships are more difficult to support, but mean annual temperatures have been correlated with the combined 10–11 and 18.6-year solar cycles. Satellite measurements during the 1980s, the latest solar cycle, show a small decrease in solar output as sunspot number approaches its *minimum*, and a subsequent recovery. Although sunspot areas are cool spots, they are surrounded by bright areas of activity known as faculae, which have higher temperatures; the net effect is for solar output to vary in parallel with the number of sunspots. Thus, the solar 'irradiance' decreases by

about 1.5 Wm^{-2} from sunspot maximum to minimum. In the long term, assuming that the earth behaves as a black body, a long-continued difference of 2 per cent in the solar constant could change the effective mean temperature of the earth's surface by as much as 1.2 °C; however, the observed fluctuations of about 0.1 per cent would change the mean global temperature by ≤0.06 °C, based on calculations of radiative equilibrium.

2 Distance from the sun

The annually changing distance of the earth from the sun produces seasonal variations in our receipt of solar energy. Owing to the eccentricity of the earth's orbit around the sun, the receipt of solar energy on a surface normal to the beam is 7 per

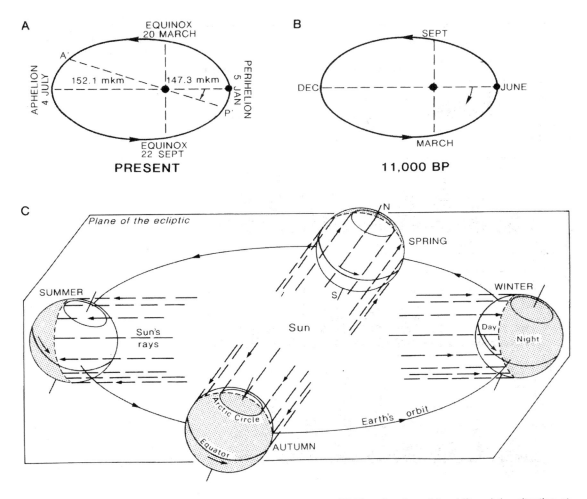

Figure 2.3 Perihelion shifts. (A) The present timing of perihelion; (B) The direction of its shift and the situation at 11,000 years BP; (C) The geometry of the present seasons (northern hemisphere).
Source: Partly after Strahler 1965.

cent more on 3 January at the perihelion than on 4 July at the aphelion (Figure 2.3). In theory (that is, discounting the interposition of the atmosphere and the difference in degree of conductivity between large land and sea masses), this difference should produce an increase in the effective January world surface temperatures of about 4 °C, over those of July. It should also make northern winters warmer than those in the southern hemisphere, and southern summers warmer than those in the northern hemisphere. In practice, atmospheric heat circulation and the effects of continentality substantially mask this global tendency, and the actual seasonal contrast between the hemispheres is reversed. Moreover, the

northern summer half-year (21 March–22 September) is five days longer than the southern hemisphere summer (22 September–21 March). This difference slowly changes; about 10,000 years ago the aphelion occurred in the northern hemisphere winter, and northern summers received 3–4 per cent more radiation than today (Figure 2.3B). This same pattern will return about 10,000 years from now.

Figure 2.4 graphically illustrates the seasonal variations of energy receipt with latitude. Actual amounts of radiation received on a horizontal surface outside the atmosphere are given in Table 2.1. The intensity on a horizontal surface (I_h) is determined from:

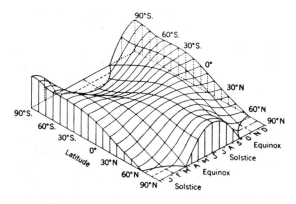

Figure 2.4 The variations of solar radiation with latitude and season for the whole globe, assuming no atmosphere. This assumption explains the abnormally high amounts of radiation received at the poles in summer, when daylight lasts for 24 hours each day.
Source: After W. M. Davis.

$$I_b = I_0 \sin d$$

where I_0 = the solar constant and d = the angle between the surface and the solar beam.

3 Altitude of the sun

The altitude of the sun (i.e. the angle between its rays and a tangent to the earth's surface at the point of observation) also affects the amount of solar radiation received at the surface of the earth. The greater the sun's altitude, the more concentrated is the radiation intensity per unit area at the earth's surface and the longer is the path length of the beam through the atmosphere, which increases the atmospheric absorption. There are, in addition, important variations with solar altitude of the proportion of radiation reflected by the surface, particularly in the case of a water surface (see B.5, this chapter). The principal factors that determine the sun's altitude are, of course, the latitude of the site, the time of

day and the season (see Figure 2.3). At the June solstice, the sun's altitude is a constant 23½° throughout the day at the North Pole and the sun is directly overhead at noon at the Tropic of Cancer (23½ °N).

4 Length of day

The length of daylight also affects the amount of radiation that is received. Obviously, the longer the time that the sun shines the greater is the quantity of radiation that a given portion of the earth will receive. At the equator, for example, the day length is close to 12 hours in all months, whereas at the poles it varies between 0 and 24 hours from winter (polar night) to summer (see Figure 2.3).

The combination of all these factors produces the pattern of receipt of solar energy at the top of the atmosphere shown by Figure 2.4. The polar regions receive their maximum amounts of solar radiation during their summer solstices, which is the period of continuous day. The amount received during the December solstice in the southern hemisphere is theoretically greater than that received by the northern hemisphere during the June solstice, due to the previously mentioned elliptical path of the earth around the sun (see Table 2.1). The equator has two radiation maxima at the equinoxes and two minima at the solstices, due to the apparent passage of the sun during its double annual movement between the northern and southern hemispheres.

B SURFACE RECEIPT OF SOLAR RADIATION AND ITS EFFECTS

1 Energy transfer within the earth–atmosphere system

So far, we have described the distribution of solar radiation as if it were all available at the earth's surface. This is, of course, an unreal view because

Table 2.1 Daily solar radiation on a horizontal surface outside the atmosphere: Wm^{-2}.

Date	90 °N	70	50	30	0	30	50	70	90 °S
Dec 21	0	0	86	227	410	507	514	526	559
Mar 21	0	149	280	378	436	378	280	149	0
June 22	524	492	482	474	384	213	80	0	0
Sept 23	0	147	276	373	430	372	276	147	0

Source: After Berger 1996.

of the effect of the atmosphere on energy transfer. Heat energy can be transferred by the three following mechanisms:

1 *Radiation*: Electromagnetic waves transfer energy (both heat and light) between two bodies, without the necessary aid of an intervening material medium, at a speed of 300×10^6 m s^{-1} (i.e. the speed of light). This is so with solar energy through space, whereas the earth's atmosphere allows the passage of radiation only at certain wavelengths and restricts that at others.

Radiation entering the atmosphere may be absorbed by atmospheric gases in certain wavelengths but, as shown in Figure 2.1, most short-wave radiation is transmitted without absorption. Scattering occurs if the direction of a photon of radiation is changed by interaction with atmospheric gases and aerosols. Two types of scattering are distinguished. For gas molecules smaller than the radiation wavelength (λ), *Rayleigh scattering* occurs in all directions and is proportional to $(1/\lambda^4)$. As a result, the scattering of blue light ($\lambda = 0.4$ μm) is an order of magnitude (i.e. $\times$ 10) greater than that of red light ($\lambda \simeq 0.7$ μm), thus creating the daytime blue sky. However, when water droplets or aerosol particles, with similar sizes (0.1–0.5 μm radius) to the radiation wavelength, are present, most of the light is scattered forward. This *Mie scattering* gives the greyish appearance of polluted atmospheres.

Within a cloud, or between low clouds and a snow-covered surface, radiation undergoes multiple scattering. In the latter case, the 'white out' conditions typical of polar regions in summer (and mid-latitude snowstorms) are experienced, when surface features and the horizon become indistinguishable.

2 *Conduction*: By this mechanism, the heat passes through a substance from point to point by means of the transfer of adjacent molecular motions. Since air is a poor conductor, this type of heat transfer can be virtually neglected in the atmosphere, but it is important in the ground.

3 *Convection*: This occurs in fluids (including gases), which are able to circulate internally and distribute heated parts of the mass. The low viscosity of air and its consequent ease of motion makes this the chief method of atmospheric heat transfer. It should be noted that *forced convection* (mechanical turbulence) occurs due to the development of eddies as air flows over uneven

surfaces, even when there is no surface heating to set up *free* (thermal) *convection*.

Convection transfers energy in two forms. The first is the *sensible heat* content of the air (called enthalpy by physicists), which is transferred directly by the rising and mixing of warmed air. It is defined as $c_p T$, where T is the temperature and c_p (= 1,004 J kg^{-1} K^{-1}) is the specific heat at constant pressure (the heat absorbed by unit mass for unit temperature increase). Sensible heat is also transferred by conduction. The second form of energy transfer by convection is indirect, involving *latent heat*. Here, there is a phase change but no temperature change. Whenever water is converted into water vapour by evaporation (or boiling), heat is required. This is referred to as the latent heat of vaporization (L). At 0 °C, L is 2.50×10^6 J kg^{-1} of water. More generally,

$$L \ (10^6 \ J \ kg^{-1}) \approx (2.5 - 0.00235T)$$

where T is in °C. When water condenses in the atmosphere (see Chapter 3D), the same amount of latent heat is given off as is used for evaporation *at the same temperature*. Similarly, for melting ice at 0 °C, the latent heat of fusion is required, which is 0.335×10^6 J kg^{-1}. If ice evaporates without melting, the latent heat of this sublimation process is 2.83×10^6 J kg^{-1} at 0 °C (i.e. the sum of the latent heats of melting and vaporization). In all of these phase changes of water there is an energy transfer. We shall return to other aspects of these processes in Chapter 3.

2 Effect of the atmosphere

Solar radiation is virtually all in the short-wavelength range, less than 4 μm (see Figure 2.1). About 18 per cent of the incoming energy is absorbed directly by ozone and water vapour. Ozone absorption is concentrated in three solar spectral bands (0.20–0.31, 0.31–0.35 and 0.45–0.85 μm), while water vapour absorbs to a lesser degree in several bands between 0.9 and 2.1 μm (see Figure 2.1). Solar wavelengths shorter than 0.285 μm scarcely penetrate below 20 km altitude, whereas those >0.295 μm reach the surface. Thus, the 3 mm (equivalent) column of stratospheric ozone attenuates ultraviolet radiation almost entirely, except for a partial window around 0.20 μm, where radiation reaches the lower stratosphere. About 30 per cent is immediately reflected back into space from the atmosphere, clouds and the earth's surface, leaving approximately 70 per cent to heat the earth and its atmosphere. Of this, the greater part eventually heats the atmosphere, but much of this heat is received secondhand by the atmosphere via the earth's surface. The ultimate retention of this energy by the atmosphere is of prime importance, because if it did not occur the average temperature of the earth's surface would fall by some 40 °C, obviously making most life impossible. The surface absorbs almost half of the incoming energy available at the top of the atmosphere and reradiates it outwards as long (infrared) waves of greater than 3 μm (see Figure 2.1). Much of this reradiated long-wave energy can be absorbed by the water vapour, carbon dioxide and ozone in the atmosphere, the rest escaping through atmospheric *windows* back into outer space, principally between 8 and 13 μm (see Figure 2.1). Figure 2.5 illustrates the relative roles of the atmosphere, clouds and the earth's surface in reflecting and absorbing solar radiation at different latitudes. (A more complete analysis of the total heat budget of the earth–atmosphere system is given in D, this chapter.)

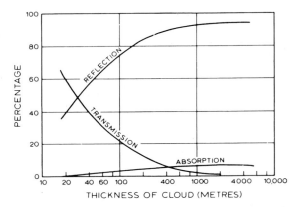

Figure 2.6 Percentage of reflection, absorption and transmission of solar radiation by cloud layers of different thickness.
Source: From Hewson and Longley 1944.

3 Effect of cloud cover

Cloud cover can, if it is thick and complete enough, form a significant barrier to the penetration of radiation. The drop in surface temperature often experienced on a sunny day when a cloud temporarily cuts off the direct solar radiation illustrates our reliance upon the sun's radiant energy. How much radiation is actually reflected depends on the amount of cloud cover and its thickness (Figure 2.6). The proportion of incident radiation that is reflected is termed the *albedo*, or reflection coefficient (expressed as a fraction or percentage). Cloud type affects the albedo. Aircraft measurements show that the albedo of a complete overcast ranges from 44 to 50 per cent for cirrostratus to 90 per cent for cumulonimbus. Average albedos, as determined by satellites, aircraft and surface measurements, are summarized in Table 2.2 (see Note 2).

The total (or global) solar radiation (direct, Q, and diffuse, q) received at the surface on cloudy days is

$$Q + q = (Q + q)_0 \, [b + (1 - b) \, (1 - c)]$$

where $(Q + q)_0$ = global solar radiation for clear skies;

c = cloudiness (fraction of sky covered);

b = a coefficient depending on cloud type and thickness; and the depth of atmosphere through which the radiation must pass.

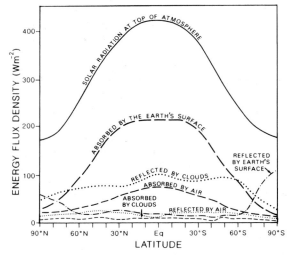

Figure 2.5 The average annual latitudinal disposition of solar radiation in W m^{-2}. Of 100 per cent radiation entering the top of the atmosphere, about 20 per cent is reflected back to space by clouds, 3 per cent by air (plus dust and water vapour), and 8 per cent by the earth's surface. Three per cent is absorbed by clouds, 18 per cent by the air, and 48 per cent by the earth.
Source: After Sellers 1965.

Table 2.2 The average (integrated) albedo of various surfaces (0.3–0.4 µm).

Planet earth	0.31
Global surface	0.14–0.16
Global cloud	0.23
Cumulonimbus	0.9
Stratocumulus	0.6
Cirrus	0.4–0.5
Fresh snow	0.8–0.9
Melting snow	0.4–0.6
Sand	0.30–0.35
Grass, cereal crops	0.18–0.25
Deciduous forest	0.15–0.18
Coniferous forest	0.09–0.15
Tropical rainforest	0.07–0.15
Water bodies*	0.06–0.10

Note: *Increases sharply at low solar angles.

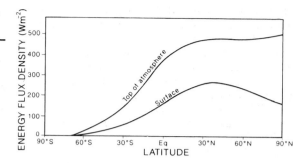

Figure 2.7 The average receipt of solar radiation with latitude at the top of the atmosphere and at the earth's surface during the June solstice.

For mean monthly values for the United States, $b \approx 0.35$, so that

$$(Q + q) \approx (Q + q)_0 [1 - 0.65c]$$

The effect of cloud cover also operates in reverse, since it serves to retain much of the heat that would otherwise be lost from the earth by radiation throughout the day and night. This largely negative role of clouds means that their presence appreciably lessens the daily temperature range by preventing high maxima by day and low minima by night. As well as interfering with the transmission of radiation, clouds act as temporary thermal reservoirs because they absorb a certain proportion of the energy they intercept. The small effect of cloud reflection and absorption of solar radiation is illustrated in Figures 2.6 and 2.7.

Global cloudiness is not yet accurately known. Ground-based observations are mostly at land stations and refer to a small ($\sim$250 km^2) area. Satellite estimates are derived from the reflected short-wave radiation and infrared irradiance measurements, with various threshold assumptions for cloud presence/absence; typically they refer to a grid area of 2,500 km^2 to 37,500 km^2. Surface-based observations tend to be about 10 per cent greater than satellite estimates due to the observer's perspective. Average winter and summer distributions of total cloud amount from surface observations are shown in Figure 2.8. The cloudiest areas are the Southern Ocean and the mid- to high-latitude North Pacific and North Atlantic storm tracks. Lowest amounts are over the Saharan–Arabian desert area (see

Plate 1). Total global cloud cover is just over 60 per cent in January and July.

4 Effect of latitude

As Figure 2.4 has already shown, different parts of the earth's surface receive different amounts of solar radiation. The time of the year is one factor controlling this, more radiation being received in summer than in winter because of the higher altitude of the sun and the longer days. Latitude is a very important control because this will determine both the duration of daylight and the distance travelled through the atmosphere by the oblique rays from the sun. However, actual calculations show the effect of the latter to be negligible in the Arctic, apparently due to the low vapour content of the air limiting tropospheric absorption. Figure 2.7 shows that in the upper atmosphere over the North Pole there is a marked maximum of solar radiation at the June solstice, yet only about 30 per cent is absorbed at the surface. This may be compared with the global average of 48 per cent of solar radiation being absorbed at the surface. The explanation lies in the high average cloudiness over the Arctic in summer and also in the high reflectivity of the snow and ice surfaces. This example illustrates the complexity of the radiation budget and the need to take into account the interaction of several factors.

A special feature of the latitudinal receipt of radiation is that the maximum temperatures experienced at the earth's surface do not occur at the equator, as one might expect, but at the tropics. A number of factors need to be taken into account. The apparent migration of the vertical sun is relatively rapid during its passage over the equator, but its rate slows

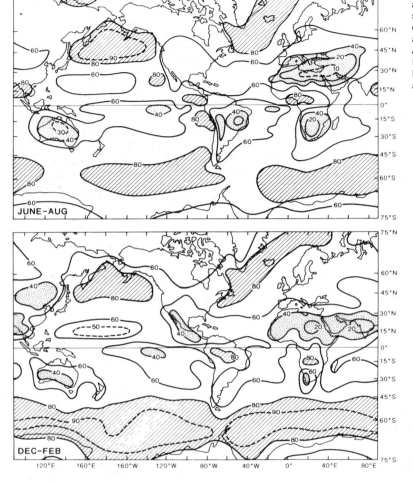

Figure 2.8 The global distribution of total cloud amount (per cent) derived from surface-based observations during the period 1971–81, averaged for the months June–August (above) and December–February (below). High percentages are shaded and low percentages are stippled.

Source: From London *et al.* 1989.

down as it reaches the tropics. Between 6 °N and 6 °S the sun's rays remain almost vertically overhead for only 30 days during each of the spring and autumn equinoxes, allowing little time for any large build-up of surface heat and high temperatures. On the other hand, between 17.5 and 23.5° latitude the sun's rays shine down almost vertically for 86 consecutive days during the period of the solstice. This longer sustained period, combined with the fact that the tropics experience longer days than at the equator, makes the maximum zones of heating occur nearer the tropics than the equator. In the northern hemisphere, this poleward displacement of the zone of maximum heating is emphasized by the effect of *continentality* (see B.5, this chapter), while low

cloudiness associated with the subtropical high-pressure belts is an additional factor. The clear skies are particularly effective in allowing large annual receipts of solar radiation in these areas. The net result of these influences is shown in Figure 2.9 in terms of the average annual solar radiation on a horizontal surface at ground level, and by Figure 2.10 in terms of the average daily maximum shade temperatures. Over the continents, the highest values occur at about 23 °N and 10–15 °S. In consequence, the mean annual *thermal equator* (i.e. the zone of maximum temperature) is located at about 5 °N. Nevertheless, the mean air temperatures, reduced to mean sea level, are very broadly related to latitude (see Figures 2.11A and B).

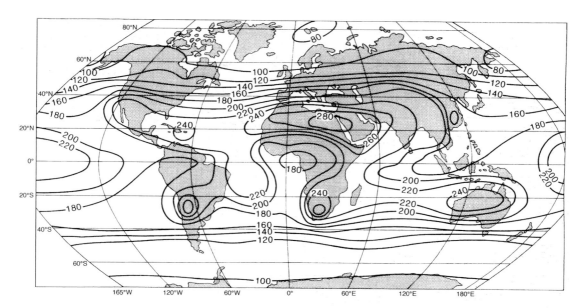

Figure 2.9 The mean annual global solar radiation (*Q* + *q*)(W m^{-2})(i.e. on a horizontal surface at ground level). Maxima are found in the world's hot deserts, where as much as 80 per cent of the solar radiation annually incident on the top of the unusually cloud-free atmosphere reaches the ground.

Source: After Budyko et al. 1962.

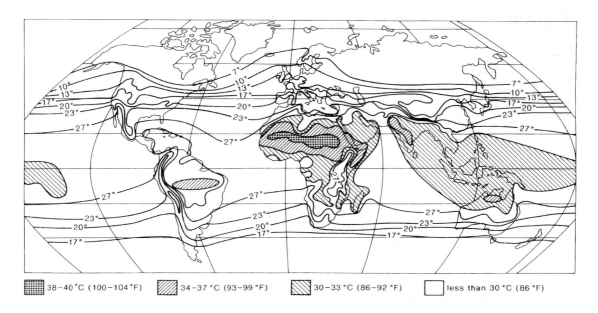

Figure 2.10 Mean daily maximum shade air temperatures (°C).

Source: After Ransom 1963.

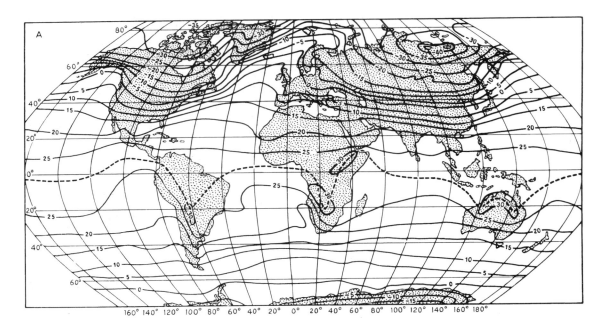

Figure 2.11A Mean sea-level temperatures (°C) in January. The position of the thermal equator is shown approximately by the dashed line.

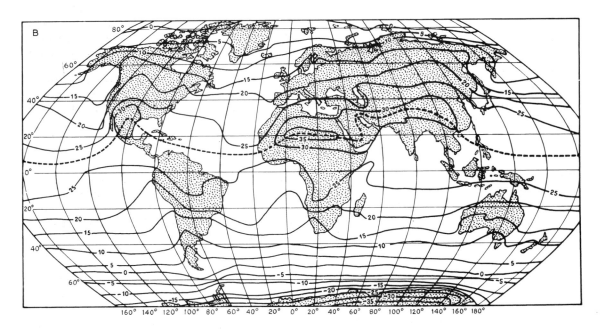

Figure 2.11B Mean sea-level temperatures (°C) in July. The position of the thermal equator is shown approximately by the dashed line.

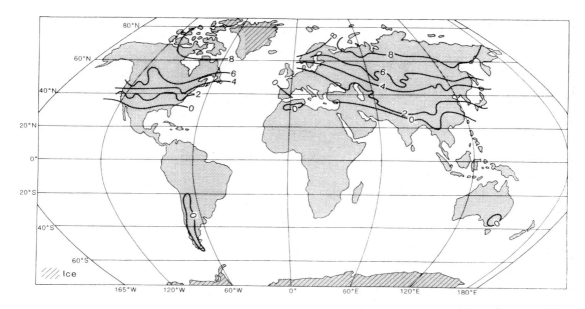

Figure 2.12 Average annual snow cover duration (months).
Source: Henderson-Sellers and Wilson 1983.

5 Effect of land and sea

Another important control on the effect of incoming solar radiation stems from the different ways in which land and sea are able to profit from it. Whereas water has a tendency to store the heat it receives, land, in contrast, quickly returns it to the atmosphere. There are several reasons for this.

A large proportion of the incoming solar radiation is reflected back into the atmosphere without heating the earth's surface at all. The proportion depends upon the type of surface (see Table 2.2). A sea surface reflects very little unless the angle of incidence of the sun's rays is large. The albedo for a calm water surface is only 2 to 3 per cent for a solar elevation angle exceeding 60°, but is more than 50 per cent when the angle is 15°. For land surfaces, the albedo is generally between 8 and 40 per cent of the incoming radiation. The figure for forests is about 9 to 18 per cent according to the type of tree and density of foliage (see Chapter 10C), for grass approximately 25 per cent, for cities 14 to 18 per cent, and for desert sand 30 per cent. Fresh snow may reflect as much as 90 per cent of solar radiation, but snow cover on vegetated, especially forested, surfaces is much less reflective (30–50 per cent). The long duration of snow cover on the northern continents (see Figure 2.12 and

Plate A) causes much of the incoming radiation to be reflected in winter, although global distribution of annual average surface albedo (Figure 2.13A) shows mainly the influence of the snow-covered Arctic sea ice and Antarctic ice sheet (see Figure 2.13B for planetary albedo).

The global solar radiation absorbed at the surface is determined from measurements of radiation incident on the surface and its albedo (a). It may be expressed as

$$(Q + q)(100 - a)$$

where the albedo is a percentage. A snow cover will absorb only about 15 per cent of the incident radiation, whereas for the sea the figure generally exceeds 90 per cent. The ability of the sea to absorb the heat received also depends upon its transparency. As much as 20 per cent of the radiation penetrates as far down as 9 m (30 ft). Figure 2.14 provides some indication of how much energy is absorbed by the sea at different depths. However, the heat absorbed by the sea is carried down to considerable depths by the turbulent mixing of water masses by the action of waves and currents. Figure 2.15, for example, illustrates the mean monthly variations with depth in the upper 100 metres of the waters of the eastern North Pacific (around 50 °N, 145 °W), showing the development of the seasonal thermocline under

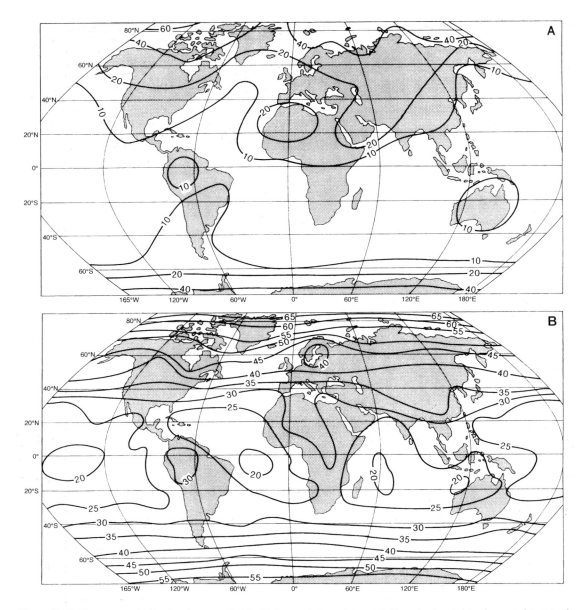

Figure 2.13 Mean annual albedos (per cent): (A) At the earth's surface; (B) On a horizontal surface at the top of the atmosphere.

Source: After Hummel and Reck; from Henderson-Sellers and Wilson 1983, and Stephens *et al.* 1981.

the influences of surface heating, vertical mixing and surface conduction.

A measure of the difference between the subsurfaces of land and sea is given in Figure 2.16, which shows ground temperatures at Kaliningrad (Königsberg) and sea temperature deviations from the annual mean at various depths in the Bay of Biscay.

Heat transmission in the soil is carried out almost wholly by conduction, and the degree of conductivity varies with the moisture content and porosity of each particular soil.

Air is an extremely poor conductor and for this reason a loose, sandy soil surface heats up rapidly by day, as the heat is not conducted away. Increased

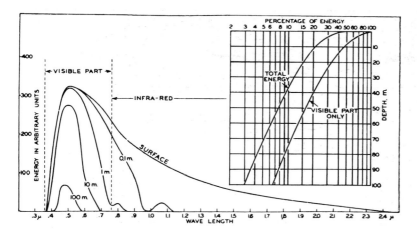

Figure 2.14 Schematic representation of the energy spectrum of the sun's radiation (in arbitrary units) that penetrates the sea surface to depths of 0.1, 1, 10 and 100 m. This illustrates the absorption of infrared radiation by water, and also shows the depths to which visible (light) radiation penetrates.

Source: From Sverdrup 1945.

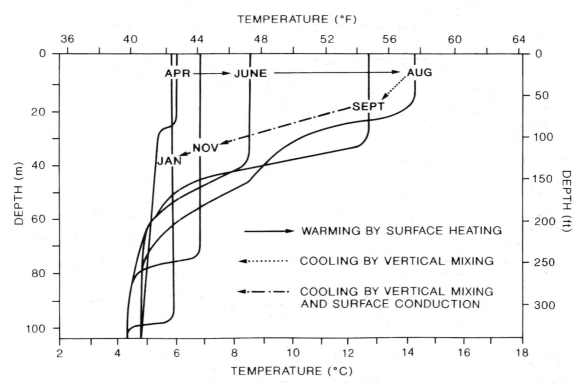

Figure 2.15 Mean monthly variations of temperature with depth in the surface waters of the eastern North Pacific. The layer of rapid temperature change is termed the thermocline.

Source: Tully and Giovando 1963; from Trenberth 1994. Reproduced by permission of the Royal Society of Canada.

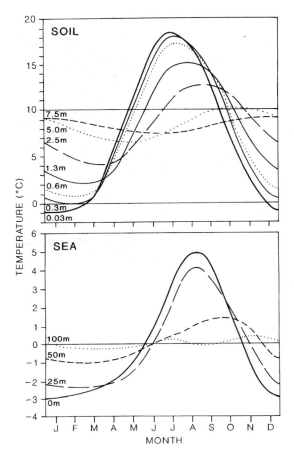

Figure 2.16 Annual variation of temperature at different depths in soil at Kaliningrad, European Russia (above) and in the water of the Bay of Biscay (at approximately 47 °N, 12 °W) (below), illustrating the relatively deep penetration of solar energy into the oceans as distinct from that into land surfaces. The bottom figure shows the temperature deviations from the annual mean for each depth.

Sources: Geiger 1965 and Sverdrup 1945.

soil moisture tends to raise the conductivity by filling the soil pores, but too much moisture increases the soil's heat capacity, thereby reducing the temperature response. The relative depths over which the annual and diurnal temperature variations are effective in wet and dry soils are roughly as follows:

	Diurnal variation	Annual variation
Wet soil	0.5 m	9 m
Dry sand	0.2 m	3 m

However, the *actual* temperature change is greater in dry soils. For example, the following values of diurnal temperature range have been observed during clear summer days at Sapporo, Japan:

	Sand	Loam	Peat	Clay
Surface	40 °C	33 °C	23 °C	21 °C
5 cm	20	19	14	14
15 cm	7	6	2	4

The different heating qualities of land and water are also partly accounted for by their different *specific heats*. The specific heat (*c*) of a substance can be represented by the number of thermal units required to raise a unit mass of it through 1 °C (4,184 J kg^{-1} K^{-1}). The specific heat of water is much greater than for most other common substances, and water must absorb five times as much heat energy to raise its temperature by the same amount as a comparable mass of dry soil. Thus for dry sand, *c* = 840 J kg^{-1} K^{-1}.

If unit volumes of water and soil are considered, the heat capacity, ρc, of the water, where ρ = density (ρc = 4.18 × 10^6 J m^{-3} K^{-1}), exceeds that of the sand approximately threefold (ρc = 1.3 × 1.6 J m^{-3} K^{-1}) if the sand is dry and twofold if it is wet. When this water is cooled the situation is reversed, for then a large quantity of heat is released. A metre-thick layer of sea water being cooled by as little as 0.1 °C will release enough heat to raise the temperature of an approximately 30 m thick air layer by 10 °C. In this way, the oceans act as a very effective reservoir for much of the world's heat. Similarly, evaporation of sea water causes a large heat expenditure because a great amount of energy is needed to evaporate even a small quantity of water (see Chapter 3C).

The thermal role of the ocean is an important and complex one. The ocean has three thermal layers:

1 The seasonal boundary, or upper mixed, layer, lying above the thermocline. This is less than 100 m deep in the tropics but is hundreds of metres deep in the subpolar seas. It is subject to annual thermal mixing from the surface.

2 The warm water sphere or lower mixed layer. This underlies '1' and slowly exchanges heat with it down to many hundreds of metres.

3 The deep ocean. This contains some 80 per cent of the total oceanic water volume and exchanges heat with '1' in polar seas.

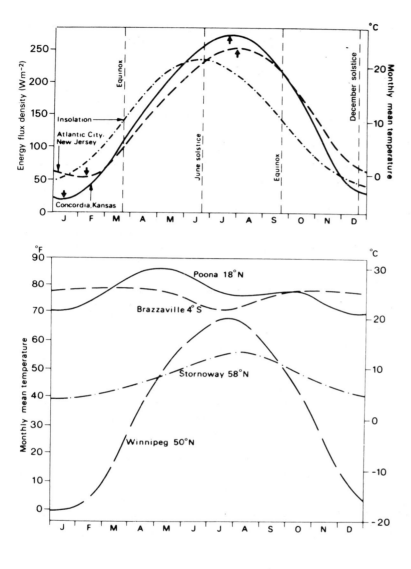

Figure 2.17 Mean annual temperature regimes in various climates and the relationships with solar radiation. Above: temperatures at maritime (Atlantic City, New Jersey, USA) and continental (Concordia, Kansas) locations in the middle latitudes. A curve of representative solar radiation is also given. Maximum and minimum points are indicated on the temperature curves, illustrating the respective time lags behind the radiation curve. †Below: mean annual temperature regimes for Pune (Poona), India (monsoon), Brazzaville, Congo Republic (equatorial), Stornoway, Scotland (temperate maritime) and Winnipeg, Canada (temperate continental).

Source: After Strahler 1965. †Data from Trewartha

This vertical thermal circulation allows global heat to be conserved in the oceans, thus damping down the global effects of climatic change produced by thermal forcing (see Chapter 11). The time for heat energy to diffuse within the upper mixed layer is two–seven months, within the lower mixed layer seven years, and within the deep ocean upwards of 300 years. The comparative figure for the outer thermal layer of the solid earth is only 11 days.

These differences between land and sea help to produce what is termed *continentality*. Continentality implies, first, that a land surface heats and cools much quicker than that of an ocean. Over the land, the lag between maximum and minimum periods of radiation and the maximum and minimum surface temperatures is only one month, but over the ocean and at coastal stations the lag is as much as two months (Figure 2.17). Second, the annual and diurnal ranges of temperature are greater in continental than in coastal locations. Figure 2.17 illustrates the annual variation of temperature at Winnipeg, Canada and Stornoway, north-western Scotland, while diurnal temperature ranges experienced in continental and maritime areas are described below (see pp. 42–4). The third effect of continentality results from the global distribution

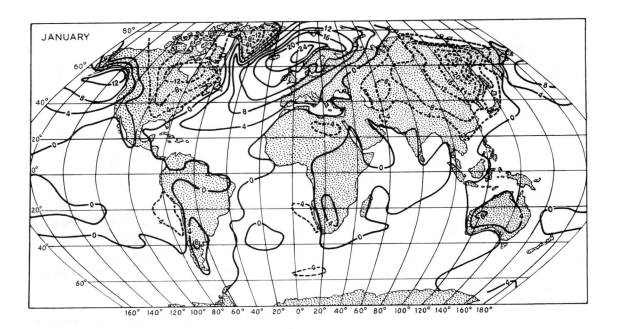

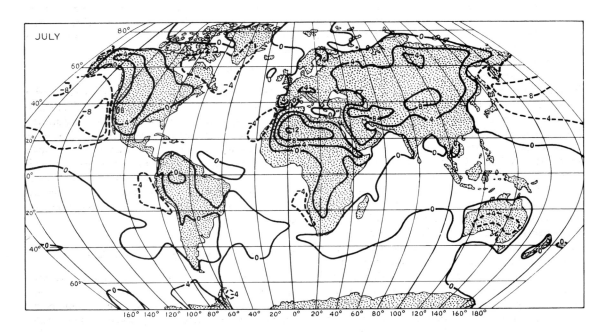

Figure 2.18 World temperature anomalies (i.e. the difference between recorded temperatures °C and the mean for that latitude) for January and July. Solid lines indicate positive, and dashed lines negative, anomalies.

of the land masses. The small sea area of the northern hemisphere causes the northern hemisphere summer to be warmer but its winters colder on the average than those of the southern hemisphere (summer, 22.4 °C versus 17.1 °C; winter, 8.1 °C versus 9.7 °C). Heat storage in the oceans causes them to be warmer in winter and cooler in summer than land in the same latitude, although ocean currents give rise to some local departures from this rule. The distribution of temperature anomalies for the latitude in January and July (Figure 2.18) illustrates the significance of continentality and also the influence of the warm drift currents in the North Atlantic and the North Pacific in winter (compare Figure 6.36).

Sea-surface temperatures can now be estimated by the use of infrared satellite imagery (see C, this chapter). Plate B is a false-colour satellite thermal image of the western North Atlantic showing the relatively warm, meandering Gulf Stream. From such images, maps of sea-surface temperatures are now routinely constructed.

6 Effect of elevation and aspect

When we come down to the local scale, even differences in the elevation of the land and its *aspect* (that is the direction that the surface faces) strongly control the amount of solar radiation received.

Obviously, some slopes are more exposed to the sun than others, whereas really high elevations that have a much smaller mass of air above them (see Figure 1.13) receive considerably more direct solar radiation under clear skies than locations near sea level, particularly below 2,000–3,000 m due to the concentration of water vapour in the lower troposphere (Figure 2.19). On the average in middle latitudes the intensity of incident solar radiation increases by 5–15 per cent for each 1,000 m increase in elevation in the lower troposphere. The difference between sites at 200 and 3,000 m in the Alps, for instance, can amount to 70 W m^{-2} on cloudless summer days. However, there is also a correspondingly greater net loss of terrestrial radiation at higher elevations because the low density of the overlying air results in a smaller fraction of the outgoing radiation being absorbed. The overall effect is invariably complicated by the greater cloudiness associated with most mountain ranges, and it is therefore impossible to generalize from the limited data at present available.

Figure 2.20 illustrates the effect of aspect and slope angle on theoretical maximum solar radiation

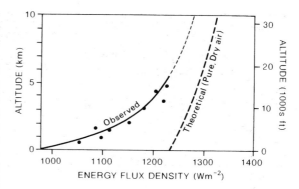

Figure 2.19 Direct solar radiation as a function of altitude observed in the European Alps. The absorbing effects of water vapour and dust, particularly below about 3,000 m, are shown by comparison with a theoretical curve for an ideal atmosphere.

Source: After Albetti, Kastrov, Kimball and Pope; from Barry 1992.

receipts at two locations in the northern hemisphere. The general effect of latitude on insolation amounts is clearly shown, but it is also apparent that increasing latitude causes a relatively greater radiation loss for north-facing slopes, as distinct from south-facing ones. The radiation intensity on a sloping surface (I_s) is

$$I_s = I_o \cos i$$

where i = the angle between the solar beam and a beam normal to the sloping surface. Relief may also affect the quantity of insolation and the duration of direct sunlight when a mountain barrier screens the sun from valley floors and sides at certain times of day. In many Alpine valleys, settlement and cultivation are noticeably concentrated on southward-facing slopes (the adret or sunny side), whereas northward slopes (ubac or shaded side) remain forested.

7 Variation of temperature with height

The last section described the gross characteristics of the vertical temperature profile in the atmosphere. Now we examine in more detail some of the features of the temperature gradient at low levels.

Vertical temperature gradients are determined in part by energy transfers and in part by vertical motion of the air. The various factors interact in a highly complex manner. The energy terms are the release of latent heat by condensation, radiational

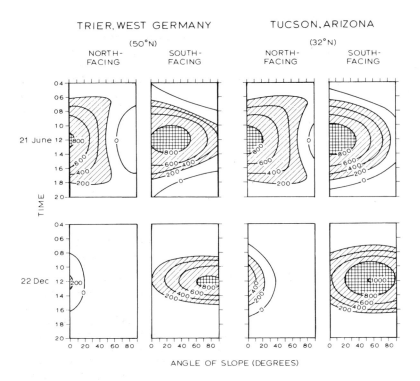

TRIER, WEST GERMANY   TUCSON, ARIZONA

(50°N)   (32°N)

NORTH-   SOUTH-   NORTH-   SOUTH-
FACING   FACING   FACING   FACING

TIME

ANGLE OF SLOPE (DEGREES)

Figure 2.20 Average direct beam solar radiation (W m⁻²) incident at the surface under cloudless skies at Trier, West Germany, and Tucson, Arizona, as a function of slope, aspect, time of day and season of year.

Source: After Geiger 1965 and Sellers 1965.

cooling of the air and sensible heat transfer from the ground. Horizontal temperature advection, by the motion of cold and warm air masses, may also be important. Vertical motion is dependent on the type of pressure system. High-pressure areas are generally associated with descent and warming of deep layers of air, hence decreasing the temperature gradient and frequently causing temperature inversions in the lower troposphere. In contrast, low-pressure systems are associated with rising air, which cools upon expansion and increases the vertical temperature gradient. This is only part of the story, since moisture is an additional complicating factor (see Chapter 3E). It remains true, however, that the middle and upper troposphere is relatively cold above a surface low-pressure area, leading to a steeper temperature gradient.

The overall vertical decrease of temperature, or *lapse rate*, in the troposphere is, as has been stated, about 6.5 °C/km. However, this is by no means constant with height, season or location. Average global values calculated by C. E. P. Brooks for July show increasing lapse rate with height: 5 °C/km in the lowest 2 km, 6 °C/km between 4 and 5 km, and

7 °C/km between 6 and 8 km. Winter values are generally smaller and in continental areas, such as central Canada or eastern Siberia, may even be negative (i.e. temperatures increase with height in the lower layer) as a result of excessive radiational cooling over a snow surface (Figure 2.21). A similar situation occurs when dense, cold air accumulates in mountain basins on calm, clear nights. On such occasions, the mountain tops may be many degrees warmer than the valley floor below (see Chapter 5C.1). For this reason, the adjustment of average temperature of upland stations to mean sea level may produce misleading results. Observations in Colorado at Pike's Peak (4,301 m) and Colorado Springs (1,859 m) show the mean lapse rate to be 4.1 °C/km in winter and 6.2 °C/km in summer! It should be noted that such topographic lapse rates may bear little relation to free lapse rates in nocturnal radiation conditions, and the two must be carefully distinguished.

Table 2.3 summarizes the seasonal characteristics of lapse rates in six major climatic zones, and examples of five of these are illustrated in Figure 2.21. The seasonal regime is very pronounced in

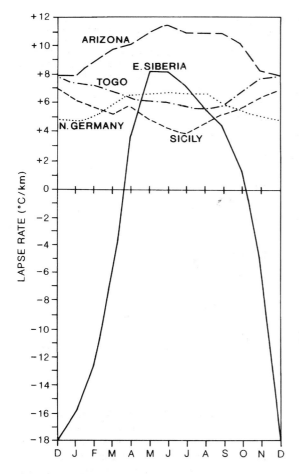

Figure 2.21 The annual variation of lapse rate in five climatic zones: tropical rainy climate (Togo); tropical desert (Arizona); mediterranean (Sicily); mid-latitude, cold winter climate (north Germany); boreal continental (eastern Siberia).

Source: Hastenrath 1968.

continental areas, with cold winters, whereas inversions persist for much of the year in the Arctic. In winter, the Arctic inversion is due to intense radiational cooling, but in summer it is the result of the surface cooling of advected warmer air. The winter lapse rate is greater than the summer one only in mediterranean climates. In these regions, there is more likelihood of rising air associated with low-pressure areas in winter. In contrast, subsidence is predominant in the desert zones in winter. The tropical and subtropical deserts have very steep lapse rates in summer, when there is considerable heat transfer from the surface and generally ascending motion.

C TERRESTRIAL INFRARED RADIATION AND THE GREENHOUSE EFFECT

Radiation from the sun is predominantly short wave, whereas that leaving the earth is long wave, or infrared, radiation (see Figure 2.1). The infrared emission from the surface is slightly less than that from a black body at the same temperature and, accordingly, Stefan's equation (see p. 20) is modified by an emissivity coefficient (ε), which is generally between 0.90 and 0.95, i.e. $F = \varepsilon\sigma T^4$. Figure 2.1 shows that the atmosphere is highly absorbent to infrared radiation (due to the effects of water vapour, carbon dioxide and other trace gases), except between about 8.5 and 13.0 μm – the 'atmospheric window'. The opaqueness of the atmosphere to infrared radiation, relative to its transparency to short-wave radiation, is commonly referred to as the *greenhouse effect*. However, in the case of an actual greenhouse, the effect of the glass roof is probably as significant in reducing cooling by restricting the turbulent heat loss as it is in retaining the infrared radiation.

The total 'greenhouse' effect results from the net infrared absorption capacity of water vapour, carbon

Table 2.3 Temperature lapse rates in the lowest 1,000–15,000 metres.

Climate	Season of maximum	Rate°C/km	Season of minimum	Rate °C/km
Tropical rainy	dry season	>5	rainy season	>4.5
Tropical and subtropical deserts	summer	>8	winter	>5
Mediterranean	winter	>5	summer	<5
Mid-latitudes (cold winter)	summer	>6	winter	0–5
Boreal continental	summer	>5	winter	<0
Arctic	summer	<0	winter	<0

Source: After Lautensach and Bögel 1956.

Table 2.4 Influence of greenhouse gases on atmospheric temperature.

Gas	Centres of main absorption bands (μm)	Temperature increase (K) for ×2 present concentration	Global warming potential on a weight basis (kg⁻¹ of air)†
Water vapour (H_2O)	6.3–8.0, >15 (8.3–12.5)*		
Carbon dioxide (CO_2)	(5.2), (10), 14.7	3.0 ± 1.5	1
Methane (CH_4)	6.52, 7.66	0.3–0.4	11
Ozone (O_3)	4.7, 9.6, (14.3)	0.9	
Nitrous oxide (N_2O)	7.78, 8.56, 17.0	0.3	270
Chlorofluoromethanes			
($CFCl_3$)	4.66, 9.22, 11.82 ⎫		3,400
(CF_2Cl_2)	8.68, 9.13, 10.93 ⎭	0.1	7,100

Sources: After Campbell; Ramanathan; Lashof and Ahuja; Luther and Ellingson; IPCC 1992.
Notes: *Important in moist atmospheres.
             †Refers to direct annual radiative forcing for the surface-troposphere system.

dioxide and other trace gases – methane (CH_4), nitrous oxide (N_2O) and tropospheric ozone (O_3) – which absorb strongly at wavelengths within the atmospheric window region, in addition to their other absorbing bands (see Figure 2.1 and Table 2.4). Moreover, because concentrations of these trace gases are low, their radiative effects increase approximately linearly with concentration, whereas the effect of CO_2 is related to the logarithm of the concentration. In addition, because of the long atmospheric residence time of nitrous oxide (132 years) and CFCs (65–140 years), the cumulative effects of human activities will be substantial. It is estimated that between 1765 and 1990, the radiative effect of increased CO_2 concentration was 1.5 W m⁻², and of all trace gases 2.5 W m⁻² (cf. the solar constant value of 1,368 W m⁻²).

The net warming contribution of the natural (non-anthropogenic) greenhouse gases to the mean 'effective' planetary temperature of 255 K (corresponding to the emitted infrared radiation) is approximately 33 K; of this, water vapour accounts for 21 K, carbon dioxide 7 K, ozone 2 K, and other trace gases (nitrous oxide, methane) about 3 K. The present global mean surface temperature is 288 K, but the surface was considerably warmer during the early evolution of the earth, when the atmosphere contained large quantities of methane, water vapour and ammonia. The largely carbon dioxide atmosphere of Venus creates a 500 K greenhouse effect on that planet.

Stratospheric ozone absorbs significant amounts of both incoming ultraviolet radiation, harmful to life, and outgoing terrestrial long-wave reradiation, so that its overall thermal role is a complex one. Its net effect on earth surface temperatures depends on the elevation at which the absorption occurs, being to some extent a trade-off between short- and long-wave absorption in that:

1 An increase of ozone above about 30 km absorbs relatively more incoming short-wave radiation, causing a net *decrease* of surface temperatures.
2 An increase of ozone below about 25 km absorbs relatively more outgoing long-wave radiation, causing a net *increase* of surface temperatures.

It is worth emphasizing that long-wave radiation is not merely terrestrial in the narrow sense. The atmosphere radiates to space, and clouds are particularly effective since they act as black bodies. For this reason, cloudiness and cloud-top temperature can be mapped from satellites by day and night using infrared sensors (see Plates 2 and 19, where high clouds appear cold). Radiative cooling of cloud layers averages about 1.5 °C per day.

For the globe as a whole, satellite measurements show that in cloud-free conditions the mean absorbed solar radiation is approximately 285 W m⁻², whereas the emitted terrestrial radiation is 265 W m⁻². Including cloud-covered areas, the corresponding global values are 241 W m⁻² and 235 W m⁻², respectively. Clouds reduce the absorbed solar radiation by 50 W m⁻², but reduce the emitted radiation by only 30 W m⁻². Hence, global cloud cover causes a net radiative loss of about 20 W m⁻², due to the dominance of cloud albedo reducing short-wave radiation absorption. In lower latitudes, this effect is much larger (up to –50 to –100 W m⁻²)

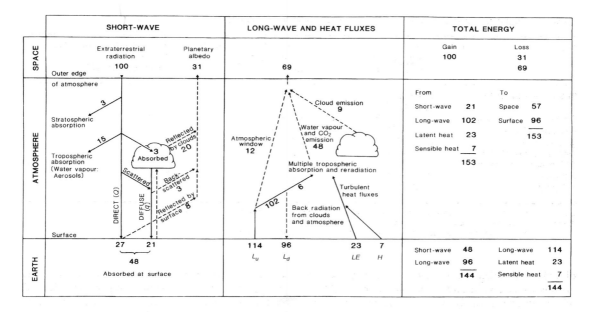

Figure 2.22 The balance of the atmospheric energy budget. The transfers are explained in the text. Solid lines indicate energy gains by the atmosphere and surface in the left-hand diagram and the troposphere in the right-hand diagram. The exchanges are referred to 100 units of incoming solar radiation at the top of the atmosphere (equal to 342 W m⁻²).

Source: Data after Trenberth *et al.* 1996.

whereas in high latitudes the two factors are close to balance, or the increased infrared absorption by clouds may lead to a small positive value. These results are important in terms of changing concentrations of greenhouse gases, since the net radiative forcing by cloud cover is four times that expected from CO_2 doubling (see Chapter 11).

D HEAT BUDGET OF THE EARTH

We can now summarize the net effect of the transfers of energy in the earth–atmosphere system averaged over the globe and over an annual period.

The incident solar radiation averaged over the globe is

$$\text{Solar constant} \times \pi r^2 / 4\pi r^2$$

where r = radius of the earth and $4\pi r^2$ is the surface area of a sphere. This figure is approximately 342 W m⁻², or 11×10^9 J m⁻² yr⁻¹ (10^9 J = 1GJ); for convenience we will regard it as 100 units. Referring to Figure 2.22, incoming radiation is absorbed in the stratosphere (3 units), by ozone mainly, and 18 units are absorbed in the troposphere by carbon dioxide (1), water vapour (12), dust (2) and water droplets

in clouds (3). Twenty units are reflected back to space from clouds, which cover about 62 per cent of the earth's surface, on average. A further 8 units are similarly reflected from the surface and 3 units are returned by atmospheric scattering. The total reflected radiation is the *planetary albedo* (31 per cent or 0.31). The remaining 48 units reach the earth either directly (27) or as diffuse radiation (21) transmitted via clouds or by downward scattering.

The pattern of outgoing terrestrial radiation is quite different (see Figure 2.22). The black-body radiation, assuming a mean surface temperature of 288 K, is equivalent to 114 units of infrared (long-wave) radiation. This is possible because most of the outgoing radiation is reabsorbed by the atmosphere; the *net* loss of infrared radiation is only 18 units. These exchanges represent a time-averaged state for the whole globe. Recall that solar radiation affects only the sunlit hemisphere, where the incoming radiation exceeds 342 W m⁻². Conversely, on the night-time hemisphere no solar radiation is received. Infrared exchanges continue, however, due to the accumulated heat in the ground. Only about 12 units escape through the atmospheric window directly from the surface, but the atmosphere radiates 57 units to space (48 from the emission by water vapour and CO_2 in the atmosphere and 9 from cloud emission), giving a total of 69 units, as well as reradiating 96 units back to the surface (L_d); $L_u + L_d = L_n$ is negative.

These radiation transfers can be expressed symbolically:

$$R_n = (Q + q) (1 - a) + L_n$$

where R_n = net radiation, $(Q + q)$ = global solar radiation, a = albedo and L_n = net long-wave radiation. At the surface, R_n = 30 units. This surplus is conveyed to the atmosphere by the turbulent transfer of sensible heat, or enthalpy (7 units), and latent heat (23 units).

$$R_n = LE + H$$

where H = sensible heat transfer and LE = latent heat transfer. There is also a flux of heat into the ground (B.5, this chapter), but for annual averages this is approximately zero.

Figure 2.22 summarizes the total balances at the surface (±144 units) and for the atmosphere (±152 units). The total absorbed solar radiation and emitted radiation for the entire earth–atmosphere system is estimated to be ±7 GJ m^{-2} yr^{-1} (±69 units). Various uncertainties are still to be resolved in these estimates. The surface short-wave and long-wave radiation budgets have an uncertainty of about 20 W m^{-2}, and the turbulent heat fluxes of about 10 W m^{-2}.

Satellite measurements now provide global views of the energy balance at the top of the atmosphere. The incident solar radiation is almost symmetrical about the equator in the annual mean (cf. Table 2.1). The mean annual totals on a horizontal surface at the top of the atmosphere are approximately 420 W m^{-2} at the equator and 180 W m^{-2} at the poles. The distribution of the planetary albedo (see Figure 2.13B) shows the lowest values over the low-latitude oceans compared with the more persistent areas of cloud cover over the continents. The highest values are over the polar ice caps. The resulting planetary short-wave radiation ranges from 340 W m^{-2} at the equator to 80 W m^{-2} at the poles. The net (outgoing) long-wave radiation (Figure 2.23B) shows the smallest losses where the temperatures are lowest and largest losses over the largely clear skies of the Saharan desert surface and over low-latitude oceans. The difference between Figure 2.23A and 2.23B represents the net radiation of the earth–atmosphere system which achieves balance about latitude 30 °N. The consequences of a low-latitude energy surplus and a high-latitude deficit are examined below.

The annual and diurnal variation of temperature are directly related to the local energy budget. Under clear skies, in middle and lower latitudes, the diurnal regime of radiative exchanges generally shows a midday maximum of absorbed solar radiation (see Figure 2.24A). A maximum of infrared (long-wave) radiation (see Figure 2.1) is also emitted by the heated ground surface at midday, when it is warmest. The atmosphere reradiates infrared radiation downward, but there is a net loss at the surface (L_n). The difference between the absorbed solar radiation and L_n is the net radiation, R_n; this is generally positive between about an hour after sunrise and an hour or so before sunset, with a midday maximum. The delay in the occurrence of the maximum air temperature until about 1400 hours local time (Figure 2.24B) is caused by the gradual heating of the air by convective transfer from the ground. Minimum R_n occurs in the early evening, when the ground is still warm; there is a slight increase thereafter. The temperature decrease after midday is slowed by heat supplied from the ground. Minimum air temperature occurs shortly after sunrise due to the lag in the transfer of heat from the surface to the air. The annual pattern of the net radiation budget and temperature regime is closely analogous to the diurnal one.

There are marked latitudinal variations in the diurnal and annual ranges of temperature. Broadly, the annual range is a maximum in higher latitudes, with extreme values about 65 °N related to the effects of continentality in Asia and North America. In contrast, in low latitudes (Figure 2.25) the boundary between land and sea is scarcely apparent due to the thermal similarity between tropical rainforests and tropical oceans. The diurnal range reaches a maximum at the tropics over land areas, but it is in the equatorial zone that the diurnal variation of heating and cooling exceeds the annual one (see Figure 2.24B). This is of course related to the small seasonal change in solar elevation angle at the equator. From the point of view of the total energy budget, the atmosphere, oceans and upper crustal skin of the earth form a complex system of storages and transfers. The global climate as a whole is dominated by the relatively small energy storages provided by the crustal skin and the atmosphere, compared with that of the oceans (see Chapter 6).

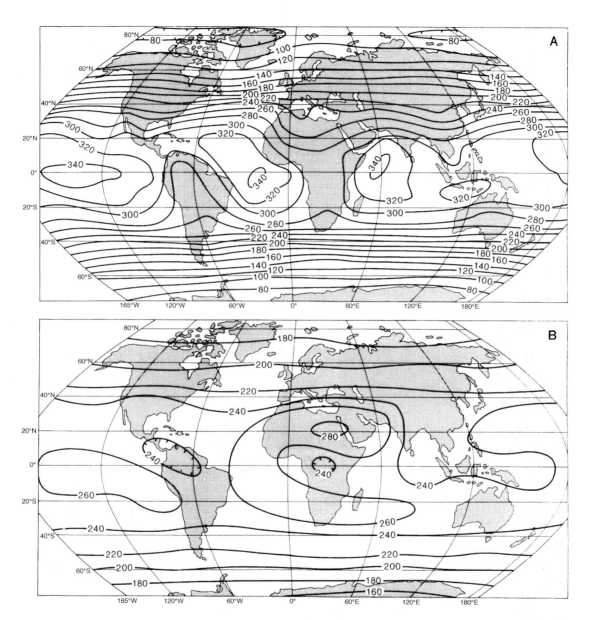

Figure 2.23 Planetary short- and long-wave radiation (W m^{-2}): (A) Mean annual absorbed short-wave radiation for the period April 1979 to March 1987. (B) Mean annual net planetary long-wave radiation (L_n) on a horizontal surface at the top of the atmosphere.

Sources: (A) Ardanuy *et al*. 1992 and Kyle *et al*. 1993. From *Bulletin of the American Meteorological Society*, by permission of the American Meteorological Society.
   (B) Stephens *et al*. 1981.

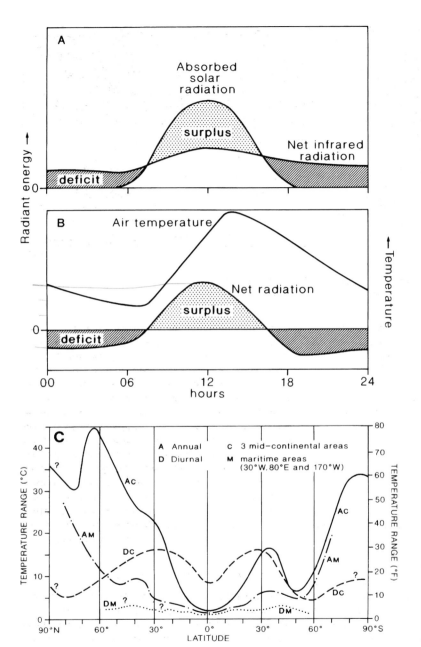

Figure 2.24 Curves showing diurnal variations of radiant energy and temperature. (A) Diurnal variations in absorbed solar radiation and infrared radiation in the middle and low latitudes. (B) Diurnal variations in net radiation and air temperature in the middle and low latitudes. (C) Annual (A) and diurnal (D) temperature ranges as a function of latitude and of continental (C) or maritime (M) location.

Source: From Paffen 1967.

E ATMOSPHERIC ENERGY AND HORIZONTAL HEAT TRANSPORT

So far, we have given an account of the earth's heat budget and its components. We have already referred to two forms of energy: internal (or heat)

energy, due to the motion of individual air molecules, and latent energy, which is released by condensation of water vapour. Two other forms of energy are important: geopotential energy due to gravity and height above the surface, and kinetic energy associated with air motion.

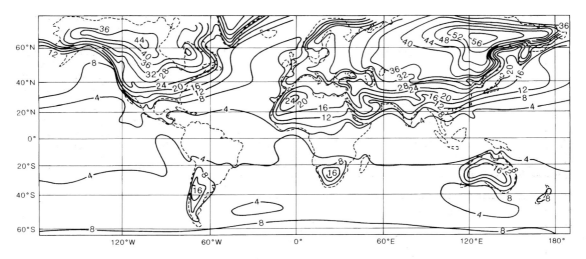

Figure 2.25 The mean annual temperature range (°C) at the earth's surface.
Source: Monin and Crowley and North 1991. Courtesy World Meteorological Organization.

Geopotential and internal energy are interrelated, since the addition of heat to an air column not only increases its internal energy but also adds to its geopotential as a result of the vertical expansion of the air column. In a column extending to the top of the atmosphere, the geopotential is approximately 40 per cent of the internal energy. These two are therefore usually considered together and termed the total potential energy (*PE*). For the whole atmosphere

potential energy ≈ 10^{24} J
kinetic energy ≈ 10^{10} J

In a later section (Chapter 6C), we shall see how energy is transferred from one form to another, but here we need only be concerned with heat energy. It is apparent that the receipt of heat energy is very unequal geographically and that this must lead to great lateral transfers of energy across the surface of the earth. Much present-day meteorological research is focused on these transfers, since undoubtedly they give rise, at least indirectly, to the observed patterns of global weather and climate.

The amounts of energy received at different latitudes vary substantially, the equator on the average receiving 2.5 times as much annual solar energy as the poles. Clearly, if this process were not modified in some way the variations in receipt would cause a massive accumulation of heat within the tropics (associated with gradual increases of temperature) and a corresponding deficiency at the poles. Yet this

does not seem to happen, and the earth as a whole is roughly in a state of thermal equilibrium in so far as no one region is obviously gaining heat at the expense of another. Some authors believe the Ice Ages to have been an exception to this rule. One explanation of this equilibrium could be that for each region of the world there is an equalization between the amount of incoming and outgoing radiation. However, observation shows that this is not so (Figure 2.26), because, whereas incoming radiation varies appreciably with changes in latitude, being highest at the equator and declining to a minimum at the poles, outgoing radiation has a more even latitudinal distribution owing to the rather small variations in atmospheric temperature. Some other explanation therefore becomes necessary.

1 The horizontal transport of heat

If the net radiation for the whole earth–atmosphere system is calculated, it is found that there is a positive budget between 35 °S and 40 °N, as shown in Figure 2.27C. The latitudinal belts in each hemisphere separating the zones of positive and negative net radiation budgets oscillate dramatically with season (Figure 2.27A and B). As the tropics do not get progressively hotter or the high latitudes colder, a redistribution of world heat energy must occur constantly, taking the form of a continuous movement of energy from the tropics to the poles. In this way, the tropics shed their excess heat and the poles,

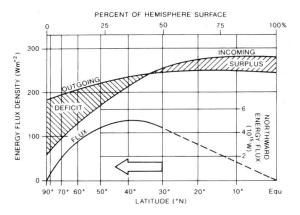

PERCENT OF HEMISPHERE SURFACE

Figure 2.26 A meridional illustration of the balance between incoming solar radiation and outgoing radiation from the earth and atmosphere* in which the zones of permanent surplus and deficit are maintained in equilibrium by a poleward energy transfer.†

Sources: *Data from Houghton; after Newell 1964.

†After Gabites.

being global heat sinks, are not allowed to reach extremes of cold. If there were no meridional interchange of heat, a radiation balance at each latitude would be achieved only if the equator were 14 °C warmer and the North Pole 25 °C colder than now. This poleward heat transport takes place within the atmosphere and oceans, and it is estimated that the former accounts for approximately two-thirds of the required total. The horizontal transport (*advection* of heat) occurs in the form of both latent heat (that is water vapour, which subsequently condenses) and sensible heat (that is warm air masses) (Figure 2.28B). It varies in intensity according to the latitude and the season. Figure 2.28B shows the mean annual pattern of energy transfer by the three mechanisms. The latitudinal zone of maximum total transfer rate is found between latitudes 35 and 45° in both hemispheres, although the patterns for the individual components are quite different from one another. The latent heat transport, which occurs almost wholly in the lowest 2 or 3 km, reflects the global wind belts on either side

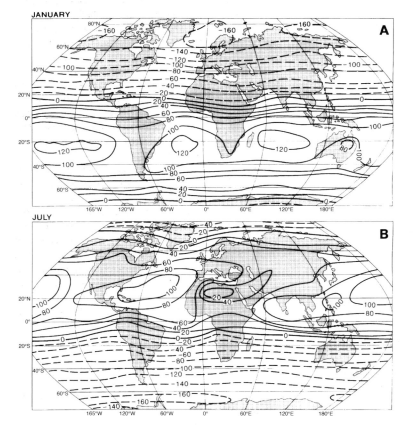

Figure 2.27 Mean net planetary radiation budget (R_n) (W m⁻²) for a horizontal surface at the top of the atmosphere (i.e. for the earth–atmosphere system). (A) January (B) July (C) Annual.

Sources: Ardanuy *et al.* 1992 and Kyle *et al.* 1993. Stephens *et al.* 1981. (C: From *Bulletin of the American Meteorological Society*, by permission of the American Meteorological Society.)

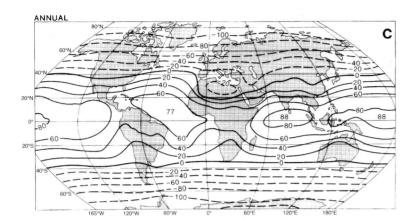

ANNUAL

Figure 2.27 (C)

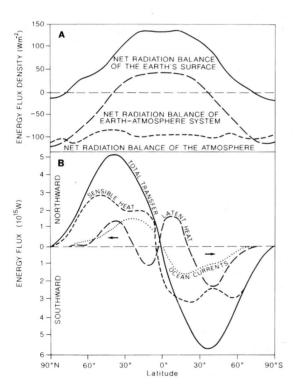

Figure 2.28 (A) Net radiation balance for the earth's surface of 101 W m⁻² (incoming solar radiation of 156 W m⁻², minus outgoing long-wave energy to the atmosphere of 55 W m⁻²); for the atmosphere of −101 W m⁻² (incoming solar radiation of 84 W m⁻², minus outgoing long-wave energy to space of 185 W m⁻²); and for the whole earth–atmosphere system of zero. (B) The average annual latitudinal distribution of the components of the poleward energy transfer (in 10¹⁵W) in the earth–atmosphere system.

Source: From Sellers 1965.

of the subtropical high-pressure zones (see Chapter 6B). The more important meridional transfer of sensible heat has a double maximum not only latitudinally but also in the vertical plane, where there are maxima near the surface and at about 200 mb. The high-level transport is particularly significant over the subtropics, whereas the primary latitudinal maximum about 50 to 60 °N is related to the travelling low-pressure systems of the westerlies.

The intensity of the poleward energy flow is closely related to the meridional (that is, north–south) temperature gradient. In winter, this temperature gradient is at a maximum and in consequence the hemispheric air circulation is most intense. The nature of the complex transport mechanisms will be discussed in Chapter 6C.

As shown in Figure 2.28B, ocean currents account for a significant proportion of the poleward heat transfer in low latitudes. Indeed, recent satellite estimates of the required total poleward energy transport indicate that the previous figures are too low. The ocean transport may be 47 per cent of the total at 30–35 °N and as much as 74 per cent at 20 °N; the Gulf Stream and Kuro Shio currents are particularly important. In the southern hemisphere, poleward transport is mainly in the Pacific and Indian Oceans (see Figure 6.30). The energy budget equation for an ocean area must be expressed as

$$R_n = LE + H + G + \Delta A$$

where ΔA = horizontal advection of heat by currents and G = the heat transferred into or out of storage in the water. The latter is more or less zero for annual averages.

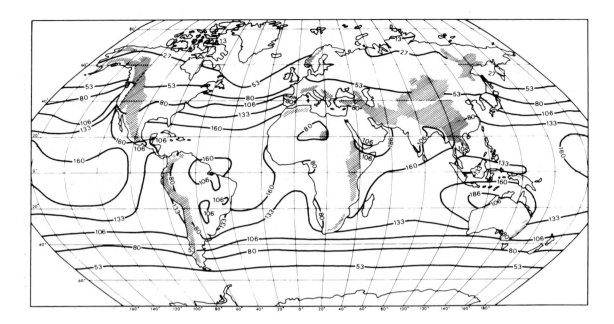

Figure 2.29 Global distribution of the annual net radiation at the surface, in W m^{-2}.
Source: After Budyko *et al.* 1962.

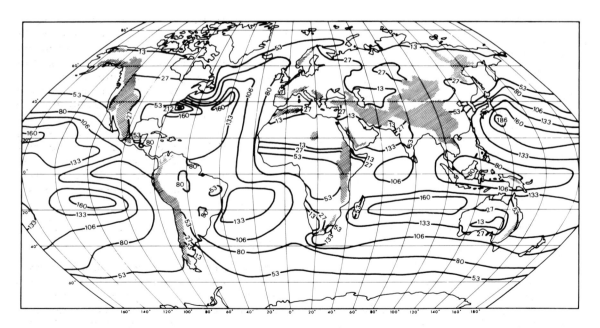

Figure 2.30 Global distribution of the vertical transfer of latent heat, in W m^{-2}.
Source: After Budyko *et al.* 1962.

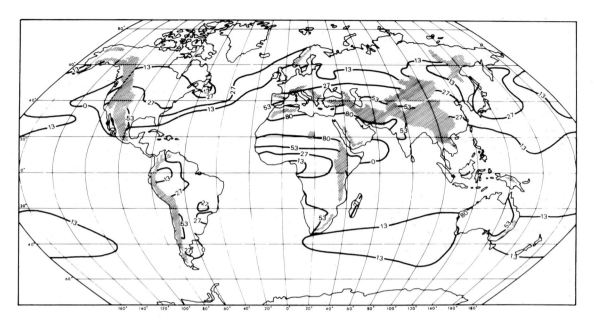

Figure 2.31 Global distribution of the vertical transfer of sensible heat, in W m⁻².
Source: After Budyko *et al*. 1962.

2 Spatial pattern of the heat budget components

The mean latitudinal values of the heat budget components discussed above conceal great spatial variations. Figure 2.29 shows the global distribution of the annual net radiation at the surface. Broadly, its magnitude decreases poleward from about 25 ° latitude, although as a result of the considerable absorption of solar radiation by the sea, the net radiation is greater over the oceans – exceeding 160 W m⁻² in latitudes 15–20 ° – than over land areas, where it is about 80–105 W m⁻² in the same latitudes. Net radiation is also rather lower in arid continental areas than in humid ones, because in spite of the increased insolation receipts under clear skies there is at the same time greater net loss of terrestrial radiation.

Figures 2.30 and 2.31 show the annual vertical transfers of latent and sensible heat to the atmosphere. Both maps show that the fluxes are distributed very differently over land and seas. Heat expenditure for evaporation is at a maximum in tropical and subtropical ocean areas, where it exceeds 160 W m⁻². It is less near the equator, where wind speeds are somewhat lower and the air has a vapour pressure close to the saturation value (see Chapter 5A). It is clear from Figure 2.30 that

the major warm currents considerably augment the evaporation rate. On land, the latent heat transfer is greatest in hot, humid regions. It is least in arid areas due to the low precipitation and in high latitudes, where there is little available energy.

The largest exchange of sensible heat occurs in the tropical deserts, where more than 80 W m⁻² is transferred to the atmosphere (see Figure 2.31). In contrast to latent heat, the sensible heat flux is generally small over the oceans, only reaching 25–40 W m⁻² in areas of warm currents. Indeed, negative values occur (transfer *to* the ocean) where warm continental air masses move offshore over cold currents.

SUMMARY

Almost all energy affecting the earth is derived from solar radiation, which is of short wavelength (<4 μm) due to the high temperature of the sun (~6,000 K) (i.e. Wien's Law). The solar constant has a value of approximately 1,370 W m⁻². The sun and the earth radiate almost as black bodies (Stefan's Law, $F = \sigma T^4$), whereas
continued

the atmospheric gases do not. Terrestrial radiation, from an equivalent black body, amounts to only about 270 W m^{-2} due to its low radiating temperature (263 K), and it is infrared (long-wave) radiation between 4 and 100 μm. Water vapour and carbon dioxide are the major absorbing gases for infrared radiation, whereas the atmosphere is largely transparent to solar radiation (the greenhouse effect). Trace gas increases are now augmenting the 'natural' greenhouse effect (33 K). Solar radiation is lost by reflection, mainly from clouds, and by absorption (largely by water vapour). The planetary albedo is 31 per cent; 48 per cent of the extraterrestrial radiation reaches the surface. The atmosphere is heated primarily from the surface by the absorption of terrestrial infrared radiation and by turbulent heat transfer. Temperature usually decreases with height at an average rate of about 6.5 °C/km in the troposphere. In the stratosphere and thermosphere, it increases with height due to the presence of radiation absorbing gases.

The excess of net radiation in lower latitudes leads to a poleward energy transport from tropical latitudes by ocean currents and by the atmosphere. This is in the form of sensible heat (warm air masses/ocean water) and latent heat (atmospheric water vapour). Air temperature at any point is affected by the incoming solar radiation and other vertical energy exchanges, surface properties (slope, albedo, heat capacity), land and sea distribution and elevation, and also by horizontal advection due to air mass movements and ocean currents.

3

Atmospheric moisture budget

A THE GLOBAL HYDROLOGICAL CYCLE

The global hydrosphere can be regarded as a series of reservoirs interconnected by water cycling in various phases. These reservoirs are the oceans; ice sheets and glaciers; terrestrial water (i.e. rivers, soil moisture, lakes and ground water); the biosphere (i.e. water in plants and animals); and the atmosphere. The oceans, with a mean depth of 3.8 km and covering 71 per cent of the earth's surface, hold 97 per cent of *all* the earth's water (1.31×10^{18} m^3). Approximately 75 per cent of the total *fresh* water is locked up in ice sheets and glaciers, while almost all of the remainder is ground water. It is an astonishing fact that at any instant rivers and lakes hold only 0.33 per cent of all fresh water and the atmosphere a mere 0.035 per cent (about $12 \times 10^{12} m^3$ (Figure 3.1). Water cycling is accomplished by evaporation, the transport of water vapour in the atmosphere, condensation, precipitation and terrestrial runoff. The average residence time of water within these reservoirs varies from hundreds or thousands of years for the oceans and polar ice to only about 10 days for the atmosphere.

Because of its large latent heat, the global occurrence and transport of water is closely linked to global energy. Atmospheric water vapour is responsible for the overriding percentage of total global energy lost into space by infrared radiation. Over 75 per cent of the energy input from the surface into the atmosphere comes from the liberation of latent heat by condensation and, principally, the production of rainfall.

For short-term processes, the water balance of the atmosphere may be assumed to be in equilibrium; however, over periods of tens of years, global warming may increase its water storage capacity.

The average storage of water vapour in the atmosphere (Table 3.1), termed the precipitable water content (about 2.5 cm), is sufficient for only some 10 days supply of rainfall over the earth as a whole. However, intense (horizontal) influx of moisture into the air over a given region makes possible short-term rainfall totals greatly in excess of 3 cm. The phenomenal record total of 187 cm fell on the island of Réunion, off Madagascar, during 24 hours in March 1952, and much greater intensities have been observed over shorter periods (see E.2a, this chapter).

B HUMIDITY

1 Moisture content

The atmospheric moisture content, comprising water vapour and water droplets and ice crystals in clouds, is determined by local evaporation, air temperature and the horizontal atmospheric transport of moisture. Cloud water, on average, amounts to only 4 per cent of atmospheric moisture. The moisture content of the atmosphere can be expressed in several ways, apart from the vapour pressure (p. 15), depending on which aspect the user wishes to emphasize. The total mass of water in a given volume of air, i.e. the density of the water vapour, is one such measure. This is termed the *absolute humidity* (ρ_w) and is measured in grams

Table 3.1 Mean water content of the atmosphere (in cm of rainfall equivalent).

	Northern hemisphere	Southern hemisphere	World
January	1.9	2.5	2.2
July	3.4	2.0	2.7

Source: After Sutcliffe 1956.

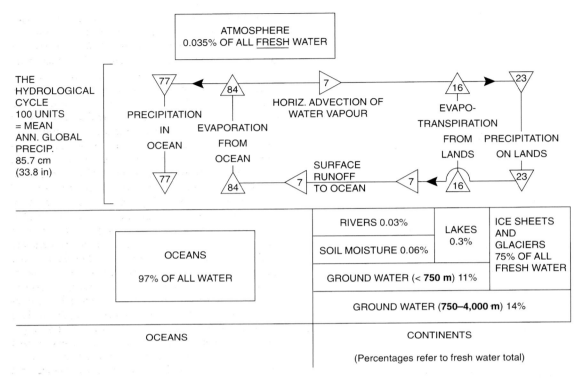

Figure 3.1 The hydrological cycle and water storage of the globe. The exchanges in the cycle are referred to 100 units, which equal the mean annual global precipitation of 85.7 cm. The percentage storage figures for atmospheric and continental water are percentages of all *fresh* water. The saline ocean waters make up 97 per cent of *all* water. The horizontal advection of water vapour indicates the *net* transfer.

Source: From More 1967.

per cubic metre (g m^{-3}). Volumetric measurements are seldom used in meteorology and more convenient is the *mass mixing ratio* (x). This is the mass of water vapour in grams per kilogram of dry air. For most practical purposes, the *specific humidity* (q) is identical, being the mass of vapour per kilogram of air, including its moisture.

More than 50 per cent of atmospheric moisture content is below 850 mb (approximately 1,450 m) and more than 90 per cent below 500 mb (5,575 m), as Figure 3.2 clearly shows. It is also apparent that the seasonal effect is most marked in the lowest 3,000 m or so, that is below about 700 mb. The global distribution of atmospheric vapour content in January and July is illustrated in Figure 3.3. Mean precipitable water generally decreases from the equator to the poles due to the direct relation between atmospheric moisture content and air temperature. There are expecially low values over deserts, where there is strong air subsidence. The greatest seasonal

variations occur in the belt 20–30 °N, the location of the Asian and African monsoons.

Air temperature sets an upper limit to water vapour pressure – the saturation value (i.e. 100 per cent relative humidity); consequently we may expect the distribution of mean vapour content to reflect this control. In January, minimum values of 0.1–0.2 cm (equivalent depth of water) occur in northern continental interiors and high latitudes, with secondary minima of 0.5–1 cm in tropical desert areas. Maximum vapour contents of 5–6 cm are over southern Asia during the summer monsoon and over equatorial latitudes of Africa and South America.

Another important measure is *relative humidity* (r), which expresses the actual moisture content of a sample of air as a percentage of that contained in the same volume of saturated air at the same temperature. The relative humidity is defined with reference to the mixing ratio, but it can be determined approximately in several ways:

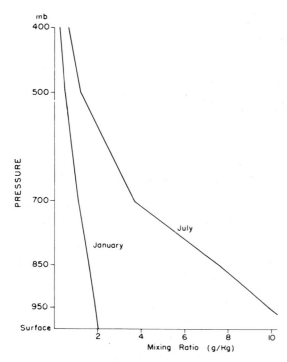

Figure 3.2 The average vertical variation of atmospheric vapour content at Portland, Maine, between 1946 and 1955.

Source: Data from Reitan 1960.

$$r = \frac{x}{x_s} \times 100 \approx \frac{q}{q_s} \times 100 \approx \frac{e}{e_s} \times 100$$

where the subscript s refers to the respective saturation values at the same temperature; e denotes vapour pressure.

A further index of humidity is the dew-point temperature. This is the temperature at which saturation occurs if air is cooled at constant pressure without addition or removal of vapour. When the air temperature and dew point are equal the relative humidity is 100 per cent, and it is evident that relative humidity can also be determined from

$$\frac{e_s \text{ at dew point}}{e_s \text{ at air temperature}} \times 100$$

The relative humidity of a parcel of air will obviously change if either its temperature or its mixing ratio is changed. In general, the relative humidity varies inversely with temperature during the day, tending to be lower in the early afternoon and higher at night.

Atmospheric moisture can be measured by at least five types of instrument. The most common for routine measurement is the *wet-bulb thermometer* installed in a louvred instrument shelter (Stevenson screen). The bulb of the standard thermometer is wrapped in muslin, which is kept moist by a wick from a reservoir of pure water. The evaporative cooling of this wet bulb gives a reading that can be used in conjunction with a simultaneous dry-bulb temperature reading to calculate the dew-point temperature. A similar portable device, called an aspirated *psychrometer*, uses a forced flow of air at a fixed rate over the dry and wet bulbs. A sophisticated instrument for determining the dew point, based on a different principle, is the *dew-point hygrometer*. This detects when condensation first occurs on a cooled surface. Two other types of instrument are used to determine relative humidity. The *hygrograph* utilizes the expansion/contraction of a bundle of human hair, in response to humidity, to record relative humidity continuously by a mechanical coupling to a pen arm marking on a rotating drum. This has an accuracy of about ± 5–10 per cent. For upper air measurements, a *lithium chloride* element is used to detect changes in its electrical resistance to vapour pressure differences. Relative humidity changes are accurate to about ± 3 per cent.

2 Moisture transport

It is sometimes overlooked that the atmosphere transports moisture horizontally as well as vertically. Figure 3.1 shows a net transport from oceans to land areas, while Figure 3.4B illustrates the quantities that must be transported meridionally in order to maintain the required moisture balance at a given latitude (i.e. evaporation – precipitation = net horizontal transport of moisture into the air column) (Figure 3.4A). Comparison of annual average precipitation and evaporation totals for latitude zones shows that in low and middle latitudes $P > E$, whereas in the subtropics $P < E$. These regional imbalances are maintained by net moisture transport into (convergence) and out of (divergence) the respective zones (ΔD, where divergence is positive).

$$E - P = \Delta D$$

A prominent feature is the equatorward transport into low latitudes and the poleward transport in middle latitudes. The reader should inspect this diagram again in the light of the discussion of winds belts in Chapter 6B. Atmospheric moisture is

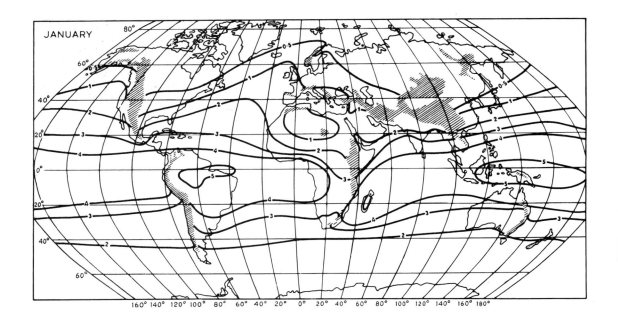

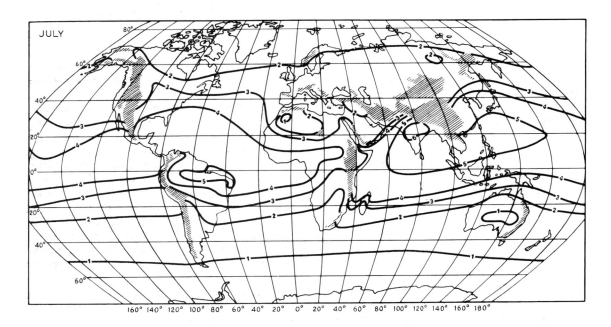

Figure 3.3 Mean atmospheric water vapour content in January and July, 1951–5, in cm of precipitable water.
Source: After Bannon and Steele 1960 (Crown Copyright Reserved).

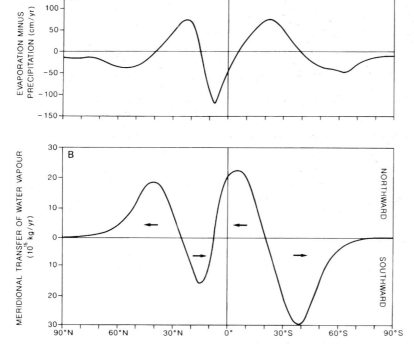

Figure 3.4 Meridional aspects of global moisture.
(A) Estimates of annual evaporation minus precipitation (in cm) as a function of latitude; (B) Annual meridional transfer of water vapour (in 10^{15} kg).

Source: A: After J. Dodd. From Browning 1993. By permission NERC.
B: From Sellers 1965.

transported by the global westerly wind systems of middle latitudes towards higher latitudes and by the easterly trade wind systems towards the equatorial region (Figure 3.4B) (see Chapter 7). There is, however, significant exchange of moisture between the hemispheres. During June to August there is a moisture transport northwards across the equator of 18.8×10^{8} kg s^{-1}; during December to February the southward transport is 13.6×10^{8} kg s^{-1}. The net annual south to north transport is 3.2×10^{8} kg s^{-1}, giving an annual excess of precipitation over evaporation in the northern hemisphere of 39 mm. This balance is returned by terrestrial runoff into the oceans.

At this point, it is necessary to stress emphatically the fact that local evaporation is, in general, not the major source of local precipitation. For example, only 24 per cent of the precipitation over the Mississippi River basin and between 25 and 35 per cent of that over the Amazon basin is of local origin, the remainder being transported into these areas (i.e. moisture advection). Even when moisture is available in the atmosphere over a region, only a small portion of it is usually precipitated. This

depends on the efficiency of the condensation and precipitation mechanisms, both microphysical and large-scale.

Using atmospheric sounding data on winds and moisture content, maps can be prepared showing seasonal and annual water vapour flux divergence (i.e. $E - P > 0$) or convergence (i.e. $E - P < 0$) (Figure 3.5), and these distributions of atmospheric moisture 'sources' (i.e. $P < E$) and 'sinks' (i.e. $P > E$) form an important basis for global climatic modelling. Figure 3.5 shows the global atmospheric vapour flux divergence for the extreme seasons. Strong divergence (outflow) of moisture occurs over the northern Indian Ocean in summer, providing moisture for the monsoon. Subtropical divergence zones are associated with the high-pressure areas. The oceanic ones are evaporation sources; those over land may reflect underground water use or are artefacts of sparse data.

C EVAPORATION

Evaporation (including transpiration) provides the moisture input into the atmospheric part of the

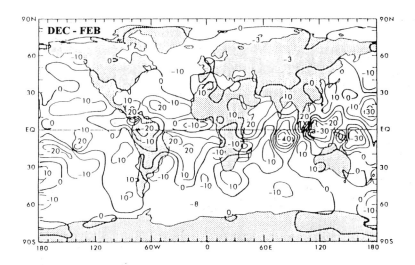

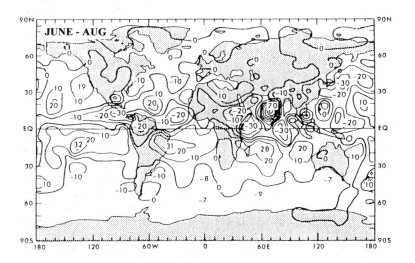

Figure 3.5 Global distribution of the horizontal divergence of atmospheric water vapour (0.1 m yr^{-1}); (above) mean December–February, (below) mean June–August. Negative values denote convergence of moisture (i.e. P > E).

Source: Peixoto and Oort 1983. From Variations in the Global Water Budget, A Street-Perrott, M. Beran and R. Ratcliffe (eds) 1983, Figs 13b and c. Copyright © D. Reidel, Dordrecht, by kind permission of Kluwer Academic Publishers.

global hydrological cycle, of which the oceans provide 84 per cent and the continents 16 per cent.

The highest annual values (150 cm), averaged zonally around the globe, occur over the tropical oceans, associated with trade wind belts, and over equatorial land areas in response to high solar radiation receipts and luxuriant vegetation growth (Figure 3.6A). The larger oceanic evaporative losses in winter, for each hemisphere (Figure 3.6B), represent the effect of outflows of cold continental air over warm ocean currents in the western North Pacific and North Atlantic (Figure 3.7) and stronger trade winds in the cold season of the southern hemisphere.

The relationship of saturation vapour pressure and temperature (see Figure 1.14) means that evaporation processes limit low-latitude ocean surface temperature (i.e. where evaporation is at a maximum) to values of about 29 °C. This factor plays an important role in regulating the temperature of ocean surfaces and overlying air in tropical latitudes. Evaporation occurs whenever energy is transported to an evaporating surface if the vapour pressure in the air is below the saturated value (e_s). As illustrated in Figure 1.14, the saturation vapour pressure increases with temperature. The change in state from liquid to vapour requires energy to be

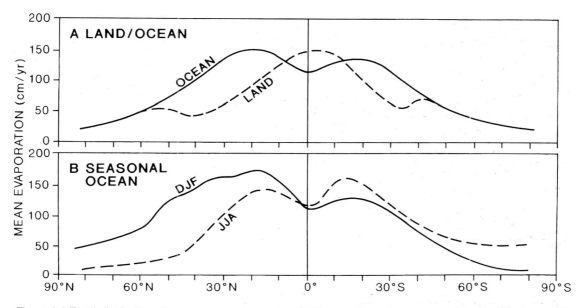

Figure 3.6 Zonal distribution of mean evaporation (cm/year) (A) annually for the ocean and land surfaces, and (B) over the oceans for December to February and June to August.

Sources: Peixoto and Oort 1983. From Variations in the Global Water Budget, A Street-Perrott, M. Beran and R. Ratcliffe (eds) 1983, Fig 22. Copyright © D. Reidel, Dordrecht, by kind permission of Kluwer Academic Publishers.
Also partly from Sellers 1965.

expended in overcoming the intermolecular attractions of the water particles. This energy is often acquired by the removal of heat from the immediate surroundings, causing an apparent heat loss (*latent heat*), as discussed on p. 25, and a consequent drop in temperature. The latent heat of vaporization to evaporate 1 kg of water at 0 °C is 2.5×10^6 J. Conversely, condensation releases this heat, and the temperature of an air mass in which condensation is occurring is increased as the water vapour reverts to the liquid state. The diurnal range of temperature is often moderated by damp air conditions, when evaporation takes place during the day and condensation at night.

Viewed another way, evaporation implies an addition of kinetic energy to individual water molecules and, as their velocity increases, so the chance of individual surface molecules escaping into the atmosphere becomes greater. As the faster molecules will generally be the first to escape, so the average energy (and therefore temperature) of those composing the remaining liquid will decrease, and the quantities of energy required for their continued release become correspondingly greater. In this way, evaporation decreases the temperature of the remaining liquid by an amount proportional to the latent heat of vaporization.

The rate of evaporation depends on a number of factors. The two most important are the difference between the saturation vapour pressure at the water surface and the vapour pressure of the air, and the existence of a continual supply of energy to the surface. Wind velocity also affects the evaporation rate, because the wind is generally associated with the importation of fresh, unsaturated air, which will absorb the available moisture.

Water loss from plant surfaces, chiefly leaves, is a complex process termed *transpiration*. It occurs when the vapour pressure in the leaf cells is greater than the atmospheric vapour pressure, and is vital as a life function in that it causes a rise of plant nutrients from the soil and cools the leaves. The cells of the plant roots can exert an osmotic tension of up to about 15 atmospheres upon the water films between the adjacent soil particles. As these soil water films shrink, however, the tension within them increases. If the tension of the soil films exceeds the osmotic root tension, the continuity of the plant's water supply is broken and wilting occurs. Transpiration is controlled by the atmospheric factors that

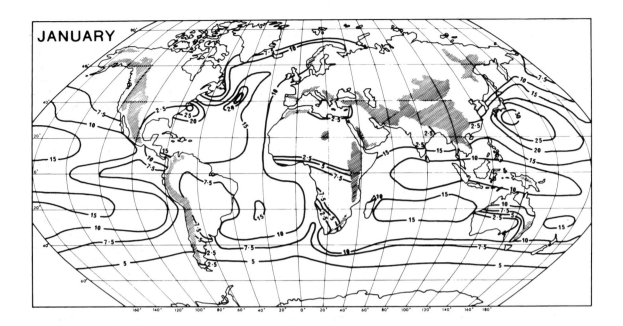

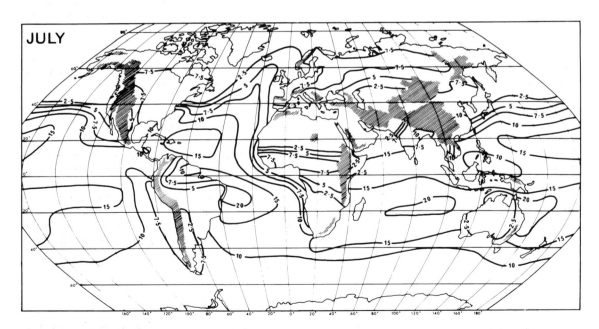

Figure 3.7 Mean evaporation (cm) for January and July.

determine evaporation as well as by plant factors such as the stage of plant growth, leaf area and leaf temperature, and also by the amount of soil moisture (see Chapter 10C). It occurs mainly during the day, when the *stomata* (i.e. small pores in the leaves), through which transpiration takes place, are open. This opening is determined primarily by light intensity. Transpiration naturally varies greatly with season, and during the winter months in mid-latitudes conifers lose only 10–18 per cent of their total annual transpiration losses and deciduous trees less than 4 per cent.

In practice, it is difficult to separate water evaporated from the soil, *intercepted moisture* remaining on vegetation surfaces after precipitation and subsequently evaporated, and transpiration. For this reason, evaporation is sometimes applied as a general term for all these, or, more correctly, the composite term *evapotranspiration* may be used.

Evapotranspiration losses from natural surfaces cannot be measured directly. There are, however, various indirect methods of assessment, as well as theoretical formulae. One approximate means of indirect measurement is based on the moisture balance equation:

precipitation = runoff + evapotranspiration + soil moisture storage change

Essentially, the method is to measure the percolation through an enclosed block of soil with a vegetation cover (usually grass) and to record the rainfall upon it. The block, termed a *lysimeter*, is weighed regularly so that weight changes unaccounted for by rainfall or runoff can be ascribed to evapotranspiration losses, provided the grass is kept short! The technique allows the determination of daily evapotranspiration amounts.

If the soil block is regularly 'irrigated' so that the vegetation cover is always yielding the maximum possible evapotranspiration, the water loss is called the *potential evapotranspiration* (or PE). More generally, PE can be defined as the water loss corresponding to the available energy. Potential evapotranspiration forms the basis for the climate classification developed by C. W. Thornthwaite (see Appendix 1).

In regions where snow cover is long lasting, evaporation/sublimation from the snowpack can be estimated by lysimeters sunk into the snow that are weighed regularly. Except in dry, sunny and windy environments, however, evaporation from snow is of minor importance compared with melt.

A meteorological solution to the calculation of evaporation uses sensitive instruments to measure the net effect of eddies of air transporting moisture upward and downward near the surface. In this 'eddy correlation' technique, the vertical component of wind and the atmospheric moisture content are measured simultaneously at the same level (say, 1.5 m) every few seconds. The product of each pair of measurements is then averaged over some time interval to determine the evaporation (or condensation). This method requires delicate rapid-response instruments, so it cannot be used in very windy conditions.

Theoretical methods for determining evaporation rates have followed two lines of approach. The first relates average monthly evaporation (E) from large water bodies to the mean wind speed (u) and the mean vapour pressure difference between the water surface and the air ($e_w - e_d$) in the form:

$$E = Ku(e_w - e_d)$$

where K is an empirical constant. This is termed the aerodynamic approach because it takes account of the factors responsible for removing vapour from the water surface. The second method is based on the energy budget. The *net balance* of solar and terrestrial radiation at the surface (R_n) is used for evaporation (E) and the transfer of heat to the atmosphere (H). A small proportion also heats the soil by day, but since nearly all of this is lost at night it can be disregarded. Thus:

$$R_n = LE + H$$

where L is the latent heat of evaporation (2.5×10^6 J kg^{-1}). R_n can be measured with a net radiometer and the ratio $H/LE = \beta$, referred to as Bowen's ratio, can be estimated from measurements of temperature and vapour content at two levels near the surface. β ranges from <0.1 for water to ≥10 for a desert surface. The use of this ratio assumes that the vertical transfers of heat and water vapour by turbulence take place with equal efficiency. Evaporation is then determined from an expression of the form:

$$E = \frac{R_n}{L(1 + \beta)}$$

The most satisfactory climatological method so far devised combines the energy budget and aerodynamic approaches. In this way, H. L. Penman succeeded in expressing evaporation losses in terms of four meteorological elements that are regularly measured, at least in Europe and North America. These are net radiation (or an estimate based on duration of sunshine), mean air temperature, mean air humidity and mean wind speed (which limit the losses of heat and vapour from the surface).

The relative roles of the factors that have been mentioned are illustrated by the global pattern of evaporation (see Figure 3.7). Losses decrease sharply in high latitudes, where there is little available energy. In middle and lower latitudes, there are appreciable differences between land and sea. Rates are naturally high over the oceans in view of the unlimited availability of water, and on a seasonal basis the maximum rates occur in January over the western Pacific and Atlantic, where cold continental air blows across warm ocean currents. On an annual basis, maximum oceanic losses occur about 15–20 °N and 10–20 °S, in the belts of the constant trade winds (see Figures 3.6B and 3.7). The highest annual losses, estimated to be about 200 cm, are in the western Pacific and central Indian Ocean near 15 °S (cf. Figure 2.30); 2,460 MJ m^{-2} yr^{-1} (78 W m^{-2} over the year) are equivalent to an evaporation of 100 cm of water/cm^2). There is a subsidiary equatorial minimum over the oceans, mainly as a result of the lower wind speeds in the doldrum belt and the proximity of the vapour

pressure in the air to its saturation value, but the land maximum occurs more or less at the equator because of the relatively high solar radiation receipts and the large transpiration losses from the luxuriant vegetation of this region. The secondary maximum over land in mid-latitudes is related to the strong prevailing westerly winds.

The annual evaporation over Britain, calculated by Penman's formula, ranges from about 38 cm in Scotland to about 50 cm in parts of south and southeast England. The average annual moisture budget can be calculated approximately by a book-keeping method devised by C.E. Thornthwaite, where potential evapotranspiration is estimated from mean temperature. Figure 3.8 illustrates this for stations in western, central and eastern Britain (compare Figure 8.25). In the winter months, there is an excess of precipitation over evaporation: this goes to recharging the soil moisture, and further surplus runs off. In summer, when evaporation exceeds precipitation, soil moisture is initially used to maintain evaporation at the potential value, but when this store is depleted, there is a water deficiency. The annual potential evapotranspiration determined by Thornthwaite's method is over 64 cm in most of south-eastern England. Since this loss is concentrated

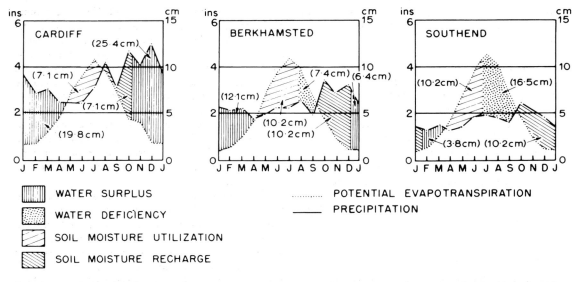

Figure 3.8 The average annual moisture budget for stations in western, central and eastern Britain determined by Thornthwaite's method. When potential evaporation exceeds precipitation, soil moisture is used; at Berkhamsted in central England and Southend on the east coast, this is depleted by July–August. Autumn precipitation excess over potential evaporation goes into replenishing the soil moisture until field capacity is reached.
Source: From Howe 1956.

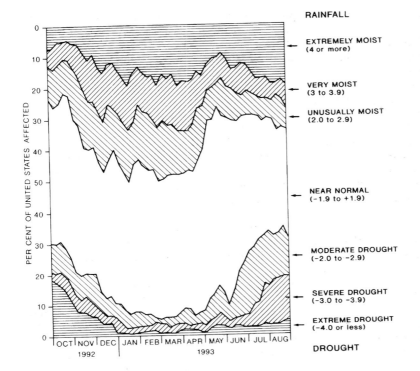

RAINFALL

EXTREMELY MOIST
(4 or more)

VERY MOIST
(3 to 3.9)

UNUSUALLY MOIST
(2.0 to 2.9)

NEAR NORMAL
(-1.9 to +1.9)

MODERATE DROUGHT
(-2.0 to -2.9)

SEVERE DROUGHT
(-3.0 to -3.9)

EXTREME DROUGHT
(-4.0 or less)

DROUGHT

Figure 3.9 Percentage of the continental USA affected by wet spells or drought, based on the Palmer Index (see scale on right), during the period October 1992–August 1993.

Sources: US Climate Analysis Center and Lott 1994. Reprinted from *Weather* by permission of the Royal Meteorological Society. Crown copyright ©.

in the period May–September, there may be seasonal water deficits of 12–15 cm in these parts of the country (as shown in Figure 3.8 for Southend), necessitating considerable use of irrigation water by farmers.

In the United States, monthly moisture conditions are commonly evaluated on the basis of the Palmer Drought Severity Index (PDSI). This is determined from accumulated weighted differences between actual precipitation and the calculated amount required for evapotranspiration, soil recharge and runoff. Accordingly, it takes account of the persistence effects of drought situations. The PDSI has a range from ≥4 (extremely moist) to ≤–4 (extreme drought). Figure 3.9 indicates an oscillation between drought and unusually moist conditions in the continental USA during the period October 1992 to August 1993.

D CONDENSATION

Condensation, the direct cause of all the various forms of precipitation, occurs under varying conditions, which in one way or another are associated with change in one of the linked parameters of air volume, temperature, pressure and humidity. Thus,

condensation takes place (1) when the temperature of the air is reduced but its volume remains constant and the air is cooled to dew point; (2) if the volume of the air is increased without addition of heat; this cooling occurs because adiabatic expansion causes energy to be consumed through work (see D, this chapter); (3) when a joint change of temperature and volume reduces the moisture-holding capacity of the air below its existing moisture content; or (4) by evaporation adding moisture to the air. The key to the understanding of condensation clearly lies in the fine balance that exists between these variables. Whenever the balance between one or more of them is disturbed beyond a certain limit, condensation may result.

The most common circumstances favourable to the production of condensation are those producing a drop in air temperature, namely contact cooling, mixing of air masses of different temperatures and dynamic cooling of the atmosphere. Contact cooling is produced, for example, within warm, moist air passing over a cold land surface (Plate 4). On a clear winter's night, strong radiation will cool the surface very quickly, and this surface cooling will gradually extend to the moist lower air, reducing the temperature to a point where condensation

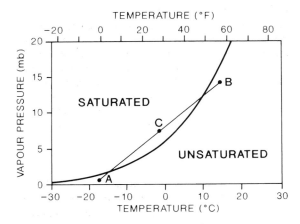

Figure 3.10 The effect of air-mass mixing. The horizontal mixing of two unsaturated air masses A and B will produce one supersaturated air mass C. The saturation vapour pressure curve is shown (cf. Figure 1.14B, which is a semi-logarithmic plot).

Source: From Petterssen 1941 [1969].

occurs in the form of dew, fog or frost, depending on the amount of moisture involved, the thickness of the cooling air layer and the dew-point value. When the latter is below 0 °C, it is referred to as the hoar frost point if the air is saturated with respect to ice.

The mixing of the differing layers within a single air mass or of two different air masses can also produce condensation. Figure 3.10 indicates how the horizontal mixing of two air masses (A and B), of given temperature and moisture characteristics, may produce an air mass (C) that is oversaturated at the intermediate temperature and consequently forms cloud. Vertical mixing of an air layer, which is discussed in Chapter 4 (see Figure 4.7), can have the same effect. Fog, or low stratus, with drizzle – known as 'crachin' – which is common along the coasts of south China and the Gulf of Tonkin in February–April, can develop as a result of either air-mass mixing or warm advection over a colder surface.

The addition of moisture into the air near the surface by evaporation occurs when cold air moves out over a warm water surface. This can cause steam fog, which is common in arctic regions, to form. Attempts at fog dispersal are one area where some progress has been made in local weather modification. Cold fogs can be dissipated locally by the use of dry ice (frozen CO_2) or the release of propane gas through expansion nozzles to produce freezing

and the subsequent fallout of ice crystals (cf. p. 87). Warm fogs (i.e. having drops above freezing temperatures) present bigger problems, but attempts at dissipation have shown some limited success in evaporating droplets by artificial heating, the use of large fans to draw down dry air from above, the sweeping out of fog particles by jets of water, and the injection of electrical charges into the fog to produce coagulation.

However, the most effective cause of condensation is undoubtedly the dynamic process of adiabatic cooling. This is considered in some detail in the next chapter.

E PRECIPITATION CHARACTERISTICS

1 Forms of precipitation

Strictly, *precipitation* refers to all liquid and frozen forms of water. The primary ones are:

Rain –   falling water drops with a diameter of at least 0.5 mm and typically 2 mm; droplets of less than 0.5 mm are termed *drizzle*. Rainfall has an accumulation rate of ≥1 mm/hour. Rain (or drizzle) that falls on a surface at a subzero temperature forms a glazed ice layer and is termed *freezing rain*.

Snow –   ice crystals falling in branched clusters as snowflakes. Wet snow has crystals bonded by liquid water in interior pores and crevices. Individual crystals have a hexagonal form (needles or platelets). At low temperatures (–30 °C), crystals may float in the air, forming 'diamond dust'.

Hail –   hard pellets, balls, or irregular lumps of ice, at least 5 mm across, formed of alternating shells of opaque and clear ice. The core of a hailstone is a frozen water drop (ice pellet) or an ice particle (graupel).

Graupel – snow pellets – opaque conical or rounded ice particles 2–5 mm in diameter formed by aggregation.

Sleet –   refers in the UK to a rain–snow mixture; in North America, to small translucent ice pellets (frozen raindrops) or snowflakes that have melted and refrozen.

Dew –   condensation droplets on the ground surface or grass, deposited when the surface temperature is below the air's dew-point temperature. *Hoar frost* is

Rime –   the frozen form, when ice crystals are deposited on a surface.

clear crystalline or granular ice deposited when supercooled fog or cloud droplets encounter a vertical structure, trees, or suspended cable. The rime deposit grows into the wind in a triangular form related to the wind speed. It is common in cold, maritime climates and on mid-latitude mountains in winter.

In general, only rain and snow make significant contributions to precipitation totals. In many parts of the world, the term rainfall can be used interchangeably with precipitation. The data in the following section refer to rainfall, since snowfall is less easily measured with the same degree of accuracy.

It must be emphasized that precipitation records are only *estimates*. Factors of site location, gauge height, large- and small-scale turbulence in the air flow, splash-in and evaporation all introduce errors in the catch. Also, more than fifty types of rain gauge are in use by national meteorological services; design differences affect the airflow over the gauge aperture and evaporation losses from the container and inner side of the funnel. Falling snow is particularly subject to wind effects, which can result in under-representation of the true amount by 50 per cent or more. Hence, corrections to gauge data need to take account of the proportion of precipitation falling in liquid and solid form, wind speeds during precipitation events, and precipitation intensity. Studies in Switzerland suggest that observed totals underestimate the true amounts by 7 per cent in summer and 11 per cent in winter below 2,000 m, but by as much as 15 per cent in summer and 35 per cent in winter in the Alps between 2,000 and 3,000 m.

Precipitation data are also limited by the density of gauge networks. On a national basis, per 10,000 km² area, this ranges from 245 gauges in Britain to ten in the United States and only three in Canada and Asia. In mountain areas and polar regions, the coverage is usually much worse. For ocean areas, there exist only island stations and ship observations of precipitation frequency and relative intensity, but satellite remote sensing has helped to provide independent estimates of ocean precipitation.

2 Precipitation characteristics

There are many measures by which the various attributes of precipitation can be described. These include mean annual precipitation, annual variability and year-to-year trends. However, hydrologists are interested in the characteristics and relationships of individual rainstorms, and it is possible here to point to some of their commonly observed features. Weather observations usually indicate the amount, duration and frequency of precipitation, and these enable other derived characteristics to be determined. Three of these are discussed below.

a Rainfall intensity

The intensity (= amount/duration) of rainfall during an individual storm, or a still shorter period, is of vital interest to hydrologists and water engineers concerned with flood forecasting and prevention, as well as to conservationists dealing with soil erosion. Chart records of the rate of rainfall (*hyetograms*) are necessary to assess intensity, which varies markedly with the time interval selected. Average intensities for short periods (thunderstorm-type downpours) are much greater than those for longer time intervals as Figure 3.11 illustrates for Milwaukee, USA, in 1973. In the case of extreme rates at different points over the earth (see Figure 3.12), the record intensity over 10 minutes is approximately three times that for 100 minutes, and the latter exceeds by as much again the record intensity over 1,000 minutes (i.e. 16.5 hours). High-intensity rain is associated with increased drop size rather than an increased number of drops. For example, with precipitation intensities of 0.1, 1.3 and 10.2 cm/hr (i.e. 0.05, 0.5 and 4.0 in/hr), the most frequent raindrop diameters are 0.1, 0.2 and 0.3 cm, respectively. The occurrence of daily amounts exceeding 1.3 cm is considered to be important for gully erosion in North America. Such falls account for 90 per cent of the annual rainfall on the Gulf Coast, compared with only 20 per cent in the Great Basin.

b Areal extent of a rainstorm

The rainfall totals received in a given time interval vary according to the size of the area that is considered, showing a relationship analogous to that of rainfall duration and intensity.

Figure 3.13 illustrates the average precipitation for a given storm area in the contrasting climatic regions of the USA, and Figure 3.14 shows maximum expected precipitation for storms of different durations and frequencies.

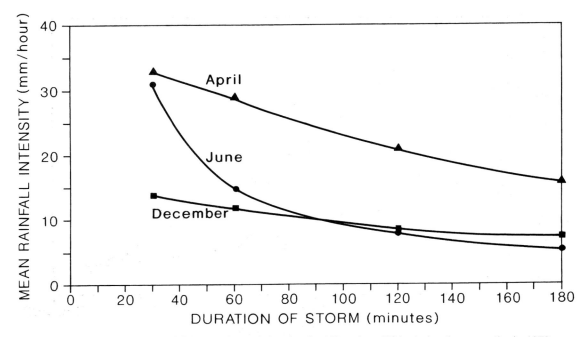

Figure 3.11 Relation between rainfall intensity and duration for Milwaukee, USA, during three months in 1973.
Source: US Environmental Data Service 1974. Courtesy US Environmental Data Service.

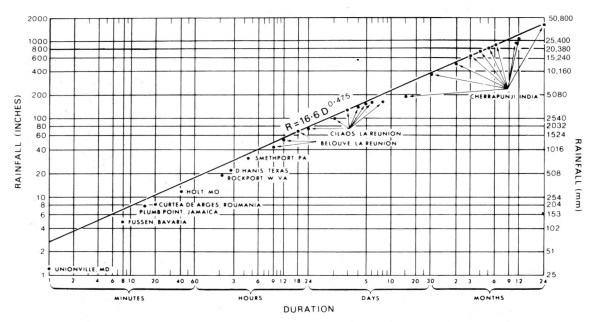

Figure 3.12 World record rainfalls (mm) together with the world envelope lines prior to 1948 and prior to 1967. The equation of the envelope line is given, together with the state or country where each record was established.
Source: From Rodda 1970.

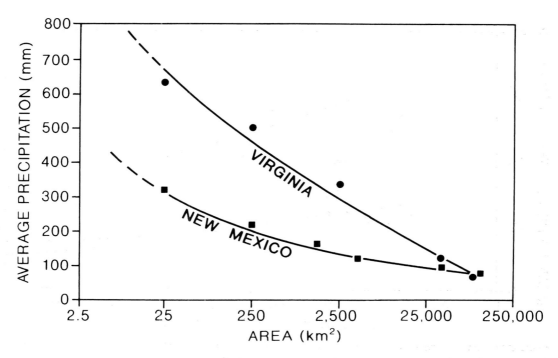

Figure 3.13 Average precipitation per area for a 24-hour storm in New Mexico in September 1941 and for an 18-hour storm in Virginia in August 1969.

Source: D.H. Miller, Water at the Surface of the Earth. © Academic Press, 1977.

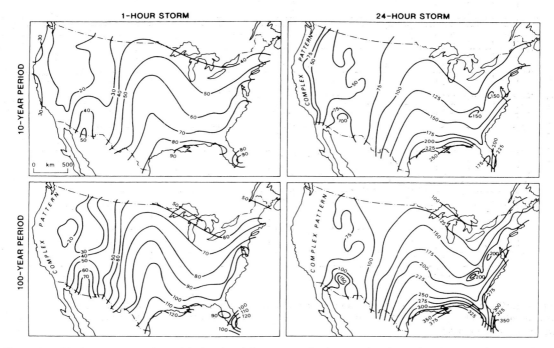

Figure 3.14 Maximum expected precipitation (mm) for storms of 1-hour and 24-hour duration occurring once in 10 and 100 years over the continental United States, calculated from records prior to 1961.

Source: US National Weather Service. Courtesy US Weather Bureau.

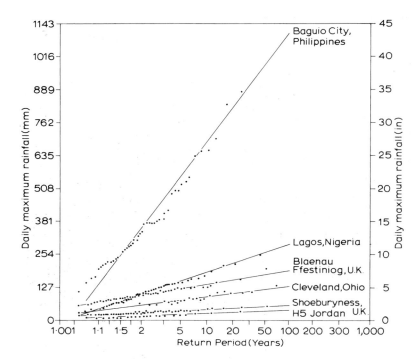

Figure 3.15
Rainfall/duration/frequency plots
for daily maximum rainfalls in
respect of a range of stations
from the Jordan desert to an
elevation of 1,482 m in the
monsoonal Philippines.

Source: After Rodda 1970; Linsley
and Franzini 1964; and Ayoade
1976.

c Frequency of rainstorms

Another useful item is the average time period within which a rainfall of specified amount or intensity can be expected to occur once. This is known as the *recurrence interval* or *return period*. Figure 3.15 gives this type of information on rainfall amount and duration for six contrasting stations. From this, for example, it would appear that on average, each 20 years, a 24-hour rainfall of at least 95 mm is likely to occur at Cleveland and 216 mm at Lagos. However, this *average* return period does not mean that such falls necessarily occur in the twentieth year of a selected period. Indeed, they might occur in the first! These estimates require long periods of observational data, but the approximately linear relationships shown by such graphs are of great practical significance for the design of flood-control systems.

Numerous case studies of rainstorm events have been carried out in different climatic areas. Two examples are shown for south-west England and

China in Figures 3.16A and B. The former was a 24-hour storm with an estimated 150–200-year return period; the latter was a 100-year storm, and the figure shows the 1-hour rainfall. Despite less orographic assistance and the shorter recurrence interval, the Hong Kong storm produced three times the maximum hourly intensity and ten times the core area of the English storm.

3 The world pattern of precipitation

A glance at the maps of precipitation amount for December–February and June–August (Figure 3.17) indicates that the distributions are considerably more complex than those, for example, of mean temperature (see Figure 2.11). Comparison of Figure 3.17 with the meridional profile of average precipitation for each latitude (Figure 3.18) brings out the marked longitudinal variations that are superimposed on the zonal pattern. The latter has a number of significant features:

Figure 3.16 (opposite) Distribution of rainfall in mm. (A) Over Exmoor, England, during a 24-hour period on 15 August 1952. 75 per cent of this rainfall occurred during a 7-hour period and 18 per cent in one hour;[*] (B) over Hong Kong during 0630 to 0730 hours on 12 June 1966.[†]

Sources: [*]From Dobbie and Wolf 1953.
[†]From So 1971.

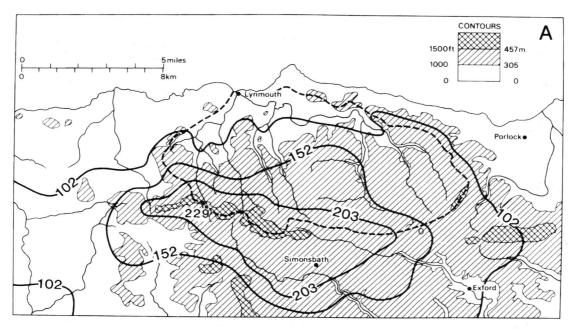

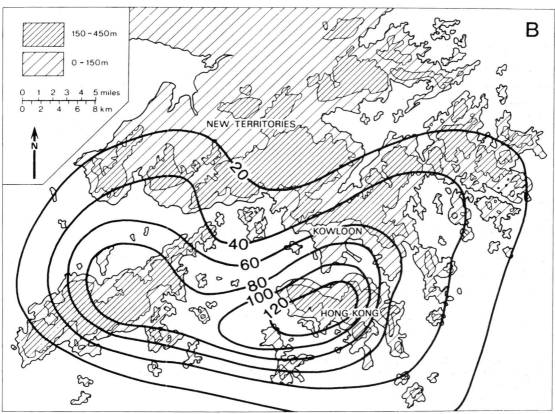

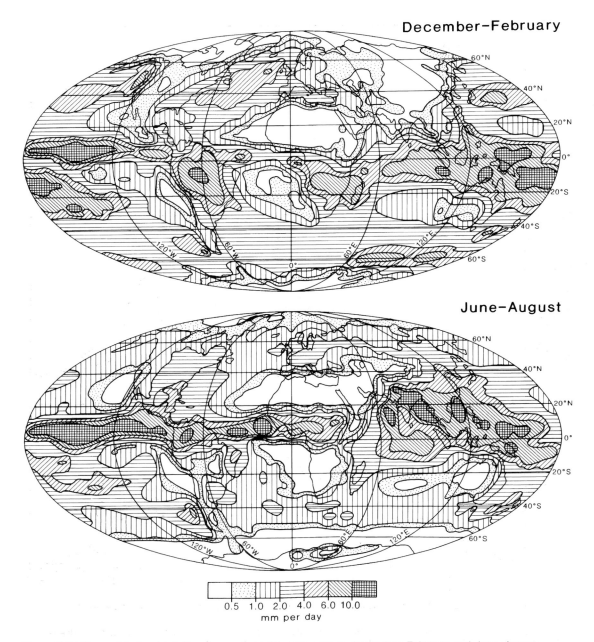

December–February

June–August

0.5 1.0 2.0 4.0 6.0 10.0
mm per day

Figure 3.17 Mean global precipitation (mm per day) for the periods December–February and June–August.

Source: From Legates 1995. From *International Journal of Climatology*, copyright © John Wiley & Sons Ltd. Reproduced with permission.

1   The 'equatorial' maximum, which is displaced into the northern hemisphere. This is related primarily to the converging trade wind systems and monsoon regimes of the summer hemisphere,

particularly in South Asia and West Africa. Annual totals over large areas are of the order of 200–250 cm or more.

2   The west coast maxima of middle latitudes

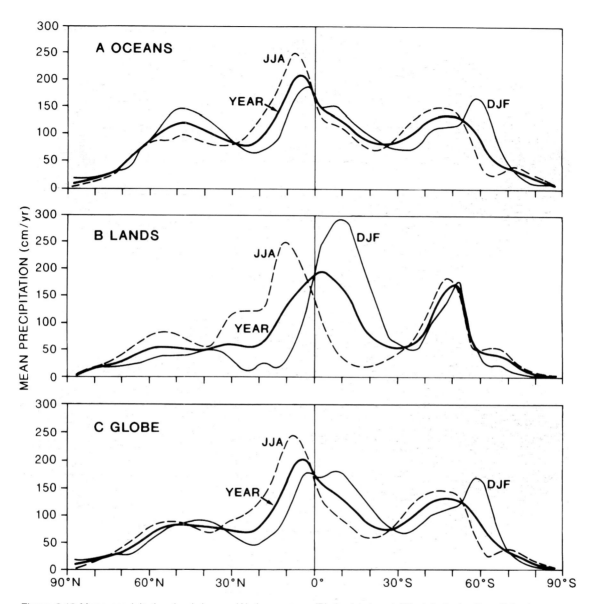

Figure 3.18 Mean precipitation (cm/yr) over (A) the oceans, (B) the land and (C) globally for Dec–Feb, June –Aug and annually.

Source: Peixoto and Oort 1983. From *Variations in the Global Water Budget*, A Street-Perrott, M. Beran and R. Ratcliffe (eds) 1983, Fig. 23. Copyright © D. Reidel, Dordrecht, by kind permission of Kluwer Academic Publishers.

associated with the belt of travelling disturbances in the westerlies. The precipitation in these areas has a very high degree of reliability.

3 The dry areas of the subtropical high-pressure cells, which include not only many of the world's major deserts but also vast oceanic expanses. In the northern hemisphere, the remoteness of the continental interiors extends these dry conditions into middle latitudes. In addition to very low average annual totals, often less than 15 cm, these regions are subject to considerable year-to-year variability.

4 Low precipitation in high latitudes and in winter over the continental interiors of the northern hemisphere. This reflects the low vapour contents of the extremely cold air. Most of this precipitation occurs in solid form.

Figure 3.17 demonstrates why the subtropics do not appear as particularly dry on the meridional profile in spite of the known aridity of the subtropical high-pressure areas (see Chapter 8). In these latitudes, the eastern sides of the continents receive considerable rainfall in summer.

In view of the complex controls involved, no brief explanation of these precipitation distributions can be very satisfactory. Various aspects of selected precipitation regimes are examined in Chapters 8 and 9, after consideration of the fundamental ideas about atmospheric motion, air masses and frontal zones. A classification of wind belts and precipitation characteristics is outlined in Appendix 1.C. It must suffice at this stage simply to note the factors that have to be taken into account in studying Figures 3.17 and 3.18.

1 The limit imposed on the maximum moisture content of the atmosphere by air temperature. This is important in high latitudes and in winter in continental interiors.
2 The major latitudinal zones of moisture influx due to atmospheric advection. This in itself is a reflection of the global wind systems and their disturbances (i.e. the converging trade wind systems and the cyclonic westerlies, in particular).
3 The distribution of the land masses. It is noteworthy that the southern hemisphere lacks the vast, arid, mid-latitude continental interiors of the northern. The oceanic expanses of the southern hemisphere allow the mid-latitude storms to increase the zonal precipitation average for 45 °S by about one-third compared with that of the northern hemisphere for 50 °N. Another major non-zonal feature is the occurrence of the monsoon regimes, especially in Asia.
4 The orientation of mountain ranges with respect to the prevailing winds.

4 Regional variations in the altitudinal maximum of precipitation

The increase of precipitation with height on mountain slopes is a worldwide characteristic, although actual profiles of precipitation differ regionally and seasonally. In middle latitudes, an increase may be

observed up to at least 3,000–4,000 m, as is the case in the Rocky Mountains in Colorado and in the Alps. In western Britain, with mountains of about 1,000 m, the maximum falls are recorded to leeward of the summits. This probably reflects the general tendency of air to go on rising for a while after it has crossed the crestline and the time lag involved in the precipitation process after condensation. Over narrow uplands, the horizontal distance may allow insufficient time for maximum cloud build-up and the occurrence of precipitation. However, a further factor may be the effect of eddies, set up in the air flow by the mountains, on the catch of rain gauges. Studies in Bavaria at the Hohenpeissenberg Observatory show that standard rain gauges may overestimate amounts by about 10 per cent on the lee slopes and underestimate them by 14 per cent on windward slopes.

In the tropics and subtropics, maximum precipitation occurs below the higher mountain summits, from which level it decreases upwards towards the crest. Observations are generally sparse in the tropics, but numerous records from Java show that the average elevation of greatest precipitation is approximately 1,200 m. Above about 2,000 m, the decrease in amounts becomes quite marked. Similar features are reported from Hawaii and, at a rather higher elevation, on mountains in East Africa (see Chapter 9H.2). Figure 3.19A shows that, despite the wide range of records for individual stations, this effect is clearly apparent along the Pacific flank of the Guatemalan highlands. Further north along the coast, the occurrence of a precipitation maximum below the mountain crest is observed in the Sierra Nevada, despite some complication introduced by the shielding effect of the Coast Ranges (Figure 3.19B), but in the Olympic Mountains of Washington precipitation increases right up to the summits. As has been previously mentioned, precipitation catches on mountain crests may underestimate the actual precipitation due to the effect of eddies, and this is particularly true where much of the precipitation falls in the form of snow, which is very susceptible to blowing by the wind.

One explanation of this orographic difference between tropical and temperate rainfall is based on the concentration of moisture in a fairly shallow layer of air near the surface in the tropics (see Chapter 9). Much of the orographic precipitation seems to derive from warm clouds (particularly cumulus congestus), composed of water droplets, which commonly have an upper limit at about 3,000 m. It is probable that

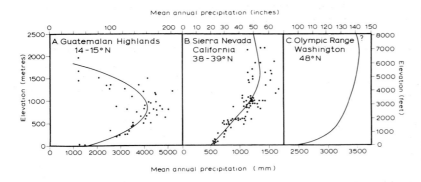

Figure 3.19 Generalized curves showing the relationship between elevation and mean annual precipitation for west-facing mountain slopes in Central and North America. The dots give some indication of the wide scatter of individual precipitation readings.

Source: Adapted from Hastenrath 1967, and Armstrong and Stidd 1967.

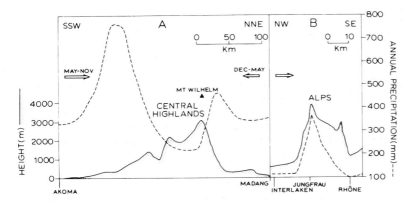

Figure 3.20 The relationship between precipitation (broken line) and relief in the tropics and mid-latitudes. (A) The highly saturated air masses over the Central Highlands of Papua New Guinea give seasonal maximum precipitations on the windward slopes of the mountains with changes in the monsoonal circulation;* (B) Across the Jungfrau massif in the Swiss Alps the precipitation is much less than in (A) and is closely correlated with the topography on the windward side of the mountains.† The arrows show the prevailing air-flow directions.

Sources: *After Barry 1981.
           †After Maurer and Lütschg (from Barry).

the height of the maximum precipitation zone is close to the mean cloud base, since the maximum size and number of falling drops will occur at that level. Thus, stations located above the level of mean cloud base will receive only a proportion of the orographic increment. In temperate latitudes, much of the precipitation, especially in winter, falls from stratiform cloud, which commonly extends through a considerable depth of the troposphere. In this case, there tends to be a smaller fraction of the total cloud depth below the station level. These differences according to cloud type and depth are apparent even on a day-to-day basis in middle latitudes. Seasonal variations in the altitude of the mean condensation level and zone of maximum precipitation are similarly observed. In the mountains of Central Asia (the

Pamirs and Tian Shan), for instance, the maximum is reported to occur at about 1,500 m in winter and at 3,000 m or more in summer. A further difference between orographic effects on precipitation in the tropics and the mid-latitudes relates to the great instability of many tropical air masses. Where high mountains obstruct the flow of very moist maritime tropical air masses, the upwind turbulence may be sufficient to trigger off sufficient precipitation to give a rainfall maximum at low elevations on the windward side of the range, as illustrated in Figure 3.20A for Papua New Guinea, where there are seasonally alternating winds. By contrast, in more stable mid-latitude air flows the rainfall maximum is more closely related to the topography (see Figure 3.20B).

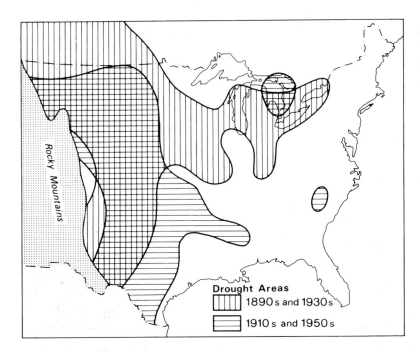

Figure 3.21 Drought areas of the central USA, based on areas receiving less than 80 per cent of the normal July–August precipitation.

Source: After Borchert 1971.

5 Drought

The term 'drought' implies an absence of significant precipitation for a period that is long enough to cause moisture deficits in the soil due to evapotranspiration and decreases in stream flow, thereby disrupting normal biological and human activities. Thus, a drought condition may obtain after only three weeks without rain in parts of Britain, whereas some areas of the tropics regularly experience many successive dry months. There is then no precise, universally applicable definition of drought. At least 150 different definitions have been suggested by specialists in meteorology, agriculture, hydrology, and socio-economic studies, who have differing perspectives! All regions suffer the temporary but irregularly recurring condition of drought, but particularly those with marginal climates alternately influenced by differing climatic mechanisms. Drought conditions are especially associated with:

1 Increases in the area and persistence of subtropical high-pressure cells. Drought in southern Israel has been shown to be significantly related to this mechanism. The major droughts in the African Sahel (see Figure 11.11) have also been attributed to an eastward and southward expansion of the Azores anticyclone.

2 Changes in the summer monsoonal circulation, causing a postponement or failure of incursions of maritime tropical air, such as may occur in Nigeria or the Punjab of India.

3 Lower ocean-surface temperatures produced by changes in currents or increased upwelling of cold waters. Rainfall in California and Chile may be affected by such mechanisms (see p. 142), and adequate rainfall in the drought-prone region of north-east Brazil appears to be strongly dependent on high sea-surface temperatures in the 0–15 °S belt of the South Atlantic.

4 Displacement of mid-latitude storm tracks associated with an expansion of the circumpolar westerlies into lower latitudes, or with the development of persistent blocking patterns of circulation in middle latitudes (see Figure 6.25). It has been suggested that droughts on the Great Plains east of the Rockies in the 1890s and 1930s were due to such changes in the general circulation. However, the droughts of the 1910s and 1950s in this area were caused by persistent high pressure in the south-east and the northward displacement of storm tracks (Figure 3.21).

Clearly, most severe and prolonged droughts involve combinations of several mechanisms. The

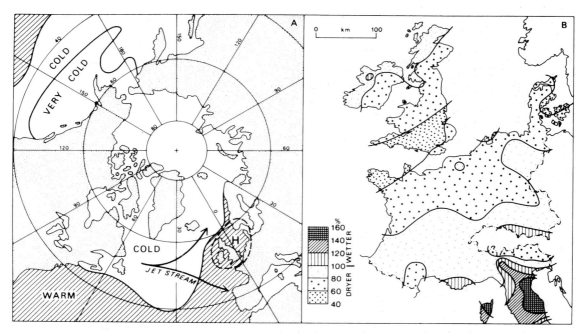

Figure 3.22 The drought of north-west Europe during May 1975 to August 1976. A: Conditions of a blocking high pressure over Britain, jet stream bifurcation and low sea-surface temperatures; B: Rainfall over Western Europe between May 1975 and August 1976 expressed as a percentage of a 30-year average.
Source: From Doornkamp and Gregory 1980.

prolonged drought in the Sahel – a 3,000 × 700 km zone stretching along the southern edge of the Sahara from Mauretania to Chad – which began in 1969 and has continued with interruptions up to the present (see Figure 11.11), has been attributed to several factors. These include an expansion of the circumpolar westerly vortex, shifting of the subtropical high-pressure belt towards the equator, and lower sea-surface temperatures in the eastern North Atlantic. There is no evidence that the subtropical high pressure was further south, but dry easterly air flow has been stronger across Africa during the drought years.

The meteorological definition of regional or global drought has, of recent years, become clouded by the subject of *desertification*, particularly since the UN conference on the subject in 1977 in Nairobi. This concern was sparked by the protracted drought in the Sahel. The removal of vegetation, increasing the surface albedo and lowering evapotranspiration, is thought to result in decreased rainfall. The problem for climatologists is that desertification is primarily marked by changes in surface cover involving the areas of savanna and

steppe surrounding the major desert regions. These areas have always been subject to climatic *oscillation*, as distinct from climatic *change* and, just as important, human activities (e.g. deforestation, mismanagement of irrigation, overgrazing, etc.) may be responsible for changes in surface cover, which modify the moisture budget.

From May 1975 to August 1976, parts of north-west Europe from Sweden to western France experienced severe drought conditions. Southern England received less than 50 per cent of average rainfall, the most severe and prolonged drought since records began in 1727 (Figure 3.22). The immediate causes of this regime were the establishment of a persistent blocking ridge of high pressure over the area, displacing depression tracks 5–10° latitude northward over the eastern North Atlantic. Farther afield, the circulation over the North Pacific had changed earlier, with the development of a stronger high-pressure cell and stronger upper-level westerlies, perhaps associated with a cooler than average sea surface. The westerlies were displaced northward over both the Atlantic and the Pacific. Over Europe, the dry conditions at the surface increased the stability of the atmosphere, further lessening the possibility of precipitation. Rainfall for April–August 1995 over England and Wales was only 46 per cent of average (compared with 47 per cent in 1976), again associated with a northward extension of the Azores anticyclone. This deficit has an estimated return period in excess of 200 years! Nevertheless, the 15-year period 1983–95 included six summers that were dry and also warm relative to the central England temperature record for 1659–1975.

F ACID PRECIPITATION

Acid precipitation includes both acid rain and snow (wet deposition) and dry deposition. The acidity of precipitation represents an excess of positive hydrogen ions [H^+] in a water solution. Acidity is measured on the pH scale ($1 - \log[H^+]$) ranging from 0 to 14, where 7 is neutral, i.e. the hydrogen cations are balanced by anions of sulphate, nitrate and chloride. Over the oceans, the main anions are Cl^- and SO_4^{2-} from sea salt. The background level of acidity in rainfall is about pH 4.8 to 5.6, because atmospheric CO_2 reacts with water to form carbonic acid. Acid solutions in rainwater are enhanced by reactions involving both gas-phase and aqueous-phase chemistry with sulphur dioxide and nitrogen dioxide. For sulphur dioxide, rapid pathways are provided by:

$$HOSO_2 + O_2 \rightarrow HO_2 + SO_3$$
$$H_2O + SO_3 \rightarrow H_2SO_4 \text{ (gas phase)}$$
$$\text{and } H_2O + HSO_3 \rightarrow H^+ + SO_4^{2-} + H_2O$$
$$\text{(aqueous phase)}$$

The OH radical is an important catalyst in gas-phase reaction and hydrogen peroxide (H_2O_2) in the aqueous phase.

Acid deposition depends on emission concentrations, atmospheric transport and chemical activity, cloud type and cloud microphysical processes, and type of precipitation. Observations in northern Europe and eastern North America in the mid-1970s, compared with the mid-1950s, showed a twofold to threefold increase in hydrogen ion deposition and rainfall acidity. Sulphate concentrations in rainwater in Europe increased over this 20-year period by 50 per cent in southern Europe and 100 per cent in Scandinavia, although there has been a subsequent decrease, apparently associated with

reduced sulphur emissions in both Europe and North America. The emissions from coal and fuel oil in these regions have high sulphur contents (2–3 per cent) and, since major SO_2 emissions occur from elevated stacks, SO_2 is readily transported by the boundary layer winds. NO_x emissions, by contrast, are primarily from automobiles and thus NO_3^- is mainly deposited locally. SO_2 and NO_x have atmospheric resident times of 1–3 days. SO_2 is not readily dissolved in cloud or raindrops unless oxidized by OH or H_2O_2, but dry deposition is quite rapid. NO is insoluble in water, but it is oxidized to NO_2 by reaction with ozone, and ultimately to HNO_3 (nitric acid), which readily dissolves.

In the western United States, where there are few major sources of emission, H^+ ion concentrations in rainwater are only 15–20 per cent of levels in the east, while sulphate and nitrate anion concentrations are one-third to one-half of those in the east. In China, where high-sulphur coal is the main energy source and there are few automobiles, rainwater sulphate concentrations are high; observations in south-west China show levels six times those in New York City. In winter, in Canada, snow has been found to have more nitrate and less sulphate than rain, apparently because falling snow scavenges nitrate faster and more effectively. Consequently, nitrate accounts for about half of the snowpack acidity. In spring, snow melt runoff causes an acid flush that may be harmful to fish populations in rivers and lakes, especially at the egg or larval stages.

In areas subject to frequent fog, or hill cloud, acidity may be greater than with rainfall; North American data indicate pH values averaging 3.4 in fog. This is a result of several factors. Small fog or cloud droplets have a large surface area, higher levels of pollutants provide more time for aqueous-phase chemical reactions, and the pollutants may act as nuclei for fog droplet condensation. In Los Angeles, fog water usually has high nitrate concentrations, due to automobile traffic during the morning rush hour, and pH values as low as 2.2 to 4.0 are observed.

The impact of acid precipitation depends on the vegetation cover, soil and bedrock type. Neutralization may occur by addition of cations in the vegetation canopy or on the surface. Such buffering is greatest if there are carbonate rocks (Ca, Mg cations); otherwise the increased acidity augments normal leaching of bases from the soil.

SUMMARY

Atmospheric humidity can be described by the absolute mass of moisture in unit mass (or volume) of air, as a proportion of the saturation value, or in terms of the water vapour pressure. When cooled at constant pressure, air becomes saturated at the dew-point temperature.

The components of the surface moisture budget are total precipitation (including condensation on the surface), evaporation, storage change of water in the soil or in a snow cover, and runoff (on the surface or in the ground). Evaporation rate is determined by the available energy, the surface–air difference in vapour pressure, and the wind speed, assuming the moisture supply is unlimited. If the moisture supply is limited, the rate is affected by soil water tension and plant factors. Evapotranspiration is best determined with a lysimeter. Otherwise, it may be calculated by formulae based on the energy budget, or on the aerodynamic profile method using the measured gradients of wind speed, temperature and moisture content near the ground.

Condensation in the atmosphere may occur by continued evaporation into the air; by mixing of air of different temperatures and vapour pressures, such that the saturation point is reached; or by adiabatic cooling of the air through lifting until the condensation level is reached.

Rainfall is described statistically by the intensity, areal extent and frequency (or recurrence interval) of rainstorms. Orography generally intensifies the precipitation on windward slopes, but there are geographical differences in this altitudinal effect. The global patterns of precipitation amount and annual regime are determined by the regional atmospheric circulation, the proximity to ocean areas, sea-surface temperatures and the atmospheric moisture budget. Droughts may occur in many different climatic regions due to various causal factors. In mid-latitudes, blocking anticyclones are a major cause. The causes of the protracted drought in the African Sahel (changes in the climate system or human-induced desertification) remain unresolved.

Acid precipitation (by wet or dry deposition) results from the reaction of cloud droplets with emissions of SO_2 and NO_x. There are large geographical variations in acid deposition.

4

Atmospheric instability, cloud formation and precipitation processes

To understand how clouds form and precipitation occurs, we begin by considering atmospheric stability/instability and the mechanisms that cause air to rise.

A ADIABATIC TEMPERATURE CHANGES

The displacement of an air parcel to an environment of lower pressure (without heat exchange with surrounding air) causes an increase in its volume and a consequent lowering of its temperature. A volume increase involves work and the consumption of energy, thus reducing the heat available per unit volume and hence the temperature. Such a temperature change, involving no subtraction or addition of heat, is termed *adiabatic*. Vertical displacements of air are obviously a major cause of adiabatic temperature changes.

Near the earth's surface, most processes of change are non-adiabatic (sometimes termed *diabatic*) because of the tendency of air to mix and modify its characteristics by lateral movement, turbulence and related physical processes. When a parcel of air moves vertically, the changes that take place often follow an adiabatic pattern because air is fundamentally a poor thermal conductor, and the air parcel as a whole tends to retain its own thermal identity, which distinguishes it from the surrounding air masses. In some circumstances, on the other hand, mixing of air with its surroundings must be taken into account.

Let us consider the changes that occur when an air parcel rises and a decrease of pressure (and density) is accompanied by volume increase and temperature decrease (see Chapter 1B). The rate at which temperature decreases in a rising, expanding air parcel is called the *adiabatic lapse rate*. If the upward movement of air does not produce condensation, then the energy expended by expansion will cause the temperature of the mass to fall at what is called the *dry adiabatic lapse rate* or DALR (9.8 °C/km). However, prolonged reduction of the temperature invariably produces condensation, and when this happens latent heat is liberated, counteracting the dry adiabatic temperature decrease to a certain extent. It is therefore a distinguishing feature of rising and saturated (or precipitating) air that it cools at a slower rate (i.e. the *saturated adiabatic lapse rate* or SALR) than air that is unsaturated. Another difference between the dry and saturated adiabatic rates is that whereas the former remains constant the latter varies with temperature. This is because air masses at higher temperatures are able to hold more moisture and on condensation therefore release a greater quantity of latent heat. For high temperatures, the saturated adiabatic lapse rate may be as low as 4 °C/km, but this rate increases with decreasing temperatures, approaching 9 °C/km at −40 °C.

In all, three different lapse rates can be differentiated, two dynamic and one static. There is the environmental (or static) rate, which is the actual temperature decrease with height on any occasion, such as an observer ascending in a balloon would record (see Chapter 1C.1). This is not an adiabatic rate, therefore, and may assume any form depending on local air temperature conditions. There are also the dynamic adiabatic dry and saturated lapse rates (or cooling rates), which apply to rising parcels of air moving through their environment. Close to the surface, the vertical temperature gradient sometimes greatly exceeds the dry adiabatic lapse rate, that is, it is superadiabatic. This is particularly common in arid areas in summer (see Table 1.7). Over most ordinary dry surfaces, the lapse rate

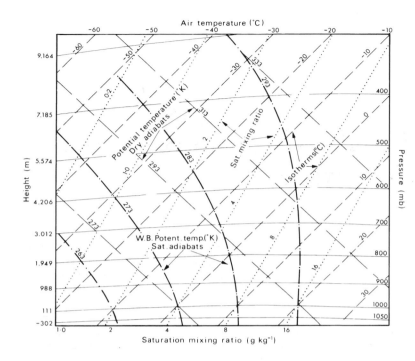

approaches the dry adiabatic value at an elevation of 100 m or so.

The changing properties of moving air parcels can be conveniently expressed by plotting them as *path curves* on suitably constructed graphs. One such adiabatic diagram in common use is the tephigram (Figure 4.1). This displays five sets of lines representing properties of the atmosphere:

1 Isotherms – i.e. lines of constant temperature (parallel lines from bottom left to top right).
2 Dry adiabats (parallel lines from bottom right to top left).
3 Isobars – i.e. lines of constant pressure (slightly curved nearly horizontal lines).
4 Saturated adiabats (curved lines sloping up from right to left).
5 Saturation mixing ratio lines (those at a slight angle to the isotherms).

The dry adiabats are also lines of constant potential temperature, θ (or isentropes). Potential temperature is the temperature of an air parcel brought dry adiabatically to a pressure of 1,000 mb. Mathematically,

$$\theta = T \left(\frac{1,000}{p} \right)^{0.286}$$

where θ and T are in °K, and p = pressure (mb).

Figure 4.2 shows schematically the relationship between T and θ; also between T and θ_w, the wet-bulb potential temperature (where the air parcel is brought to a pressure of 1,000 mb by a saturated adiabatic process).

Figure 4.2 illustrates that on an aerological diagram (such as the tephigram or adiabatic chart), a line along a dry adiabat (θ) through the dry-bulb

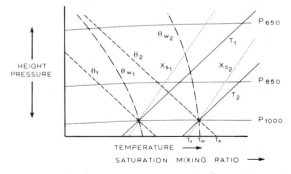

Figure 4.2 Graph showing the relationships between temperature (T), potential temperature (θ), wet-bulb potential temperature (θ_w) and saturation mixing ratio (X_s). T_d = dew point, T_w = wet-bulb temperature and T_A = air temperature.

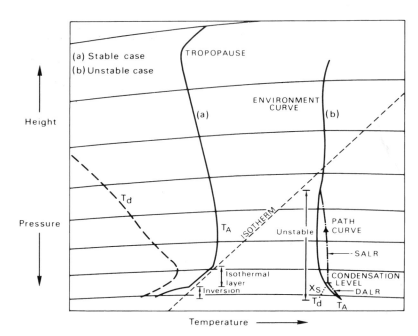

Figure 4.3 Tephigram showing (a) stable air case – T_A is the air temperature and T_d the dew point; and (b) unstable air case. The lifting condensation level is shown, together with the path curve (arrowed) of a rising air parcel. X_s is the saturation humidity mixing ratio line through the dew-point temperature (see text).

temperature of the surface air (T_A), an isopleth of saturation mixing ratio (x_s) through the dew point (T_d) and a saturated adiabat (θ_w) through the wet-bulb temperature (T_w) all intersect at a point corresponding to saturation for the air mass. This relationship, known as Normand's theorem, is used to estimate the *lifting condensation level* (see Figures 4.3 and 4.5). For example, with an air temperature of 20 °C and a dew point of 10 °C at 1,000 mb surface pressure, the lifting condensation level is at 860 mb with a temperature of 8 °C (see Figure 4.1). The altitude of this 'characteristic point' can be estimated roughly by

$$h \ (m) = 120(T - T_d)$$

where T = air temperature and T_d = dew-point temperature at the surface in °C.

The lifting condensation level (LCL) formulation does not take account of vertical mixing. A modified calculation defines a *convective condensation level* (CCL). In the near-ground layer, surface heating may establish a superadiabatic lapse rate, but convection modifies this to the DALR profile. Daytime heating steadily raises the surface air temperature from T_0 to T_1, T_2 and T_3 (Figure 4.4). Convection also equalizes the humidity mixing ratio, assumed equal to the value for the initial temperature. The CCL is located at the intersection of the environment temperature curve with a saturation mixing ratio line corresponding to the average mixing ratio in the surface layer (1,000–1,500 m). Expressed in another way, the surface air temperature is the minimum that will allow cloud to form as a result of free convection. Because the air near the surface is often well mixed, the CCL and LCL, in practice, are commonly nearly identical.

Experimentation with a tephigram shows that both the convective and the lifting condensation levels rise as the surface temperature increases, with little change of dew point. This is commonly observed in the early afternoon, when the base of cumulus clouds tends to be at higher levels.

Potential temperature provides an important yardstick for air mass characteristics, since if the air is affected only by dry adiabatic processes the potential temperature remains constant. This helps to identify different air masses and indicates when latent heat has been released through saturation of the air mass or when non-adiabatic temperature changes have occurred.

B AIR STABILITY AND INSTABILITY

The important characteristic of stable air is that if it is forced up or down it has a tendency to return to its former position once the motivating force

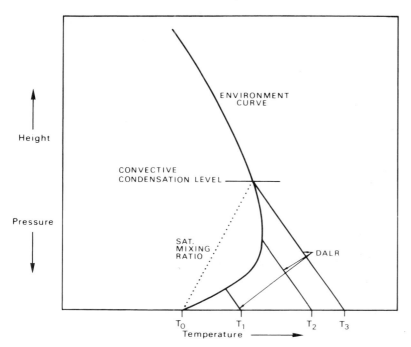

Figure 4.4 Schematic adiabatic chart used to determine the convective condensation level (see p. 78). T_0 represents the early morning temperature: T_1, T_2 and T_3 illustrate daytime heating of the surface air.

ceases. Figure 4.3 shows the reason for this, in that the environment temperature curve (a) lies to the right of any *path curve* representing the lapse rate of an unsaturated air parcel cooling dry adiabatically when forced to rise. At any level, the rising parcel is cooler and more dense than its surroundings and therefore tends to revert to its former level. Similarly, if the air is forced downwards it will gain in temperature at the dry adiabatic rate, always be warmer and less dense than the surrounding air, and tend to return to its former position (unless prevented from doing so). However, if local surface heating causes the environmental lapse rate near the surface to exceed the dry adiabatic lapse rate (b), then the adiabatic cooling of a convective air parcel allows it to remain warmer and less dense than the surrounding air, so it continues to rise through buoyancy. Similarly, if an air parcel is impelled downwards under these same conditions from a higher level it will become colder than its surroundings and there will be no check on its downward progress until it reaches the surface. The characteristic of unstable air is a tendency to continue moving away from its original level when set in motion.

A further possibility is illustrated in Figure 4.5. The air is stable in the lower layers, but if the air is forced to rise, for example by passage over a mountain range, or by local surface heating, until the path curve crosses the environment curve (the level of free convection) then the air, being warmer than its surroundings, is free to rise. This is termed *conditional instability*, as the development of instability is dependent on the air mass becoming saturated. Since the environmental lapse rate is frequently between the dry and saturated adiabatic rates, a state of conditional instability is a common one.

The environment curve in Figure 4.5 intersects the path curve at a higher level. Above the level of this intersection the atmosphere is stable, but the buoyant energy gained by the rising parcel enables it to move some distance into this region. The theoretical upper limit of cloud development can be estimated from the tephigram by determining an area (B) above the intersection of the environment and path curves equal to that between the two curves from the level of free convection to the intersection (A in Figure 4.5); the tephigram is so constructed that equal areas represent equal energy.

These examples assume that a small air parcel is being displaced without any compensating air motion or mixing of the parcel with its surroundings. These assumptions are rather unrealistic, since dilution of an ascending air parcel by mixing of the

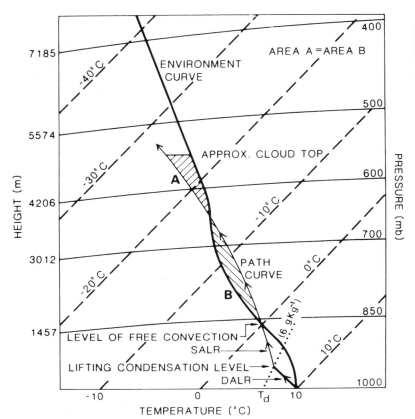

Figure 4.5 Schematic tephigram illustrating the conditions associated with the conditional instability of an air mass that is forced to rise. The saturation mixing ratio is a broken line and the lifting condensation level (cloud base) is below the level of free convection.

surrounding air with it through *entrainment* will reduce its buoyant energy. However, the method is generally satisfactory for routine forecasting, especially perhaps since the assumptions approximate to conditions in the updraught of cumulonimbus clouds.

A further consideration is that a deep layer of air may be displaced by vertical motion over an extensive topographic barrier. Figure 4.6 shows a case where the air in the upper levels is less moist than that at lower levels. If the whole layer is forced bodily upwards, the drier air at B follows the dry adiabatic rate, and so, for a while, will the air about A, but there eventually comes a time when the condensation level is reached, after which the lower layers of the rising air mass cool at the saturated adiabatic rate. This has the final effect of increasing the actual lapse rate of the total thickness of the raised layer, and, if this new rate is greater than that of the saturated adiabatic, the air layer becomes unstable, and may overturn. This is termed *convective* (or *potential*) *instability*.

Vertical mixing of air was referred to earlier as a possible cause of condensation. This is best illustrated by use of a tephigram. Figure 4.7 shows an initial distribution of temperature and dew point. Vertical mixing has the effect of averaging these conditions through the layer affected. Thus, the *mixing condensation level* is determined from the intersection of the average values of saturation humidity mixing ratio and potential temperature. The areas above and below the points where these average-value lines cross the initial environment curves are equal.

Subsidence usually results from either radiational cooling or an excess of horizontal convergence of air in the upper troposphere. Subsiding air generally moves with a vertical velocity of only 1–10 cm s^{-1}, although convection conditions provide an exception (see G, this chapter). Subsidence can produce substantial changes in the atmosphere; for instance, if an air mass subsides about 300 m, all average-size cloud droplets will usually be evaporated.

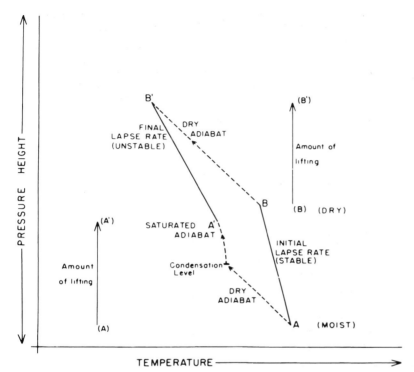

Figure 4.6 Convective instability. AB represents the initial
state of an air column; moist at
A, dry at B. After uplift of the
whole air column the temperature gradient A′ B′ exceeds the
saturated adiabatic lapse rate,
so the air column is unstable.

C CLOUD FORMATION

The formation of clouds is dependent on atmospheric instability and vertical motion but it is also
controlled by microscale processes. These are now
discussed before we examine larger-scale aspects of
cloud development and type.

1 Condensation nuclei

It must be emphasized that condensation occurs with
the utmost difficulty in *clean* air; moisture must generally find a suitable surface upon which it can
condense. If clean air is reduced in temperature
below its dew point it becomes *supersaturated* (i.e.
relative humidity exceeding 100 per cent). To maintain a pure water drop of radius 10^{-7} cm (0.001 μm)
requires a relative humidity of 320 per cent, and for
one of 10^{-5} cm (0.1 μm) radius 101 per cent.

Usually, condensation occurs on a foreign surface,
which can be a land or plant surface, as is the case
for dew or frost, while in free air condensation
begins around so-called *hygroscopic nuclei*. These
are microscopic particles – aerosols – the surfaces
of which (like the weather enthusiast's seaweed!)

have the property of *wettability*. Aerosols include
dust, smoke, salts and chemical compounds. Sea
salts, which are particularly hygroscopic, enter the
atmosphere principally by the bursting of air
bubbles in foam and are a major component of the
aerosol load near the surface of the oceans. Other
large contributions are from fine soil particles and
various natural, industrial and domestic combustion
products raised by the wind. A further source is the

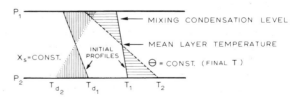

Figure 4.7 Graph illustrating the effects of vertical
mixing in an air mass. The horizontal lines are
pressure surfaces (P_2, P_1). The initial temperature (T_1)
and dew-point temperature (T_{d1}) gradients are modified
by turbulent mixing to T_2 and T_{d2}. The condensation
level occurs where the dry adiabat (θ) through T_1
intersects the saturation humidity mixing ratio line (X_s)
through T_{d2}.

conversion of atmospheric trace gas to particles through photochemical reactions, particularly over urban areas. Nuclei range in size from those with a radius of 0.001 μm, which are ineffective because of the high supersaturations required for their activation, to *giants* of over 10 μm, which do not remain airborne for very long (see pp. 2–3). On average, oceanic air contains 1 million condensation nuclei per litre (i.e. dm³), and land air holds some 5 or 6 million. Near the surface, marine aerosols are predominantly sea salts, but these tend to be removed rapidly due to their size. In the marine troposphere, there are fine particles (mainly ammonium sulphate). A photochemical origin associated with anthropogenic activities accounts for about half of these in the northern hemisphere. Dimethyl sulphide (DMS), associated with algal decomposition, also undergoes oxidation to sulphate. Over the tropical continents, aerosols are produced biogenically, by forest vegetation and surface litter, and through biomass burning during the dry season; particulate organic carbon predominates. In mid-latitudes, remote from anthropogenic sources, coarse particles are mostly of crustal origin (calcium, iron, potassium and aluminium) whereas crustal, organic and sulphate particles are almost equally represented in the fine aerosol load.

Hygroscopic aerosols are soluble. This is very important since the saturation vapour pressure is less over a solution droplet (for example, sodium chloride or sulphuric acid) than over a pure water drop of the same size and temperature (Figure 4.8). Indeed, condensation begins on hygroscopic particles before the air is saturated: in the case of sodium chloride nuclei at 78 per cent relative humidity. Figure 4.8 illustrates droplet radii for three sets of solution droplets of sodium chloride (a common sea salt) in relation to their equilibrium relative humidity, called Kohler curves. Droplets in an environment where values are below/above the appropriate line will evaporate/grow. Each curve has a maximum beyond which the droplet can grow in air with less supersaturation.

Once they are formed initially, the process of growth of water droplets is far from simple and much remains to be explained. In the early stages, the solution effect is predominant and small drops grow more quickly than large ones, but as the size of a droplet increases, its growth rate by condensation decreases (Figure 4.9). The radial growth rate obviously slows down as the drop size increases, because there is a greater surface area to add to

with every increment of radius. However, the condensation rate is limited by the speed with which the released latent heat can be lost from the drop by conduction to the air, and this heat reduces the vapour gradient. Moreover, competition between droplets for the available moisture increasingly tends to reduce the degree of supersaturation.

Supersaturation in clouds very rarely exceeds 1 per cent and, because the saturation vapour pressure is greater over a curved droplet surface than over a plane water surface, very small droplets (<0.1 μm radius) are readily evaporated (see Figure 4.8). In the early stages, the nucleus size is important; for supersaturation of 0.05 per cent a droplet of 1 μm radius with a salt nucleus of mass 10^{-13} g reaches 10 μm in 30 minutes, whereas one with a salt nucleus of 10^{-14} g would take 45 minutes. Later, when the dissolved salt has ceased to have significant effect, the radial growth rate becomes slow as a result of decreasing supersaturation.

Figure 4.9 illustrates the very slow growth of water droplets by condensation – in this case, at 0.2 per cent supersaturation from an initial radius of 10 μm. As there is an immense size difference between cloud droplets (<1 to 50 μm radius) and raindrops (>1 mm diameter), it is very clear that the gradual process of condensation is inadequate to explain the rates of formation of raindrops that are often observed. For example, in most clouds precipitation develops within an hour. The coalescence mechanism illustrated in Figure 4.9 is described below (p. 89). It must be remembered too that falling raindrops undergo evaporation in the unsaturated air below the cloud base. A droplet of 0.1 mm radius evaporates after falling only 150 m at a temperature of 5 °C and 90 per cent relative humidity, but a drop of 1 mm radius would fall 42 km before evaporating. It seems likely then that cloud droplets are not necessarily the immediate source of raindrops. This point is taken up again in Section D.

2 Cloud types

The great variety of cloud forms necessitates a classification for purposes of weather reporting. The internationally adopted system is based upon (1) the general shape, structure and vertical extent of the clouds, and (2) their altitude.

These primary characteristics are used to define the ten basic groups (or genera) as shown in Figure 4.10. High cirriform cloud is composed of ice

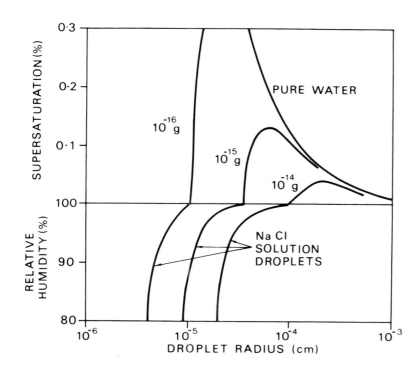

Figure 4.8 Kohler curves showing the variation of equilibrium relative humidity or supersaturation (%) with droplet radius for pure water and NaCl solution droplets. The numbers show the mass of sodium chloride (a similar family of curves is obtained for sulphate solutions). The pure water droplet line illustrates the curvature effect.

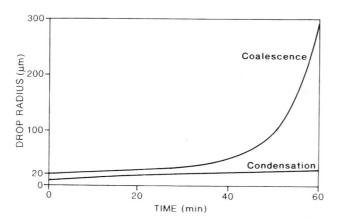

Figure 4.9 Droplet growth by condensation and coalescence.

Source: Jonas 1994. Reprinted from Weather by permission of the Royal Meteorological Society. Crown copyright ©.

crystals, giving a generally fibrous appearance. Stratiform clouds are layer-shaped, while cumuliform ones have a heaped appearance and usually show progressive vertical development. Other prefixes are *alto-* for middle-level (medium) clouds and *nimbo-* for low clouds of considerable thickness, which appear dark grey and from which continuous rain is falling.

The height of the cloud base may show a considerable range of any of these types and varies with latitude. The approximate limits in thousands of metres for different latitudes are shown in Table 4.1.

Following taxonomic practice, the classification subdivides the major groups into species and varieties with Latin names according to details of their appearance. The World Meteorological Organization has produced an *International Cloud Atlas* illustrating all of these types.

Other possible classifications of clouds take into account their mode of origin. For instance, a broad genetic grouping can be made according to the

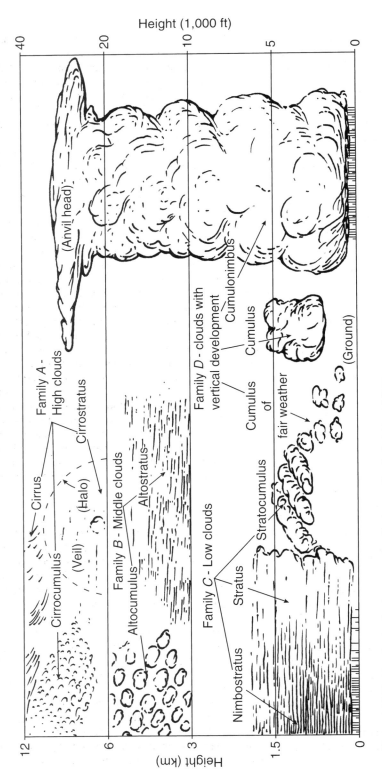

Figure 4.10 The ten basic cloud groups classified according to height and form.
Source: Modified after Strahler 1965.

Table 4.1 Cloud base height (in 000s m).

	Tropics	Middle latitudes	High latitudes
High cloud	Above 6	Above 5	Above 3
Medium cloud	2–7.5	2–7	2–4
Low cloud	Below 2	Below 2	Below 2

mechanism of vertical motion that produces condensation. Four categories are:

1 gradual uplift of air over a wide area in association with a low-pressure system;
2 thermal convection (on the local cumulus scale);
3 uplift by mechanical turbulence (*forced convection*);
4 ascent over an orographic barrier.

Group 1 includes a wide range of cloud types and will be discussed more fully in Chapter 7D.2. In connection with thermal convection, which forms cumuliform clouds (group 2), it is worth noting that upward convection currents (thermals) form plumes of warm air that, as they rise, expand and are carried downwind. Towers in cumulus and other clouds (Plate 5) are caused not by thermals of surface origin, but by ones set up *within* the cloud as a result of the release of latent heat by condensation. Thermals gradually lose their impetus as mixing of cooler, drier air from the surroundings dilutes the more buoyant warm air. Cumulus towers also tend to evaporate as updraughts diminish, leaving a shallow oval-shaped 'shelf' cloud (*stratocumulus cumulogenitus*), which may amalgamate with others to produce a high overcast. Group 3 includes fog, stratus or stratocumulus and is important whenever air near the surface is cooled to dew point by conduction or night-time radiation and the air is stirred by irregularities of the ground. The final group (4) could include stratiform or cumulus clouds produced by forced uplift of air over mountains (see Plate 6). Hill fog is simply stratiform cloud enveloping high ground. A special and important category is the wave (lenticular) cloud (Plate 7), which develops when air flows over hills, setting up a wave motion in the air current downwind of the ridge (see Chapter 3C.2). Clouds form in the crest of these waves if the air reaches its condensation level.

A great deal of information on cloud amounts, especially in remote areas, and on cloud patterns in relation to weather systems is now being provided by operational weather satellites. These supply direct-readout imagery and data that cannot be obtained by ground observations. Special classifications of cloud elements and patterns have been devised in order to analyse satellite imagery. The most common patterns seen on satellite photographs are cellular, or honeycomb-like, with a typical diameter of 30 km. These develop from the movement of cold air over a warmer sea surface. An open cellular pattern, where cumulus clouds are along the cell sides, forms where there is a large air–sea temperature difference, whereas closed polygonal cells occur if this difference is small. In both cases, there is subsidence above the cloud layer. Open (closed) cells are more common over warm (cool) ocean currents to the east (west) of the continents. The honeycomb pattern has been attributed to mesoscale convective mixing, but the cells have a width–depth ratio of about 30:1, whereas laboratory thermal convection cells have a corresponding ratio of only 3:1. Thus the true explanation may be more complicated. Occasionally, subsidence over the tropical oceans leads to a modification of the cellular pattern referred to as 'actiniform' (or radiating), illustrated in Plate 8. Another common pattern over oceans and uniform terrain is provided by linear cumulus cloud 'streets'. Helical motion in these two-dimensional cloud cells develops with surface heating, particularly when outbreaks of polar air move over warm seas (see Plate 9) and there is a capping inversion. Cloud patterns related to cyclonic systems are discussed in Chapter 7D.

3 Global cloud cover

Clouds represent an important sink for radiative energy in the earth–atmosphere system by absorption, as well as a source by reflection and reradiation (see Chapter 2B). Although clouds contain only 4 per cent of the total water in the atmosphere at any one time on average, their mean annual net forcing effect (i.e. that leading to net atmospheric warming or cooling) is very significant (Figure 4.11). However, cloud forcing is very complex; for example, more cloud in general implies more absorption of outgoing terrestrial radiation (positive forcing, leading to warming) whereas more high cloud produces increased reflection of incoming solar radiation (negative forcing, leading to cooling).

The difficulties in predicting future cloud amounts present problems for those hoping to estimate future global temperatures with the aid of computer

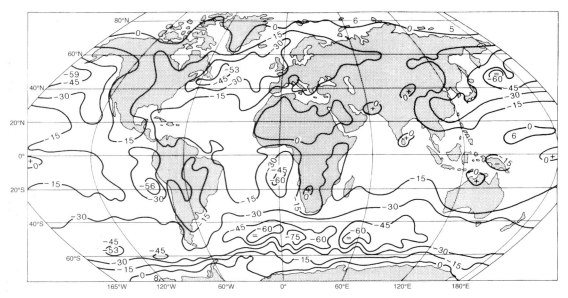

Figure 4.11 Mean annual net cloud forcing (W m⁻²) observed by the *Nimbus-7* ERB satellite for the period June 1979 to May 1980.

Source: Kyle *et al.* 1993. From *Bulletin of the American Meteorological Society*, by permission of the American Meteorological Society.

modelling. Figure 4.12, for example, shows the striking increase in cloud cover over the United States during the present century (especially between 1940 and 1950), perhaps mainly from a rise in atmospheric sulphate concentrations due to an increase in coal burning. In recent years, satellites have provided large quantities of cloud data but, as noted earlier, surface-based estimates of cloud amounts are some 10 per cent greater than those derived from satellites, the greatest discrepancies occurring in summer in the subtropics and in the highest latitudes.

Total cloud percentage amounts show characteristic geographical, latitudinal and seasonal distributions (see Figures 2.8 and 4.13). During the northern summer, there are high percentages over West Africa, north-west South America and southeast Asia, with minima over the southern hemisphere continents, southern Europe, North Africa and the Near East. During the northern winter, there are high percentages over tropical land areas in the southern hemisphere, partly due to convection along the Intertropical Convergence Zone, and in subpolar ocean areas due to the advection of moist air. Minimal cloud cover is naturally associated with the subtropical high-pressure regions throughout the year, whereas persistent maximum

cloud cover occurs over the open ocean belt at 50–70 °S and over large parts of the northern hemisphere oceans north of about 45 °N.

D FORMATION OF PRECIPITATION

The puzzle of raindrop formation has already been briefly mentioned. The simple growth of cloud droplets is apparently an inadequate mechanism by itself, but if this is the case then more complex processes have to be envisaged.

Various early theories of raindrop growth can be discounted. Proposals were that differently charged droplets could coalesce by electrical attraction, but it appears that distances between drops are too great and the difference between the electrical charges too small for this to happen; that large drops might grow at the expense of small ones, but observations show that the distribution of droplet size in a cloud tends to maintain a regular pattern, with the average radius between about 10 and 15 μm, with a few larger than 40 μm; that if atmospheric turbulence brought warm and cold cloud droplets into close conjunction, the supersaturation of the air with reference to the cold drop surfaces and the undersaturation with reference to the warm drop surfaces would cause the warm drops to evaporate and the

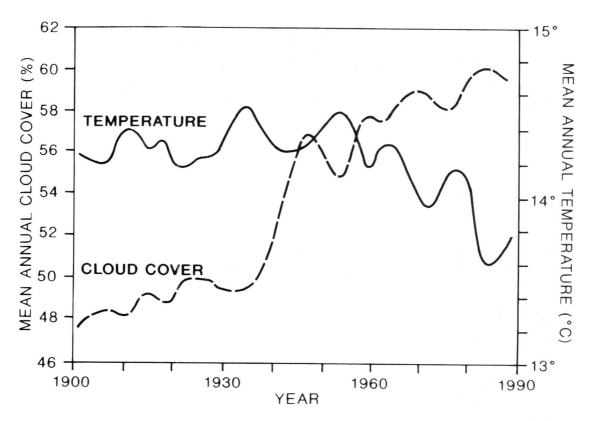

Figure 4.12 The relation between mean annual cloud cover and mean annual temperature over the United States during the present century.
Source: Pearce 1994. Copyright © *New Scientist*.

cold ones to develop at their expense. However, except perhaps in some tropical clouds, the temperature of cloud droplets is too low for this differential mechanism to operate. Figure 1.14 shows that below about −10 °C the slope angle of the saturation vapour pressure curve is low. Another theory was that raindrops grow around exceptionally large condensation nuclei (such as have been observed in some tropical storms). Large nuclei do experience a more rapid rate of initial condensation, but after this stage they are subject to the same limiting rates of growth that apply to all other atmospheric water drops.

The two current theories that attempt to explain the rapid growth of raindrops involve the growth of ice crystals at the expense of water drops, and the coalescence of small water droplets by the sweeping action of falling drops.

1 Bergeron–Findeisen theory

This widely accepted theory is based on the fact that at subzero temperatures the atmospheric vapour pressure decreases more rapidly over an ice surface than over water (Figure 1.14). This results in the saturation vapour pressure over water becoming greater than that over ice, especially between temperatures of −5 and −25 °C, where the difference exceeds 0.2 mb. If ice crystals and supercooled water droplets exist together in a cloud, the latter tend to evaporate and direct deposition takes place from the vapour on to the ice crystals.

Freezing nuclei are necessary before ice particles can form – usually at very low temperatures (about −15 to −25 °C). Small water droplets can, in fact, be supercooled in pure air to −40 °C before spontaneous freezing occurs, but ice crystals generally predominate in clouds where temperatures are below about −22 °C. Freezing nuclei are far less numerous

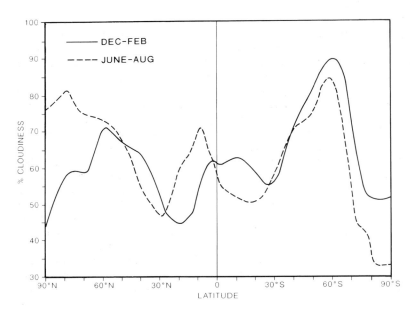

Figure 4.13 The average zonal distribution of total cloud amount (%), derived from surface observations over the total global surface (i.e. land plus water) for the months of December–February and June–August during the period 1971–81.

Source: From London *et al.* 1989.

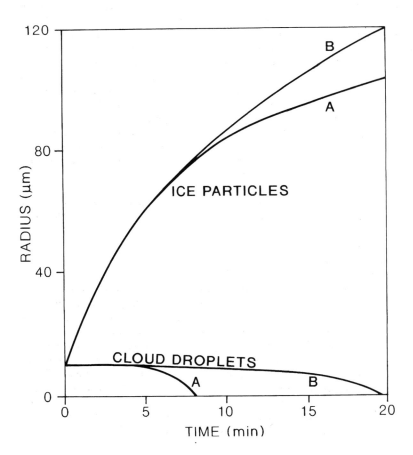

Figure 4.14 The effect of a small proportion of initially frozen droplets on the relative increase/decrease in the sizes of cloud ice and water particles. The initial droplets were at a temperature of −10 °C and at water saturation. A: a density of 100 drops per cc, 1% of which were assumed to be frozen. B: a density of 1,000 drops per cc, 0.1% of which were assumed to be frozen.

Source: Jonas 1994. Reprinted from *Weather* by permission of the Royal Meteorological Society. Copyright ©.

than condensation nuclei; there may be as few as 10 per litre at −30 °C and probably rarely more than 1,000. However, some become active at higher temperatures. Kaolinite, a common clay mineral, initially becomes active at −9 °C and on subsequent occasions at −4 °C. The origin of freezing nuclei has been the subject of much debate but it is generally considered that very fine soil particles are a major source. Common biogenic aerosols emitted by decaying plant litter, in the form of complex chemical compounds, also serve as freezing nuclei. In the presence of certain associated bacteria, ice nucleation can take place at only −2 to −5 °C.

Tiny ice crystals grow readily by deposition from vapour, with different hexagonal forms of crystal developing at different temperature ranges. The number of ice crystals also tends to increase progressively because small splinters become detached by air currents during growth and act as fresh nuclei. The freezing of supercooled water drops may also produce ice splinters (see F, this chapter). Figure 4.14 shows that a small density of ice particles is capable of rapid growth in an environment of cloud water droplets, resulting in a slower decrease in the average size of the much larger number of cloud droplets; the latter, nevertheless, take place within a timescale of 10^1 minutes. Ice crystals readily aggregate upon collision, due to their branched (dendritic) shape, and tens of crystals may form a single snowflake. Temperatures between about 0 and −5 °C are particularly favourable to aggregation, because fine films of water on the crystals' surfaces freeze when two crystals touch, binding them together. When the fall speed of the growing ice mass exceeds the existing velocities of the air upcurrents the snowflake falls, melting into a raindrop if it falls about 250 m below the freezing level.

This theory can account for most precipitation in middle and higher latitudes, yet it is not completely satisfactory. Cumulus clouds over tropical oceans can give rain when they are only some 2,000 m deep and the cloud-top temperature is 5 °C or more. In middle latitudes in summer, precipitation may fall from cumuli that have no subfreezing layer (*warm clouds*). A suggested mechanism in such cases is that of 'droplet coalescence', discussed below.

Practical *rainmaking* has been based on the Bergeron theory with some success. The basis of such experiments is the freezing nucleus. Supercooled (water) clouds between −5 and −15 °C are *seeded* with especially effective materials, such as silver iodide or 'dry ice' (CO_2) from aircraft or ground-based silver iodide generators, promoting the growth of ice crystals and encouraging precipitation. The seeding of some cumulus clouds at these temperatures probably produces a mean increase of precipitation of 10–15 per cent from clouds that are already precipitating or are 'about to precipitate'. Increases of up to 10 per cent have resulted from seeding winter orographic storms. However, the seeding of depressions has produced no apparent precipitation increases, and it appears likely that clouds with an abundance of natural ice crystals or with above-freezing temperatures throughout are not susceptible to rainmaking. Premature release of precipitation may destroy the updraughts and cause dissipation of the cloud, however. This explains why some seeding experiments have actually *decreased* the rainfall! In other instances, cloud growth and precipitation have been achieved, with some individually spectacular results being obtained by such methods in Australia and the United States. Programmes aimed at increasing winter snowfall on the western slopes of the Sierra Nevada and Rocky Mountains by seeding cyclonic storms have been carried out for a number of years with mixed results. Their success depends in any case on the presence of suitable supercooled clouds. When several cloud layers are present in the atmosphere, natural seeding may be important. For example, if ice crystals fall from high-level cirrostratus or altostratus (a 'releaser' cloud) into nimbostratus (a 'spender' cloud) composed of supercooled water droplets, the latter can grow rapidly by the Bergeron process and such situations may lead to extensive and prolonged precipitation. This is a frequent occurrence in cyclonic systems in winter and is important in orographic precipitation (see E3, this chapter).

2 Coalescence theories

Alternative raindrop theories use collision, coalescence and 'sweeping' as the drop growth generator. It was originally thought that atmospheric turbulence by making cloud particles collide would cause a significant proportion to coalesce. However, particles just as easily break up if subjected to collisions. Langmuir offered a variation of this simple collision theory by pointing out that falling drops have terminal velocities (typically 1–10 cm s^{-1}) directly related to their diameters, such that the larger drops might overtake and absorb small droplets and that the latter might also be swept into the wake of the former and be absorbed by them. Figure 4.9 gives experimental results of the rate of growth of water drops by coalescence from an initial radius of 20 μm in a cloud having a water content of 1 g/m³ (assuming maximum efficiency). Although coalescence is initially rather slow, the drop can reach 100–200 μm radius in 50 minutes; moreover, the growth rate is rapid for droplets with radii greater than 40 μm. Calculations show that drops must exceed 19 μm in radius before they can coalesce with other droplets, smaller droplets being swept aside without colliding. The initial presence of a few very large cloud droplets calls for the availability of giant nuclei (e.g. salt particles) if the cloud top does not reach above the freezing level. Observations show that maritime clouds do have relatively few large condensation nuclei (10–50 μm radius) and a high liquid water content, whereas continental air tends to contain many small nuclei (~1 μm) and less liquid water. Hence, rapid onset of showers is feasible by the coalescence mechanism in maritime clouds. Alternatively, if a few ice crystals are present at higher levels in the cloud (or if seeding occurs with ice crystals coming from higher cloud layers) they may eventually fall through the cloud as drops and the coalescence mechanism comes into action. Turbulence, especially in cumuliform clouds, serves to encourage collisions in the early stages. Thus, the coalescence process allows a more rapid growth than simple condensation can provide and is, in fact, common in 'warm' clouds in tropical maritime air masses, even in temperate latitudes.

3 Solid precipitation

Rain has been discussed at length because it is the most common form of precipitation. Snow occurs when the freezing level is so near the surface that aggregations of ice crystals do not have time to melt before reaching the ground. Generally, this means that the freezing level must be below 300 m. Mixed snow and rain ('sleet' in British usage) is especially likely when the air temperature at the surface is about 1.5 °C. Snow rarely occurs with an air temperature exceeding 4 °C.

Soft hail pellets (roughly spherical, opaque grains of ice with much enclosed air) occur when the Bergeron process operates in a cloud with a small liquid water content and ice particles grow mainly by deposition of water vapour. Limited accretion of small supercooled droplets forms an aggregate of soft, opaque ice particles 1 mm or so in radius. Showers of such pellets are quite common in winter and spring from cumulonimbus clouds.

Ice pellets may develop if the soft hail falls through a region of large liquid water content above the freezing level. Accretion forms a casing of clear ice around the pellet. Alternatively, an ice pellet consisting entirely of transparent ice may result from the freezing of a raindrop or the refreezing of a melted snowflake.

True hailstones are roughly concentric accretions of clear and opaque ice. The embryo is a raindrop carried aloft in an updraught and frozen. Successive accretions of opaque ice (rime) occur due to impact of supercooled droplets, which freeze instantaneously, whereas the clear ice (glaze) represents a wet surface layer, developed as a result of very rapid collection of supercooled drops in parts of the cloud with large liquid water content, which has subsequently frozen. A major difficulty in early theories was the necessity to postulate violently fluctuating upcurrents to give the observed banded hailstone structure, but a new thunderstorm model successfully accounts for this by demonstrating that the growing hailstones are recycled by the moving storm (see Chapter 4G). On occasions, hailstones may reach giant size, weighing up to 0.76 kg each (recorded in September 1970 at Coffeyville, Kansas). In view of their rapid fall speeds, hailstones may fall considerable distances with little melting. Hailstorms are a cause of severe damage to crops and property when large hail falls.

E PRECIPITATION TYPES

It is usual to identify three main types of precipitation – convective, cyclonic and orographic – according to the primary mode of uplift of the air. Essential to this analysis is some knowledge of storm systems. These are treated in later chapters, and the newcomer to the subject may prefer to read the following in conjunction with them (Chapter 7).

1 'Convective type' precipitation

This is associated with towering cumulus (cumulus congestus) and cumulonimbus clouds. Three subcategories can be distinguished according to their degree of spatial organization.

1 Scattered convective cells develop through strong heating of the land surface in summer (Plate 10),

especially when low upper tropospheric temperatures facilitate the release of conditional or convective instability (see B, this chapter). Precipitation, often including hail, is of the thunderstorm type, although thunder and lightning do not necessarily occur. Small areas (20 to 50 km²) are affected by individual heavy downpours, which generally last for about 30 minutes to 1 hour.

2 Showers of rain, snow or soft hail pellets may form in cold, moist unstable air passing over a warmer surface. Convective cells moving with the wind can produce a streaky distribution of precipitation parallel to the wind direction, although over a period of several days the variable paths and intensities of the showers tend to obscure this pattern. Two locations in which these cells may occur are parallel to a surface cold front in the warm sector of a depression (sometimes as a squall line) or parallel to and ahead of the warm front (see Chapter 7D). Hence the precipitation is widespread, although of limited duration at any locality.

3 In tropical cyclones, cumulonimbus cells become organized about the centre in spiralling bands (see Chapter 9B.2). Particularly in the decaying stages of such cyclones, typically over land, the rainfall can be very heavy and prolonged, affecting areas of thousands of square kilometres.

2 'Cyclonic type' precipitation

Precipitation characteristics vary according to the type of low-pressure system and its stage of development, but the essential mechanism is ascent of air through horizontal convergence of airstreams in an area of low pressure (see Chapter 5B). In extra-tropical depressions, this is reinforced by uplift of warm, less dense air along an air-mass boundary (see Chapter 7D.2). Such depressions give moderate and generally continuous precipitation over very extensive areas as they move, usually eastward, in the westerly wind belts between about 40 and 65° latitude. The precipitation belt in the forward sector of the storm can affect a locality in its path for 6 to 12 hours, whereas the belt in the rear gives a shorter period of thunderstorm-type precipitation. These sectors are, therefore, sometimes distinguished in precipitation classifications, and a more detailed breakdown is illustrated in Table 8.2. Polar lows (see Chapter 7H.3) combine the effects of airstream convergence and convective activity of

category 2 (previous section), whereas troughs in the equatorial low-pressure area give convective precipitation as a result of airstream convergence in the tropical easterlies (see Chapter 9B.1).

3 Orographic precipitation

Orographic precipitation is commonly regarded as a distinct type, but this requires careful qualification. Mountains are not especially efficient in causing moisture to be removed from airstreams crossing them, yet because precipitation falls repeatedly in more or less the same locations, the cumulative totals are large. Orography, dependent on the alignment and size of the barrier, may cause (1) forced ascent on a smooth mountain slope, producing adiabatic cooling, condensation and precipitation; (2) triggering of conditional or convective instability by blocking of the airflow and upstream lifting; (3) triggering of convection by diurnal heating of slopes and upslope winds; (4) precipitation from low-level cloud over the mountains through 'seeding' of ice crystals or droplets from an upper-level feeder cloud (Figure 4.15); and (5) increased frontal precipitation by retarding the movement of cyclonic systems and fronts. West coast mountains with onshore flow, such as the Western Ghats, India, during the southwest summer monsoon; the west coasts of Canada, Washington and Oregon; or coastal Norway, in winter months, supposedly illustrate smooth forced ascent, yet most of the other processes seem to be involved. The limited width of the coastal ranges, with average wind speeds, generally allow insufficient time for the basic mechanisms of precipitation growth to operate (see Figure 4.9). In view of the complexity of processes involved, Tor Bergeron proposed using the term 'orogenic', rather than orographic, precipitation, i.e. an origin related to various orographically produced effects. An extreme example of orographic precipitation is provided by the Godzone Wetzone on the western side of the mountains of the South Island of New Zealand, where mean annual totals exceed 1,000 cm (Figure 4.16).

In mid-latitude areas where precipitation is predominantly of cyclonic origin, orographic effects tend to increase both the frequency and intensity of winter precipitation, whereas during summer and in continental climates with a higher condensation level the main effect of relief is the occasional triggering of intense thunderstorm-type precipitation. The orographic influence occurs only in the proximity of high ground in the case of a stable

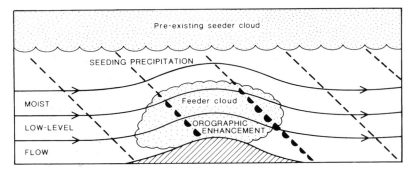

Figure 4.15 Schematic diagram of T. Bergeron's 'seeder–feeder' cloud model of orographic precipitation over hills.

Source: After Browning and Hill 1981, by permission of the Royal Meteorological Society.

Note: This process may also operate in deep nimbostratus layers.

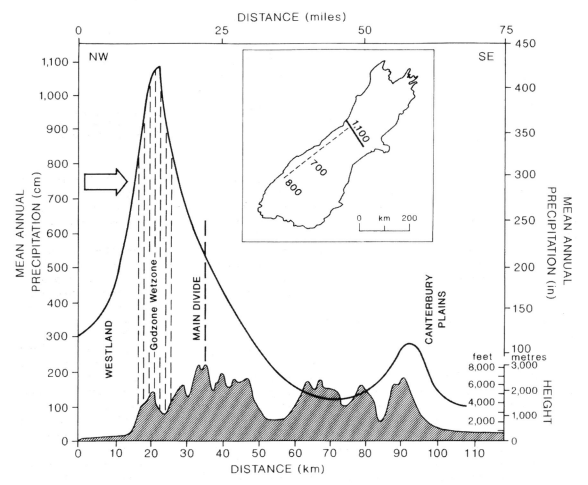

Figure 4.16 Mean annual precipitation (1951–80) along a transect of the South Island of New Zealand, shown as the solid line in the inset map. On the latter, the dashed line indicates the position of the Godzone Wetzone and the figures give the precipitation peaks (cm) at three locations along the Godzone Wetzone.

Sources: Chinn 1979 and Henderson 1993, by permission of the New Zealand Alpine Club Inc. and T.J. Chinn.

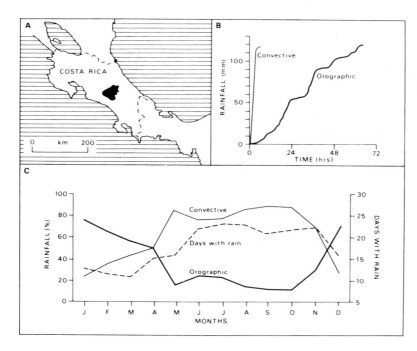

Figure 4.17 Orographic and convective rainfall in the Cachi region of Costa Rica for the period 1977–80. A: the Cachi region, elevation 500–3,000 m. B: typical accumulated rainfall distributions for individual convective (duration 1–6 hours, high intensity) and orographic (1–5 days, lower intensity except during convective bursts) rainstorms. C: monthly rainfall divided into percentages of convective and orographic, plus days with rain, for Cachi (1,018 m).

Source: From Chacon and Fernandez 1985, by permission of the Royal Meteorological Society.

atmosphere. Recent radar studies show that the main effect in this case is one of redistribution, whereas in the case of an unstable atmosphere precipitation appears to be increased, or at least redistributed on a larger scale, since the orographic effects may extend well downwind due to the activation of mesoscale rain bands (see Figure 7.13).

In tropical highland areas, there is a clearer distinction between orographic and convective contributions to total rainfall than in the mid-latitude cyclonic belt, although elements of tropical orographic rainfall are of convective origin due to the orographic release of potential instability. Figure 4.17 shows that in the mountains of Costa Rica the temporal character of convective and orographic rainfalls and their seasonal occurrences are quite distinguishable. Convective rain occurs mainly in the May–November period, when 60 per cent of the rain falls in the afternoons between 1200 and 1800 hours; orographic rain predominates between December and April, with a secondary maximum in June and July coinciding with an intensification of the trades.

Even low hills may have an orographic effect. Research in Sweden shows that wooded hills, rising only 30–50 m above the surrounding lowlands, increase precipitation amounts locally by 50–80 per

cent during cyclonic spells. Until Doppler radar studies of the motion of falling raindrops became feasible, the processes responsible for such effects were unknown. A principal cause is the 'seeder–feeder' ('releaser–spender') cloud mechanism, proposed by Tor Bergeron (described above). This is illustrated in Figure 4.15. In moist stable airflows, shallow cap clouds form over hill tops. Precipitation falling from an upper layer of altostratus (the seeder cloud) grows rapidly by the wash out of droplets in the lower (feeder) cloud. The seeding cloud may release ice crystals, which subsequently melt, or droplets. Precipitation from the upper cloud layer alone would not give significant amounts at the ground, as the droplets would have insufficient time to grow in the airflow, which may traverse the hills in only 15–30 minutes. Most of the precipitation intensification takes place in the lowest kilometre of the atmosphere in moist, fast-moving airflows.

Two special cases of orographic effects may be mentioned. One is the general influence of surface friction, which by turbulent uplift may assist the formation of stratus or stratocumulus layers when other conditions are suitable (see C.2, this chapter). Only light precipitation (drizzle, light rain or snow grains) is to be expected under these circumstances. The other case arises through frictional slowing

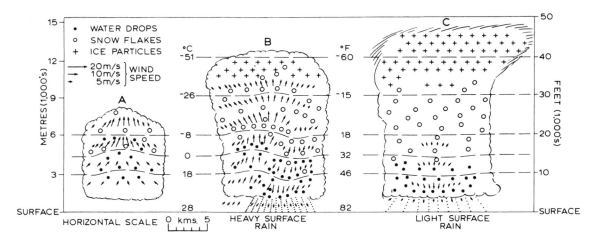

Figure 4.18 The cycle of a local thunderstorm. The arrows indicate the direction and speed of air currents. A: the developing stage of the initial updraught. B: the mature stage with updraughts and downdraughts. C: the dissipating stage, dominated by cool downdraughts.
Source: After Byers and Braham; adapted from Petterssen 1958 [1969].

down of an airstream moving inland over the coast. A particular instance of the convergence and uplift that this may initiate has been reported by Bergeron. During a 24-hour period in October 1945, a west-south-westerly airstream over the Netherlands produced a belt of precipitation (3 cm or more) – a result of frictional convergence and uplift – in crossing the narrow zone of coastal sand dunes only a few metres high. Over the remainder of that virtually flat country a series of lee waves developed in the tropospheric airflow downwind from the coast (see Figure 5.11, for example), and these gave a series of transverse (north–south) bands of precipitation up to 2 cm in amount. On the following day, the surface flow had changed little, but a temperature decrease from –20 to -28 °C at the 500 mb level altered the vertical stability, so that the lee waves broke down and the precipitation distribution showed convective streaks, up to 4 cm per day, parallel to the surface wind direction.

F THUNDERSTORMS

In temperate latitudes probably the most spectacular example of moisture changes and associated energy releases in the atmosphere is the thunderstorm. Extreme upward and downward movements of air are both the principal ingredients and motivating machinery of such storms. They occur: (1) as rising cells of excessively heated moist air; (2) in association with the triggering-off of conditional

instability by uplift over mountains; or (3) along the *squall line* in association with an air-mass discontinuity (see p. 97).

The life-cycle of a local storm lasts only 1–2 hours and begins when a parcel of air is either warmer than the air surrounding it or actively undercut by colder encroaching air. In both instances, the air begins to rise and the embryo thunder cell forms as an unstable updraught of warm air (Figure 4.18). As condensation begins to form cloud droplets, latent heat is released and the initial upward impetus of the air parcel is augmented by an expansion and a decrease in density until the whole mass becomes completely out of thermal equilibrium with the surrounding air. At this stage, updraughts may increase from 3–5 m s^{-1} at the cloud base to 8–10 m s^{-1} some 2–3 km higher, and they can exceed 30 m s^{-1}. The constant release of latent heat continuously injects fresh supplies of energy, which accelerate the updraught. The air mass will continue to rise as long as its temperature remains greater (or, in other words, its density less) than that of the surrounding air. Cumulonimbus clouds tend to form where the air is already moist, as a result of previous penetrating towers from a cluster of clouds, and there is persistent ascent.

Raindrops begin to develop rapidly when the ice stage (or freezing stage) is reached by the vertical build-up of the cell, allowing the Bergeron process to operate. They do not immediately fall to the ground, because the updraughts are able to support

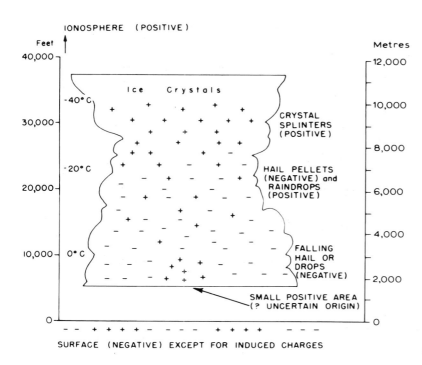

Figure 4.19 The distribution of electrostatic charges in a thundercloud.

Source: After Mason 1962.

them. The minimum cumulus depth for showers over ocean areas seems to be between 1 and 2 km, but 4–5 km is more typical inland. The corresponding minimum time intervals needed for showers to fall from growing cumulus are about 15 minutes over ocean areas and ≥30 minutes inland. Falls of hail require the special cloud processes, described in the last section, involving phases of 'dry' (rime accretion) and 'wet' growth on hail pellets. The mature stage of a storm (see Figure 4.18B) is usually associated with precipitation downpours and lightning (see Plate C). The precipitation causes frictional downdraughts of cold air. As these gather momentum, cold air may eventually spread out below the thunder cell in a wedge. Gradually, as the moisture of the cell is expended, the supply of released latent heat energy diminishes, the downdraughts progressively gain in power over the warm updraughts, and the cell dissipates.

To simplify the explanation, a thunderstorm with only one cell was illustrated. Usually storms are far more complex in structure and consist of several cells arranged in clusters of 2–8 km across, 100 km or so in length and extending up to 10 km or more (see Plate 11). Such systems are known as squall lines (see section G, this chapter).

Two general hypotheses have been developed to account for thunderstorm electrification. One involves the induction mechanism, the other non-inductive charge transfer. As an example of the first category, since the ionosphere at 30–40 km altitude is positively charged (owing to the action of cosmic and solar ultraviolet radiation in ionization) and the earth's surface is negatively charged during fine weather, cloud droplets can acquire an induced positive charge on their lower side and negative charge on their upper side. Non-inductive charge transfer requires contact between cloud or precipitation particles.

The typically observed distribution of charges in a thundercloud is shown in Figure 4.19. The separation of electrical charges of opposite sign may involve several mechanisms: raindrop break-up (large droplets retaining positive charge, the surface spray carrying negative ions), or the selective capture of negative atmospheric ions by falling cloud particles are possible factors, but they do not appear to create sufficiently large charges. A third mechanism is the splintering of ice crystals during the freezing of cloud droplets. This operates as follows. A supercooled droplet freezes inwards from its surface and this leads to a negatively charged warmer core (OH⁻ ions) and a positively charged colder surface due to the migration of H⁺ ions outwards down the temperature gradient. When this soft hailstone ruptures during freezing, small ice splinters carrying a positive charge are ejected by the ice shell and preferentially lifted to the top of the convection cell in updraughts. This theory can explain the charge distribution in Figure 4.19, which shows that the upper part of the cloud (above about the −20 °C isotherm) is positively charged. Equally, the negatively charged hail pellets fall towards the cloud base. However, the ice-splintering mechanism appears to work only for a narrow range of temperature conditions, and the charge transfer is small.

According to J. Latham, the major factor in cloud electrification is non-inductive charge transfer. This involves collisions between splintered ice crystals and warmer pellets of soft hail. Previous accretion of supercooled droplets on the hail pellets produces an irregular surface, which is warmed as the droplets release latent heat on freezing. The impacts of ice crystals on this irregular surface generate negative charge, whereas the crystals acquire positive charge. Negative charge is usually concentrated between about −10 and −20 °C in a thundercloud, where ice crystal concentrations are large, due to splintering at about the −5 °C level and then ascent of the crystals in upcurrents. Radar studies show that lightning is associated with both ice particles in clouds and rising air currents causing upward motion of small hail. The origin of small positive areas near the cloud base (see Figure 4.19) is still under discussion. They could arise through the action of convective updraughts carrying positive charge. It is likely that the very varied electrical properties of thunderclouds (from cloud to cloud and within individual clouds as they develop) cannot be explained by any single theory of charge generation.

Lightning commonly begins more or less simultaneously with precipitation downpours. It may occur between the lower part of the cloud and the ground (which locally has an induced positive charge). The first (leader) stage of the flash bringing down negative charge from the cloud is met near the ground by a return stroke, which rapidly takes positive charge upwards along the already formed channel of ionized air. Just as the leader is neutralized by the return stroke, so the latter is neutralized in turn by the cloud. Subsequent leaders and return strokes drain higher regions of the cloud until its supply of negative charge is temporarily exhausted. The total

flash, with about eight return strokes, typically lasts only about 0.5 seconds. Other, more frequent, flashes occur within a cloud or between clouds. The extreme heating and explosive expansion of air immediately around the path of the lightning sets up intense sound waves, causing thunder to be heard. The sound travels at about 300 m s⁻¹.

Lightning is only one aspect of the atmospheric electricity cycle. During fine weather, the earth's surface is negatively charged, the ionosphere positively charged. The potential gradient of this vertical electrical field in fine weather is about 100 V m⁻¹ near the surface, decreasing to about 1 V m⁻¹ at 25 km, whereas beneath a thundercloud it reaches 10,000 V m⁻¹ just before a discharge. The 'breakdown potential' for lightning to occur in dry air is 3×10^6 V m⁻¹, but this is ten times the largest observed potential in thunderclouds. Hence, the necessity for localized cloud droplet/ice crystal charging processes, as already described, to initiate flash leaders.

Atmospheric ions conduct electricity from the ionosphere down to the earth, and hence a return supply must be forthcoming to maintain the observed electrical field. One source is the slow point discharge, from objects such as buildings and trees, of ions carrying positive charge induced by the negative thundercloud base. Similar upward currents occur above thunderstorm clouds. The other source (estimated to be smaller in its effect over the earth as a whole) is the instantaneous upward transfer of positive charge by lightning strokes, leaving the earth negatively charged. The joint operation of these supply currents, in approximately 1,800 thunderstorms over the globe at any instant, is thought to be sufficient to balance the air–earth leakage, and this number seems to agree reasonably well with observations.

Globally, thunderstorms are most frequent between 1200 and 2100 local time, with a minimum around 0300. An analysis of lightning on visible satellite imagery at local midnight shows a predominance of flashes over tropical land areas between 15 °N and 30 °S (Plate C). In the austral summer, lightning signatures are along the equatorial trough and south to about 30 °S over the Congo, South Africa, Brazil, Indonesia and northern Australia, with activity along cyclone paths, in the northern hemisphere. In the boreal summer, activity is concentrated in Central and northern South America, West Africa – the Congo, northern India and South-east Asia and the southeastern United

States. Lightning is a significant environmental hazard. In the United States alone there are 100–150 deaths per year, on average, as a result of lightning accidents.

G MESOSCALE CONVECTIVE SYSTEMS

Mesoscale convective systems are intermediate in size and lifespan between synoptic disturbances and individual cumulonimbus cells (see Figure 4.20). They occur seasonally in both middle latitudes (the central United States, eastern China and South Africa) and the tropics (India, West and Central Africa and northern Australia) as either nearly circular clusters of convective cells or linear squall lines. The *squall line* consists of a narrow line of thunderstorm cells, which may extend for hundreds of kilometres. It is marked by a sharp veer of wind direction and very gusty conditions. The squall line often occurs ahead of a cold front, maintained either as a self-propagating disturbance or by thunderstorm downdraughts. It may form a pseudo-cold front between rain-cooled air and a rainless zone within the same air mass. In frontal cyclones, cold air in the rear of the depression may overrun air in the warm sector. The intrusion of this nose of cold air sets up great instability, and the subsiding cold wedge tends to act as a scoop forcing up the slower-moving warm air (see Plate 11).

Figure 4.21 shows the movement of clusters of convective cells, each cell about 1 km in diameter, as they crossed southern Britain with a cold front. Each individual cell may be short-lived, but cell

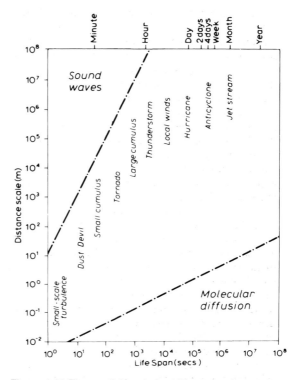

Figure 4.20 The spatial scale and lifespans of mesoscale and other meteorological systems.
Source: After Smagorinsky 1974.

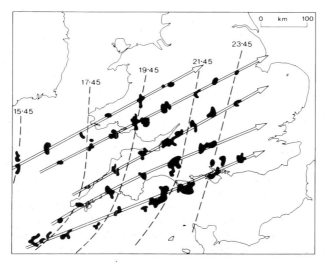

Figure 4.21 Successive positions of individual clusters of middle-tropospheric convective cells moving across southern Britain at about 50 km hr^{-1} with a cold front. Cell location and intensity were determined by radar.

Source: After Browning 1980.

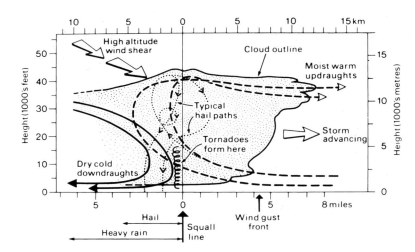

Figure 4.22 Thunder cell structure with hail and tornado formation.

Source: After Hindley 1977.

clusters may persist for hours, strengthening or weakening due to orographic and other factors.

Figure 4.22 shows that the *relative* motion of the warm air is towards the squall line. Such conditions generate severe frontal thunderstorms like that which struck Wokingham, England, in September 1959. This moved from the south-west at about 20 m s⁻¹, steered by strong south-westerly flow aloft. The cold air subsided from high levels as a violent squall, and the updraught ahead of this produced an intense hailstorm. Hailstones grow by accretion in the upper part of the updraught, where speeds in excess of 50 m s⁻¹ are not uncommon, are blown ahead of the storm by strong upper winds, and begin to fall. This causes surface melting, but the stone is caught up again by the advancing squall line and reascends. The melted surface freezes, giving glazed ice as the stone is carried above the freezing level, and further growth occurs by the collection of supercooled droplets (see also Chapter 4, pp. 87 and 89).

The *mesoscale convective complex* is common over the central United States in spring and summer, where it brings widespread severe weather. It develops from initially isolated cumulonimbus cells. The sequence, illustrated in Figure 4.21, appears to be as follows. As rain falls from the thunderstorm clouds, evaporative cooling of the air beneath the cloud bases sets up cold downdraughts (see Figure 4.18B), and when these become sufficiently extensive they create a local high pressure of a few millibars intensity. The downdraughts trigger the ascent of displaced warm air, and a general warming

of the middle troposphere results from latent heat release. Inflow develops towards this warm region, above the cold outflow, causing additional convergence of moist, unstable air. In at least some cases, this inflow is provided by a low-level jet. As individual cells become organized in a cluster along the leading edge of the surface high, new cells tend to form on the right flank (in the n. hemisphere) through interaction of cold downdraughts with the surrounding air. Through this process and the decay of older cells on the left flank, the storm system tends to move 10–20° to the right of the mid-tropospheric wind direction. As the thunderstorm high intensifies, a 'wake low', associated with clearing weather, develops to the rear of it. The system is now producing violent winds, and intense downpours of rain and hail accompanied by thunder. During the triggering of new cells, tornadoes may form, as discussed later in this section. As the complex reaches maturity, during the evening and night hours over the Great Plains of the United States, the mesoscale circulation is capped by an extensive (>100,000 km²) cold upper-cloud shield, readily identified on infrared satellite images. Statistics for forty-three systems over the Great Plains in 1978 showed that the systems lasted on average 12 hours, with initial mesoscale organization occurring in the early evening (1800–1900 LST) and maximum extent seven hours later. During their life-cycle, systems may travel from the Colorado–Kansas border to the Mississippi River or the Great Lakes, or from the Missouri–Mississippi river valley to the east coast. The complex usually decays when

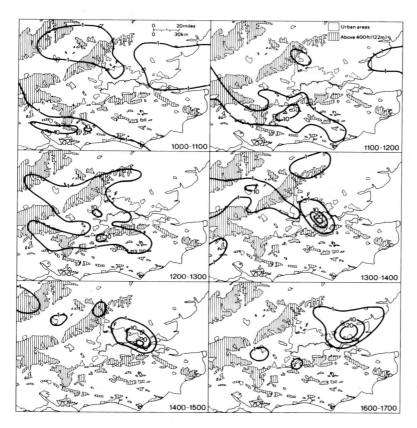

Figure 4.23 Hourly rainfalls (mm) over south-east England during the periods 1000–1500 hours and 1600–1700 hours on 14 September 1968.

Source: After Jackson 1977.

synoptic-scale features inhibit its self-propagation. The production of cold air is shut off when new convection ceases, so that the meso-high and -low are steadily weakened and the rainfall becomes light and sporadic, eventually stopping altogether.

On 14 September 1968, a belt of intense thunderstorms moved northwards into south-east England. Strong convergence of air from the south and north-east occurred along this trough in association with jet streams aloft, allowing intense surface convection of hot, moist air. The intense rainstorms during the period 1000–1700 hours (Figure 4.23) were the first phase of a two-day period of instability, yielding 200 mm (7.87 in) of rainfall at Tilbury, just east of London, which proved to be the outstanding rainfall event of the century in south-east England.

Particularly severe thunderstorms are associated with great potential vertical instability (e.g. hot, moist air underlying dryer air, with colder air aloft). This was the case with a severe storm in the vicinity of Sydney, Australia, on 21 January 1991 (Figure 4.24). This storm formed in a hot, moist, low-level airstream flowing north-east on the eastern side of the Blue Mountains escarpment. This flow was overlain by a hot, dry, northerly airstream at an elevation of 1,500–6,000 metres, which, in turn, was capped by cold air associated with a nearby cold front.

On occasions, so-called supercell thunderstorms may develop (Figure 4.25). These are about the same size as thundercell clusters but are dominated by one giant updraught and localized strong downdraughts. They are often associated with the production of large hailstones and tornadoes.

Tornadoes, which may develop from such squall-line thunderstorms, are common over the Great Plains of the United States, especially in spring and early summer (see Figure 4.26). During this period, cold, dry air from the high plateaux may override maritime tropical air (see Note 1). Subsidence beneath the upper troposphere westerly jet (Figure 4.27) caps the low-level moist air, forming an inversion at about 1,500–2,000 m. The moist air is

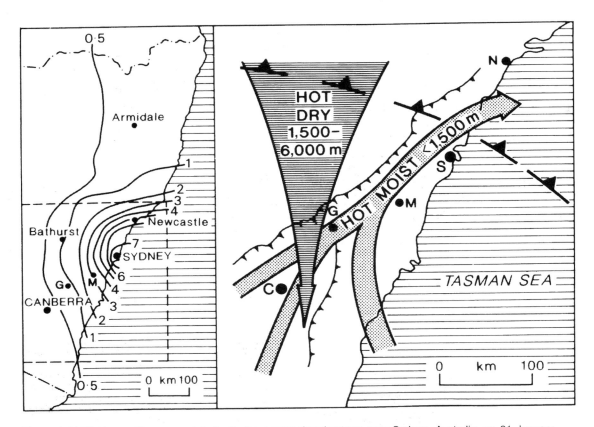

Figure 4.24 Right: conditions associated with the severe thunderstorm near Sydney, Australia, on 21 January 1991. Left: the average annual number of severe thunderstorms (per 25,000 km²) over eastern New South Wales for the period 1950–89. C–Canberra; G–Goulburn; M–Moss Vale; N–Newcastle; S–Sydney.

Sources: Eyre 1992 and Griffiths *et al.* 1993. Left: *Climatic Change* **25**. Copyright © Kluwer Academic Publishers, Dordrecht. Reproduced also by kind permission of the NSW Bureau of Meteorology. Right: reprinted from *Weather* by permission of the Royal Meteorological Society. Copyright ©.

extended northwards by a low-level southerly jet (cf. p. 102) and through continuing advection the air beneath the inversion becomes progressively more warm and moist. Eventually, the general convergence and ascent in the depression trigger the potential instability of the air, generating large cumulus clouds, which penetrate the inversion. The convective trigger is sometimes provided by the approach of the cold front towards the western edge of the moist tongue. Tornadoes may also occur in association with tropical cyclones (see p. 246) and in other synoptic situations if the necessary vertical contrast is present in the temperature, moisture and wind fields.

The exact tornado mechanism is still not fully understood because of the observational problems involved. Tornadoes generally develop in a meso-

low on the periphery of severe rotating thunderstorm systems, where horizontal convergence increases the vorticity and rising air is replenished by moist air from progressively lower levels as the vortex descends and intensifies (see Figure 4.22). Such generating thunderstorms are identifiable, when seen in plan view on a radar display, by a 'hook echo' pattern representing spiral cloud bands about a small central eye. Pressure in the mesoscale thunderstorm low is only 2–5 mb lower than in the surrounding environment. The tornado funnel has been observed to originate in the cloud base and extend towards the surface, and one idea is that convergence beneath the base of cumulonimbus clouds, aided by the interaction between cold precipitation downdraughts and neighbouring updraughts, may initiate rotation. Other observations suggest that the funnel forms

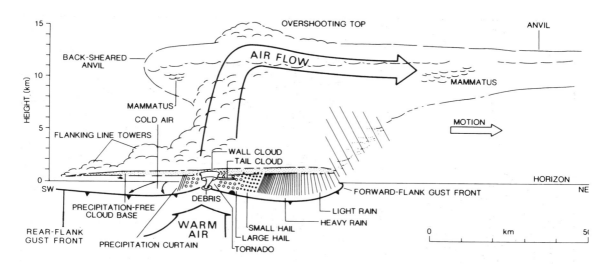

Figure 4.25 A supercell thunderstorm.

Source: After the National Severe Storms Laboratory, USA and H. Bluestein; from Houze and Hobbs 1982 (copyright © Academic Press, reproduced by permission).

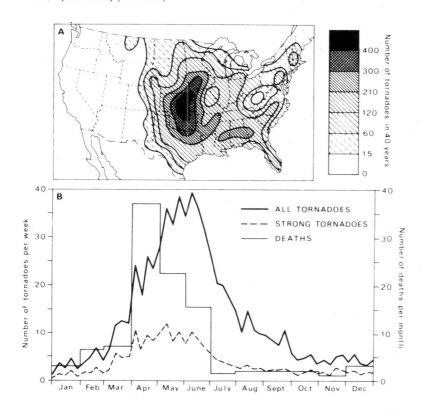

Figure 4.26 Incidence of tornadoes in the United States. A: number of tornadoes reported over a 40-year period. B: weekly average number of tornadoes (1950–82) and monthly averages of resulting deaths (1953–82).

Sources: A: from Hindley 1977 (this first appeared in *New Scientist*, London). B: from 'The tornado' by J. T. Snow (copyright © 1984 by Scientific American Inc. All rights reserved).

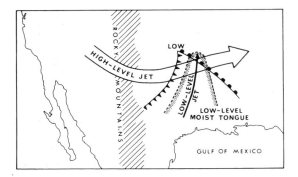

Figure 4.27 The synoptic conditions favouring severe storms and tornadoes over the Great Plains.

simultaneously throughout a considerable depth of cloud, usually a towering cumulus. It appears that the upper portion of the tornado spire in this cloud may become linked to the main updraught of a neighbouring cumulonimbus, thereby causing rapid removal of air from the spire and allowing a violent pressure decrease at the surface. The pressure drop is estimated to exceed 200–250 mb in some cases, and it is this that makes the funnel visible by causing air entering the vortex to reach saturation. Over water, tornadoes are termed waterspouts; the majority rarely attain extreme intensities. The tornado vortex is usually only a few hundred metres in diameter (see Plate 12), and in an even more restricted band around the core the winds can attain speeds of 50–100 m s^{-1}. The fastest tornadoes often split

into multiple vortices rotating anticlockwise with respect to the main tornado axis, each following a cycloidal path and the whole tornado system giving a complex pattern of destruction (Figure 4.28), with maximum wind speeds on the right-side boundary (in the northern hemisphere), where the translational and rotational speeds are combined. Destruction results not only from the high winds, because buildings near the path of the vortex may explode outwards owing to the pressure reduction outside.

Intense tornadoes present problems as to their energy supply, and it has recently been suggested that the release of heat energy by lightning and other electrical discharges may be a necessary additional source of energy.

Tornadoes commonly occur in families and move in rather straight paths (typically between 10 and 100 km long and 100 m to 2 km wide) at velocities dictated by the low-level jet. Thirty-year averages indicate some 750 tornadoes per year in the United States, with 60 per cent of these in April, May and June (see Figure 4.26B). They cause about 100 fatalities and 1,800 injuries each year, on average, although most of the deaths and destruction result from a few enduring mature tornadoes, making up only 1.5 per cent of the total reported. For example, the most severe recorded tornado travelled 200 km in 3 hours across Missouri, Illinois and Indiana on 18 March 1925, killing 689 people.

Tornadoes are not unknown even in the British Isles. During 1960–82 there were 14 days per year with tornado occurrences. Most are minor outbreaks,

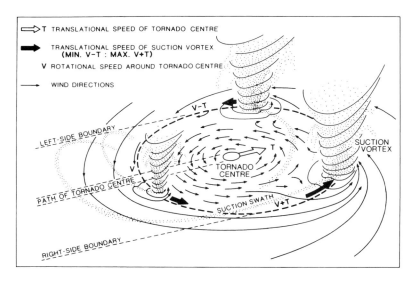

Figure 4.28 Schematic diagram of a complex tornado with multiple suction vortices.

Source: After Fujita, J. Atmos. Sci. 38, 1981, Fig. 15, p. 1,251.

but on 23 November 1981, 102 were reported during south-westerly flow ahead of a cold front. They are most common in autumn, when cold air moves over relatively warm seas.

SUMMARY

Air may be lifted through instability due to surface heating or mechanical turbulence, ascent of air at a frontal zone, or forced ascent over an orographic barrier. Instability is determined by the actual rate of temperature decrease with height in the atmosphere relative to the appropriate adiabatic rate. The dry adiabatic lapse rate is 9.8 °C/km; the saturated adiabatic rate is less than the DALR due to latent heat released by condensation. It is least (around 5 °C/km) at high temperatures, but approaches the DALR at subzero temperatures.

Condensation requires the presence of hygroscopic nuclei such as salt particles in the air. Otherwise, supersaturation occurs. Similarly, ice crystals only form naturally in clouds containing freezing nuclei (clay mineral particles). Otherwise, water droplets may supercool to −39 °C. Both supercooled droplets and ice crystals may be present at cloud temperatures of −10 to −20 °C.

Clouds are classified in ten basic types, according to altitude and cloud form. Satellites are providing new information on spatial patterns of cloudiness, revealing cellular (honeycomb) areas and linear cloud streets, as well as large-scale storm patterns.

Precipitation drops do not form directly by growth of cloud droplets through condensation. Two processes may be involved – coalescence of falling drops of differing sizes, and the growth of ice crystals by vapour deposition (the Bergeron–Findeisen process). Low-level cloud may be seeded naturally by ice crystals from upper cloud layers, or by introducing artificial nuclei. There is no single cause of the orographic enhancement of precipitation totals, and at least four contributing processes can be distinguished.

Thunderstorms are generated by convective uplift, which may result from daytime heating, orographic ascent or squall lines. Several cells may be organized in a mesoscale convective complex and move with the large-scale flow. Such thunderstorms provide an environment for hailstone growth and for the generation of tornadoes.

The freezing process appears to be a major element of cloud electrification in thunderstorms. Lightning plays a key role in maintaining the electrical field between the surface and the ionosphere.

5

Atmospheric motion: principles

The atmosphere is in constant motion on scales ranging from short-lived, local wind gusts to storm systems spanning several thousand kilometres and lasting about a week, and to the more or less constant global-scale wind belts circling the earth. Before considering the global aspects, however, it is important to look at the immediate controls on air motion. The downward-acting gravitational field of the earth sets up the observed decrease of pressure away from the earth's surface that is represented in the vertical distribution of atmospheric mass (see Figure 1.13). This mutual balance between the force of gravity and the vertical pressure gradient is referred to as *hydrostatic equilibrium* (p. 14). This state of balance, together with the general stability of the atmosphere and its shallow depth, greatly limits vertical air motion. Average horizontal wind speeds are of the order of one hundred times greater than average vertical movements, although individual exceptions occur – particularly in convective storms.

A LAWS OF HORIZONTAL MOTION

There are four controls on the horizontal movement of air near the earth's surface: the pressure-gradient force, the Coriolis force, centripetal acceleration and frictional forces. The primary cause of air movement is the development of a horizontal pressure gradient, and the fact that such a gradient can persist (rather than being destroyed by air motion towards the low pressure) results from the effect of the earth's rotation in giving rise to the Coriolis force.

1 The pressure-gradient force

The pressure-gradient force has vertical and horizontal components but, as already noted, the vertical component is more or less in balance with the force of gravity. Horizontal differences in pressure can be due to thermal or mechanical causes (often not easily distinguishable), and these differences control the horizontal movement of an air mass. In effect, the pressure gradient serves as the motivating force that causes the movement of air away from areas of high pressure and towards areas where it is lower, although other forces prevent air from moving directly across the isobars (lines of equal pressure). The pressure-gradient force per unit mass is expressed mathematically as

$$-\frac{1}{\rho}\frac{\mathrm{d}p}{\mathrm{d}n}$$

where ρ = air density and $\mathrm{d}p/\mathrm{d}n$ = the horizontal gradient of pressure. Hence the closer the isobar spacing the more intense is the pressure gradient and the greater the wind speed. The pressure-gradient force is also inversely proportional to air density, and this relationship is of particular importance in understanding the behaviour of upper winds.

2 The earth's rotational deflective (Coriolis) force

The Coriolis force arises from the fact that the movement of masses over the earth's surface is usually referred to a moving co-ordinate system (i.e. the latitude and longitude grid, which 'rotates' with the earth). The simplest way to begin to visualize the manner in which this deflecting force operates is to picture a rotating disc on which moving objects are deflected. Figure 5.1 shows the effect of such a deflective force operating on a mass moving outward from the centre of a spinning disc. The body follows a straight path in relation to a fixed frame of reference (for instance, a box that contains the spinning disc), but viewed relative to co-ordinates rotating with the disc the body swings to the right of its initial

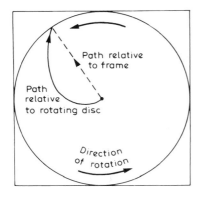

Figure 5.1 The Coriolis deflecting force operating on an object moving outward from the centre of a rotating turntable.

line of motion. This effect is readily demonstrated if a pencil line is drawn across a white disc on a rotating turntable. Figure 5.2 illustrates a case where the movement is not from the centre of the turntable and the object possesses an initial momentum in relation to its distance from the axis of rotation. In the analogous case of the rotating earth (with rotating reference co-ordinates of latitude and longitude), there is apparent deflection of moving objects to the right of their line of motion in the northern hemisphere and to the left in the southern hemisphere, as viewed by observers on the earth. The deflective force (per unit mass) is expressed by:

$$-2\ \omega\ V \sin \phi$$

where ω = the angular velocity of spin ($15°$ hr^{-1} or $2\pi/24$ rad hr^{-1} for the earth = 7.29×10^{-5} rad s^{-1});

ϕ = the latitude and V = the velocity of the mass. $2\omega \sin \phi$ is referred to as the Coriolis parameter (f).

The magnitude of the deflection is directly proportional to: (1) the horizontal velocity of the air (i.e. air moving at 11 m s^{-1} having half the deflective force operating on it as on that moving at 22 m s^{-1}); and (2) the sine of the latitude (sin $0° = 0$; sin $90° = 1$). The effect is thus a maximum at the poles (i.e. where the plane of the deflecting force is parallel to the earth's surface) and decreases with the sine of the latitude, becoming zero at the equator (i.e. where there is no component of the deflection in a plane parallel to the surface). Values of f vary with latitude as follows:

Latitude	0°	10°	20°	43°	90°N
$f(10^{-4}$ s$^{-1})$	0	0.25	0.50	1.00	1.458

The Coriolis force always acts at right angles to the direction of the air motion, to the right in the northern hemisphere (f positive) and to the left in the southern hemisphere (f negative).

The earth's rotation also produces a vertical component of rotation about a horizontal axis. This is a maximum at the equator (zero at the poles), but it is much less important to atmospheric motions due to the existence of hydrostatic equilibrium.

3 The geostrophic wind

Observations in the *free atmosphere* (above the level affected by surface friction at about 500 to 1,000 m) show that the wind blows more or less at right angles

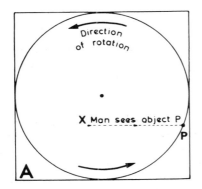

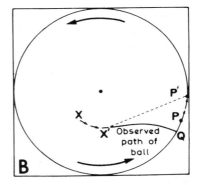

Figure 5.2 The Coriolis deflecting force on a rotating turntable. A: a man at X sees the object P and attempts to throw a ball towards it. Both locations are rotating anticlockwise. B: the man's position is now X′ and the object is at P′. To the man, the ball appears to follow a curved path and lands at Q. The man overlooked the fact that P was moving to his left and that the path of the ball would be affected by the initial impetus due to the man's own rotation.

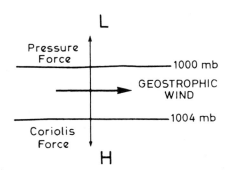

Figure 5.3 The geostrophic wind case of balanced motion (northern hemisphere).

to the pressure gradient (i.e. parallel to the isobars) with, for the northern hemisphere, the high-pressure core on the right and low pressure on the left when viewed downwind. This implies that for steady motion the pressure-gradient force is exactly balanced by the Coriolis deflection acting in the diametrically opposite direction (Figure 5.3). The wind in this idealized case is called a *geostrophic wind*, the velocity (V_g) of which is given by the following formula:

$$V_g = \frac{1}{2\omega \sin \phi \rho} \cdot \frac{dp}{dn}$$

where dp/dn = the pressure gradient. The velocity is thus inversely dependent on latitude, such that the same pressure gradient associated with a geostrophic wind speed of 15 m s^{-1} at latitude 43° will produce a velocity of only 10 m s^{-1} at latitude 90°. Except in low latitudes, where the Coriolis deflection approaches zero, the geostrophic wind is a close approximation to the observed air motion in the free atmosphere. Since pressure systems are rarely stationary, this fact implies that air motion must continually change towards a new balance. In other words, mutual adjustments of the wind and

pressure fields are constantly taking place. The common 'cause-and-effect' argument that a pressure gradient is formed and air begins to move towards low pressure before coming into geostrophic balance is an unfortunate oversimplification of reality.

4 The centripetal acceleration

For a body to follow a curved path there must be an inward acceleration (*c*) towards the centre of rotation. This is expressed by:

$$c = -\frac{mV^2}{r}$$

where m = the moving mass, V = its velocity and r = the radius of curvature. This factor is sometimes regarded for convenience as a centrifugal force operating radially outwards (see Note 1). In the case of the earth itself, this is valid. The centrifugal effect due to rotation has in fact resulted in a slight bulging of the earth's mass in low latitudes and a flattening near the poles. The small decrease in apparent gravity towards the equator (see Note 2) reflects the effect of the centrifugal force working against the gravitational attraction directed towards the earth's centre. It is, therefore, only necessary to consider the forces involved in the rotation of the air about a local axis of high or low pressure. Here the curved path of the air (parallel to the isobars) is maintained by an inward-acting, or centripetal, acceleration.

Figure 5.4 shows (for the northern hemisphere) that in a low-pressure system balanced flow is maintained in a curved path (referred to as the *gradient wind*) by the Coriolis force being weaker than the pressure force. The difference between the two gives the net centripetal acceleration inwards. In the high-pressure case, the inward acceleration is provided by the Coriolis force exceeding the pressure force. Since the pressure gradients are assumed to be equal, the different contributions of the Coriolis force in

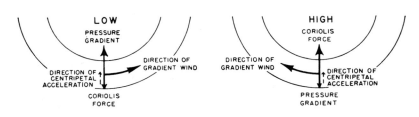

(a) Low pressure   (b) High pressure

Figure 5.4 The gradient wind case of balanced motion around a low pressure (a) and a high pressure (b) in the northern hemisphere.

each case imply that the wind speed around the low pressure must be lower than the geostrophic value (*subgeostrophic*), whereas in the case of high pressure it is *supergeostrophic*. In reality, this effect is obscured by the fact that the pressure gradient in a high is usually much weaker than in a low. Moreover, the fact that the earth's rotation is cyclonic imposes a limit on the speed of anticyclonic flow. The maximum occurs when the angular velocity is $f/2$ ($= \omega \sin \phi$), at which value the absolute rotation of the air (viewed from space) is just cyclonic. Beyond this point anticyclonic flow breaks down ('dynamic instability'). There is no maximum speed in the case of cyclonic rotation.

The magnitude of the centripetal acceleration is generally small, but it becomes important where high-velocity winds are moving in very curved paths (i.e. around an intense low-pressure system). Two cases are of meteorological significance: first, in intense cyclones near the equator, where the Coriolis force is negligible; and, second, in a narrow vortex such as a tornado. Under these conditions, when the large pressure-gradient force provides the necessary centripetal acceleration for balanced flow parallel to the isobars, the motion is called *cyclostrophic*.

The above arguments all assume steady conditions of balanced flow. This simplification is useful, but it must be noticed that two factors prevent a continuous state of balance. Latitudinal motion changes the Coriolis parameter, and the movement or changing intensity of a pressure system leads to acceleration or deceleration of the air, causing some degree of cross-isobaric flow. Pressure change itself depends on air displacement through the breakdown of the balanced state. If air movement were purely geostrophic there would be no growth or decay of pressure systems. The acceleration of air in moving at upper levels from a region of cyclonic isobaric curvature (subgeostrophic wind) to one of anticyclonic curvature (supergeostrophic wind) causes a fall of pressure at lower levels in the atmosphere through the removal of air aloft. The significance of this fact will be discussed in Chapter 7G. The interaction of horizontal and vertical air motions is outlined in B.2, this chapter.

5 Frictional forces and the planetary boundary layer

The last force that has an important effect on air movement is that due to friction with the earth's surface. Towards the surface (i.e. below about 500 m for flat terrain), friction begins to reduce the wind velocity below its geostrophic value. This layer of frictional influence is known as the *planetary boundary layer* (PBL). The temporal variability of PLB structure can now be routinely measured by atmospheric profilers (lidar and radar). Its depth varies over land from a few hundred metres at night, when the air is stable as a result of nocturnal surface cooling, to 1–2 km during afternoon convective conditions. Exceptionally, over hot dry surfaces, convective mixing may extend to 4–5 km. Over the oceans, it is more consistently near 1 km deep and in the tropics especially is often capped by an inversion due to sinking air. The boundary layer is typically either stable or unstable. Yet, for theoretical convenience, it is often treated as being neutrally stable (i.e. the lapse rate is that of the DALR, or the potential temperature is constant with height, see Figure 4.1). Under this ideal state, the wind turns clockwise (veers) with increased height above the surface, setting up a wind spiral (Figure 5.5). This spiral profile was first demonstrated in the turning of ocean currents with depth (see Chapter 6D.1a) by V. W. Ekman; both are referred to as *Ekman spirals*. The slowing of the wind towards the surface modifies the deflective force, which is dependent on velocity causing it also to decrease. At low levels, the wind consequently blows obliquely across the

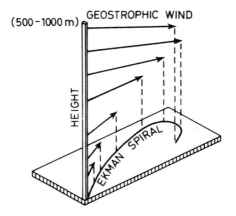

Figure 5.5 The Ekman spiral of wind with height, in the northern hemisphere. The wind attains the geostrophic velocity at between 500 and 1,000 m in the middle and higher latitudes as frictional drag effects become negligible. This is a theoretical profile of wind velocity under conditions of mechanical turbulence.

isobars in the direction of the pressure gradient. The angle of obliqueness increases with the growing effect of frictional drag due to the earth's surface averaging about 10–20° at the surface over the sea and 25–35° over land. The inflow of air towards the low-pressure centre generates upward motion at the top of the PBL, known as *Ekman pumping*.

Wind velocity decreases exponentially close to the earth's surface due to the frictional effects of the surface. These involve 'form drag' over obstacles (buildings, forests, hills) and the stress exerted by the air at the surface interface. The mechanism of *form drag* involves the creation of locally higher pressure on the windward side of an obstacle and a lateral pressure gradient. Wind stress arises from, first, the molecular resistance of the air to the vertical wind shear (i.e. increasing air motion with height above the surface); this molecular viscosity operates in a laminar sub-layer only millimetres thick. Second, turbulent eddies, a few metres to tens of metres in size, act to slow the air motion on a larger scale (eddy viscosity). The aerodynamic roughness of terrain is characterized by the *roughness length* (z_0), defined by Oke as 'the height at which the neutral wind profile extrapolates to zero wind speed'. The concept of roughness length involves the lifting of airflow over a rough surface close to which the airflow has low velocity. Typical roughness lengths are given in Table 5.1.

In summary, the surface wind (neglecting any curvature effects) represents a balance between the pressure-gradient force and the Coriolis force perpendicular to the air motion and friction parallel to the air motion.

Table 5.1 Typical roughness lengths (m) associated with terrain surface characteristics.

Terrain surface characteristics	Roughness length (m)
Groups of high buildings	1–10
Temperate forest	0.8
Groups of medium buildings	0.7
Suburbs	0.5
Trees and bushes	0.2
Farmland	0.05–0.1
Grass	0.008
Bare soil	0.005
Snow	0.001
Smooth sand	0.0003
Water	0.0001

Source: After Troen and Petersen 1989.

B DIVERGENCE, VERTICAL MOTION AND VORTICITY

These three terms essentially hold the key to a proper understanding of modern meteorological studies of wind and pressure systems on a synoptic and global scale. Mass uplift or descent of air occurs primarily in response to dynamic factors related to horizontal airflow and is only secondarily affected by air-mass stability. Hence the significance of these factors for weather processes.

1 Divergence

Different types of horizontal flow are shown in Figure 5.6A. The first panel shows that air may accelerate (decelerate), leading to velocity divergence (convergence). When streamlines (lines of instantaneous air motion) spread out or squeeze together, this is termed diffluence or confluence, respectively. If the streamline pattern is strengthened by that of the isotachs (lines of equal wind speed), as shown in the third panels of Figure 5.6A, then there may be mass divergence or convergence at a point (Figure 5.6B). In this case, the compressibility of the air causes the density to decrease or increase, respectively. Usually, however, confluence is associated with an increase in air velocity and diffluence with a decrease. In the intermediate case, confluence is balanced by an increase in wind velocity and diffluence by a decrease in velocity. Hence, convergence (divergence) may give rise to vertical stretching (shrinking), as illustrated in Figure 5.6C. It is important to note that if all winds were geostrophic, there could be no convergence or divergence and hence no weather!

Other ways in which convergence or divergence can occur are the result of surface friction effects. Onshore winds undergo convergence at low levels when the air slows down on crossing the coastline owing to the greater friction overland, whereas offshore winds accelerate and become divergent. Frictional differences can also set up coastal convergence (or divergence) if the geostrophic wind is parallel to the coastline with, for the northern hemisphere, land to the right (or left) of the air current, viewed downwind.

2 Vertical motion

Horizontal inflow or outflow near the surface has to be compensated by vertical motion, as illustrated

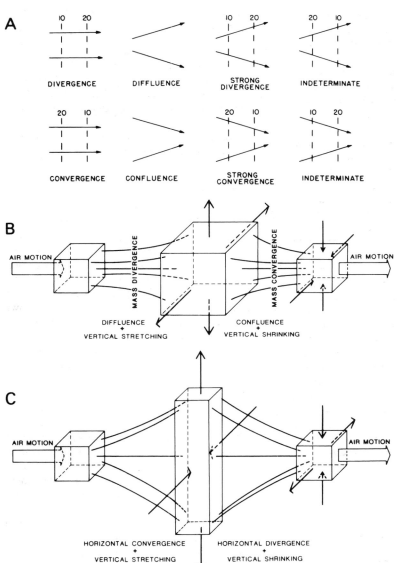

Figure 5.6 Convergence and divergence. A: plan view of horizontal flow patterns producing divergence and convergence – the broken lines are schematic isopleths of wind speed (isotachs). B: schematic illustration of local mass divergence and convergence, assuming density changes. C: typical convergence-stretching and divergence-shrinking relationships in atmospheric flow.

in Figure 5.7, if the low- or high-pressure systems are to persist and there is to be no continuous density increase or decrease. Air rises above a low-pressure cell and subsides over high pressure, with compensating divergence and convergence, respectively, in the upper troposphere. In the middle troposphere, there must clearly be some level at which horizontal divergence or convergence is effectively zero; the mean 'level of non-divergence' is generally at about 600 mb. Large-scale vertical

motion is extremely slow compared with convective up- and downdraught currents in cumulus, for example. Typical rates in large depressions and anticyclones are of the order of ± 5–10 cm s^{-1}, whereas updraughts in cumulus may exceed 10 m s^{-1}.

3 Vorticity

Vorticity implies the rotation or angular velocity of minute (imaginary) particles in any fluid system. The

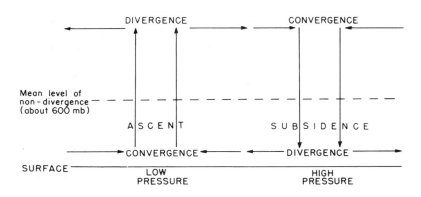

Figure 5.7 Cross-section of the patterns of vertical motion associated with (mass) divergence and convergence in the troposphere, illustrating mass continuity.

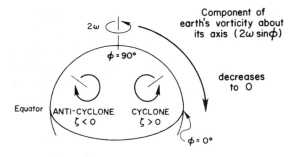

Figure 5.8 Sketch of the relative vertical vorticity (ζ) about a cyclone and an anticyclone in the northern hemisphere. The component of the earth's vorticity about its axis of rotation (or, the Coriolis parameter, f), is equal to twice the angular velocity (ω) times the sine of the latitude (ϕ). At the pole $f = 2\omega$, diminishing to 0 at the equator. Cyclonic vorticity is in the same sense as the earth's rotation about its own axis, viewed from above, in the northern hemisphere: this cyclonic vorticity is defined as positive ($\zeta > 0$).

air within a depression can be regarded as comprising an infinite number of small air parcels, each rotating cyclonically about an axis vertical to the earth's surface (Figure 5.8). Vorticity has three elements – magnitude (defined as *twice* the angular velocity, ω) (see Note 3), direction (the horizontal or vertical axis about which the rotation occurs) and the sense of rotation. Rotation in the same sense as the earth's rotation – cyclonic in the northern hemisphere – is defined as positive. Cyclonic vorticity may result from cyclonic curvature of the streamlines, from cyclonic shear (stronger winds on the right side of the current, viewed downwind in the northern hemisphere), or a combination of the two (Figure 5.9). Lateral shear (see Figure 5.9B) results from changes in isobar spacing. Anticyclonic

vorticity occurs with the corresponding anticyclonic situation. The component of vorticity about a vertical axis is referred to as the vertical vorticity. This is generally the most important, but near the ground surface frictional shear causes vorticity about an axis parallel to the surface and normal to the wind direction.

Vorticity is related not only to air motion around a cyclone or anticyclone (*relative vorticity*), but also to the location of that system on the rotating earth. The vertical component of *absolute vorticity* consists of the relative vorticity (ζ) and the latitudinal value of the Coriolis parameter, $f = 2\omega \sin \phi$ (see Chapter 6A). At the equator, the local vertical is at right angles to the earth's axis, so $f = 0$, but at the North Pole cyclonic relative vorticity and the earth's rotation act in the same sense (see Figure 5.8).

C LOCAL WINDS

To the practising meteorologist, local controls over air movement often provide more problems than the effects of the major planetary forces just discussed. Diurnal tendencies are superimposed upon both the large- and the small-scale patterns of wind velocity. These are particularly noticeable in the case of local winds and therefore are examined before we consider the major types of local wind regime.

In normal conditions, there is a general tendency for wind velocities to be least about dawn, at which time there is little vertical thermal mixing and the lower air does not therefore partake of the velocity of the more freely moving upper air (see Chapter 6A). Conversely, velocities of some local winds are greatest between 1300 and 1400 hours, because this is the time when the air suffers its greatest tendency

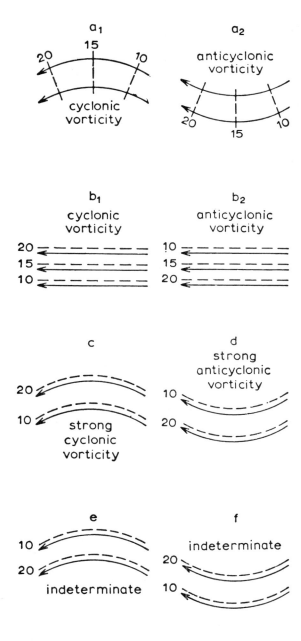

Figure 5.9 Streamline models illustrating in plan view the flow patterns with cyclonic and anticyclonic vorticity in the northern hemisphere. In c and d, the effects of curvature (a_1 and a_2) and lateral shear (b_1 and b_2) are additive, whereas in e and f they more or less cancel out. Dashed lines are schematic isopleths of wind speed.

Source: After Riehl *et al.* 1954.

to move vertically due to terrestrial heating, allowing it, subject to surface frictional effects, to join in the freer upper-air movement. Upper air always moves more freely than air at surface levels, because it is not subject to the retarding effects of friction and obstruction.

1 Mountain and valley winds

Terrain irregularities give rise to their own special meteorological conditions. On warm sunny days, the heated air in a valley is laterally constricted, compared with that over an equivalent area of lowland, and so tends to expand vertically. The volume ratio of lowland:valley air is typically about 2 or 3:1 and this difference in heating sets up a density and pressure differential, which causes air to flow from the lowland up the axis of the valley. This valley wind is generally very light and requires a weak regional pressure gradient in order to develop. This flow along the main valley develops more or less simultaneously with *anabatic* (upslope) winds, which result from greater heating of the valley sides compared with the valley floor. These slope winds rise above the ridge lines and feed an upper return current along the line of the valley to compensate for the valley wind. This feature may be obscured, however, by the regional airflow. Speeds reach a maximum around 1400 hours. At night, there is a reverse process as the denser cold air at higher elevations drains into depressions and valleys; this is known as a *katabatic* wind (Figure 5.10).

If the air drains downslope into an open valley, a 'mountain wind' develops more or less simultaneously along the axis of the valley. This flows towards the plain, where it replaces warmer, less dense air. The maximum velocity occurs just before sunrise at the time of maximum diurnal cooling. Like the valley wind, the mountain wind is also overlain by an upper return current, in this case up-valley.

Katabatic drainage is usually cited as the cause of frost pockets in hilly and mountainous areas. It is argued that greater radiational cooling on the slopes, especially if they are snow-covered, leads to a gravity flow of cold, dense air into the valley bottoms. Observations in California and elsewhere, however, suggest that the valley air remains colder than the slope air from the onset of nocturnal cooling, so that the air moving downslope slides over the denser air in the valley bottom. Moderate

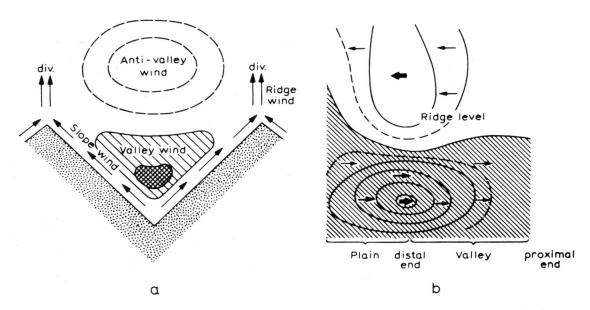

Figure 5.10 Valley winds in an ideal V-shaped valley. a: Section across the valley. The valley wind and anti-valley wind are directed at right angles to the plane of the paper. The arrows show the slope and ridge wind in the plane of the paper, the latter diverging (div.) into the anti-valley wind system. b: Section running along the centre of the valley and out on to the adjacent plain, illustrating the valley wind (below) and the anti-valley wind (above).

Source: After Buettner and Thyer 1965.

drainage winds will also act to raise the valley temperatures through turbulent mixing. One suggestion is that cold air pockets in valley bottoms and hollows may result from the cessation of turbulent heat transfer to the surface in sheltered locations rather than by cold air drainage. Clearly, the problem requires further study.

2 Winds due to topographic barriers

Mountain ranges have important effects on the air-flow across them. The displacement of air upwards over the obstacle may trigger instability if the air is conditionally unstable and buoyant (see Chapter 4B), whereas stable air returns to its original level in the lee of the barrier as the gravitational effect counteracts the initial displacement. This descent often forms the first of a series of *lee waves* (or *standing waves*) downwind, as shown in Figure 5.11. The wave form remains more or less stationary relative to the barrier, with the air moving quite rapidly through it. Below the crest of the waves, there may be circular air motion in a vertical plane, which is termed a *rotor*. The formation of such features is naturally of vital interest to pilots. The development

of lee waves is commonly disclosed by the presence of lenticular clouds (see Plate 7), and on occasion a rotor causes reversal of the surface wind direction in the lee of high mountains (see Plate 13).

Winds on mountain summits are usually strong, at least in middle and higher latitudes. Average speeds on summits in the Rocky Mountains in winter months are around 12–15 m s^{-1}, for example, and on Mount Washington, New Hampshire, an extreme value of 103 m s^{-1} has been recorded. Peak monthly speeds in excess of 40–50 m s^{-1} are typical in both these areas in winter. Airflow over a mountain range causes the air to be constricted and thus accelerated particularly at and near the crest line (the Venturi effect), but friction with the ground also retards the flow, compared with free air at the same level. The net result is predominantly one of retardation, but the outcome depends on the topography, wind direction and stability.

In the case of low hills, the boundary layer is displaced upward and an acceleration occurs just over the summit. Figure 5.12 shows instantaneous airflow conditions across Askervein Hill (relief *c.* 120 m) on the island of South Uist in the Scottish Hebrides, where the near-crest wind speed at a

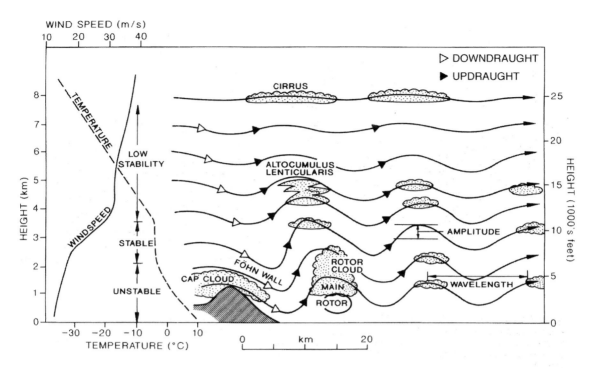

Figure 5.11 Lee waves and rotors are produced by airflow across a long mountain range. The first wave crest usually forms less than one wavelength downwind of the ridge. There is a strong surface wind down the lee slope. Wave characteristics are determined by the wind speed and temperature relationships, shown schematically on the left of the diagram. The existence of an upper stable layer is particularly important.
Source: After Ernst 1976.

height of 10 m above the surface approaches 80 per cent more than the undisturbed upstream velocity. In contrast, there was a 20 per cent decrease on the initial run-up to the hill and a 40 per cent decrease on the lee side, probably due to horizontal divergence. Knowledge of such local variability is critical in the siting of wind-energy systems.

A wind of local importance in mountain areas is the föhn, or chinook. It is a strong, gusty, dry and warm wind that develops on the lee side of a mountain range when stable air is forced to flow over the barrier by the regional pressure gradient. Sometimes, there is a loss of moisture by precipitation on the windward side of the mountains (Figure 5.13) and the air, having cooled at the saturated adiabatic lapse rate above the condensation level, subsequently warms at the greater dry adiabatic lapse rate as it descends on the lee side, with a consequent lowering of both the relative and the absolute humidity. Other investigations show that in many instances there is no loss of moisture over the mountains. In such cases, the föhn effect is the result of the blocking of air to windward of the mountains by a summit-level temperature inversion. This forces air from higher levels to descend and warm adiabatically. Föhn winds are common along the northern flanks of the Alps and the mountains of the Caucasus and Central Asia in winter and spring, when the accompanying rapid temperature rise may help to trigger avalanches on the snow-covered slopes. At Tashkent in Central Asia, where the mean winter temperature is about freezing point, temperatures may rise to more than 21°C during a föhn. In the same way, the chinook is a significant feature at the eastern foot of the New Zealand Alps, the Andes in Argentina and the Rocky Mountains. At Pincher Creek, Alberta, a temperature rise of 21°C (38°F) occurred in four minutes with the onset of a chinook on 6 January 1966. Less spectacular effects are also noticeable in the lee of the Welsh mountains, the Pennines and the Grampians in Great Britain, where the importance of föhn winds lies mainly in the dispersal of cloud by the subsiding dry air (see also Chapters 8A.6

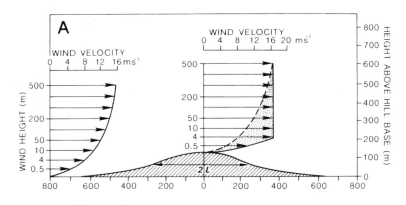

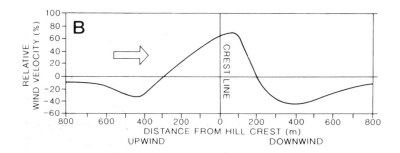

Figure 5.12 Airflow over Askervein Hill, South Uist, off the west coast of Scotland. A: vertical airflow profiles (not true to scale) measured simultaneously 800 m upwind of the crest line and at the crest line. L is the *characteristic length* of the obstruction (i.e. one-half the hill width at mid-elevation, here 500 m) and is also the height above ground level to which the flow is increased by the topographic obstruction (shaded). The maximum speed-up of the airflow due to vertical convergence over the crest is to about 16.5 m s⁻¹ at a height of 4 m. B: the relative speed-up (%) of airflow upwind and downwind of the crest line measured 14 m above ground level.

Source: After Taylor, Teunissen and Salmon *et al.* From Troen and Petersen 1989.

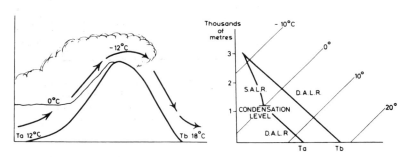

Figure 5.13 The föhn effect when an air parcel is forced to cross a mountain range. Ta refers to the temperature at the windward foot of the range and Tb to that at the leeward foot.

and 8B.3). This is an important component of so-called 'rain shadow' effects.

In some parts of the world, winds descending on the lee slope of a mountain range are cold. The type example of such 'fall-winds' is the *bora* of the northern Adriatic, although similar winds occur on the northern Black Sea coast, in northern Scandinavia, in Novaya Zemlya and in Japan. These winds occur when cold continental air masses are forced across a mountain range by the pressure gradient and, despite adiabatic warming, displace warmer air. They are therefore primarily a winter phenomenon.

On the eastern slope of the Rocky Mountains in Colorado (and probably also in other similar continental locations), winds of either bora or chinook type can occur depending on the initial airflow characterics. Locally, at the foot of the mountains, such winds may attain hurricane force, with gusts exceeding 45 m s⁻¹ (100 mph). A few downslope storms of this type have caused millions of dollars of property damage in Boulder, Colorado, and the immediate vicinity. These windstorms develop when a stable layer close to the mountain-crest level prevents airflow to windward from crossing over

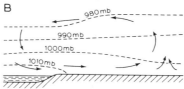
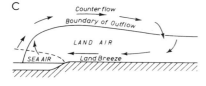
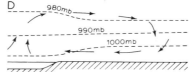

Figure 5.14 Diurnal land and sea breezes. A and B: sea breeze circulation and pressure distribution in the early afternoon during anticyclonic weather. C and D: land breeze circulation and pressure distribution at night during anticyclonic weather.

Source: A and C after Oke 1978.

the mountains. Extreme amplification of a lee wave (see Figure 5.11) drags air from above the summit level (about 4,000 m) down to the plains (1,700 m) in a very short distance, so that the high velocities occur. However, the flow is not simply 'downslope'; the storm may affect the mountain slopes but not the foot of the slope, or vice versa, depending on the location of the lee wavetrough. The high winds are caused by the acceleration of the air towards this local pressure minimum.

3 Land and sea breezes

Another thermally induced type of air movement is the land and sea breeze (see Figure 5.14). The vertical expansion of the air column that occurs daily during the hours of heating over the more rapidly heated land (see Chapter 2B.5) tilts the isobaric surfaces downwards at the coast, causing onshore winds at the surface and a compensating offshore movement aloft. Typical land–sea pressure differences are of the order of 2 mb. At night, the air over the sea is warmer and the situation is reversed, although this reversal is also the effect of downslope winds blowing off the land. Figure 5.15 shows that these local winds can have a decisive effect on coastal temperature and humidity. The advancing cool sea air may form a distinct line (or *front*, see Chapter 7D) marked by cumulus cloud development, behind which there is a distinct wind velocity maximum. This often develops in summer, for example, along the Gulf Coast of Texas. On a smaller scale, such features can also be observed in Britain, particularly along the south and east coasts. The sea breeze has a depth of about 1 km, although it thins towards the advancing edge, and may pene-

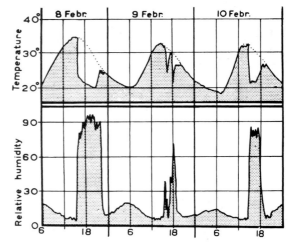

Figure 5.15 The effect of the afternoon sea breeze on the temperature (°C) and relative humidity (per cent) at Joal on the Senegal coast, 8–10 February 1893.

Source: After Angot and De Martonne; from Kuenen 1955.

trate 50 km or more inland by 2100 hours. Typical wind speeds in such sea breezes are 4–7 m s^{-1}, although these may be greatly increased where a well-marked low-level temperature inversion produces a 'Venturi effect' in constricting and accelerating the flow. The much shallower land breezes are usually only about 2 m s^{-1}. The counter-currents aloft are generally less evident and may be obscured by the regional airflow, but recent work along the Oregon coast has suggested that under certain conditions this upper return flow may be very closely related to the lower sea breeze conditions, even to the extent of mirroring the surges in the latter. It is worth noting that in middle latitudes the Coriolis deflection causes turning of a well-developed onshore sea breeze (clockwise in the northern hemisphere) so that eventually it may blow more or less parallel to the shore. Analogous 'lake breeze' systems develop adjacent to large inland water bodies such as the Great Lakes.

SUMMARY

Air motion is described by its horizontal and vertical components; the latter are much smaller than the horizontal velocities. Horizontal motions compensate for vertical imbalances between gravitational acceleration and the vertical pressure gradient.

Horizontal wind velocity is determined by the horizontal pressure gradient, the earth's rotational effect (Coriolis force), and the curvature of the isobars (centripetal acceleration). All three factors are accounted for in the gradient wind equation, but this can be approximated in large-scale flow by the geostrophic wind relationship. Below 1,500 m, the wind speed and direction are affected by surface friction.

Air ascends (descends) in association with surface convergence (divergence) of air. Air motion is also subject to relative vertical vorticity as a result of curvature of the streamlines and/or lateral shear; this, together with the earth's rotational effect, makes up the absolute vertical vorticity.

Local winds occur as a result of diurnally varying thermal differences setting up local pressure gradients (mountain–valley winds and land–sea breezes) or due to the effect of a topographic barrier on airflow crossing it (examples are the lee-side föhn and bora winds).

Plate 1 Visible image of Africa, Europe and the Atlantic Ocean taken by METEOSAT on 19 August 1978 at 1155 hours GMT. An anticyclone is associated with clear skies over Europe and the Mediterranean, while frontal-wave cyclones are evident in the North Atlantic. Cloud clusters appear along the oceanic Intertropical Convergence Zone (ITCZ) and there are extensive monsoon cloud masses over equatorial West Africa. Less-organized cloud cover is present over East Africa. The subtropical anticyclone areas are largely cloud-free but possess trade-wind cumulus, particularly in the South-East Trade Wind belt of the South Atlantic. The highly reflective desert surfaces of the Sahara are prominent (*METEOSAT image supplied by the European Space Agency*).

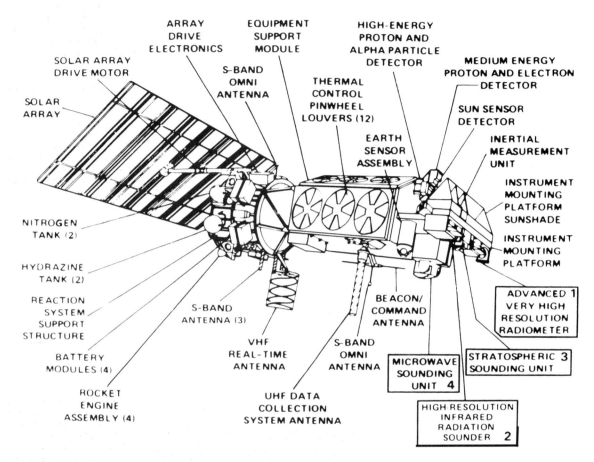

ARRAY DRIVE ELECTRONICS

EQUIPMENT SUPPORT MODULE

HIGH-ENERGY PROTON AND ALPHA PARTICLE DETECTOR

MEDIUM ENERGY PROTON AND ELECTRON DETECTOR

SOLAR ARRAY DRIVE MOTOR

S-BAND OMNI ANTENNA

SUN SENSOR DETECTOR

SOLAR ARRAY

THERMAL CONTROL PINWHEEL LOUVERS (12)

INERTIAL MEASUREMENT UNIT

EARTH SENSOR ASSEMBLY

INSTRUMENT MOUNTING PLATFORM SUNSHADE

NITROGEN TANK (2)

INSTRUMENT MOUNTING PLATFORM

HYDRAZINE TANK (2)

REACTION SYSTEM SUPPORT STRUCTURE

S-BAND ANTENNA (3)

BEACON/ COMMAND ANTENNA

ADVANCED 1 VERY HIGH RESOLUTION RADIOMETER

BATTERY MODULES (4)

VHF REAL-TIME ANTENNA

S-BAND OMNI ANTENNA

STRATOSPHERIC 3 SOUNDING UNIT

MICROWAVE SOUNDING UNIT 4

ROCKET ENGINE ASSEMBLY (4)

UHF DATA COLLECTION SYSTEM ANTENNA

HIGH-RESOLUTION INFRARED RADIATION SOUNDER 2

Plate 2 The TIROS-N spacecraft, having a length of 3.71 m and a weight of 1421 kg. The four instruments of particular meteorological importance are shown in the numbered boxes: 1 Visible and infra-red detector – discerns clouds, land–sea boundaries, snow and ice extent and temperatures of clouds, earth's surface and sea surface. 2 Infra-red detector – permits calculation of temperatures profile from the surface to the 10-mb level, as well as the water vapour and ozone contents of the atmosphere in cloud-free areas. 3 Device for measuring temperatures in the stratosphere. 4 Device for measuring microwave radiation from the earth's surface which supplements unit 2 in cloudy areas (*NOAA: National Oceanic and Atmospheric Administration*).

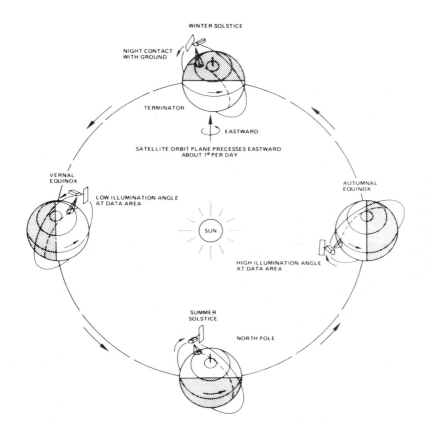

WINTER SOLSTICE

NIGHT CONTACT
WITH GROUND

TERMINATOR

EASTWARD

SATELLITE ORBIT PLANE PRECESSES EASTWARD
ABOUT 1° PER DAY

VERNAL
EQUINOX

LOW ILLUMINATION ANGLE
AT DATA AREA

SUN

AUTUMNAL
EQUINOX

HIGH ILLUMINATION ANGLE
AT DATA AREA

SUMMER
SOLSTICE

NORTH POLE

Plate 3 The TIROS-N satellite system consisting of two spacecraft in polar orbit at 833 and 870 km, respectively. The orbital plane of the second satellite lags 90° longitude behind that of the first and the orbital plane of each processes eastward at about 1° longitude per day. Each satellite transmits data from a circular area of the earth's surface 6200 km in diameter. The satellites make 14.18 and 14.07 orbits of the earth per day, respectively, such that each point on the earth is sensed for 13–14 minutes at a time (*NOAA*).

Plate 4 Cold, fog-laden air draining over the southern rim of the Grand Canyon, Arizona, elevation 2075 m (6800 ft) in the early morning (*photograph Ernst Haas; courtesy Time/Life Publications*).

Plate 5 Cumulus towers with powerful thunderstorms along the ITCZ over Zaire photographed in April 1983 from the Space Shuttle at an elevation of 280 km. The largest tower shows a double mushroom cap reaching to more than 15,240 m and the symmetrical form of the caps indicates a lack of pronounced airflow at high levels (*courtesy NASA*).

Plate 6 Cumulus orographic clouds developed over the dip-slope of the South Downs in Sussex, England. To the west (*right*), southern Hampshire is covered by stratiform clouds. The English Channel is in the upper left. This infra-red photograph was taken from an elevation of about 12,000 m (40,000 ft) (B = Burgess Hill; Br = Brighton; H = Haywards Heath; S = Shoreham; W = Worthing) (*P.A.–Reuters Ltd*).

Plate 7 View north along the eastern front of the Colorado Rockies, showing lee-wave clouds (*NCAR photograph by Robert Bumpas*).

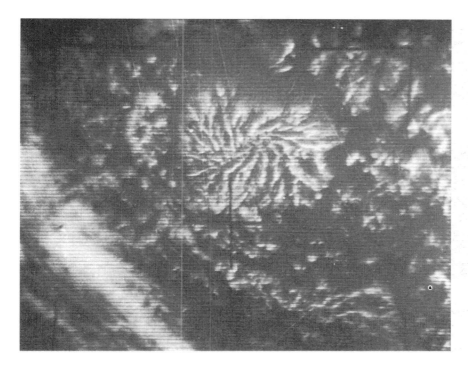

Plate 8 Radiating or dendritic cellular (actiniform) cloud pattern. These complex convective systems, some 150–250 km (90–150 miles) in diameter, were only discovered as a result of satellite photography. They usually occur in groups over areas of subsidence inversions intensified by cold ocean currents (e.g. in the low latitudes of the eastern Pacific) (*Environmental Science Service Administration*).

Plate 9 DMSP visible image of the coastal area off New England at 1433 hours GMT, 17 February 1979. A northerly airflow averaging 10 m s^{-1}, with surface air temperatures of about −15°C, moves offshore where sea-surface temperatures increase to 9°C within 250 km of the coast. Convective cloud streets are visible, also ice in James Bay (*upper left*), and in the Gulf of St Lawrence (*image courtesy of National Snow and Ice Data Center, University of Colorado, Boulder.*) (See *Monthly Weather Review III*, 1983, p. 245).

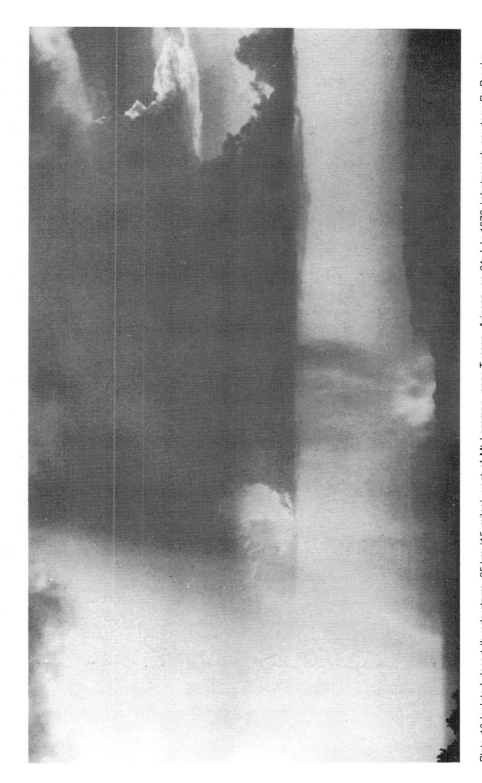

Plate 10 Isolated desert thunderstorm, 25 km (15 miles) west of Mt Lemmon, near Tucson, Arizona, on 31 July 1973 (*photograph courtesy Dr Brooks Martner, University of Wyoming*).

Plate 11
Thunderstorm
approaching
Östersund, Sweden,
during late afternoon
on 23 June 1955.
Ahead of the region of
intense precipitation
there are rings of
cloud formed over the
squall-front (*copyright
F.H. Ludlam;
originally published in*
Weather, *vol. XV.
no. 2, 1960, p. 63*).

Plate 12 View from 5 km distance of a tornado north-east of Tracy, Minnesota, on 13 June 1968 (*photograph courtesy of Eric Lantz and Associated Press*). A convectively unstable tongue of warm air extended north from Texas and by mid-afternoon its temperature had risen to 32°C, and severe thunderstorms had set in ahead of a cold front lying to its west. Surface pressure continued to drop in this belt, which was supported by a trough at 500 mb and surmounted by a jet over 45 m s^{-1} at the tropopause extending from Oregon to eastern Canada. Thunderstorm activity reached a maximum at about 1800 hours as the unstable belt moved into Minnesota, and individual cells were shown by radar to have built up to over 15,000 m. This combination of conditions was ideal for tornado inception and on that afternoon thirty-four funnels were sighted within 480 km of Minneapolis. The Tracy tornado appeared 13 km south-west of the town at 1900 hours and moved north-east at 13 m s^{-1} for 21 km, cutting a 90 to 150-m wide belt of total destruction through Tracy, killing nine people, injuring 125 and causing $3 million worth of damage. Unlike most tornadoes, it did not lift off the ground on encountering the rough urban surface but 'dug in' for its whole course until it suddenly dissolved a few seconds after the photograph was taken (*description courtesy of the Director, National Severe Storms Forecast Center, Kansas City*).

Plate 13 View looking south-south-east from about 9000 m (30,000 ft) along the Owen's Valley, California, showing a roll cloud developing in the lee of the Sierra Nevada mountains. The lee-wave crest is marked by the cloud layer, and the vertical turbulence is causing dust to rise high into the air (W = Mount Whitney, 4418 m (14,495 ft); I = Independence) (*photograph by Robert F. Symons: courtesy R.S. Scorer*).

Plate 14 Photograph by an astronaut from Gemini XII manned spacecraft from an elevation of some 180 km (112 miles) looking south-east over Egypt and the Red Sea. The bank of cirrus clouds is associated with strong upper winds, possibly concentrated as a jet stream (*NASA photograph*).

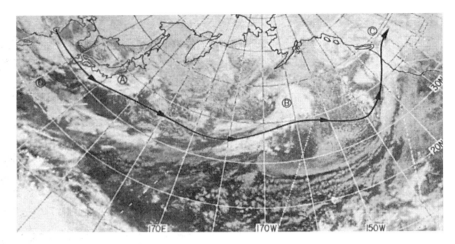

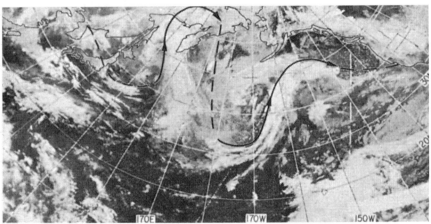

Plate 15 Infra-red photographs of the North Pacific, with the 200-mb jet stream inserted. *Above*: general zonal flow associated with a high zonal index, 12 March 1971: three major cloud systems (A, B, C) occur along the belt of zonal flow, and the largest east–west belt of cloud (D) to the south of Japan is also characteristic of accentuated zonal flow. *Below*: large-amplitude flow regime associated with a lower zonal index, 23 April 1971 (*World Meterological Organization 1973*).

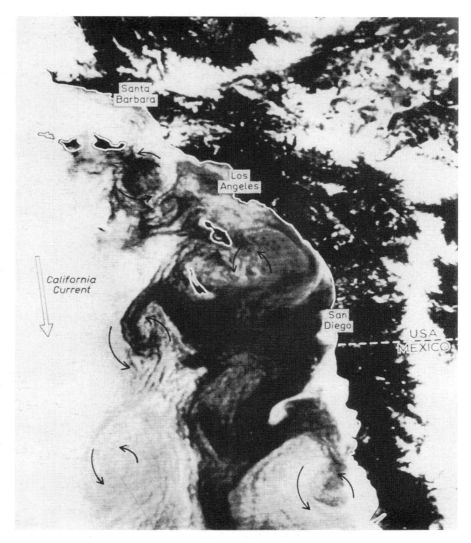

Plate 16 Infra-red photograph showing large vortices of cold water (*light*) up-welling in the warmer surface coastal waters (*dark*) off southern California. The colder offshore California Current is clearly shown (*NASA photograph*).

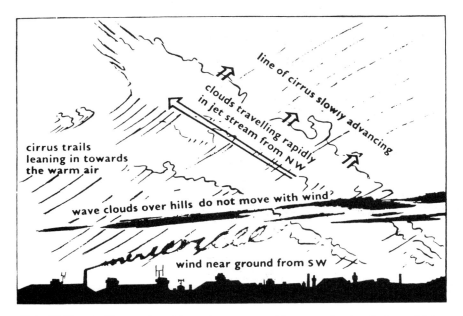

Plate 17 View looking westward towards an approaching warm front, with lines of jet-stream clouds extending from the north-west, from which trails of ice crystals are falling. In the middle levels are dark wave clouds formed in the lee of small hills by the south-westerly airflow, whereas the wind direction at the surface is more southerly – as indicated by the smoke from the chimney (*photograph copyright by F.H. Ludlam: diagram by R.S. Scorer: both published in* Weather, *vol. 18, no. 8. 1963. pp. 266–7).*

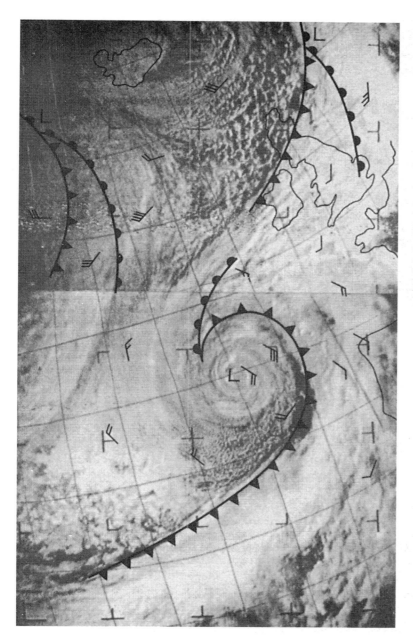

Plate 18 Photograph of a depression to the south-west of the British Isles taken by the ESSA 2 satellite at 1018 hours GMT on 12 November 1966, showing a well-marked spiral cloud structure and open cell cloud areas behind the cold fronts. The depression began as hurricane 'Lois' in the eastern Caribbean and moved north-eastwards until its winds fell below gale force on 10 November. Thereafter it began to deepen again and the pressure at the centre had dropped to 962 mb when this photograph was taken. The frontal zones are well developed, although the 1000–500-mb wind shear was unusually weak for such a well-defined system (*courtesy* Weather, *vol. XXIV, no. 6, 1969, p. 222; Crown Copyright Reserved*).

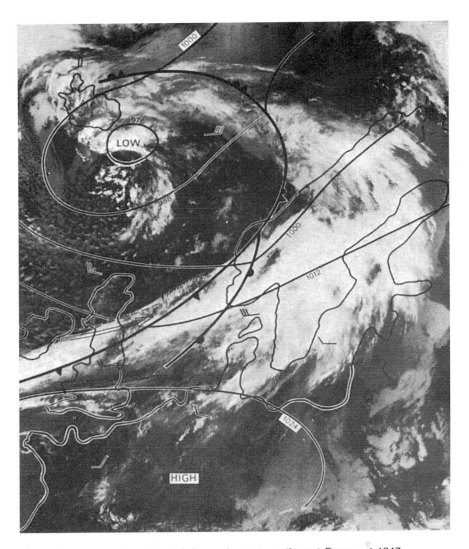

Plate 19 A partly occluded frontal depression over north-west Europe at 1347
hours GMT on 19 October 1979. This TIROS infra-red image was processed to show
the coldest surfaces (e.g. cloud tops) as white and the warmest (e.g. land surfaces)
as black. Surface fronts, isobars and winds are shown, together with the upper
tropospheric jet stream (arrowed), lying on the warm side of the cold front aloft
(see Figure 7.13) (maximum velocity 65 m s^{-1} at about 250 mb; 11.3 km); also note
the break (dashed) between the mid-latitude tropopause (200 mb; 11.8 km) and the
polar tropopause (270 mb; 10.2 km). This added synoptic information relates to 1200
hours GMT (*image courtesy of P.E. Baylis, Department of Electrical Engineering and
Electronics, University of Dundee; synoptic interpretation courtesy of Ross Reynolds,
Department of Meteorology, University of Reading*).

6

Planetary-scale motions in the atmosphere and ocean

In this chapter, we examine global-scale motions in the atmosphere and their role in redistributing energy, momentum and moisture. As we saw in Chapter 2 (p. 47), there are also close links between the atmosphere and oceans with the latter making a major contribution to poleward energy transport. Thus, we also discuss ocean circulations and the coupling of the atmosphere–ocean system.

The atmosphere acts rather like a gigantic heat engine in which the constantly maintained difference in temperature existing between the poles and the equator provides the energy supply necessary to drive the planetary atmospheric and ocean circulation. The conversion of the heat energy into kinetic energy to produce motion must involve rising and descending air, but vertical movements are generally much less in evidence than horizontal ones, which may cover vast areas and persist for periods of a few days to several months. We begin by examining the relationships between winds and pressure patterns in the troposphere and those at the surface.

A VARIATION OF PRESSURE AND WIND VELOCITY WITH HEIGHT

Changes of height reveal variations of both pressure and wind characteristics. Above the level of surface frictional effects (about 500–1,000 m), the wind increases in speed and becomes more or less geostrophic. With further height increase, the reduction of air density leads to a general increase in wind speed (see Chapter 5A.1). At 45°N, a geostrophic wind of 14 m s^{-1} at 3 km is equivalent to one of 10 m s^{-1} at the surface for the same pressure gradient. There is also a seasonal variation in wind speeds aloft, these being much greater in the northern hemisphere during winter months, when the meridional temperature gradients are at a maximum. In addition, the persistence of these gradients tends to cause the upper winds to be more constant in direction.

1 The vertical variation of pressure systems

The air pressure at the surface, or at any level in the atmosphere, depends on the weight of the overlying air column. In Chapter 1B, we noted that air pressure is proportional to air density and that density varies inversely with air temperature. Accordingly, increasing the temperature of an air column between the surface and, say, 3 km will reduce the air density and therefore lower the air pressure at the surface without affecting the pressure at 3 km altitude. Correspondingly, if we compare the heights of the 1,000 and 700 mb pressure surfaces, warming of the air column will lower the height of the 1,000 mb surface but not affect the height of the 700 mb surface (i.e. the thickness of the 1,000–700 mb layer increases).

The general relationships between surface and tropospheric pressure conditions are illustrated by the models of Figure 6.1. A low-pressure cell at sea level with a cold core will intensify with elevation, whereas one with a warm core will tend to weaken and may be replaced by high pressure. A warm air column of relatively low density causes the pressure surfaces to bulge upwards, and conversely a cold, more dense air column leads to downward contraction of the pressure surfaces. Thus, a surface high-pressure cell with a cold core (a *cold anticyclone*), such as the Siberian winter anticyclone, weakens with increasing elevation and is replaced by low pressure aloft. Cold anticyclones are shallow and rarely extend their influence above about 2,500 m.

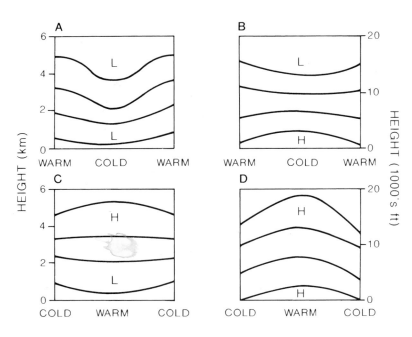

Figure 6.1 Models of the vertical pressure distribution in cold and warm air columns. A: a surface low pressure intensifies aloft in a cold air column. B: a surface high pressure weakens aloft and may become a low pressure in a cold air column. C: a surface low pressure weakens aloft and may become a high pressure in a warm air column. D: a surface high pressure intensifies aloft in a warm air column.

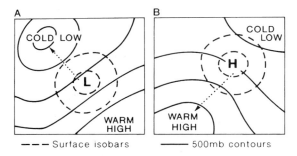

--- Surface isobars   —— 500mb contours

Figure 6.2 The characteristic slope of the axes of low- and high-pressure cells with height in the northern hemisphere.

By contrast, a surface high with a warm core (a *warm anticyclone*) intensifies with height (Figure 6.1D). This is characteristic of the large subtropical cells, which maintain their warmth through dynamic subsidence. The warm low (Figure 6.1C) and cold high (Figure 6.1B) are consistent with the vertical motion schemes illustrated in Figure 5.7, whereas the other two types are primarily produced by dynamic processes. The high surface pressure in a warm anticyclone is linked hydrostatically with cold, relatively dense air in the lower stratosphere. Conversely, a cold depression (Figure 6.1A) is associated with a warm lower stratosphere.

Mid-latitude low-pressure cells have cold air in the rear, and in consequence the axis of low pressure slopes with height towards the colder air to the west. High-pressure cells slope towards the warmest air (Figure 6.2), and in this manner the northern hemisphere subtropical high-pressure cells are displaced 10–15° south in latitude at the 3,000 m level, as well as towards the west. Even so, this slope of the high-pressure axes is not constant through time, and stations located between the cells may experience widely fluctuating upper winds associated with variations in the inclination of the axes.

2 Mean upper-air patterns

It is helpful to begin by considering the patterns of pressure and wind in the middle troposphere. These are less complicated in appearance than surface maps as a result of the diminished effects of the land masses. Rather than using pressure maps at a particular height, it is convenient to depict the height of a selected pressure surface; this is termed a *contour chart* by analogy with topographic relief map (see Note 1). Figure 6.3 shows that in the middle troposphere of the southern hemisphere there is a vast circumpolar cyclonic vortex poleward of latitude 30°S in summer and winter. The vortex is more or less symmetrical about the pole, although

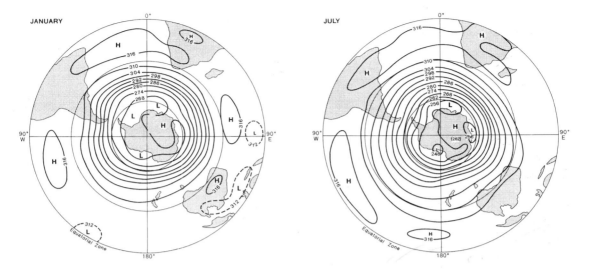

Figure 6.3 The mean contours (g.p. decametres) of the 700 mb pressure surface in January and July for the southern hemisphere, 1949–60.
Source: After Taljaard *et al.* 1969.

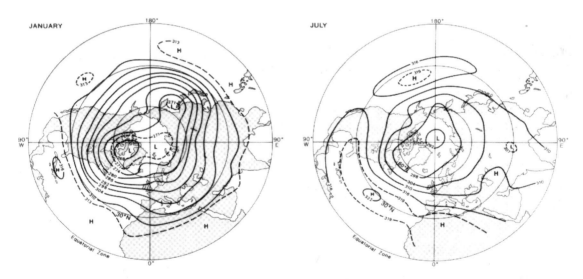

Figure 6.4 The mean contours (g.p. decametres) of the 700 mb pressure surface in January and July for the northern hemisphere, 1950–9.
Source: Adapted from O'Connor 1961.

the low centre is towards the Ross Sea sector. Corresponding charts for the northern hemisphere (Figure 6.4) also show an extensive cyclonic vortex, but one that is markedly more asymmetric with a primary centre over the eastern Canadian Arctic and a secondary one over eastern Siberia. The major troughs and ridges form what are referred to as *long waves* (or *Rossby waves*) in the upper flow. It is worth considering first the reason why the hemispheric westerlies show this large-scale wave motion. The key to this problem lies in the rotation of the earth and the latitudinal variation of the Coriolis parameter (Chapter 5A.2). It can be shown that for large-scale motion the absolute vorticity

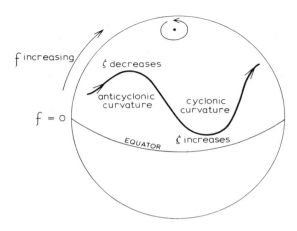

Figure 6.5 A schematic illustration of the mechanism of long-wave development in the tropospheric westerlies.

and tend to be steered by the quasi-stationary long waves.

The two major troughs at about 70°W and 150°E are thought to be induced by the combined influence on upper-air pressure and winds of large orographic barriers, like the Rockies and the Tibetan Plateau, and heat sources such as warm ocean currents (in winter) or land masses (in summer). It is worth noting that land surfaces occupy over 50 per cent of the northern hemisphere between latitudes 40 and 70°N. The subtropical high-pressure belt has only one clearly distinct cell in January over the eastern Caribbean, whereas in July cells are well developed over the North Atlantic and North Pacific. In addition, the July map shows greater prominence of the subtropical high over the Sahara and southern North America. The northern hemisphere shows a marked summer to winter intensification of the mean circulation, which is explained below.

In the southern hemisphere, the predominance of ocean surface (which comprises 81 per cent of the hemisphere) considerably reduces the development of long waves in the upper westerlies. Nevertheless, asymmetries are initiated by the effects on the atmosphere of such geographical features as the Andes, the elevated and extensive dome of eastern Antarctica, and ocean currents, particularly the Humboldt and Benguela Currents (see Figure 6.31 and the associated cold coastal upwellings).

about a vertical axis $(f + \zeta)$ tends to be conserved, i.e.

$$d\,(f + \zeta)\,dt = 0$$

The symbol d/dt denotes a rate of change following the motion (a total differential). Consequently, if air moves polewards so that f increases, the cyclonic vorticity tends to decrease. The curvature thus becomes anticyclonic and the current returns towards lower latitudes. If the air moves equatorward of its original latitude, f tends to decrease (Figure 6.5), requiring ζ to increase, and the resulting cyclonic curvature again deflects the current polewards. In this manner, large-scale flow tends to oscillate in a wave pattern.

Rossby related the motion of these waves to their wavelength (L) and the speed of the zonal current (U). The speed of the wave (or phase speed, c), is:

$$c = U - \beta \left(\frac{L}{2\pi} \right)^2$$

where β = df/dy, i.e. the variation of the Coriolis parameter with latitude (a local, partial differential). For stationary waves, where c = 0, $L = 2\pi\sqrt{u/\beta}$. At 45° latitude, this stationary wavelength is 3,120 km for a zonal velocity of 4 m s^{-1}, increasing to 5,400 km at 12 m s^{-1}. The wavelengths, at 60° latitude for zonal currents of 4 and 12 m s^{-1} are, respectively, 3,170 and 6,430 km. Long waves tend to remain stationary, or even to move westward against the current, so that $c \leq 0$. Shorter waves travel eastward with a speed close to that of the zonal current

3 Upper wind conditions

It is a common observation that clouds at different levels move in different directions. The wind speeds at these levels may also be markedly different, although this is not so evident to the casual observer. The gradient of wind velocity with height is referred to as the (vertical) *wind shear*, and in the free air, above the friction level, the amount of shear depends upon the temperature structure of the air. This important relationship is illustrated in Figure 6.6. The diagram shows hypothetical contours of the 1,000 and 500 mb pressure surfaces. As discussed in section A1, this chapter, the *thickness* of the 1,000–500 mb layer is proportional to its mean temperature – low thickness values correspond to cold air, high thickness values to warm air. This relationship is apparent in the vertical sections of Figure 6.1. The theoretical wind vector (V_T) blowing parallel to the thickness lines, with a velocity proportional to their gradient, is termed the

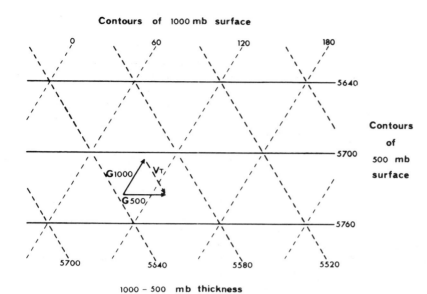

Contours of 1000 mb surface

1000 - 500 mb thickness

Figure 6.6 Schematic map of superimposed contours of isobaric height and thickness of the 1,000–500 mb layer (in metres). G_{1000} is the geostrophic velocity at 1,000 mb, G_{500} that at 500 mb, V_T is the resultant 'thermal wind' blowing parallel to the thickness lines.

thermal wind. The geostrophic wind velocity at 500 mb (G_{500}) is the vector sum of the 1,000 mb geostrophic wind (G_{1000}) and the thermal wind (V_T), as shown in Figure 6.6.

Since the thermal wind component blows with cold air (low thickness) to the left in the northern hemisphere, when viewed downwind, it is readily apparent that in the troposphere the poleward decrease of temperature should be associated with a large westerly component in the upper winds. Furthermore, since the meridional temperature gradient is steepest in winter (in the northern hemisphere), the zonal westerlies are most intense at this time.

The total result of the above influences is that in both hemispheres the mean upper geostrophic winds are dominantly westerly between the subtropical high-pressure cells (centred aloft at about 15° latitude) and the polar low-pressure centre aloft. Between the subtropical high-pressure cells and the equator they are easterly. This dominant, westerly circulation reaches maximum speeds of 45–67 m s^{-1}, which even increase to 135 m s^{-1} in winter. These maximum speeds are concentrated in a narrow band, often situated at about 30° latitude between 9,000 and 15,000 m, called the *jet stream* (see Note 2). Plate 14 shows bands of cirrus cloud that may have been related to jet-stream systems.

The jet stream, which is essentially a fast-moving ribbon of air, is connected with the zone of maximum slope, folding or fragmentation of the tropopause, which in turn coincides with the latitude of maximum poleward temperature gradient and energy transfer. The thermal wind, as described above, is a major component of the jet stream, but the basic reason for the concentration of the meridional temperature gradient in a narrow zone (or zones) is still uncertain. One theory is that the temperature gradient becomes accentuated when the upper wind pattern is confluent (see Chapter 5B.1). Figure 6.7 gives a generalized view of the wind and temperature distribution in the northern hemisphere troposphere in winter. Figure 6.8 shows that there are three westerly jet streams. The more northerly ones, termed the *Polar Front* and *Arctic Front Jet Streams* (Chapter 7E), are associated with the steep temperature gradient where polar and tropical air and polar and arctic air, respectively, interact (see Figure 6.8), but the *Subtropical Jet Stream* is related to a temperature gradient confined to the upper troposphere. The Polar Front Jet Stream is very irregular in its longitudinal location and is commonly discontinuous, whereas the Subtropical Jet Stream is much more persistent. For these reasons, the location of the mean jet stream (Figure 6.9) primarily reflects the position of the Subtropical Jet Stream. The synoptic pattern of jet stream occurrence may be further complicated in some sectors by the presence of additional frontal zones (see Chapter 7E), each associated with a jet stream. This

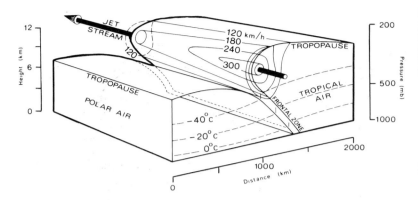

Figure 6.7 Structure of the mid-latitude frontal zone and associated jet stream showing generalized distribution of temperature, pressure and wind velocity.

Source: After Riley and Spolton 1981.

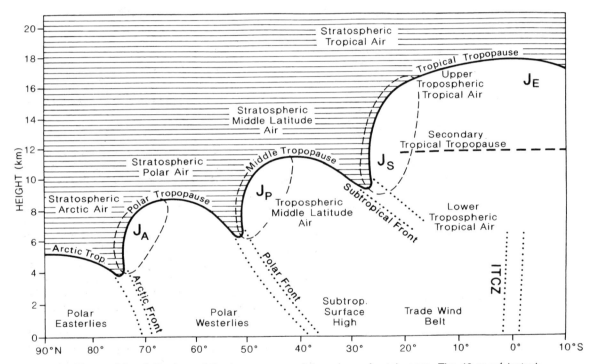

Figure 6.8 The meridional structure of the tropopause and the primary frontal zones. The 40 m s⁻¹ isotach (dashed) encloses the arctic (J_A), polar (J_P) and subtropical (J_S) jet streams. The tropical easterly (J_E) jet stream is also shown. Occasionally, the arctic and polar or the polar and subtropical fronts and jet streams may merge to form single systems in which about 50 per cent of the pole-to-equator mid-tropospheric pressure gradient is concentrated into a singe frontal zone approximately 200 km wide. The tropical easterly jet stream may be accompanied by a lower easterly jet at about 5 km elevation.

Source: Shapiro *et al.* 1987. From *Monthly Weather Review* **115**, p. 450 by permission of the American Meteorological Society.

situation is common in winter over North America. Comparison of Figures 6.4 and 6.9 indicates that the main jet stream cores are associated with the principal troughs of the Rossby long waves. In summer, an *Easterly Tropical Jet Stream* forms in the upper troposphere over India and Africa due to regional reversal of the S–N temperature gradient (p. 258). The relationships between these upper tropospheric wind systems and surface weather and climate will be considered in Chapters 7, 8 and 9.

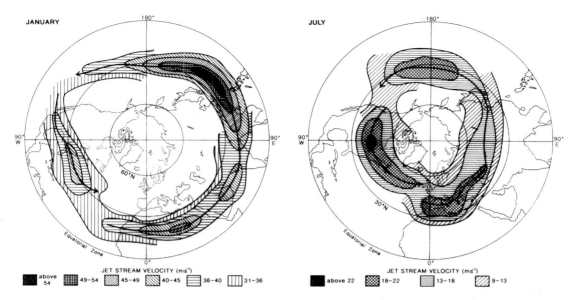

Figure 6.9 The mean location and velocities (m s⁻¹) of the westerly jet stream in the northern hemisphere in January and July.
Source: After Namias and Clapp (adapted from Petterssen 1958).

In the southern hemisphere, the mean jet stream in winter is similar in strength to its northern hemisphere winter counterpart and it weakens less in summer, because the meridional temperature gradient between 30° and 50°S is reinforced by heating over the southern continents. Upper air temperature gradients in the northern hemisphere, in contrast, are much stronger in winter, as noted above.

4 Surface pressure conditions

The most permanent features of the mean sea-level pressure maps are the oceanic subtropical high-pressure cells (Figures 6.10 and 6.11). These anticyclones are located at about 30° latitude, suggestively situated below the mean Subtropical Jet Stream. They move a few degrees equatorwards in winter and polewards in summer in response to the seasonal expansion and contraction of the two circumpolar vortices. In the northern hemisphere, the subtropical ridges of high pressure weaken over the heated continents in summer but are thermally intensified over them in winter. The principal subtropical high-pressure cells are located (1) over the Bermuda–Azores ocean region (aloft the centre of this cell lies over the east Caribbean); (2) over the south and south-west United States (the Great

Basin or Sonoran cell) – this continental cell is seasonal, being replaced by a thermal surface low in summer; (3) over the east and north Pacific – a large and powerful cell (sometimes dividing into two, especially during the summer); and (4) over the Sahara – this, like other continental source areas, is seasonally variable in both intensity and extent, being most prominent in winter. In the southern hemisphere, the subtropical anticyclones are oceanic, except over southern Australia in winter.

The latitude of the subtropical high-pressure belt depends on the meridional temperature difference between the equator and the pole and on the vertical temperature lapse rate (i.e. vertical stability). The greater the meridional temperature difference the more equatorward is the location of the subtropical high-pressure belt (Figure 6.12).

Equatorward of the subtropical anticyclones there is an equatorial trough of low pressure, associated broadly with the zone of maximum insolation and tending to migrate with it, especially towards the heated continental interiors of the summer hemisphere. Poleward of the subtropical anticyclones lies a general zone of subpolar low pressure. In the southern hemisphere, this sub-Antarctic Trough is virtually circumpolar (see Figure 6.11), whereas in the northern hemisphere the major centres are near Iceland and the Aleutians in winter and primarily

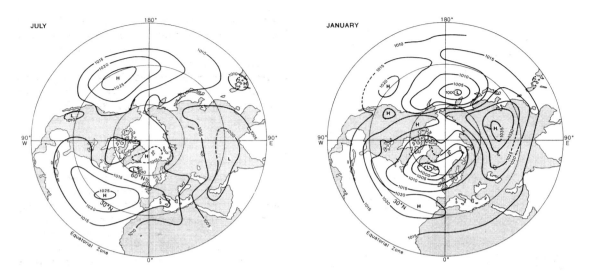

Figure 6.10 The mean sea-level pressure distribution (mb) in January and July for the northern hemisphere, 1950–9.
Source: After O'Connor 1961.

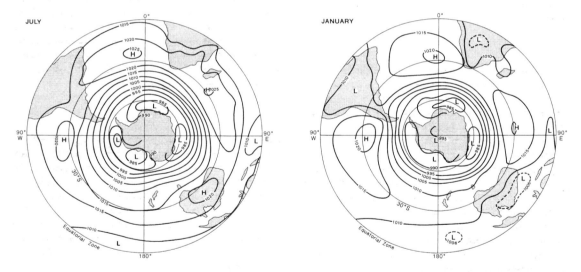

Figure 6.11 The mean sea-level pressure distribution (mb) in January and July for the southern hemisphere.
Source: From Taljaard et al. 1969.

over continental areas in summer. It is commonly stated that in high latitudes there is a surface anticyclone due to the cold polar air, but in the Arctic this is true only in spring over the Canadian Arctic Archipelago. In winter, the polar basin is affected by high- and low-pressure cells with the major semi-permanent cold air anticyclones over Siberia and, to a lesser extent, north-western Canada. The

shallow Siberian high is in part a result of the exclusion of tropical air masses from the interior by the Tibetan massif and the Himalayas. Over Antarctica, it is meaningless to speak of sea-level pressure but, on average, there is high pressure over the 3–4-km-high eastern Antarctic plateau.

The main circulation in the southern hemisphere is much more zonal at both 700 mb and sea level

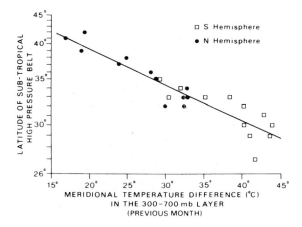

Figure 6.12 A plot of the meridional temperature difference at the 300–700 mb level in the previous month against the latitude of the centre of the subtropical high-pressure belt, assuming a constant vertical tropospheric lapse rate.

Source: After Flohn, in Flohn and Fantechi 1984 (copyright © 1980/1982 D. Reidel Publishing Company. Reprinted by permission).

than in the northern hemisphere, due to the limited area and effect of the southern land masses. There is also little difference between summer and winter circulation intensity (see Figures 6.3 and 6.11). Indeed, it is important at this point to differentiate between mean pressure patterns and the highs and lows shown on synoptic weather maps. Thus, in the southern hemisphere, the zonality of the mean circulation conceals a high degree of day-to-day variability. The synoptic map is one that shows the principal pressure systems over a very large area at a given time, local wind features, for example, being ignored. The subpolar lows over Iceland and the Aleutians (see Figure 6.10) shown on recurrent mean pressure maps represent the passage of deep depressions across these areas downstream of the upper long-wave troughs. The mean high-pressure areas, however, relate to more or less permanent highs. The intermediate areas, such as the zones about 50–55°N and 40–60°S, affected by travelling depressions and ridges of high pressure, appear on the mean maps as being of neither markedly high nor markedly low pressure. The movement of depressions is considered in Chapter 7F.

On comparing the surface and tropospheric pressure distributions for January (see Figures 6.4 and

6.10), it will be noticed that only the subtropical high-pressure cells extend to high levels. The reasons for this are evident from Figures 6.1B and D. In summer, the equatorial low-pressure belt is also evident aloft over South Asia. The subtropical cells are still discernible at 300 mb, showing them to be a fundamental feature of the global circulation and not merely a response to surface conditions.

B THE GLOBAL WIND BELTS

One fact that emerges from the preceding discussion is the importance of the subtropical high-pressure cells. Dynamic, rather than immediately thermal, in origin, and situated between 20 and 30° latitude, they seem to provide the key to the world's surface wind circulation. In the northern hemisphere, the pressure gradients surrounding these cells are strongest between October and April. In terms of actual pressure, however, oceanic cells experience their highest pressure in summer, the belt being counterbalanced at low levels by thermal low-pressure conditions over the continents. Their strength and persistence clearly mark them as the dominating factor controlling the position and activities of both the trades and the westerlies. Figure 6.13 shows the general distribution of the surface wind zones over the globe.

1 The trade winds

The trades (or tropical easterlies) are important because of the great extent of their activity; they blow over nearly half the globe (see Figure 6.14). They originate at low latitudes on the margins of the subtropical high-pressure cells, and their constancy of direction and speed (about 7 m s⁻¹) is remarkable. Trade winds, like the westerlies, are strongest during the winter half-year, which suggests they are both controlled by the same fundamental mechanism.

The two trade wind systems tend to converge in the *Equatorial Trough* (of low pressure). Over the oceans, particularly the central Pacific, the convergence of these air streams is often pronounced and in this sector the term *Intertropical Convergence Zone* (ITCZ) is applicable. Generally, however, the convergence is discontinuous in space and time (see Plate 1). Equatorward of the main belts of the trades over the eastern Pacific and eastern Atlantic are regions of light, variable winds, known traditionally as the *doldrums* and much feared in past centuries

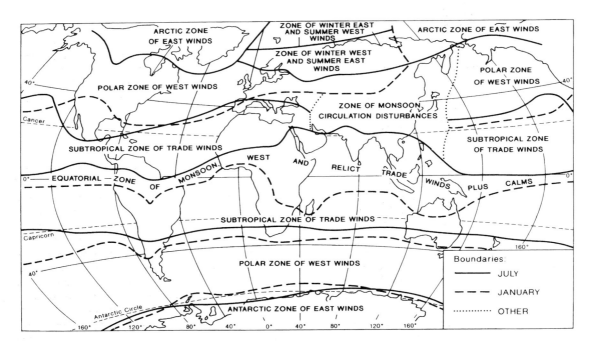

Figure 6.13 Global wind zones.

Source: After Okolowicz and Martyn. Reprinted from D. Martyn (1992) *Climates of the World* with kind permission from Elsevier Science NL, Sara Burgerhartstraat 25, KV Amsterdam, the Netherlands.

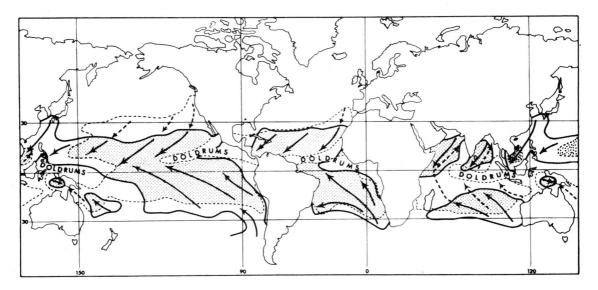

Figure 6.14 Map of the trade wind belts and the doldrums. The limits of the trades − enclosing the area within which 50 per cent of all winds are from the predominant quadrant − are shown by the solid (January) and the dashed (July) lines. The stippled area is affected by trade wind currents in both months. Schematic streamlines are indicated by the arrows − dashed (July) and solid (January, or both months).

Source: Based on Crowe 1949 and 1950.

by the crews of sailing ships. Their seasonal extent varies considerably: from July to September they spread westward into the central Pacific while in the Atlantic they extend to the coast of Brazil. A third major doldrum zone is located in the Indian Ocean and western Pacific. In March–April it stretches 16,000 km from East Africa to 180° longitude and it is again very extensive during October–December.

2 The equatorial westerlies

In the summer hemisphere, and over continental areas especially, there is a zone of generally westerly winds intervening between the two trade wind belts (Figure 6.15). This westerly system is well marked over Africa and South Asia in the northern hemisphere summer, when thermal heating over the continents assists the northward displacement of the Equatorial Trough (see Figure 6.14). Over Africa, the westerlies reach to 2–3 km and over the Indian Ocean to 5–6 km. In Asia, these winds are known as the 'Summer Monsoon', but this is now recognized to be a complex phenomenon, the cause of which is partly global and partly regional in origin (see Chapter 9C). The equatorial westerlies are not simply trades of the opposite hemisphere that recurve (due to the changed direction of the Coriolis

deflection) on crossing the equator, since there is *on average* a westerly component in the Indian Ocean at 2–3°S in June and July and at 2–3°N in December and January. Over the Pacific and Atlantic Oceans, the ITCZ does not shift sufficiently far from the equator to permit the development of this westerly wind belt.

3 The mid-latitude (Ferrel) westerlies

These are the winds of the mid-latitudes emanating from the poleward sides of the subtropical high-pressure cell (see Figure 6.13). They are far more variable than the trades in both direction and intensity, for in these regions the path of air movement is frequently affected by cells of low and high pressure, which travel generally eastwards within the basic flow. Also, in the northern hemisphere the preponderance of land areas with their irregular relief and changing seasonal pressure patterns tend to obscure the generally westerly airflow. The Isles of Scilly, off south-west England, lying in the south-westerlies, record 46 per cent of winds from between south-west and north-west, but fully 29 per cent from the opposite sector, between north-east and south-east.

The westerlies of the southern hemisphere are stronger and more constant in direction than those

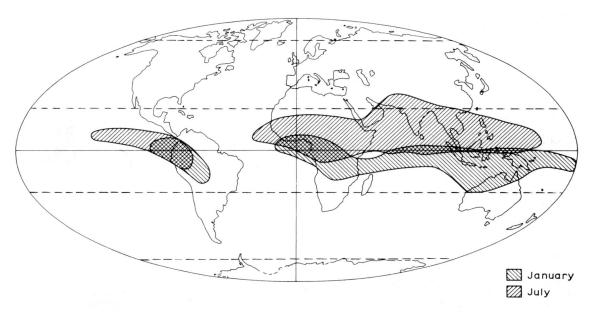

Figure 6.15 Distribution of the equatorial westerlies in any layer below 3 km (about 10,000 ft) for January and July.
Source: After Flohn in Indian Meteorological Department 1960.

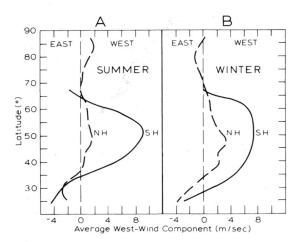

Figure 6.16 Profiles of the average west wind component (m s⁻¹) at sea level in the northern and southern hemispheres during their respective summer (A) and winter (B) seasons.

Source: After van Loon 1964.

of the northern hemisphere because the broad expanses of ocean rule out the development of stationary pressure systems (Figure 6.16). Kerguelen Island (49°S, 70°E) has an annual frequency of 81 per cent of winds from between south-west and north-west, and the comparable figure of 75 per cent for Macquarie Island (54°S, 159°E) shows that this predominance is widespread over the southern oceans. However, the apparent zonality of the southern circumpolar vortex (see Figure 6.11) conceals considerable synoptic variability of wind velocity.

4 The polar easterlies

This term is applied to winds that are supposed to occur between a polar high pressure and the belt of low pressure of the higher mid-latitudes. The polar high, as has already been pointed out, is by no means a quasi-permanent feature of the arctic circulation. Easterly winds occur mainly on the poleward sides of depressions over the North Atlantic and North Pacific, and if average wind directions are calculated for entire latitude belts in high latitudes there is found to be little sign of a coherent system of polar easterlies. The situation in high latitudes of the southern hemisphere is complicated by the presence of Antarctica, but anticyclones appear to be frequent over the high plateau of eastern Antarctica, and easterly winds prevail over the Indian Ocean sector of the Antarctic coastline. For example, in 1902–3 the expedition ship *Gauss*, at 66°S, 90°E, observed winds between north-east and south-east for 70 per cent of the time, and at many coastal stations the constancy of easterlies may be compared with that of the trades. However, westerly components predominate over the sea areas off west Antarctica.

C THE GENERAL CIRCULATION

The observed patterns of wind and pressure prompt consideration of the mechanisms maintaining the *general circulation* of the atmosphere – the large-scale patterns of wind and pressure that persist throughout the year or recur seasonally. Reference has already been made to one of the primary driving forces, the imbalance of radiation between lower and higher latitudes (see Figure 2.26), but it is also important to appreciate the significance of energy transfers in the atmosphere. Energy is continually undergoing changes of form, as shown schematically in Figure 6.17. Unequal heating of the earth and its atmosphere by solar radiation generates potential energy, some of which is converted into kinetic energy by the rising of warm air and the

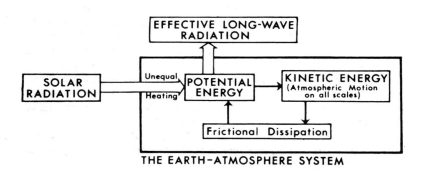

Figure 6.17 Schematic changes of energy involving the earth–atmosphere system.

sinking of cold air. Ultimately, the kinetic energy of atmospheric motion on all scales is dissipated by friction and small-scale turbulent eddies (i.e. internal viscosity). In order to maintain the general circulation, the rate of generation of kinetic energy must obviously balance its rate of dissipation. These rates are estimated to be about 2 W m^{-2}, which amounts to only 1 per cent of the average global solar radiation absorbed at the surface and in the atmosphere. In other words, the atmosphere is a highly inefficient heat engine (see Chapter 2E).

A second controlling factor is the angular momentum of the earth and its atmosphere. This is the tendency for the earth's atmosphere to move, with the earth, around the axis of rotation. Angular momentum is proportional to the rate of spin (that is the angular velocity) and the square of the distance of the air parcel from the axis of rotation. With a uniformly rotating earth and atmosphere, the total angular momentum must remain constant (in other words, there is a *conservation of angular momentum*). If, therefore, a large mass of air changes its position on the earth's surface such that its distance from the axis of rotation also changes, then its angular velocity must change in a manner so as to allow the angular momentum to remain constant. Naturally, absolute angular momentum is high at the equator (see Note 3) and decreases with latitude to become zero at the poles (that is, the axis of rotation), so air moving polewards tends to acquire progressively higher eastward velocities. For example, air travelling from 42 to 46° latitude and conserving its angular momentum would increase its speed relative to the earth's surface by 29 m s^{-1}. This is the same principle that causes an ice skater to spin more violently when her arms are progressively drawn into the body. In practice, this increase of air-mass velocity is countered or masked by the other forces affecting air movement (particularly friction), but there is no doubt that many of the important features of the general atmospheric circulation result from this poleward transfer of angular momentum.

The necessity for a poleward momentum transport is readily appreciated in terms of the maintenance of the mid-latitude westerlies (Figure 6.18). These winds continually impart westerly (eastward) relative momentum to the earth by friction, and it has been estimated that they would cease altogether due to this frictional dissipation of energy in little over a week if their momentum were not continually replenished from elsewhere. In low latitudes, the extensive tropical easterlies are gaining westerly relative momentum by friction as a result of the earth rotating in a direction opposite to their flow (see Note 4), and this excess is transferred polewards with the maximum poleward transport occurring, significantly, in the vicinity of the mean subtropical jet stream at about 250 mb at 30°N and 30°S.

1 Circulations in the vertical and horizontal planes

There are two possible ways in which the atmosphere can transport heat and momentum. One is by circulation in the vertical plane as indicated in Figure 6.19 which shows three meridional cells. The low-latitude (or Hadley) cell and its counterpart in the southern hemisphere were considered to be analogous to the convective circulations set up when a pan of water is heated over a flame and are referred to as *thermally direct* cells. Warm air near the equator was thought to rise and generate a low-level flow towards the equator, the earth's rotation deflecting these currents, which thus form the north-east and south-east trades. This explanation was put forward by G. Hadley in 1735, although in 1856 W. Ferrel pointed out that the conservation of angular momentum would be a more effective factor in causing easterlies, because the Coriolis force is small in low latitudes. The low-latitude cell, according to the above scheme, would be completed by poleward counter-currents aloft, with the air sinking at about 30° latitude as it is cooled by radiation. However, this scheme is not entirely correct, since the atmosphere does not have a simple heat source at the equator, the trades are not continuous around the globe (see Figure 6.14) and poleward upper flow is restricted mainly to the western ends of the subtropical high-pressure cells aloft (see Figures 6.4 and 6.11).

Figure 6.19 shows another thermally direct (polar) cell in high latitudes with cold dense air flowing out from a polar high pressure. The reality of this is doubtful, but it is in any case of limited importance to the general circulation in view of the small mass involved. It is worth noting at this point that a single direct cell in each hemisphere is not possible, because the easterly winds near the surface would slow down the earth's rotation. On average the atmosphere must rotate with the earth, requiring a balance between easterly and westerly winds over the globe.

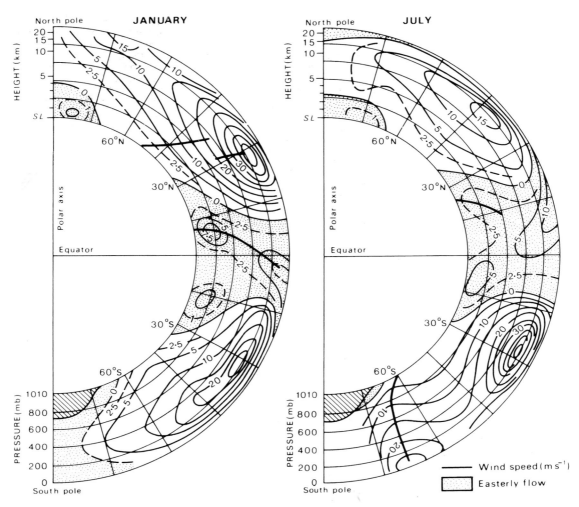

Figure 6.18 Mean zonal wind speeds (m s⁻¹) calculated for each latitude and for elevations up to more than 20 km. Note the weak mean easterly flow at all levels in low latitudes dominated by the Hadley cells, and the strong upper westerly flow in mid-latitudes, localized into the subtropical jet streams.

Source: After Mintz; from Henderson-Sellers and Robinson 1986.

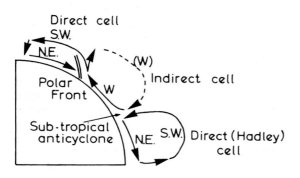

The mid-latitude (Ferrel) cell in Figure 6.19 is thermally indirect and would need to be driven by the other two. Momentum considerations indicate the necessity for upper easterlies in such a scheme, yet observations with upper-air balloons during the 1930s and 1940s demonstrated the existence of strong westerlies in the upper troposphere (see A.3, this chapter). Rossby modified the three-cell model

Figure 6.19 Three-cell model of the northern hemisphere meridional circulation.

Source: After Rossby 1941; from Barry 1967.

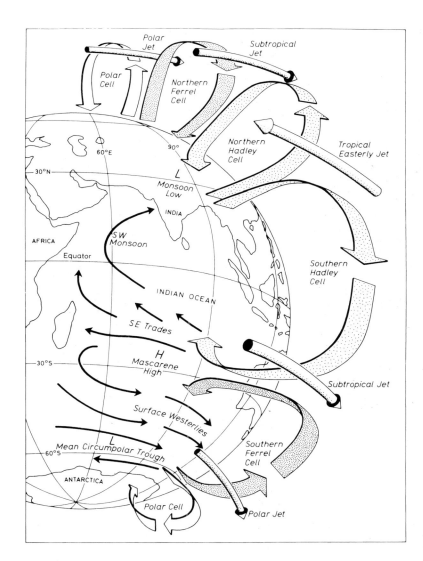

Figure 6.20 Meridional distribution of zonally averaged circulation features at 90 °E at the time of the Indian summer monsoon. The Hadley, Ferrel and polar cells are shown, together with jet streams and surface winds over the Indian Ocean.

to incorporate this fact, proposing that westerly momentum was transferred to middle latitudes from the upper branches of the cells in high and low latitudes. Such horizontal mixing could, for example, be accomplished by troughs and ridges in the upper flow. Figure 6.20 shows an idealized meridional circulation at 90°E containing northern and southern Hadley, Ferrel and polar cells, separated aloft by one easterly and four westerly jet streams.

These views underwent radical amendment from about 1948 onwards. The alternative means of transporting heat and momentum – by horizontal circulations – had been suggested in the 1920s by

A. Defant and H. Jeffreys but could not be tested until adequate upper-air data became available. Calculations for the northern hemisphere by V. P. Starr and R. M. White at the Massachusetts Institute of Technology showed that in middle latitudes horizontal cells transport most of the required heat and momentum polewards. This operates through the mechanism of the quasi-stationary highs and the travelling highs and lows near the surface acting in conjunction with their related wave patterns aloft. The importance of such horizontal eddies for energy transport is shown in Figure 6.21 (see also Figure 2.28B). The modern concept of the general

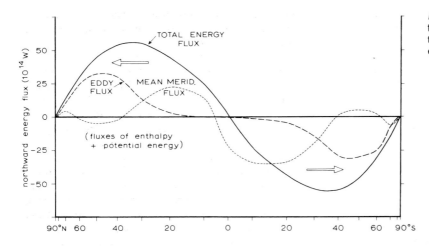

Figure 6.21 The poleward transport of energy, showing the importance of horizontal eddies in mid-latitudes.

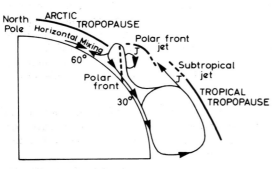

Figure 6.22 General meridional circulation model for the northern hemisphere in winter.
Source: After Palmén 1951; from Barry 1967.

circulation therefore views the energy of the zonal winds as being derived from travelling waves, not from meridional circulations. In lower latitudes, however, this mechanism may be insufficient by itself to account for the total energy transport estimated to be necessary for energy balance. For such reasons, the mean Hadley cell still features in current representations of the general circulation, as Figure 6.22 shows, but the low-latitude circulation is recognized as being complex. In particular, the vertical heat transport in the Hadley cell is effected by giant cumulonimbus clouds in disturbance systems associated with the Equatorial Trough (of low pressure), which is located on average at 5°S in January and at 10°N in July (see Chapter 9A). The Hadley cell of the winter hemisphere is by far the most important, and it gives rise to low-level transequatorial flow into the summer hemisphere. The traditional

model of global circulation with twin cells, symmetrical about the equator, is found only in spring/autumn.

Longitudinally, the Hadley cells are linked with the monsoon regimes of the summer hemisphere. Rising air over South Asia (and also South America and Indonesia) is associated with east–west (zonal) outflow, and these systems are known as 'Walker circulations' (pp. 133–4). The poleward return transport of the meridional Hadley cells takes place in troughs that extend into low latitudes from the mid-latitude westerlies. This tends to occur at the western ends of the upper tropospheric subtropical high-pressure cells. Horizontal mixing predominates in middle and high latitudes, although it is also thought that there is a weak indirect mid-latitude cell in much reduced form (Figure 6.22). The relationship of the jet streams to regions of steep meridional temperature gradient has already been noted (see Figure 6.7). A complete explanation of the two wind maxima and their role in the general circulation is still lacking, but they undoubtedly form an essential part of the story.

In the light of these theories, the origin of the subtropical anticyclones that play such an important role in the world's climates may be re-examined. Their existence has been variously ascribed to the piling up of poleward-moving air as it is increasingly deflected eastwards through the earth's rotation and the conservation of angular momentum; to the sinking of poleward currents aloft by radiational cooling; to the general necessity for high pressure near 30° latitude separating approximately equal zones of east and west winds; or to combinations of such mechanisms. An adequate theory must account not only for their permanence but also for their cellular nature and the vertical inclination of the axes. The preceding discussion shows that ideas of a simplified Hadley cell and momentum conservation are only partially correct. Moreover, recent studies rather surprisingly show no relationship, on a seasonal basis, between the intensity of the Hadley cell and that of the subtropical highs. Descent occurs near 25°N in winter, whereas North Africa and the Mediterranean are generally driest in summer, when the vertical motion is weak.

It is probable that the high-level anticyclonic cells that are evident on synoptic charts (these tend to merge on mean maps) are related to anticyclonic eddies on the equatorward side of jet streams. Theoretical and observational studies show that, as a result of the latitudinal variation of the Coriolis parameter, cyclones in the westerlies tend to move poleward and anticyclonic cells equatorward. Hence the subtropical anticyclones are constantly regenerated. There is a statistical relationship between the latitude of the subtropical highs and the mean meridional temperature gradient (see Figure 6.12); a stronger gradient causes an equatorward shift of the high pressure, and vice versa. This shift is evident on a seasonal basis. The cellular pattern at the surface clearly reflects the influence of heat sources. The cells are stationary and elongated north–south over the northern hemisphere oceans in summer, when continental heating creates low pressure and also the meridional temperature gradient is weak. In winter, on the other hand, the zonal flow is stronger in response to a greater meridional temperature gradient, and continental cooling produces east–west elongation of the cells. Undoubtedly, surface and high-level factors reinforce one another in some sectors and tend to cancel out in others. Indeed, it has been suggested that the Azores high-pressure cell, in particular, owes part of its summer intensification and its tendency to extend eastwards to air masses that rise locally in areas of high monsoonal rainfall over Africa, enter the tropical easterly jet stream circulation (see Chapter 9C.4) and then subside over the western Sahara and the eastern North Atlantic (see Figure 6.23). This idea has been extended to include a link between ascending motion over the Indian summer monsoon region and areas of descent over the eastern Mediterranean and eastern Sahara Desert and the Kyzylkum–Karakum Desert, Uzbekistan. However, whereas the ascending air originates in the tropical easterlies, Rossby waves in the mid-latitude westerlies are thought to be the source of the descending air.

Just as Hadley circulations represent major meridional (i.e. north–south) components of the atmospheric circulation, so Walker circulations represent the large-scale zonal (i.e. east–west) components of tropical airflow. These latter circulations are driven by major east–west pressure gradients set up by differences between, on the one hand, air rising over heated continents and the warmer parts of the oceans and, on the other, air subsiding over cooler parts of the oceans, over continental areas where deep high-pressure systems have become established, and in association with subtropical high-pressure cells. The Walker circulations were first identified by Sir Gilbert Walker in 1922–3 as the result of an inverse correlation that he observed between pressure over the eastern Pacific Ocean and Indonesia. The strength and phase of this so-called *Southern Oscillation* is commonly measured by the pressure difference between Tahiti (18°S, 150°W) and Darwin, Australia (12°S, 130°E). This Southern Oscillation Index (SOI) has two extreme phases (Figure 6.24):

- *positive* when there is a strong high pressure in the south-east Pacific and a low centred over Indonesia with ascending air and convective precipitation;
- *negative* (or low) when the area of low pressure and convection is displaced eastward towards the Date Line.

Positive (negative) SOI implies strong easterly trade winds (low level equatorial westerlies) over the central–western Pacific. These Walker circulations are subject to fluctuations in which an oscillation (El Niño–Southern Oscillation: ENSO) between high phases (i.e. non-ENSO events) and low phases (i.e. ENSO events) is the most striking (see Chapter 9G.1):

1 *High phase* (Figure 6.24A). This is characterized by four major zonal cells involving rising low-pressure limbs and accentuated precipitation over Amazonia, central Africa and Indonesia/India; and subsiding high-pressure limbs and decreased precipitation over the eastern Pacific, South Atlantic and western Indian Ocean. During this phase, low-level easterlies strengthen over the Pacific and subtropical westerly jet streams in both hemispheres weaken, as does the Pacific Hadley cell.

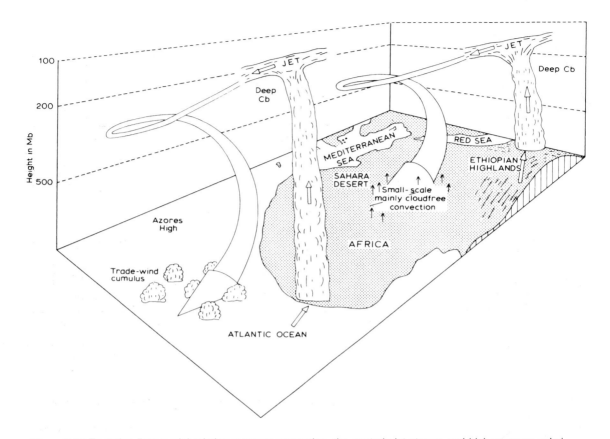

Figure 6.23 Tentative flow model relating summer convection, the easterly jet stream and high-pressure subsidence over northern Africa and the eastern North Atlantic.

Source: Adapted from Walker 1972. Reproduced from the *Meteorological Magazine* Crown Copyright © by permission of the Controller of Her Majesty's Stationery Office.

2 *Low phase* (Figure 6.24B). This is characterized by five major zonal cells involving rising low-pressure limbs and accentuated precipitation over the South Atlantic, the western Indian Ocean, the western Pacific and the eastern Pacific; and subsiding high-pressure limbs and decreased precipitation over Amazonia, central Africa, Indonesia/India and the central Pacific. During this phase, low-level westerlies and high-level easterlies dominate over the Pacific, and subtropical westerly jet streams in both hemispheres intensify, as does the Pacific Hadley cell.

2 Variations in the circulation of the northern hemisphere

The pressure and contour patterns during certain periods of the year may be radically different from those indicated by the mean maps (see Figure 6.4). Variations, of three to eight weeks duration, occur irregularly but are rather more noticeable in the winter months, when the general circulation is strongest. The nature of the changes is illustrated schematically in Figure 6.25. The zonal westerlies over middle latitudes develop waves, and the troughs and ridges become accentuated, ultimately splitting up into a cellular pattern with pronounced meridional flow at certain longitudes. The strength of the westerlies between 35 and 55°N is termed the *zonal index*; strong zonal westerlies are representative of a high index, and marked cellular patterns occur with a low index (see Plate 15). A relatively low index may also occur if the westerlies are well south of their usual latitudes and, paradoxically, such expansion of the zonal circulation pattern is associated with strong westerlies in lower

A   HIGH PHASE

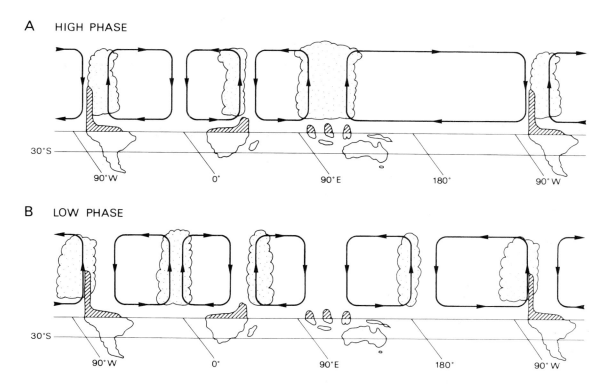

B   LOW PHASE

Figure 6.24 The Walker circulation during high (A) and low (B) phases of the Southern Oscillation. Low phases correspond to ENSO events (see text). The shading indicates the very exaggerated vertical scale.

Source: After J. A. Lindesay; from Tyson 1986, *Climatic Change and Variability in Southern Africa*, copyright © Oxford University Press, Southern Africa.

latitudes than usual. Figures 6.26 and 6.27 illustrate the mean 700 mb contour patterns and zonal wind speed profiles for two contrasting months. In December 1957, the westerlies were stronger than normal north of 40°N, and the troughs and ridges were weakly developed, whereas in February 1958 there was a low zonal index and an expanded circumpolar vortex, giving rise to strong low-latitude westerlies. The 700 mb pattern shows very weak subtropical highs, deep meridional troughs and a blocking anticyclone off Alaska (see Figure 6.25D). The cause of these variations is still uncertain, although it would appear that fast zonal flow is unstable and tends to break down. This tendency is certainly increased in the northern hemisphere by the arrangement of the continents and oceans.

Detailed studies are now beginning to show that the irregular index fluctuations, together with secondary circulation features, such as cells of low and high pressure at the surface or long waves aloft,

play a major role in redistributing momentum and energy. Laboratory experiments with rotating 'dish-pans' of water to simulate the atmosphere, and computer studies using numerical models of the atmosphere's behaviour, demonstrate that a Hadley circulation cannot provide an adequate mechanism for transporting heat polewards. In consequence, the meridional temperature gradient increases and eventually the flow becomes unstable in the Hadley mode, breaking down into a number of cyclonic and anticyclonic eddies. This phenomenon is referred to as *baroclinic instability*. In energy terms, the potential energy in the zonal flow is converted into potential and kinetic energy of the eddies. It is also now known that the kinetic energy of the zonal flow is derived *from* the eddies, the reverse of the classical picture, which viewed the disturbances within the global wind belts as superimposed detail. The significance of atmospheric disturbances and the variations of the circulation are becoming

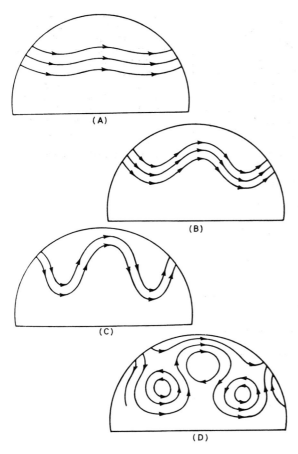

Figure 6.25 The index cycle. A schematic illustration of the development of cellular patterns in the upper westerlies, usually occupying 3–8 weeks and being especially active in February and March in the northern hemisphere. Statistical studies indicate no regular periodicity in this sequence. A: high zonal index. The jet stream and the westerlies lie north of their mean position. The westerlies are strong, pressure systems have a dominantly east–west orientation, and there is little north–south air-mass exchange. B and C: the jet expands and increases in velocity, undulating with increasingly larger oscillations. D: low zonal index. The latter is associated with a complete break-up and cellular fragmentation of the zonal westerlies, formation of stationary deep occluding cold depressions in lower mid-latitudes and deep warm blocking anticyclones at higher latitudes. This fragmentation commonly begins in the east and extends westwards at a rate of about 60° of longitude per week.

Source: After Namias; from Haltiner and Martin 1957.

increasingly evident. The mechanisms of the circulation are, however, greatly complicated by numerous interactions and *feedback* processes, one of the most important of which involves the oceanic circulation, as outlined in D, this chapter. The significance of interactions between the oceanic and atmospheric heat and moisture budgets has already been discussed in Chapters 2E and 3C.

D OCEAN STRUCTURE AND CIRCULATION

The oceans occupy 71 per cent of the earth's surface, with over 60 per cent of the global ocean area in the southern hemisphere. Three-quarters of the ocean area is between 3,000 and 6,000 m deep, whereas only 11 per cent of the land area exceeds 2,000 m altitude.

1 Above the thermocline

a Vertical

The major atmosphere–ocean interactive processes (Figure 6.28) involve heat exchanges, evaporation, density changes and wind shear. The effect of these processes is to produce a major oceanic layering that is of great climatic significance:

1 At the ocean surface, winds produce a *thermally mixed surface layer* averaging a few tens of metres deep poleward of latitude 60°, 400 m at latitude 40° and 100–200 m at the equator.
2 Below the relatively warm mixed layer is the *thermocline*, a layer in which temperature decreases and density increases (the pycnocline) markedly with depth. The thermocline layer, within which stable stratification tends to inhibit vertical mixing, acts as a barrier between the warmer surface water and the colder deep-layer water. In the open ocean between latitudes of 60° north and south the thermocline layer extends from depths of about 200 m to a maximum of 1,000 m (at the equator from about 200 to 800 m; at 40° latitude from about 400 to about 1,100 m). Poleward of 60° latitude, the colder deep-layer water approaches the surface. Within the thermocline layer, the location of the steepest temperature gradient is termed the *permanent thermocline*, which has a dynamically inhibiting effect in the ocean similar to that of a major inversion in the atmosphere. However, heat exchanges take place between the oceans and the atmosphere by turbulent mixing above the permanent

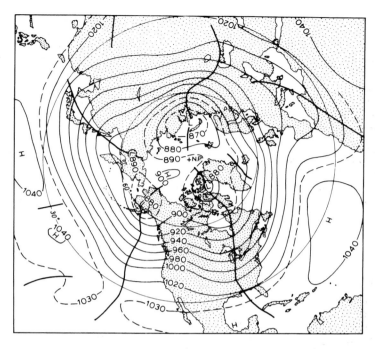

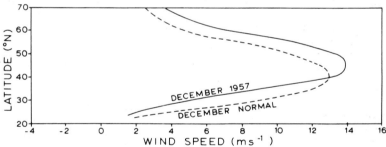

Figure 6.26 Above: mean 700 mb contours (in tens of feet) for December 1957, showing a fast, westerly, small-amplitude flow typical of a high zonal index. Below: mean 700 mb zonal wind speed profiles (m s⁻¹) in the western hemisphere for December 1957, compared with those of a normal December. The westerly winds were stronger than normal and displaced to the north.

Source: After Dunn 1957.

thermocline, as well as by upwelling and downwelling involving deep-layer water.

During spring and summer in the mid-latitudes, accentuated surface heating leads to the development of a *seasonal thermocline* occurring at depths of 50 to 100 m. Surface cooling and wind mixing tend to destroy this layer in autumn and winter.

3 Below the thermocline layer is a *deep layer* of cold, dense water. Within this, water movements are mainly driven by density variations, commonly due to salinity differences (i.e. having a *thermohaline* mechanism).

The ocean may be viewed as consisting of a large number of layers, the topmost subject to wind stress, the next layer down to frictional drag by the layer above, and so on; all layers being acted on by the Coriolis force. The surface water tends to be deflected to the right (in the northern hemisphere) by an angle averaging some 45° in the surface wind direction and to move at about 3 per cent of its velocity. This deflection increases with depth as the friction-driven velocity of the current decreases exponentially (Figure 6.29). On the equator, where there is no Coriolis force, the surface water moves in the same direction as the surface wind. This theoretical Ekman spiral was developed under assumptions of idealized ocean depth, wind constancy, uniform water viscosity and constant water pressure at a given depth. This is seldom the case in reality, and under most oceanic conditions the thickness of the wind-driven layer is about 100 to 200 m.

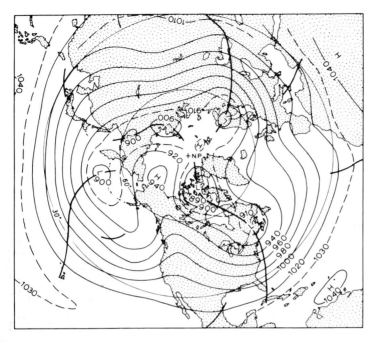

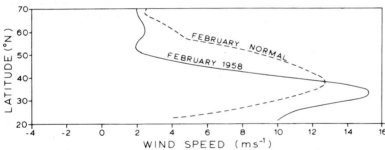

Figure 6.27 Above: mean 700 mb contours (in tens of feet) for February 1958. Below: mean 700 mb zonal wind speed profiles (m s⁻¹) in the western hemisphere for February 1958, compared with those of a normal February. The westerly winds were stronger than normal at low latitudes, with a peak at about 33 °N.

Source: After Klein 1958.

b Horizontal

(1) General

Comparisons between the structure and dynamics of the oceans and the atmosphere have long been made, particularly in respect of their behaviour above the permanent thermocline and below the tropopause – their two most significant stabilizing boundaries. Within these two zones, fluid-like circulations are maintained by meridional thermal energy gradients, dominantly directed polewards (Figure 6.30), and acted upon by the Coriolis force. Prior to the last quarter of a century, oceanography was studied in a coarsely averaged spatial–temporal framework similar to that applied in classical climatology. At the present day, however, its similarities with modern meteorology are more marked. The major behavioural differences between the oceans and the atmosphere derive from the greater density and viscosity of ocean waters and the much greater frictional constraints placed on their global movement.

Macroscale characteristics of ocean dynamics that invite comparison with atmospheric features include the general circulation, major oceanic gyres (similar to atmospheric subtropical high-pressure cells), major jet-like streams such as sections of the Gulf Stream, large-scale areas of subsidence and uplift, the stabilizing layer of the permanent thermocline, boundary layer effects, frontal discontinuities created by temperature and density contrasts, and water mass ('mode water') regions.

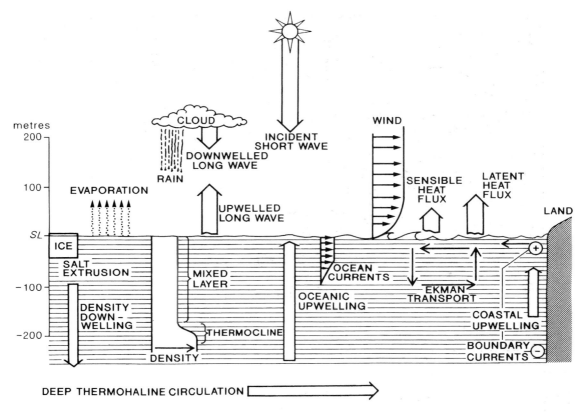

Figure 6.28 Generalized depiction of the major atmosphere–ocean interaction processes.
Source: Modified from NASA (n.d.). Courtesy NASA.

Mesoscale characteristics inviting atmospheric analogues are oceanic cyclonic and anticyclonic eddies, current meanders, cast-off ring vortices, jet filaments, and circulations produced by irregularities in the North Equatorial Current (Figure 6.31).

(2) Macroscale
The most obvious feature of the surface oceanic circulation is the control exercised over it by the low-level planetary wind circulation, especially by the subtropical oceanic high-pressure circulations and the westerlies. The oceanic circulation also displays seasonal reversals of flow in the monsoonal regions of the northern Indian Ocean, off East Africa and off northern Australia (see Figure 6.31). As water moves meridionally, the conservation of angular momentum implies changes in relative vorticity (see pp. 110 and 120), with poleward-moving currents acquiring anticyclonic vorticity and equatorward-moving currents acquiring cyclonic vorticity.

The more or less symmetrical atmospheric subtropical high-pressure cells produce oceanic gyres with centres displaced towards the west sides of the oceans. The gyres in the southern hemisphere are more symmetrically located than those in the northern, due possibly to their connection with the powerful West Wind Drift. This results, for example, in the Brazil Current being not much stronger than the Benguela Current. The most powerful southern hemisphere current, the Agulhas, possesses nothing like the jet-like character of its northern counterparts.

Equatorward of the subtropical high-pressure cells, the persistent trade winds generate the broad North and South Equatorial Currents (see Figure 6.31). On the western sides of the oceans, most of this water swings polewards with the airflow and thereafter increasingly comes under the influence of the Ekman deflection and of the anticyclonic vorticity effect. However, some water tends to pile

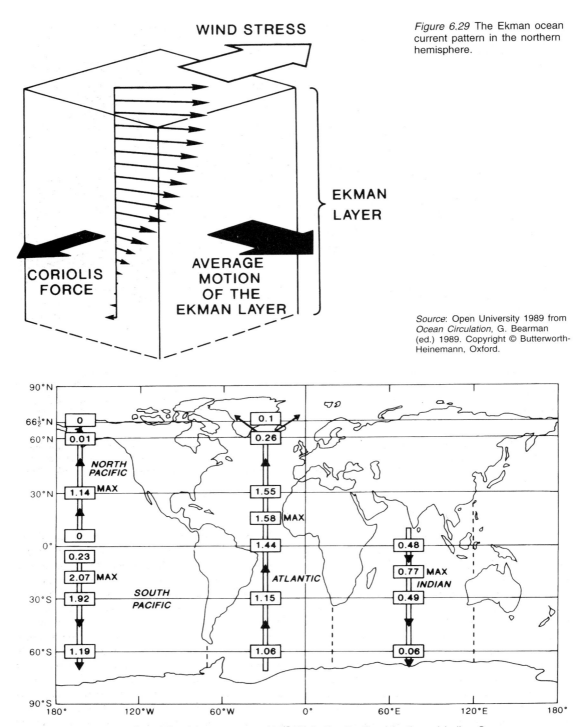

WIND STRESS

Figure 6.29 The Ekman ocean current pattern in the northern hemisphere.

EKMAN LAYER

CORIOLIS FORCE

AVERAGE MOTION OF THE EKMAN LAYER

Source: Open University 1989 from *Ocean Circulation*, G. Bearman (ed.) 1989. Copyright © Butterworth-Heinemann, Oxford.

Figure 6.30 Mean annual meridional heat transport (10¹⁵ W) in the Pacific, Atlantic and Indian Oceans, respectively (delineated by the dashed lines). The latitudes of maximum transport are indicated.

Source: Hastenrath 1980. From *Journal of Physical Oceanography* by permission of the American Meteorological Society.

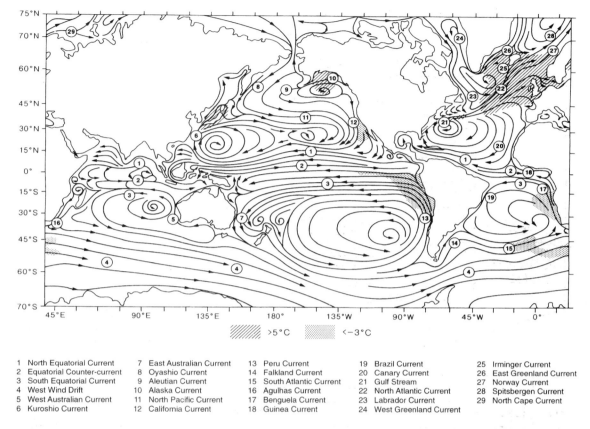

Figure 6.31 The general ocean current circulation in January. This holds broadly for the year, except that in the northern summer some of the circulation in the northern Indian Ocean is reversed by the monsoonal airflow. The shaded areas show mean annual anomalies of ocean surface temperatures (°C) of greater than +5 °C and less than –3 °C.

Sources: US Naval Oceanographic Office and Niiler 1992. Courtesy US Naval Oceanographic Office.

1 North Equatorial Current	7 East Australian Current	13 Peru Current
2 Equatorial Counter-current	8 Oyashio Current	14 Falkland Current
3 South Equatorial Current	9 Aleutian Current	15 South Atlantic Current
4 West Wind Drift	10 Alaska Current	16 Agulhas Current
5 West Australian Current	11 North Pacific Current	17 Benguela Current
6 Kuroshio Current	12 California Current	18 Guinea Current

19 Brazil Current	25 Irminger Current
20 Canary Current	26 East Greenland Current
21 Gulf Stream	27 Norway Current
22 North Atlantic Current	28 Spitsbergen Current
23 Labrador Current	29 North Cape Current
24 West Greenland Current	

up near the equator on the western sides of oceans, partly because here the Ekman effect is virtually absent, with little poleward deflection and no reverse current at depth. To this is added some of the water that is displaced northwards into the equatorial zone by the especially active subtropical high-pressure circulations of the southern hemisphere. This accumulated water flows back eastward down the hydraulic gradient as compensating narrow surface Equatorial Counter-Currents, unimpeded by the weak surface winds. Near the equator in the Pacific Ocean, upwelling raises the thermocline to only 50–100 m depth, and within this layer there exist thin, jet-like Equatorial Undercurrents flowing eastwards (under hydraulic gradients) at the considerable velocity of 1 to 1.5 m s⁻¹.

As the circulations swing polewards around the western margins of the oceanic subtropical high-pressure cells, there is the tendency for water to pile up against the continents, giving, for example, an appreciably higher sea level in the Gulf of Mexico than that along the Atlantic coast of the United States. This accumulated water cannot escape by sinking because of its relatively high temperature and resulting vertical stability, and it consequently continues polewards in the dominant direction of surface airflow, augmented by the geostrophic force acting at right angles to the ocean surface slope. As a result of this movement, the current gains anti-cyclonic vorticity, which reinforces the similar tendency imparted by the winds, leading to relatively narrow currents of high velocity (for example, the Kuroshio, Brazil, Mozambique–Agulhas and, to a less-marked extent, the East Australian Current). In the North Atlantic, the configuration of the Caribbean Sea and Gulf of Mexico especially favours this pile-up of water, which is released polewards through the Florida Straits as the particularly narrow and fast Gulf Stream. These poleward currents are opposed both by their friction with the nearby continental margins and by energy losses due to turbulent diffusion, such as those accompanying the formation and cutting off of meanders in the Gulf Stream. These poleward western boundary currents (e.g. the Gulf Stream and the Kuroshio Current) are fast, deep and narrow (i.e. approximately 100 km wide and reaching surface velocities greater than 2 m s^{-1}), contrasting with the slower, wider and more diffuse eastern boundary currents such as the Canary and California (i.e. approximately 1,000 km wide with surface velocities generally less than 0.25 m s^{-1}). The northward-flowing Gulf Stream causes a heat flux of 1.2×10^{15} W, 75 per cent of which is lost to the atmosphere and 25 per cent in heating the Greenland–Norwegian Seas area.

On the poleward sides of the subtropical high-pressure cells, westerly currents dominate, and where they are unimpeded by land masses in the southern hemisphere they form the broad and swift West Wind Drift. This strong current, driven by unimpeded winds, occurs within the zone 50 to 65°S and is associated with a southward-sloping ocean surface generating a geostrophic force, which intensifies the flow. Within the West Wind Drift, the action of the Coriolis force produces a convergence zone at about 50°S marked by westerly submarine jet streams reaching velocities of 0.5 to 1 m s^{-1}. In the northern hemisphere, a great deal of the eastward-moving current in the Atlantic swings northwards, leading to anomalously very high sea temperatures, and is compensated for by a southward flow of cold arctic water at depth. However, more than half of the water mass comprising the North Atlantic Current, and almost all that of the North Pacific Current, swings south around the east sides of the subtropical high-pressure cells, forming the Canary and California Currents. Their southern-hemisphere equivalents are the Benguela, Humboldt or Peru, and West Australian Currents.

(3) Mesoscale

Mesoscale eddies and rings in the upper ocean are generated by a number of mechanisms, sometimes by atmospheric convergence or divergence or by the casting off of vortices by jet-like currents such as the Gulf Stream (Figure 6.32). These vortices are generated by the transfer of warm water from low to high latitudes. Oceanographic eddies occur on the scale of 50–400 km diameter and are analogous to atmospheric low- and high-pressure systems. Ocean mesoscale systems are much smaller than atmospheric depressions (which average about 1,000 km diameter), travel much slower (a few kilometres per day, compared with about 1,000 km per day for a depression) and persist from one to several months (compared with a depression life of about a week). Their maximum rotational velocities occur at a depth of about 150 m, but the vortex circulation may be felt to depths of several thousands of metres. Some eddies move parallel to the main flow direction, but many move irregularly equatorwards or polewards. In the North Atlantic, for example, this produces a 'synoptic-like' situation in which up to 50 per cent of the area may be occupied by mesoscale eddies (Figure 6.33 and Plate B). Cyclonic rings are commonly three times as numerous as anticyclonic eddies, having a maximum rotational velocity of about 1.5 m s^{-1}.

2 Deep ocean water interactions

a Upwelling

In contrast with the currents on the west sides of the oceans, equatorward-flowing eastern currents acquire cyclonic vorticity, which is in opposition to the anticyclonic wind tendency, leading to relatively broad flows of low velocity. In addition, the deflection due to the Ekman effect causes the surface water to move westwards away from the coasts, leading to replacement by the upwelling of cold water from depths of 100–300 m (Figure 6.34). Average rates of upwelling are low (1–2 m/day), being about the same as the offshore surface current velocities, with which they are balanced. The rate of upwelling therefore varies with the surface wind stress. As the latter is proportional to the square of the wind speed, small changes in wind velocity can lead to marked variations in rates of upwelling. Although the band of upwelling may be quite narrow (about 200 km wide for the Benguela Current), the Ekman effect spreads this cold water westwards. On the poleward margins of these cold-water coasts, the meridional swing of the wind belts imparts a strong seasonality to the upwelling, the

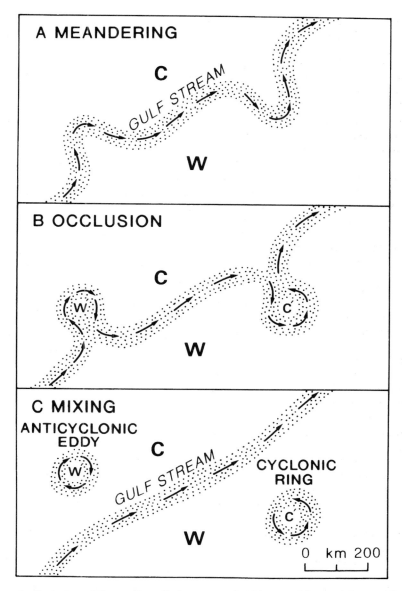

Figure 6.32 Schematic illustration of the casting off of warm- and cold-core eddies and rings as the result of Gulf Stream meandering.

Source: After *Modern Physical Geography* (4th edn), A.N. Strahler and A.H. Strahler. Copyright © 1992. Reproduced by permission of John Wiley and Sons, Inc.

California Current upwelling, for example (Plate 16), being particularly well marked during the period March–July.

A major region of deep-water upwelling is along the west coast of South America (Figure 9.52), which is normally assisted by the offshore airflow associated with the large-scale convective Walker cell (see Chapters 6C.1 and 9G) linking South-east Asia with the eastern South Pacific. Every 2–10 years or so this pressure difference is disturbed, producing an El Niño event with weakening trade winds and a pulse of warm surface water spreading eastwards over the South Pacific, raising average ocean-surface temperatures from about 24 to 30°C.

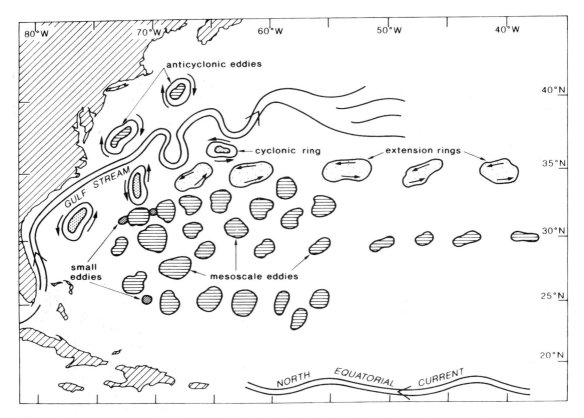

Figure 6.33 Schematic map of the western North Atlantic showing the major types of ocean surface circulation.
Source: From D. Tolmazin (1994) *Elements of Dynamic Oceanography*. Copyright © Chapman & Hall.

Coastal upwelling can also be caused by other less important mechanisms such as surface current divergence or the effect of the ocean bottom configuration (see Figure 6.34).

b Deep ocean circulation

Whereas above the permanent thermocline the ocean circulation is mainly wind driven, the deep ocean circulation is powered by density gradients due to salinity and temperature differences. These differences are mostly produced by surface processes, which feed cold, saline water to the deep ocean basins in compensation for the deep water delivered to the surface by upwelling. Although upwelling occurs chiefly in narrow coastal locations, subsidence takes place largely in two broad ocean regions – the northern North Atlantic and the Antarctic Weddell Sea.

In the North Atlantic, particularly in winter, heating and evaporation produce warm, saline

water which flows northwards both in the near-surface Gulf Stream–North Atlantic Current and at intermediate depths of around 800 m. In the Norwegian and Greenland Seas, its density is enhanced by further evaporation due to high winds, by the formation of sea ice, which expels brine during ice growth, and by cooling. Exposed to evaporation and to the chill high-latitude air masses, the surface water cools from about 10 to 2°C, releasing immense amounts of heat to the atmosphere, supplementing solar insolation there by some 25–30 per cent and heating Western Europe.

The resulting dense high-latitude water, equivalent in volume to about twenty times the combined discharge of all the world's rivers, sinks to the bottom of the North Atlantic and fuels a south-ward-flowing density (thermohaline) current, which forms part of a global deep-water conveyor belt (Figure 6.35). This broad, slow and diffuse flow, occurring at depths of greater than 1,500 m, is

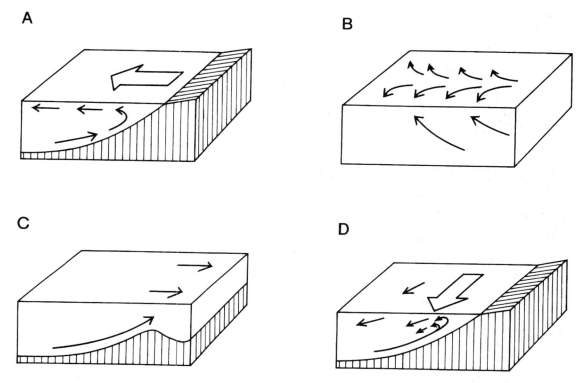

Figure 6.34 Schematic illustration of mechanisms that cause ocean upwelling. The large arrows indicate the dominant wind direction and the small arrows the currents. A: the effects of a persistent offshore wind. B: divergent surface currents. C: deep-current shoaling. D: Ekman motion with coastal blocking (northern hemisphere case).

Source: Partly modified after Stowe, *Ocean Science*, © 1983 by John Wiley and Sons, Inc. Reproduced by permission.

augmented in the South Atlantic/circum-Antarctic/ Weddell Sea region by more cold, saline, dense subsiding water. The conveyor belt then flows eastwards under the Coriolis influence, turning north into the Indian and, especially, the Pacific Ocean. The time taken for the conveyor belt circulation to move from the North Atlantic to the North Pacific has been estimated at 500–1,000 years. In the Pacific and Indian Oceans, a decrease of salinity due to water mixing causes the conveyor belt to rise and to form a less deep return flow to the Atlantic, the whole global circulation occupying some 1,500 years or so. An important aspect of this conveyor belt flow is that the western Pacific Ocean contains a deep source of warm summer water (29°C) (Figure 6.36). This heat differential with the eastern Pacific assists the high phase Walker circulation (see Figure 6.24).

The thermal significance of the conveyor belt implies that any change in it may promote climatic changes, which may be apparent at time scales of several hundred or thousand years. It has been suggested, however, that any impediment to the rise of deep conveyor belt water might cause ocean surface temperatures to drop by 6°C within 30 years at latitudes of 60°N. Changes to the conveyor belt circulation might be caused by lowering the salinity of the surface water of the North Atlantic by increased precipitation, ice melting, or fresh water inflow. However, the complex mechanisms and consequences of the deep ocean conveyor belt are still only imperfectly understood.

3 The oceans and atmospheric regulation

The atmosphere and the surface ocean waters are closely connected both in temperature and in CO_2 concentrations. The atmosphere contains less than 1.7 per cent of the CO_2 held by the oceans, and the amount absorbed by surface ocean water rapidly

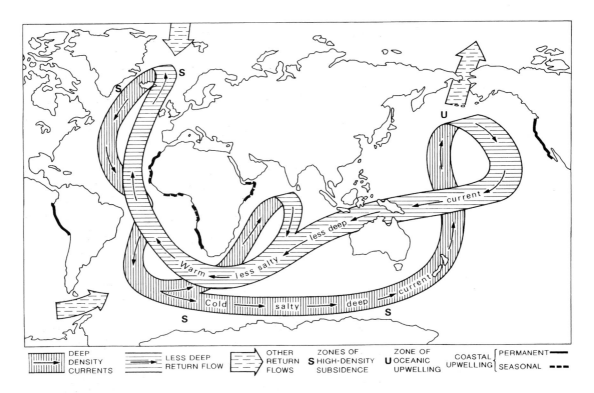

Figure 6.35 The deep ocean thermohaline circulation system leading to Broecker's concept of the oceanic conveyor belt.

Source: Kerr 1988. Reprinted with permission from *Science* **239** (1988), Fig. 259. Copyright © 1988 American Association for the Advancement of Science.

regulates the concentration in the atmosphere. The absorption of CO_2 by the oceans is greatest where the water is rich in organic matter or where it is cold. Thus the oceans are capable of regulating atmospheric CO_2, of changing the greenhouse effect, and of contributing to climate change. The most important aspect of the carbon cycle linking atmosphere and ocean is the difference between the partial pressure of CO_2 in the lower atmosphere and that in the upper oceanic layer. This results in atmospheric CO_2 being dissolved in the oceans and in some of this being subsequently converted into particulate carbon, mainly through the agency of plankton, ultimately sinking to form carbon-rich deposits in the deep ocean as part of a cycle lasting hundreds of years. Thus two of the major effects of ocean surface warming would be to increase its CO_2 equilibrium partial pressure and to decrease the abundance of plankton. Both of these effects would tend to decrease the oceanic uptake of CO_2 and therefore to increase its atmospheric concentration,

thereby producing a positive feedback (i.e. enhancing) effect on global warming. However, as will be seen in Chapter 11, the operation of the atmosphere–ocean system is sufficiently complex that, for example, global warming may so increase oceanic convective mixing that the resulting imports of cooler water and plankton into the surface layers might exert a brake (i.e. negative feedback) on the system warming.

Sea-surface temperature anomalies in the Atlantic appear to have marked effects on climate in Europe, Africa, and South America. For example, warmer sea surfaces off north-west Africa augment West African summer monsoon rainfall; and dry conditions in the Sahel have been linked to a cooler North Atlantic. There are similar links between tropical sea-surface temperatures and droughts in north-eastern Brazil. A circulation anomaly/teleconnection pattern known as the North Atlantic Oscillation (NAO) also shows strong air–sea interactions. The NAO is an oscillation in the pressure field between

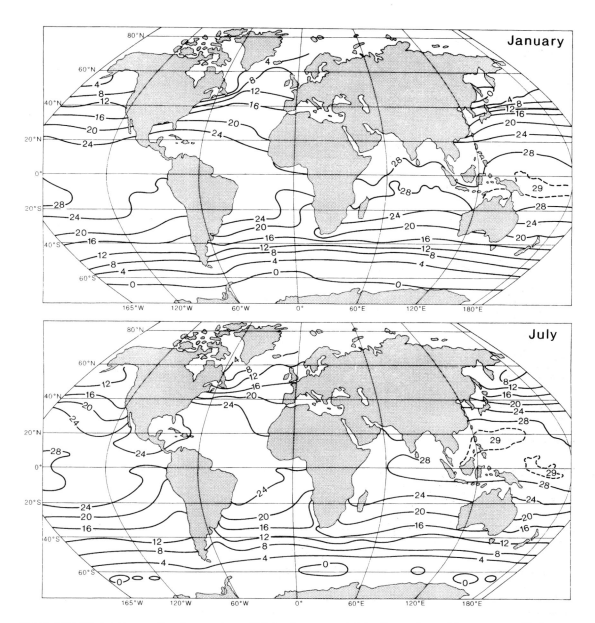

Figure 6.36 Mean ocean-surface temperatures (°C) for January and July. Comparison of these maps with those of mean sea-level air temperatures (Figure 2.11) show similarities during the summer but a significant difference in the winter.

Source: Reprinted from M. Bottomley *et al.* (1990) *Global Ocean Surface Temperature Atlas*, be permission of the Meteorological Office. Crown copyright ©.

Iceland (65°N) and a zone about 40°N across the Atlantic. Fluctuations in the NAO give rise to alternating mild/severe winters in western Greenland–Labrador and north-west Europe. Severe winters in the Greenland area also have cold northerly airflow and more extensive sea ice in the Labrador Sea. On a longer time scale, the NAO index (of S–N pressure difference) was generally low from 1925 to 1970,

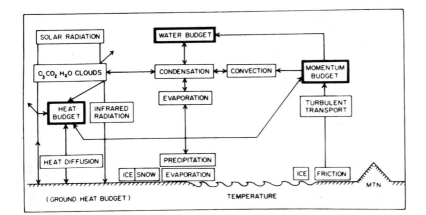

Figure 6.37 Schematic diagram of the interactions among physical processes in a general circulation model.

Source: From Druyan *et al.* 1975.

when air temperatures in the northern hemisphere were above normal and cyclones over the east coast of North America tended to be located over the ocean, thus causing longer, drier east coast summers. Prior to 1925, a regime of colder climatic conditions was associated with a higher NAO index. Since 1989, the NAO has been consistently positive.

E MODELLING THE ATMOSPHERIC CIRCULATION AND CLIMATE

Better understanding of the complex behaviour of the atmosphere and climate processes has been obtained in recent years through numerical modelling studies. Here we can only sketch the essential features of this approach. There are three basic categories of model: the general circulation model (GCM), the energy budget model (EBM) and the radiative–convective model (RCM).

In the GCM, all dynamic and thermodynamic processes and the radiative and mass exchanges that have been treated in Chapters 1–5 are modelled using five basic sets of equations. These account for the horizontal and vertical components of motion (see Chapter 5A, B), the conservation of energy (F, this chapter) and of water substance (Chapter 3), the continuity of mass (p. 109), and the equation of state (p. 14). In addition, radiative processes, including a diurnal and seasonal cycle, surface friction and cloud formation are usually represented. These are coupled in the manner shown schematically in Figure 6.37. Beginning with a set of initial atmospheric conditions, the equations are solved repeatedly for time steps of a few minutes at a large number of grid points over the earth and at many

levels in the atmosphere; 2–15 levels may be used. The horizontal grid is usually of the order of 2.5 to 5° latitude and longitude (except near the poles). Another computationally faster approach is to represent the horizontal fields by a series of two-dimensional sine and cosine functions (a spectral model). The number of two-dimensional waves that are included is described by a truncation level. The truncation procedure may be rhomboidal (*R*) or triangular (*T*); *R*15 (or *T*21) corresponds approximately to a 5° grid spacing, *R*30 (*T*42) to a 2.5° grid, and *T*102 to a 1° grid. Coastlines and mountains as well as essential elements of the surface vegetation (albedo, roughness) and soil (moisture content) may be incorporated. Sea ice extent and sea-surface temperatures are usually specified for each month (unless the GCM is coupled to an interactive ocean model where the ocean circulation and heat transports are calculated). Snow cover on the ground is often computed as part of a hydrological cycle.

The development of atmospheric GCMs for weather forecasting purposes (see Chapter 7I) began at leading meteorological institutions around the world in the 1960s, and there are now some thirty models in common use. Similar models are used to simulate climatic change by representing inputs (i.e. forcings), storage and transfers (see Figure 6.38 and Chapter 11). The periods of time shown in Figure 6.38 refer to:

1   *Forcing times.* The characteristic timespans over which solar and anthropogenic changes of input occur. In the case of the former, these are the periods of solar radiation cycles and in the case

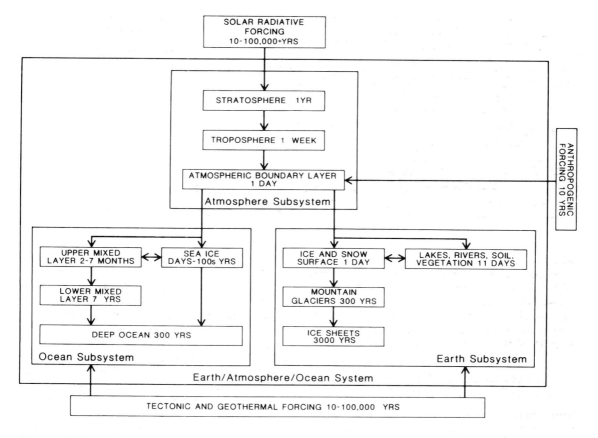

Figure 6.38 The earth–atmosphere–ocean system showing estimated equilibrium times, together with the wide time variations involving the external solar, tectonic, geothermal and anthropogenic forcing mechanisms. *Source*: After Saltzman 1983.

of the latter the average timespan over which significant changes of such anthropogenic effects as atmospheric CO_2 occur.

2 *Storage times*. For each compartment of the atmosphere and ocean subsystems these are the average times taken for an input of thermal energy to diffuse and mix within the compartment. For the earth subsystem, the average times are those required for inputs of water to move through each compartment.

It is important to evaluate how well such models perform, and an Atmospheric Model Intercomparison Program (AMIP) is designed to do this by using common procedures and standardized data (on sea-surface temperatures for example), as well as by providing extensive documentation on the model design and *parameterizations*. These are empirical and statistical relationships used to represent certain complex and usually small-scale physical processes that the model does not treat explicitly.

Model simulations of present-day climate conditions involve running a model for perhaps 5–10 years of simulated time, with time steps of only a few minutes between each iteration of the numerical equations. Figure 6.39 illustrates modelled 500 mb height fields for January from the National Center for Atmospheric Research (NCAR) Community Climate Model CCM1 compared with the observed climatological mean. The main features are well represented, although the modelled heights in the polar vortex are lower than observed. The evaluation of models requires analysis of their ability to reproduce interannual variability and synoptic-scale variability as well as mean conditions.

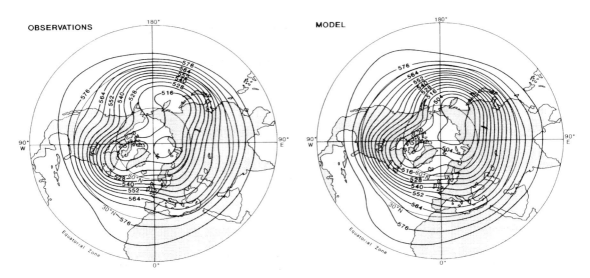

Figure 6.39 January mean 500 mb height field (*dm*) from (a) observations and (b) NCAR CCM1 simulation.
Source: Boville and Randel 1996. From *Journal of the Atmospheric Sciences* **43**, by permission of the American Meteorological Society.

Recent models incorporate improved spatial resolution and fuller treatment of some previously neglected physical processes. However, both changes may create additional problems as a result of the need to treat complex interactions between the land surface (including soil moisture, canopy structure, etc.) and the atmospheric boundary layer, or interactions between clouds, radiative exchanges and precipitation mechanisms, for example. Fine-scale spatial resolution is especially appropriate in the treatment of cloud and rain bands associated with frontal zones in mid-latitude cyclones, since the dynamics of convergence and vertical motion are key factors. However, mesoscale convective systems of continental areas and low-latitudes require representation of moisture exchanges (evaporation, condensation) and cloud microphysics.

Over the last decade, coupled atmosphere–ocean–sea ice models have been developed. The coupling of an atmospheric GCM to the ocean (Figure 6.40) may involve (1) a swamp ocean, where sea-surface temperatures are calculated through an energy budget and no annual cycle is possible; (2) a slab (mixed layer) ocean, where storage and release of energy can take place seasonally; or (3) an ocean GCM that treats all dynamic and thermodynamic processes and interactions. A major difficulty in the last case is the century time scale of deep ocean circulation, compared with the atmospheric

response to upper ocean conditions of the order of weeks to months. When the global ocean is considered, seasonal freezing/melting and the effects of sea ice on energy exchanges and salinity must also be modelled. Some models incorporate sea ice dynamics (ice drift), whereas others have only thermodynamic growth/decay of ice. Currently, there are about twenty global coupled models, most with relatively low-resolution atmospheric GCMs. Because coupled models are used in long-term (century-scale) simulations, an important concern is model 'drift' due to accumulating errors. These tendencies are sometimes constrained by using observed climatology at certain high-latitude or deep ocean boundaries, or by adjusting the net fluxes of heat and fresh water at each grid point on an annual basis, but such arbitrary procedures are the subject of controversy, especially for climate change studies.

Climate models can be used in several different ways. A common procedure is to analyse the model's sensitivity to a specified change in a single variable. This may involve changes in external forcing (increased/decreased solar radiation, atmospheric CO_2 concentrations, or a volcanic dust layer), surface boundary conditions (orography, land surface albedo, continental ice sheets) or in the model physics (modifying the convective scheme or the treatment of biospheric exchanges). In these

Coupled Model Hierarchy

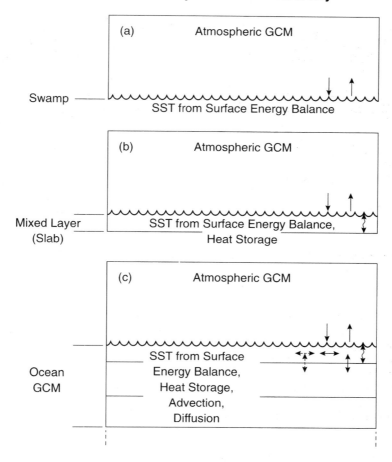

Figure 6.40 Schematic illustration of the three types of coupling of an atmospheric GCM to the ocean: (a) swamp ocean (b) mixed layer, slab ocean (c) ocean GCM.

Source: From Meehl 1992 in K. E. Trenberth (ed.) *Climate System Modeling*. Copyright © Cambridge University Press.

simulations, the model is allowed to reach a new equilibrium and the result is compared with a control experiment. A second approach is to conduct a genuine climate change experiment where, for example, the climate is allowed to evolve as atmospheric trace gas concentrations are increased at a specified annual rate (a transient experiment).

A key issue in assessments of greenhouse-gas-induced warming is the sensitivity of global climate to CO_2 doubling. Atmospheric GCM simulations for equilibrium condition changes, with a simple ocean treatment, indicate an increase in global mean surface air temperature of 2.5 to 5°C, comparing $1 \times CO_2$ and $2 \times CO_2$ concentrations in the models. The range is in part the result of a dependence of the temperature change on the temperature level simulated for the base-state $1 \times CO_2$, and in part arises from the feedback mechanisms incorporated in the models involving atmospheric water vapour, clouds, snow cover and sea ice. Use of coupled atmosphere–ocean models, however, suggest only a 1–2°C warming for century-long transient or doubled CO_2 experiments (see Chapter 11).

Because GCMs require massive computer resources, other approaches to modelling climate have developed. A variant of the GCM is the statistical–dynamical model (SDM), in which only zonally averaged features are analysed and north–south energy and momentum exchanges are not treated explicitly but are represented statistically through parameterization. Simpler still are the energy balance model (EBM) and the radiative convective model (RCM). The EBM assumes a global radiation balance and describes the integrated north–south transports of energy in terms of the poleward temperature gradients; EBMs can be one-dimensional (latitude variations only), two-dimensional (latitude–longitude, with simple land–ocean weightings or simplified geography) and even zero-dimensional (averaged for the globe). They are used particularly in climate change studies. The RCMs represent a single, globally averaged vertical column. The vertical temperature structure is analysed in terms of radiative and convective exchanges. These less complete models complement the GCMs because, for example, the RCM allows study of complex cloud-radiation interactions and atmospheric composition on lapse rates in the absence of large-scale dynamic processes.

SUMMARY

The vertical change of pressure with height depends on the temperature structure. High-(low-)pressure systems intensify with altitude in a warm (cold) air column; thus warm lows and cold highs are shallow features. This 'thickness' relationship is illustrated by the upper-level subtropical anticyclones and polar vortex in both hemispheres. The intermediate mid-latitude westerly winds thus have a large 'thermal wind' component. They become concentrated into upper tropospheric jet streams above sharp thermal gradients, such as fronts.

The upper flow displays a large-scale long-wave pattern, especially in the northern hemisphere, related to the influence of mountain barriers and land–sea differences. The surface pressure field is dominated by semi-permanent subtropical highs, subpolar lows and, in winter, shallow cold continental highs in Siberia and north-western Canada. The equatorial zone is predominantly low pressure. The associated global wind belts are the easterly trade winds and the mid-latitude westerlies. There are more variable polar easterlies and over land areas in summer a band of equatorial westerlies representing the monsoon systems. This mean zonal (west–east) circulation is intermittently interrupted by 'blocking' highs; an idealized sequence is known as the index cycle.

The atmospheric general circulation, which transfers heat and momentum polewards, is predominantly in a vertical meridional plane in low latitudes (the Hadley cell), but there are also important east–west circulations (Walker cells) between the major regions of subsidence and convective activity. Heat and momentum exchanges in middle and high latitudes are accomplished by horizontal waves and eddies (cyclones/anticyclones). Substantial energy is also carried polewards by ocean current systems. Surface currents are mostly wind driven, but the slow deep ocean circulation (global conveyor belt) is due to thermohaline forcing.

The ocean's vertical structure varies latitudinally and regionally. In general, the thermocline is deepest in mid-latitudes, thus permitting greater turbulent mixing and atmospheric heat exchanges. The oceans are important regulators of both atmospheric temperatures and CO_2 concentrations. Ocean dynamics and circulation features are analogous to those in the atmosphere on both meso- and macroscales. The wind-driven Ekman layer extends to 100–200 m. Ekman transport and coastal upwelling maintain normally cold sea surfaces off western South America and south-west Africa in particular.

Various types of numerical model are used to study the mechanisms of the atmospheric circulation and climate processes. These include vertical column models of radiative and convective processes, one- and two-dimensional energy balance models and complete three-dimensional general circulation models (GCMs). While initially developed for weather forecasting such models are now widely used to study climatic anomalies and past and future changes of global climate. These uses require coupling of atmospheric and oceanic GCMs and the representation of land surface processes.

7

Mid-latitude synoptic systems

This chapter examines the classical ideas about air masses and their role in the formation of frontal boundaries and in the development of extratropical cyclones. It also discusses the limitations of those ideas and more recent models of mid-latitude weather systems. The chapter concludes with a brief overview of weather forecasting.

A THE AIR-MASS CONCEPT

An air mass may be defined as a large body of air whose physical properties, especially temperature, moisture content and lapse rate, are more or less uniform horizontally for hundreds of kilometres. The theoretical ideal is a *barotropic* atmosphere where surfaces of constant pressure are not intersected by isosteric (constant-density) surfaces, so that in any vertical cross-section, as shown in Figure 7.1, isobars and isotherms are parallel.

Three main factors determine the nature and degree of uniformity of air-mass characteristics.

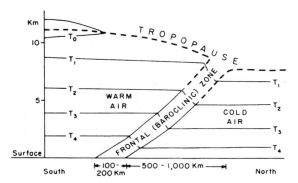

Figure 7.1 A schematic height cross-section for the northern hemisphere showing barotropic air masses and a baroclinic frontal zone (assuming that density decreases with height only).

They are: (1) the nature of the source area (from which the air mass obtains its original qualities); (2) the direction of movement and changes that occur in an air mass as it moves over long distances; and (3) the age of the air mass. The physical properties of all air masses are classified according to the way in which they compare with the corresponding properties of the underlying surface region or with those of adjacent air masses.

B NATURE OF THE SOURCE AREA

We have already observed how most of the physical processes of our atmosphere result from self-regulating attempts to equalize the major differences that arise from inequalities in the distribution of heat, moisture and pressure. On the world scale, the heat and momentum balances refer only to long-term average conditions. However, on a smaller scale, radiation and vertical mixing can produce some measure of equilibrium between the surface conditions and the properties of the overlying air mass if air remains over a given geographical region for a period of about three to seven days. Naturally, the chief source regions of air masses are areas of extensive, uniform surface type, which are normally overlain by quasi-stationary pressure systems. These requirements are fulfilled where there is slow divergent flow from the major thermal and dynamic high-pressure cells, whereas low-pressure regions are zones of convergence into which air masses move (see F, this chapter).

Air masses are classified on the basis of two primary factors. The first is the temperature, giving arctic, polar and tropical air, and the second is the type of surface in their region of origin, giving maritime and continental categories. The major cold and warm air masses will now be discussed.

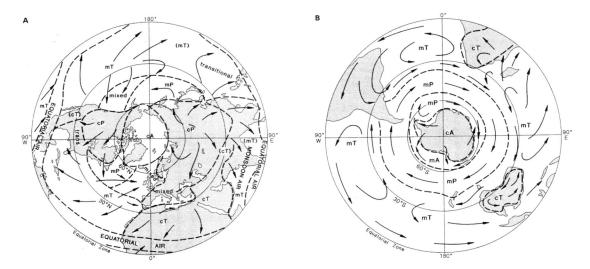

Figure 7.2 Air masses in winter. A: northern hemisphere; B: southern hemisphere.
Sources: A: After Petterssen 1950 and Crowe 1965. B: After Taljaard *et al.* 1969 and Newton 1972.

1 Cold air masses

The principal sources of cold air in the northern hemisphere are (1) the continental anticyclones of Siberia and northern Canada, which originate in continental polar (cP) air masses, and (2) the Arctic Basin, when it is dominated by high pressure (Figure 7.2). Some classifications designate air of the latter category as continental arctic (cA), but the differences between cP and cA air masses are limited mainly to the middle and upper troposphere, where temperatures are lower in the cA air.

The snow-covered source regions of these two air masses lead to marked cooling of the lower layers (Figure 7.3) and, since the vapour content of cold air is very limited, the air masses generally have a mixing ratio of only 0.1–0.5 g/kg near the surface. The stability produced by the effect of surface cooling prevents vertical mixing, so further cooling occurs more slowly by radiation losses only. The effect of this radiative cooling and the tendency for air-mass subsidence in high-pressure regions combine to produce a prominent temperature inversion from the surface up to about 850 mb in typical cA and cP air. In view of their extreme dryness, these air masses are characterized by small cloud amounts and low temperatures. In summer, continental heating over northern Canada and Siberia causes the virtual disappearance of their sources of cold air. The Arctic Basin source remains (Figure

7.4A), but the cold air here is very limited in depth at this time of year. In the southern hemisphere, the Antarctic continent and the ice shelves are a source of cA air in all seasons (see Figures 7.2B and 7.4B). There are no sources of cP air, however, due to the dominance of ocean areas in middle latitudes. At all seasons, cA or cP air is greatly modified by a passage over the ocean. Secondary types of air mass are produced by such means and these will be considered below.

2 Warm air masses

These have their origins in the subtropical high-pressure cells and, during the summer season, in the great accumulations of warm surface air that characterize the heart of large land areas.

The tropical (T) sources are either maritime (mT), originating in the oceanic subtropical high-pressure cells, or continental (cT), either originating from the continental parts of these subtropical cells (e.g. as does the North African *Harmattan*) or simply associated with regions of generally light variable winds, assisted by upper tropospheric subsidence, over the major continents in summer (e.g. Central Asia). In the southern hemisphere, the source area of mT air covers about half of the hemisphere. There is no zone of significant temperature gradient between the equator and the oceanic Subtropical Convergence at about 40°S.

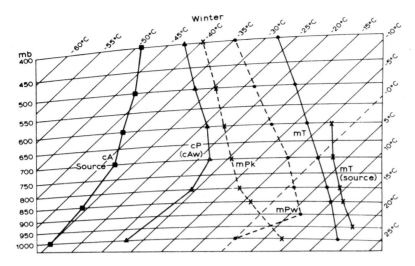

Figure 7.3 The average vertical temperature structure for selected air masses affecting North America at about 45–50 °N, recorded over their source areas or over North America in winter.

Sources: After Godson 1950, Showalter 1939, and Willett.

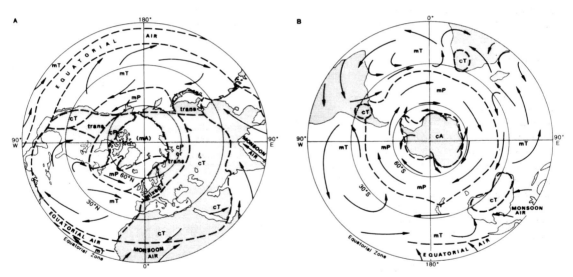

Figure 7.4 Air masses in summer. A: northern hemisphere; B: southern hemisphere.
Sources: A: After Petterssen 1950 and Crowe 1965. B: After Taljaard *et al.* 1969 and Newton 1972.

The maritime type is characterized by high temperatures (accentuated by the warming action to which the descending air is subjected), high humidity of the lower layers over the oceans, and stable stratification. Since the air is warm and moist near the surface, stratiform cloud commonly develops as the air moves polewards from its source. The continental type in winter is restricted mainly to North Africa (see Figure 7.2), where it is a warm, dry and stable air mass. In summer, warming of the lower layers by the heated land generates a steep lapse rate, but despite its instability the low relative and specific humidity prevent the development of cloud and precipitation. In the southern hemisphere, cT air is rather more prevalent in winter over the subtropical continents, except South America. In summer, much of southern Africa and northern Australia is affected by mT air, while there is a small source of cT air over Argentina (see Figure 7.4B).

The characteristics of the primary air masses are illustrated in Figures 7.3 and 7.5. In some cases, their properties have been considerably affected by

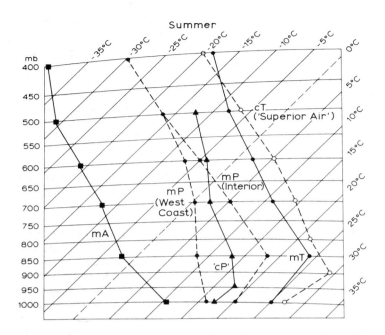

Figure 7.5 The average vertical temperature structure for selected air masses affecting North America in summer.

Sources: After Godson 1950, Showalter 1939, and Willett.

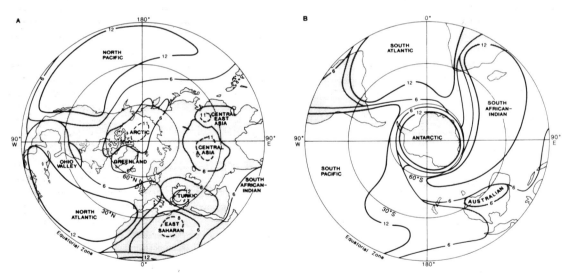

Figure 7.6 Air-mass source regions in the northern hemisphere (A) and the southern hemisphere (B). Numbers show the areas affected by each air mass in months per year.

Sources: After Wendland and Bryson 1981, and Wendland and McDonald 1986.

movement away from the source region, and this question is discussed below (see pp. 157–9).

A different view of source regions can be obtained from analysis of airstreams. Streamlines of the mean resultant winds (see Note 1) in individual months can be used to analyse areas of divergence repre-senting air-mass source regions, downstream airflow and the confluence zones between different air-streams. Figure 7.6A shows the sources in the northern hemisphere and their annual duration. Four sources are dominant: the subtropical North Pacific and North Atlantic anticyclones, and their southern hemisphere counterparts. For the entire year, air from these sources covers at least 25 per cent of the northern hemisphere; for six months of each year they affect almost three-fifths of the hemisphere. In the southern hemisphere, in contrast, the airstream climatology is much simpler as a result of the dominance of ocean surfaces (see Figure 7.6B). The source areas are associated with the three oceans, and especially their subtropical anticyclones. Antarctica is the major continental source, with another mainly in winter over Australia.

C AIR-MASS MODIFICATION

As an air mass moves away from its source region it is affected by different heat and moisture exchanges with the ground surface and by dynamic processes in the atmosphere. Thus an initially barotropic air mass is gradually changed into a moderately *baroclinic* airstream in which isosteric and isobaric surfaces intersect one another. The presence of horizontal temperature gradients means that air cannot travel as a solid block maintaining an unchanging internal structure. The trajectory (i.e. actual path) followed by an air parcel in the middle or upper troposphere will normally be quite different from that of a parcel nearer the surface, due to the increase of westerly wind velocity with height in the troposphere. The actual structure of an airstream at a given instant is determined to a large extent by the past history of air-mass modification processes. In spite of these qualifications, the air-mass concept nevertheless remains of considerable practical value.

1 Mechanisms of modification

The mechanisms by which air masses are modified are, for convenience, treated separately, although this rigid distinction is often not justified in practice.

a Thermodynamic changes

An air mass may be heated from below either by passing from a cold to a warm surface or by solar heating of the ground over which the air is located. Similarly, but in reverse, it can be cooled from below. Heating from below acts to increase air-mass instability, so the effect may be spread rapidly through a considerable thickness of air, whereas surface cooling produces a temperature inversion, which greatly limits the vertical extent of the cooling. For this reason, cooling occurs mainly

through radiative heat loss by the air, a process that takes place only very gradually.

Changes can also occur through increased evaporation, the moisture being supplied either from the underlying surface or by precipitation from an overlying air-mass layer. In reverse, the abstraction of moisture by condensation or precipitation can also cause changes. A parallel, and most important, change is the respective addition or loss of latent heat accompanying this condensation or evaporation. Annual values of latent and sensible heat transfers to the atmosphere are illustrated in Figures 2.30 and 2.31.

b Dynamic changes

Dynamic (or mechanical) changes are, superficially at any rate, different from thermodynamic changes because they involve mixing or pressure changes associated with the actual movement of the air mass. The distribution of the physical properties of air masses has been shown to be considerably modified, for example, by a prolonged period of turbulent mixing (see Figure 4.7). This process is particularly important at low levels, where surface friction intensifies the natural turbulence of airflow, providing a ready mechanism for the upward transfer of heat and moisture.

The radiative and advective exchanges discussed previously are diabatic, but the ascent or descent of air causes adiabatic changes of temperature. Large-scale lifting may result from forced ascent by a mountain barrier or from airstream convergence. Conversely, sinking may occur when high-level convergence sets up subsidence or when stable air, having been forced up over high ground by the pressure gradient, descends in its lee. Dynamic processes in the middle and upper troposphere are in fact a major cause of air-mass modification. The decrease in stability aloft, as air moves away from the areas of subsidence, is a common example of this type of mechanism.

2 The results of modification: secondary air masses

Study of the ways in which air masses change in character tells us a great deal about many common meteorological phenomena.

a Cold air

Continental polar air frequently streams out from Canada over the western Atlantic in winter, where it undergoes rapid transformation. Heating over the Gulf Stream Drift rapidly makes the lower layers unstable, and evaporation into the air leads to sharp increases of moisture content (see Figure 3.7) and cloud formation (see Plate 9). The turbulence associated with the convective instability is marked by gusty conditions. By the time the air has reached the central Atlantic, it has become a cool, moist, maritime polar (mP) air mass. Analogous processes occur with outflow from Asia over the north Pacific (see Figure 7.2). Over middle latitudes of the southern hemisphere, the circumpolar ocean gives rise to a continuous zone of mP air that, in summer, extends to the margin of Antarctica. In this season, however, a considerable gradient of ocean temperatures associated with the Antarctic Convergence makes the zone far from uniform in its physical properties. The weather in cP airstreams is typically that of bright periods and squally showers, with a variable cloud cover of cumulus and cumulonimbus. As mP air moves eastwards towards Europe, the cooler sea surface may produce a neutral or even stable stratification near the surface, especially in summer, but subsequent heating over land will again regenerate unstable conditions. Similar conditions, but with lower temperatures, result if cA air crosses sea areas in high latitudes, producing maritime arctic (mA) air.

When cP air moves southwards over land in winter, in central North America for example, it acquires a greater tendency towards instability, but there is little gain in moisture content. The cloud type is scattered shallow cumulus, which only rarely gives showers. Exceptions occur in early winter around the eastern and southern shores of Hudson Bay and the Great Lakes. Until these water bodies freeze over, cold airstreams that cross them are warmed rapidly and supplied with moisture, leading to locally heavy snowfalls (p. 207).

Over Eurasia and North America, cP air may move southwards and later recurve northwards. Some schemes of air-mass classification cater for such possibilities by specifying whether the air is colder (k), or warmer (w), than the surface over which it is passing. For example, cPk refers to cold, dry continental polar air that is moving over a warmer surface and is thereby likely to become unstable. In general, a 'k' air mass has gusty, turbulent winds, which make for good visibility as smoke and haze are dispersed. The instability leads to cumuliform clouds. Likewise, mPw indicates moist, maritime polar air that is progressively being cooled

near the surface and, hence, becoming more stable. A 'w' air mass typically has stable or inversion conditions with stratiform cloud. Limited vertical mixing allows the concentration of smoke, haze and fog at low levels.

In many parts of the world, the surface conditions and air circulation produce air masses with intermediate characteristics. Northern Asia and northern Canada fall into this category in summer. In a general sense, the air has affinities with continental polar air masses but these land areas have extensive bog and water surfaces, so the air is moist and cloud amounts are quite high. In a similar manner, meltwater pools and leads in the arctic pack-ice make it more appropriate to regard the area as a source of maritime arctic (mA) air in summer (see Figure 7.4A). This designation is also applied to air over the Antarctic pack-ice in winter that is much less cold in its lower levels than the air over the continent itself.

b Warm air

The modification of warm air masses is usually a gradual process. Air moving polewards over progressively cooler surfaces becomes increasingly stable in the lower layers. In the case of mT air with high moisture content, surface cooling may produce advection fog, and this is particularly common, for example, in the south-western approaches to the English Channel during spring and early summer, when the sea is still cool. Similar development of advection fog in mT air occurs along the South China coast in February–April, and also off Newfoundland and over the coast of northern California in spring and summer. If the wind velocity is sufficient for vertical mixing, low stratus cloud forms in the place of fog, and drizzle may result. In addition, forced ascent of the air by high ground, or by overriding of an adjacent air mass, can produce heavy rainfall.

The cT air originating in those parts of the subtropical anticyclones situated over the arid subtropics in summer is extremely hot and dry. It is typically unstable at low levels and dust storms may occur, but the dryness and the subsidence of the upper air limit cloud development. In the case of North Africa this cT air may move out over the Mediterranean rapidly acquiring moisture, with the consequent release of potential instability triggering off showers and thunderstorm activity.

Air masses in low latitudes present considerable problems of interpretation. The temperature contrasts found in middle and high latitudes are virtually absent, and what differences do exist are due principally to moisture content and, more particularly, to the presence or absence of subsidence. *Equatorial air* is usually cooler than that subsiding in the subtropical anticyclones, for example. On the equatorial sides of the subtropical anticyclones in summer, the air is moving westwards from areas with cool sea surfaces (e.g. off north-west Africa and California) towards those of higher sea-surface temperatures. Moreover, the south-western parts of the high-pressure cells are affected only by weak subsidence due to the vertical structure of the cells. As a result of these circumstances, the mT air moving westwards around the equatorward sides of the subtropical highs becomes much less stable than that on the north-eastern margin of the cells. Eventually, such air forms the very warm, moist, unstable 'equatorial air' of the Intertropical Convergence Zone (see Figures 7.2 and 7.4). *Monsoon air* is indicated separately in these figures, although there is no basic difference between it and mT air. Better approches to tropical climatology are discussed in Chapter 9.

3 The age of the air mass

Eventually, the mixing and modification necessarily accompanying the movement of an air mass away from its source will cause the rate of energy exchange with the surroundings to diminish, and the various weather phenomena associated with these changes will dissipate. This process will be associated with the loss of its original identity until, finally, its features merge with those of surrounding airstreams and the air may again become subject to the influence of a new source region.

North-west Europe is shown as an area of 'mixed' air masses in Figures 7.2 and 7.4. This is intended to refer to the variety of sources and directions from which air may invade the region. The same is also true of the Mediterranean sea in winter, although the area does impart its own particular characteristics to polar and other air masses that stagnate over it. Such air is termed *mediterranean*. In winter, it is convectively unstable (see Figure 4.6) as a result of the moisture picked up over the Mediterranean Sea.

The length of time during which an air mass retains its original characteristics depends very much on the extent of the source area and the type of pressure pattern affecting the area. In general, the lower air is changed much more rapidly than that at higher levels, although dynamic modifications aloft are no less significant in terms of weather

processes. Modern air-mass concepts must, therefore, be flexible from the point of view of both synoptic and climatological studies.

D FRONTOGENESIS

The first real advance in our detailed understanding of mid-latitude weather variations was made with the discovery that many of the day-to-day changes are associated with the formation and movement of boundaries, or *fronts*, between different air masses. Observations of the temperature, wind directions, humidity and other physical phenomena during unsettled periods showed that discontinuities often persist between impinging air masses of differing characteristics. The term 'front' for these surfaces of air-mass conflict was a logical one, proposed during the First World War by a group of meteorologists (including V. and J. Bjerknes, H. Solberg and T. Bergeron) working in Norway, and their ideas are still an integral part of most weather analysis and forecasting, particularly in middle and high latitudes.

1 Frontal waves

The typical geometry of an air-mass interface, or front, resembles a wave form (Figure 7.7). Similar wave patterns are, in fact, found to occur on the interfaces between many different media, for example, waves on the sea surface, ripples on beach sand, aeolian sand dunes, etc. Unlike these wave forms, however, the frontal waves in the atmosphere are commonly unstable; that is, they suddenly originate, increase in size, and then gradually dissipate. Numerical model calculations show that in middle latitudes waves in a baroclinic atmosphere are unstable if their wavelength exceeds a few thousand kilometres. Frontal wave cyclones are typically 1,500–3,000 km in wavelength. The initially attractive analogy between atmospheric wave systems and waves formed on the interface of other media is, therefore, an insufficient basis on which to develop explanations of frontal waves. In particular, the circulation of the upper troposphere plays a key role in providing appropriate conditions for their development and growth, as will be shown below.

2 The frontal-wave depression

A depression (also termed a low or cyclone) (see Note 2) is an area of relatively low pressure, with a more or less circular isobaric pattern. It covers an area

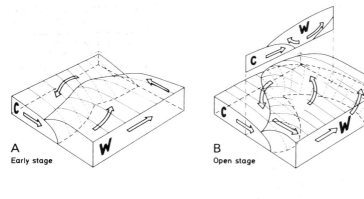

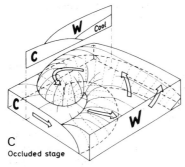

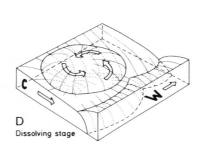

Figure 7.7 Four stages in the typical development of a mid-latitude depression. Satellite views of the cloud systems corresponding to these stages are shown in Figure 7.8.

Source: Mostly after Strahler 1965, modified after Beckinsale.

Notes: C = cold air; W = warm air.

1,500–3,000 km in diameter and usually has a life-span of four to seven days. Systems with these characteristics, which are prominent on daily weather maps, are referred to as *synoptic-scale* features. The depression, in mid-latitudes at least, is usually associated with a convergence of contrasting air masses. According to the 'Norwegian cyclone model' (see Figure 7.7), the interface between these air masses develops into a wave form with its apex located at the centre of the low-pressure area. The wave encloses a mass of warm air between modified cold air in front and fresh cold air in the rear. The formation of the wave also creates a distinction between the two sections of the original air-mass discontinuity for, although each section still marks the boundary between cold and warm air, the weather characteristics found in the neighbourhood of each section are very different. The two sections of the frontal surface are distinguished by the names *warm front* for the leading edge of the wave and *cold front* for that of the cold air to the rear (see Figure 7.7B).

The boundary between two adjacent air masses is marked by a strongly baroclinic zone of large temperature gradient 100–200 km (see C, this chapter,

and Figure 7.1). Sharp discontinuities of temperature, moisture and wind properties at fronts, especially the warm front, are rather uncommon. Such discontinuities are usually the result of a pronounced surge of fresh, cold air in the rear sector of a depression, but in the middle and upper troposphere they are often caused by subsidence and may not coincide with the location of the baroclinic zone. In meteorological centres, numerous indicators may be analysed to locate frontal boundaries, including 1,000–500 mb thickness gradients, 850 mb wet-bulb potential temperature, cloud and precipitation bands, and wind shifts. However, a skilled forecaster may have to make a subjective decision where some of these criteria give conflicting information.

On satellite imagery, active cold fronts in a strong baroclinic zone commonly show marked spiral cloud bands formed as a result of the thermal advection (Figure 7.8B, C). Warm fronts, however, are typically covered by a cirrus shield. As Figure 6.7 shows, an upper tropospheric jet stream is closely associated with the baroclinic zone, blowing roughly parallel to the line of the upper front (see Plate 17). This relationship is examined further below.

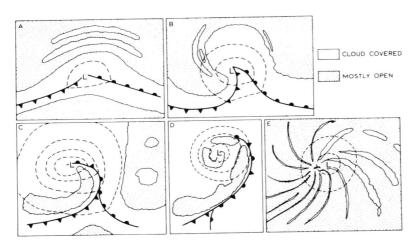

Figure 7.8 Schematic patterns of cloud cover (white) observed from satellites, in relation to surface fronts and generalized isobars. A, B, C and D correspond to the four stages in Figure 7.7.

Source: After Boucher and Newcomb 1962.

Air behind the cold front, away from the low centre, commonly has an anticyclonic trajectory and hence moves at a greater than geostrophic speed (see Chapter 5A.4), impelling the cold front to acquire a supergeostrophic speed also. The wedge of warm air is pinched out at the surface and lifted bodily off the ground. This stage of *occlusion* eliminates the wave form at the surface (see Figure 7.7). The depression usually achieves its maximum intensity 12–24 hours after the beginning of occlusion. The occlusion gradually works outwards from the centre of the depression along the warm front. Sometimes, the cold air wedge advances so rapidly that, in the friction layer close to the surface, cold air overruns the warm air and generates a *squall line* (see Chapter 4G).

By no means all frontal lows follow the idealized life-cycle discussed above (cf. the caption for Plate 18). It is generally characteristic of oceanic cyclogenesis, although the evolution of those systems has been re-examined using aircraft observations collected during North Atlantic meteorological field programmes in the 1980s. These suggest a revised view of the evolution of maritime frontal cyclones (Figure 7.9). Four stages are identified: (1) a broad (400 km) continuous frontal zone during cyclone inception, (2) frontal fracture near the centre of the low and tighter frontal gradients, (3) development of a T-bone structure and bent-back warm front, and (4) the mature cyclone showing seclusion of the warm core within the polar airstream behind the cold front. Figure 7.10 illustrates an intense cyclone system with bent-back warm front and secluded warm core. This storm gave extreme wind

speeds and caused widespread damage over southern England in October 1987.

Over central North America, cyclones forming in winter and spring depart considerably from the Norwegian model. In particular, they feature an outflow of cold arctic air east of the Rocky Mountains, forming an arctic front, a lee trough with dry air descending from the mountains, and warm, moist southerly flow from the Gulf of Mexico (Figure 7.11). The trough superposes dry air over warm, moist air, generating instability and a rain band analogous to a warm front. The arctic air flows southwards west of the low centre, causing lifting of warmer, dry air but giving little precipitation. There may also be an upper cold front advancing over the trough that forms a rain band along its leading edge. Such a system is thought to have caused a record rainstorm at Holt, Missouri, on 22 June 1947, when 305 mm fell in just 42 minutes!

E FRONTAL CHARACTERISTICS

The activity of a front in terms of weather depends upon the vertical motion in the air masses. If the air in the warm sector is rising relative to the frontal zone the fronts are usually very active and are termed *ana-fronts*, whereas sinking of the warm air relative to the cold air masses gives rise to less intense *kata-fronts* (Figure 7.12).

1 The warm front

The warm front represents the leading edge of the warm sector in the wave. The frontal zone here has

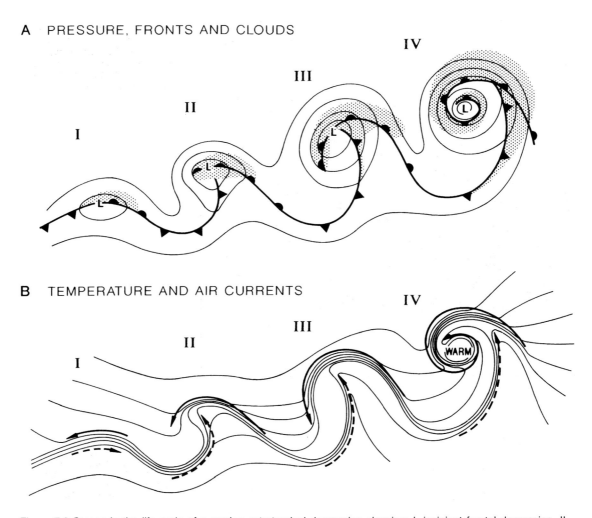

A PRESSURE, FRONTS AND CLOUDS

B TEMPERATURE AND AIR CURRENTS

Figure 7.9 Stages in the life-cycle of a marine extratropical depression showing: I. incipient frontal depression, II frontal fracture, III. bent-back warm front and frontal T-bone, IV. warm-core seclusion. A: schematic isobars of sea-level pressure, fronts and cloud cover (stippled). B: isotherms and flows of cold air (solid arrows) and warm air (dashed arrows).

Source: After Shapiro and Keyser 1990. From C. W. Newton and E. D. Holopainen (eds), *Extratropical Cyclones*, by permission of the American Meteorological Society.

a very gentle slope, of the order of 0.5–1°, so the cloud systems associated with the upper portion of the front herald its approach some 12 hours or more before the arrival of the surface front (see Plate 17). The ana-warm front, with rising warm air, has multi-layered cloud that steadily thickens and lowers towards the surface position of the front. The first clouds are thin, wispy cirrus, followed by sheets of cirrus and cirrostratus, and altostratus (Figure 7.12A). The sun is obscured as the alto-stratus layer thickens and drizzle or rain begins to

fall. The cloud often extends through most of the troposphere and with continuous precipitation occurring is generally designated as nimbostratus. Patches of stratus may also form in the cold air as rain falling through this air undergoes evaporation and quickly saturates it.

The descending warm air of the kata-warm front greatly restricts the development of medium- and high-level clouds. The frontal cloud is mainly strato-cumulus, with a limited depth as a result of the subsidence inversions in both air masses (see Figure

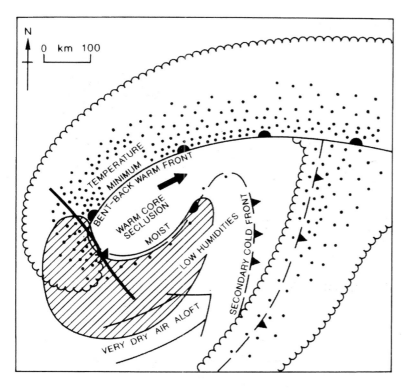

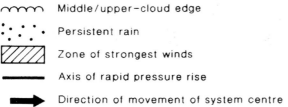

Figure 7.10 Schematic model showing the distribution of airflows and surface weather within an intense mid-latitude depression that produced very high wind velocities.

⌒⌒⌒⌒ Middle/upper–cloud edge

: ˙ : Persistent rain

▨▨▨▨ Zone of strongest winds

━━━━━ Axis of rapid pressure rise

➡ Direction of movement of system centre

Source: From Monk, 1992. Reprinted from the *Meteorological Magazine* (Crown copyright ©) by permission of the Controller of Her Majesty's Stationery Office.

7.12B). Precipitation is usually light rain or drizzle formed by coalescence, since the freezing level tends to be above the inversion level, particularly in summer.

At the passage of the warm front the wind veers, the temperature rises and the fall of pressure is checked. The rain becomes intermittent or ceases in the warm air and the thin stratocumulus cloud sheet may break up.

Forecasting the extent of rain belts associated with the warm front is complicated by the fact that most fronts are not ana- or kata-fronts throughout their length or even at all levels in the troposphere. For this reason, radar is increasingly being used to determine by direct means the precise extent of rain belts and to detect differences in rainfall intensity.

Such studies have shown that most of the production and distribution of precipitation is controlled by a broad airflow a few hundred kilometres across and several kilometres deep, which flows parallel to and ahead of the surface cold front (see Figure 7.13 and Plate 19).

Just ahead of the cold front, the flow occurs as a low-level jet with winds of up to 25–30 m s^{-1} at about 1 km above the surface. The air, which is warm and moist, rises over the warm front and turns south-eastwards ahead of it as it merges with the mid-tropospheric flow (B in Figure 7.13). This flow has been termed a 'conveyor belt' (for large-scale heat and momentum transfer in mid-latitudes). Broad-scale convective (potential) instability is generated by the overrunning of this low-level flow by potentially

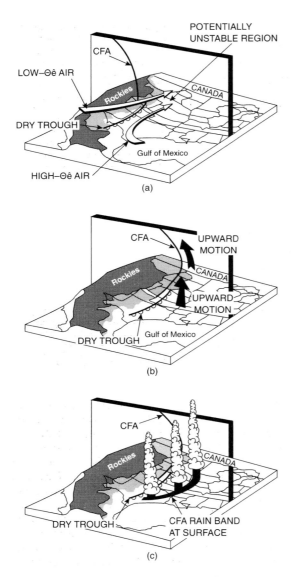

Figure 7.11 Schematic model of a dry trough and frontogenesis east of the Rocky Mountains. (a) Warm, dry air with low equivalent potential temperature (θ_e) from the Rockies overrides warm, moist, high θ_e air from the Gulf of Mexico, forming a potentially unstable zone east of the dry trough. (b) Upward motion associated with the cold front aloft (CFA). (c) Location of the CFA rain band at the surface. [Equivalent potential temperature is the potential temperature of an air parcel that is expanded adiabatically until all water vapour is condensed and the latent heat released then compressed adiabatically to 1,000 mb pressure.]

Source: After Locatelli *et al*. 1995. From *Monthly Weather Review* **123**(9), by permission of the American Meteorological Society.

colder, drier air in the middle troposphere. Instability is released mainly in small-scale convection cells that are organized into clusters, known as mesoscale precipitation areas (MPAs). These MPAs are further arranged in bands, 50–100 km wide (Figure 7.14). Ahead of the warm front, the bands are broadly parallel to the airflow in the rising section of the conveyor belt, whereas in the warm sector they parallel the cold front and the low-level jet. In some cases, cells and clusters are further arranged in bands within the warm sector and ahead of the warm front (see Figures 7.14 and 7.15). Precipitation from warm front rain bands often involves 'seeding' by ice particles falling from the upper cloud layers. It has been estimated that 20–35 per cent of the precipitation originates in the 'seeder' zone and the remainder in the lower clouds (see also Figure 4.15). Some of the cells and clusters are undoubtedly set up through orographic effects, and these influences may extend well downwind when the atmosphere is unstable.

2 The cold front

The weather conditions observed at cold fronts are equally variable, depending upon the stability of the warm-sector air and the vertical motion relative to the frontal zone. The classical cold-front model is of the ana-type, and the cloud is usually cumulonimbus. Figure 7.16 illustrates the warm conveyor belt associated with such a frontal zone and the line convection. Over the British Isles, air in the warm sector is rarely unstable, so that nimbostratus occurs more frequently at the cold front (see Figure 7.12A). With the kata-cold front the cloud is generally stratocumulus (see Figure 7.12B) and precipitation is light. With ana-cold fronts there are usually brief, heavy downpours, sometimes accompanied by thunder. The steep slope of the cold front, roughly 2°, means that the bad weather is of shorter duration than at the warm front. With the passage of the cold front, the wind veers sharply, pressure begins to rise and temperature falls (Plate 20). The sky may clear very abruptly, even before the passage of the surface cold front in some cases, although with kata-cold fronts the changes are altogether more gradual.

3 The occlusion

Occlusions are classified as either *cold* or *warm*, the difference depending on the relative states of the cold air masses lying in front and to the rear of

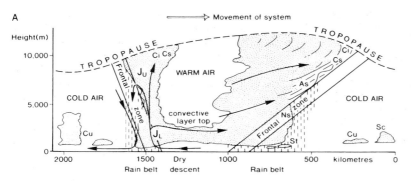

A

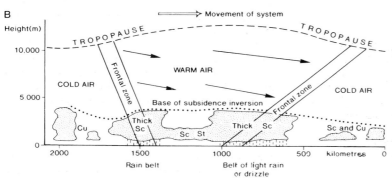

B

Figure 7.12 A: cross-sectional model of a depression with ana-fronts, where the air is rising relative to each frontal surface. Note that an ana-warm front may occur with a kata-cold front and vice versa. J_U and J_L show the locations of the upper and lower jet streams, respectively. B: model of a depression with kata-fronts, where the air is sinking relative to each frontal surface.

Sources: After Pedgley, *A Course in Elementary Meteorology*, and Bennetts *et al.* 1988 ((Crown copyright ©), reproduced by permission of the Controller of Her Majesty's Stationery Office).

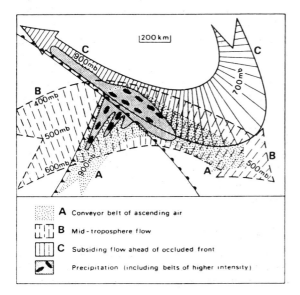

A Conveyor belt of ascending air

B Mid-troposphere flow

C Subsiding flow ahead of occluded front

Precipitation (including belts of higher intensity)

the warm sector (Figure 7.17). If the air is colder than the air following it then the occlusion is warm, but if the reverse is so (which is more likely over the British Isles) it is termed a cold occlusion. The air in advance of the depression is most likely to be coldest when depressions occlude over Europe in winter and very cold cP air is affecting the continent.

The line of the warm air wedge aloft is associated with a zone of layered cloud (similar to that found with a warm front) and often of precipitation. Hence its position is indicated separately on some weather maps and it is referred to by Canadian meteorologists as a *trowal* (trough of warm air aloft). The passage of an occluded front and trowal brings a change back to polar air-mass weather.

Figure 7.13 Model of the large-scale flow and mesoscale precipitation structure of a partially occluded depression typical of those affecting the British Isles. It shows the 'conveyor belt' (A) rising from 900 mb ahead of the cold front over the warm front. This is overlaid by a mid-tropospheric flow (B) of potentially colder air from behind the cold front. Most of the precipitation occurs in the well-defined region shown, within which it exhibits a cellular and banded structure.
Source: After Harrold 1973.

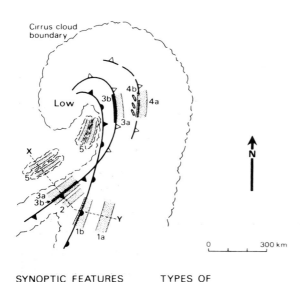

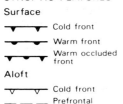

SYNOPTIC FEATURES

Surface

▼——▼ Cold front

●——● Warm front

▼——▼ Warm occluded
         front

Aloft

▽——▽ Cold front

▽– –▽ Prefrontal
         cold surge

TYPES OF
MESOSCALE RAINBANDS

1 Warm – frontal

2 Warm – sector

3 Cold – frontal

4 Prefrontal cold – surge

5 Postfrontal

Figure 7.14 Fronts and associated rain bands typical of a mature depression. The broken line X–Y shows the location of the cross-section given in Figure 7.15.

Source: After Hobbs; from Houze and Hobbs 1982.

A different process occurs when there is interaction between the cloud bands within a polar trough and the main polar front, giving rise to an *instant occlusion*. A warm conveyor belt on the polar front ascends as an upper tropospheric jet, forming a stratiform cloud band (Figure 7.18), while a low-level polar trough conveyor belt at right angles to it produces a convective cloud band and precipitation area poleward of the main polar front (see Plate 21) on the leading edge of the cold pool.

The occurrence of *frontolysis* (frontal decay) is not necessarily linked with occlusion, although it represents the final phase of a front's existence. Decay occurs when differences no longer exist between adjacent air masses. This may arise in four ways: (1) through their mutual stagnation over a similar surface, (2) as a result of both air masses moving on parallel tracks at the same speed, (3) as a result of their movement in succession along the same track at the same speed, or (4) by the system incorporating into itself air of the same temperature.

4 Frontal-wave families

Observation has shown that frontal waves, or depressions, do not generally occur as separate units but in *families* of three or four (see Figure 7.9; Plate 22), with the depressions that succeed the original one forming as *secondaries* along the trailing edge of an extended cold front. Each new member follows a course that is south of its progenitor as the polar air pushes farther south to the rear of each depression in the series. Eventually, the front trails far to the south and the cold polar air forms an extensive meridional wedge of high pressure, terminating the sequence.

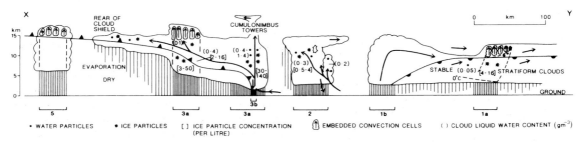

• WATER PARTICLES   • ICE PARTICLES   [ ] ICE PARTICLE CONCENTRATION
                                                (PER LITRE)   🔒 EMBEDDED CONVECTION CELLS   ( ) CLOUD LIQUID WATER CONTENT (gm⁻³)

Figure 7.15 Cross-section along the line X–Y in Figure 7.14 showing cloud structures and rain bands. The vertical hatching represents rainfall location and intensity. Raindrop and ice particle regions are shown, as are ice particle concentrations and cloud liquid water content. Numbered belts refer to those shown in Figure 7.14. Scales are approximate.

Source: After Hobbs and Matejka *et al*.; from Houze and Hobbs 1982.

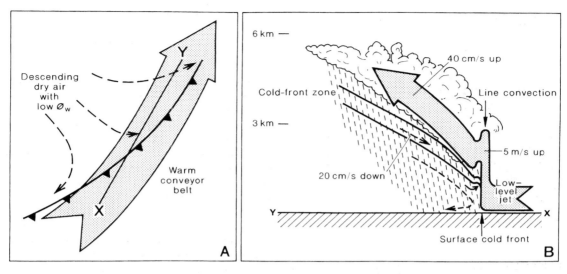

Figure 7.16 Schematic diagrams showing airflows, relative to the moving frontal system, at an ana-cold front. A warm conveyor belt (stippled) ascends above the front with cold air (dashed arrows) descending beneath it. A: plan view. B: vertical section along the line X–Y, showing rates of vertical motion.

Source: Browning 1990. From C. W. Newton and E. D. Holopainen (eds) *Extratropical Cyclones*, by permission of the American Meteorological Society.

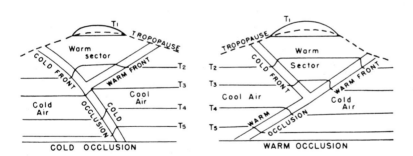

Figure 7.17 Schematic cross-sections of a cold and a warm occlusion in the classical model.

Source: After Pedgley, *A Course in Elementary Meteorology* ((Crown copyright ©), reproduced by permission of the Controller of Her Majesty's Stationery Office).

Another pattern of development may take place on the warm front, particularly at the point of occlusion, as a separate wave forms and runs ahead of the parent depression. This type of secondary is more likely with very cold (cA, mA or cP) air ahead of the warm front, and its formation is encouraged when the eastward movement of the occlusion is barred by mountains. This situation often occurs when a primary depression is situated in the Davis Strait and a breakaway wave forms south of Cape Farewell (the southern tip of Greenland), moving away eastwards. Analogous developments take place in the Skagerrak–Kattegat area when the occlusion is held up by the Scandinavian mountains.

F ZONES OF WAVE DEVELOPMENT AND FRONTOGENESIS

Fronts and associated depressions do not form everywhere, and their development is restricted to well-defined areas. Frontal formation in the temperate latitudes has been studied intensively, and knowledge of the general weather conditions to be expected is reasonably accurate. Arctic and polar fronts are caused primarily by gross differences in air-mass characteristics, whereas tropical disconti-nuities within and between somewhat similar air masses are produced mainly by the nature of the large-scale air motion and especially by confluence within an airstream or between two air currents of different humidity.

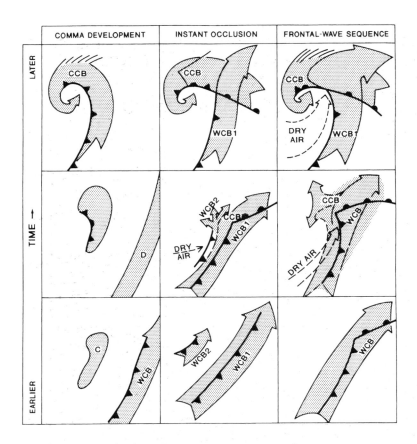

Figure 7.18 Schematic illustrations of vortex developments in satellite imagery. The sequences run from bottom to top. Left: comma cloud (c) developing in a polar airstream. Centre: Instant occlusion from the interaction of a polar trough with a wave on the polar front. Right: the classical frontal wave with cold and warm conveyor belts (CCB, WCB). C = enhanced convection; D = decaying cloud band; cloud cover stippled.

Source: After Browning 1990. From C. W. Newton and E. D. Holopainen (eds), *Extratropical Cyclones*, by permission of the American Meteorological Society.

The major zones of frontal wave development are naturally those areas which are most frequently baroclinic as a result of airstream confluence (Figure 7.19). This is the case, for instance, off East Asia and eastern North America, especially in winter, when there is a sharp temperature gradient between the snow-covered land and warm offshore currents. These zones are referred to as the Pacific Polar and Atlantic Polar Fronts, respectively (Figure 7.20). Their position is quite variable, but they show a general equatorward shift in winter, when the Atlantic Frontal Zone may extend into the Gulf of Mexico. In this area, there is convergence of air masses of different stability between adjacent subtropical high-pressure cells. Depressions developing here commonly move north-eastwards, sometimes following or amalgamating with others of the northern part of the Polar Front proper or of the Canadian Arctic Front. Frontal frequency remains high across the North Atlantic, but it decreases eastwards in the North Pacific, perhaps owing to a less pronounced gradient of sea-surface temperature. Frontal activity is most common in the central North Pacific when the subtropical high is split into two cells with converging air currents between them.

Another section of the Polar Front, often referred to as the *Mediterranean Front*, is located over the Mediterranean–Caspian Sea areas in winter. At intervals, fresh Atlantic mP air, or cool cP air from southeast Europe, converges with warmer air masses, often of North African origin, over the Mediterranean Basin and initiates frontogenesis. In summer, the area lies under the influence of the Azores subtropical high-pressure cell, and the frontal zone is absent.

The summer locations of the Polar Front over the western Atlantic and Pacific are some 10° further north than in winter (see Figure 7.20), although the frontal zone is rather weak at this time of year. There is now a frontal zone over Eurasia and a corresponding one over middle North America. These reflect the general meridional temperature gradient

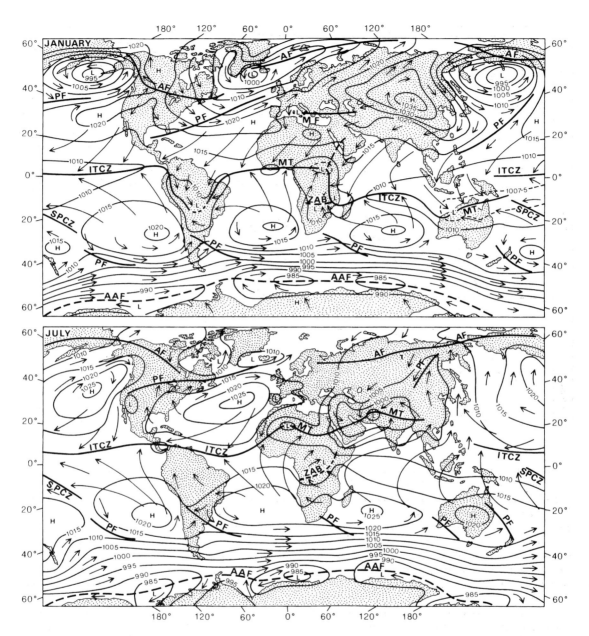

Figure 7.19 Mean pressure (mb) and surface winds for the world in January and July. The major frontal and convergence zones are shown as follows: Intertropical Convergence Zone (ITCZ), South Pacific Convergence Zone (SPCZ), Monsoon Trough (MT), Zaire Air Boundary (ZAB), Mediterranean Front (MF), northern and southern hemisphere Polar Fronts (PF), Arctic Fronts (AF) and Antarctic Fronts (AAF).

Source: Partly from Liljequist 1970.

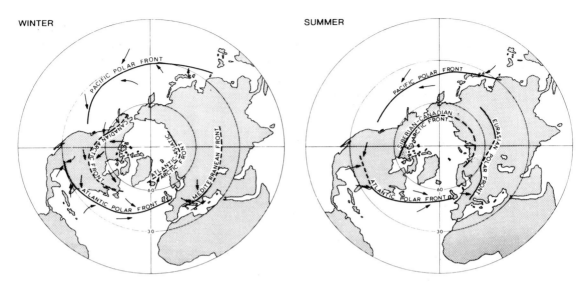

Figure 7.20 The major northern hemisphere frontal zones in winter and summer.

and probably also the large-scale influence of orography on the general circulation (see G, this chapter).

In the southern hemisphere, the Polar Front is on average about 45°S in January (summer), with branches spiralling polewards towards it from about 32°S off eastern South America and from 30°S, 150°W in the South Pacific (Figure 7.21). In July (winter), there are two Polar Frontal Zones spiralling towards Antarctica from about 20°S; one starts over South America and the other at 170°W. They terminate some 4–5° latitude further poleward than in summer. It is noteworthy that the southern hemisphere has more cyclonic activity in summer than does the northern hemisphere in its summer. This appears to be related to the seasonal strengthening of the meridional temperature gradient noted earlier (see p. 123).

The second major frontal zone is the Arctic Front, associated with the snow and ice margins of high latitudes (see Figure 7.20). In summer, this zone is developed along the tundra margin in Siberia and North America. In winter over North America, it is formed between cA (or cP) air and Pacific maritime air modified by crossing the Coast Ranges and the Rockies. There is also a less pronounced arctic frontal zone in the North Atlantic–Norwegian Sea area, extending along the Siberian coast. A similar weak frontal zone is found in winter in the southern hemisphere. It is located at 65–70°S near the edge of the Antarctic pack-ice in the Pacific

sector (see Figure 7.21), although few cyclones form there. Zones of airstream confluence in the southern hemisphere (cf. Figures 7.2B and 7.4B) are less numerous and more persistent, particularly in coastal regions, than in the northern hemisphere.

The principal tracks of depressions in the northern hemisphere in January are shown in Figure 7.22. The major ones reflect the primary frontal zones already discussed. In summer, the Mediterranean route is absent and lows move across Siberia, but the other tracks are similar although generally more zonal at this season and located in higher latitudes (around 60°N).

Between the two hemispherical belts of subtropical high pressure there is a further major world convergence zone, the Intertropical Convergence Zone (or ITCZ). Formerly this was referred to as the Intertropical Front (ITF), but air-mass contrasts only occur in limited sectors. This zone moves seasonally north and south away from the equator, as the subtropical high-pressure cell activity alternates in opposite hemispheres. The contrast between the converging air masses obviously increases with the distance of the ITCZ from the equator, and the degree of difference in their characteristics is naturally associated with considerable variation in activity along the convergence zone. Activity is most intense in June–July over South Asia and West Africa, when the contrast between the maritime and continental air masses that are involved is at a

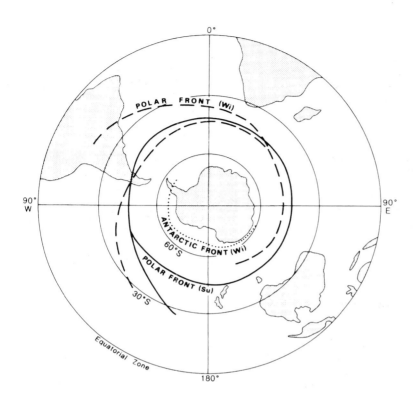

Figure 7.21 The major southern hemisphere frontal zones in winter (Wi) and summer (Su).

maximum. In these sectors, the zone merits the term Intertropical Front, although this does not imply that it behaves like a mid-latitude frontal zone. The nature of the ITCZ and its role in tropical weather are discussed in Chapter 9.

G SURFACE/UPPER-AIR RELATIONSHIPS AND THE FORMATION OF DEPRESSIONS

It has already been pointed out that a wave depression is associated with air-mass convergence, yet the barometric pressure at the centre of the low may decrease by 10–20 mb in 12–24 hours as the system intensifies. The explanation of this apparent discrepancy is that upper air divergence removes rising air more quickly than convergence at lower levels replaces it (see Figure 5.7). The superimposition of a region of upper divergence over a frontal zone is the prime motivating force of *cyclogenesis* (i.e. depression formation).

The long (or Rossby) waves in the middle and upper troposphere, which were discussed in Chapter 6A.2, are particularly important in this respect. The latitudinal circumference limits the circumpolar westerly flow to between three and six major Rossby waves, and these affect the formation and movement of surface depressions. The main stationary waves tend to be located about 70°W and 150°E in response to the influence on the atmospheric circulation of orographic barriers, such as the Rocky Mountains and the Tibetan plateau, and of heat sources. On the eastern limb of troughs in the upper westerlies of the northern hemisphere the flow is normally divergent, since the gradient wind is sub-geostrophic in the trough but supergeostrophic in the ridge (see Chapter 5A.4). Thus the sector ahead of an upper trough is a very favourable location for a surface depression to form or deepen (see Figure 7.23), and it will be noted that the mean upper troughs are significantly positioned just west of the Atlantic and Pacific Polar Front Zones in winter.

With these ideas in mind, we may now consider further the three-dimensional nature of depression development and the important links existing between upper and lower tropospheric flow. The basic theory relates to the vorticity equation, which states that, for frictionless horizontal motion, the rate of change of the vertical component of absolute

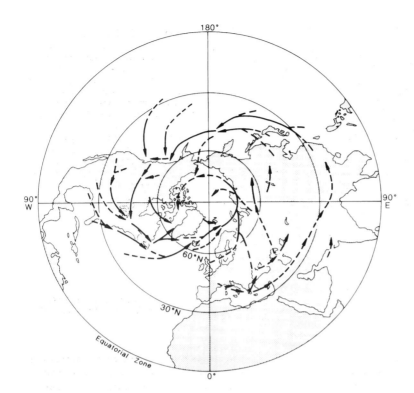

Figure 7.22 The principal northern hemisphere depression tracks in January. The full lines show major tracks, the dashed lines secondary ones that are less frequent and less well defined. The frequency of lows is a local maximum where arrowheads end. An area of frequent cyclogenesis is indicated where a secondary track changes to a primary one or where two secondary tracks merge to form a primary.

Source: After Klein 1957.

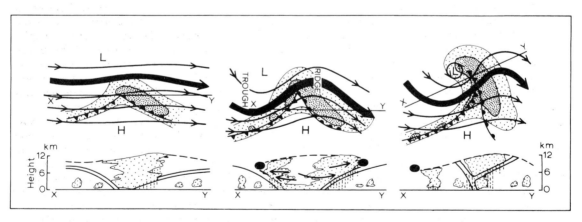

Figure 7.23 Stages in the development of an occluding depression. Above: upper winds and Polar Jet Stream in relation to surface fronts and precipitation areas (dark stipple), and cloud (lighter stipple). Below: cross-sections along the lines marked X–Y.
Source: After Flohn.

vorticity (dQ/dt or $d(f + \zeta)/dt$) is proportional to air-mass convergence ($-D$, i.e. negative divergence):

$$\frac{dQ}{dt} = DQ \text{ or } D = -\frac{1}{Q}\frac{dQ}{dt}$$

The relationship implies that a converging (diverging) air column has increasing (decreasing) absolute vorticity. The conservation of vorticity equation, which we have already discussed, is in fact a special case of this relationship.

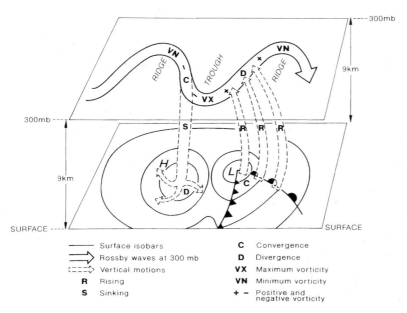

Figure 7.24 Schematic representation of the relationship between surface pressure (H and L), airflow and frontal systems, on the one hand, and the location of troughs and ridges in the Rossby waves at the 300 mb level. The locations of maximum (cyclonic) and minimum (anticyclonic) relative vorticity are shown, as are those of negative (anticyclonic) and positive (cyclonic) vorticity advection.

———	Surface isobars	**C**	Convergence
⟹	Rossby waves at 300 mb	**D**	Divergence
⊏⊏⊏⊐	Vertical motions	**VX**	Maximum vorticity
R	Rising	**VN**	Minimum vorticity
S	Sinking	**+ −**	Positive and negative vorticity

Sources: Mostly after Musk 1988, with additions from Uccellini 1990.

In the sector ahead of an upper trough, the decreasing cyclonic vorticity causes divergence (i.e. D positive), since the change in ζ outweighs that in f, thereby favouring surface convergence and low-level cyclonic vorticity (see Figure 7.24). Once the surface cyclonic circulation has become established, vorticity production is increased through the effects of thermal advection. Poleward transport of warm air in the warm sector and the eastward advance of the cold upper trough act to sharpen the baroclinic zone, strengthening the upper jet stream through the thermal wind mechanism (see p. 121). The vertical relationship between jet stream and front has already been shown (see Figure 6.8); a model depression sequence is demonstrated in Figure 7.24. The actual relationship may depart from this idealized case, although the jet is commonly located in the cold air (Plate 21). Velocity maxima (core zones) occur along the jet stream, as shown schematically in Figure 7.25, and the distribution of vertical motion upstream and downstream of these cores is known to be quite different. In the area of the jet entrance (i.e. upstream of the core), divergence causes lower-level air to rise on the equatorward (i.e. right) side of the jet, whereas in the exit zone (downstream of the core) ascent is on the poleward side. The second depression in Figure 7.25 is moving eastwards towards the area of maximum cyclogenetic tendency (upper divergence).

Figure 7.26 shows how precipitation is more often related to the position of the jet stream than to that of surface fronts; maximum precipitation areas are in the right entrance sector of the jet core. This vertical motion pattern is also of basic importance in the initial deepening stage of the depression. If the upper-air pattern is unfavourable (e.g. beneath left entrance and right exit zones, where there is convergence) the depression will fill. Note that to the rear of the second depression in Figure 7.25 upper convergence will encourage a polar outbreak through subsidence.

The development of a depression can also be considered in terms of energy transfers. A cyclone requires the conversion of potential into kinetic energy, and this is achieved by the upward (and poleward) motion of *warm* air. The rising warm air is driven by the vertical wind shear and by the superimposition of upper tropospheric divergence over a baroclinic zone. Intensification of this zone further strengthens the upper winds. The upper divergence allows surface convergence and pressure fall to occur simultaneously. Modern theory relegates the fronts to a quite subordinate role. They develop within depressions as narrow zones of intensified ascent, probably through the effects of cloud formation.

Recent research has identified a category of mid-latitude cyclones that develop and intensify rapidly, acquiring characteristics that resemble tropical hurricanes. These have been termed 'bombs' in view

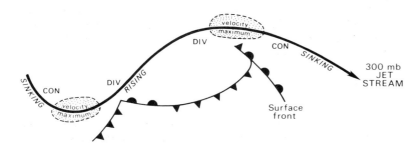

Figure 7.25 Model of the jet stream and surface fronts, showing zones of upper tropospheric divergence and convergence and the jet stream cores.

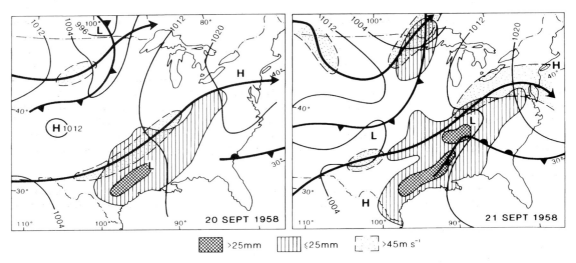

>25mm ▨ <25mm ▥ >45m s⁻¹ ⬚

Figure 7.26 The relations between surface fronts and isobars, surface precipitation (≤25 mm vertical hatching; >25 mm cross-hatching), and jet streams (wind speeds in excess of about 45 m s⁻¹ shown by stipple) over the United States on 20 September 1958 and 21 September 1958. This illustrates how the surface precipitation area is related more to the position of the jets than to that of the surface fronts. The air over the south-central United States was close to saturation, whereas that associated with the northern jet and the maritime front was much less moist.

Source: After Richter and Dahl 1958.

of their explosive rate of deepening; pressure falls of at least 24 mb/24 hr are observed. For example, the 'QE II storm', which battered the ocean liner *Queen Elizabeth II* off New York on 10 September 1978, developed a central pressure below 950 mb with hurricane-force winds and an eye-like storm centre within 24 hours (see Chapter 9C.2). These systems are observed mainly during the cold season off the east coast of the United States, off Japan, and over parts of the central and north-eastern North Pacific, in association with major baroclinic zones and close to strong gradients of sea-surface temperature. Synoptic analyses indicate that explosive cyclogenesis is favoured by an unstable lower troposphere and is often located downstream of a

travelling 500-mb-level trough. They are characterized by strong vertical motion, associated with a sharply defined level of non-divergence near 500 mb, and large-scale release of latent heat. Wind maxima in the upper troposphere, organized as jet streaks, serve to amplify the lower-level instability and upward motion. Further studies have revealed that *average* cyclonic deepening rates over the North Atlantic and North Pacific are about 10 mb/24 hr, or three times greater than over the continental United States (3 mb/24 hr). Hence, it is suggested that explosive cyclogenesis represents a more intense version of typical maritime cyclone development.

The movement of depressions is determined essentially by the upper westerlies and, as a rule of thumb, a depression centre travels at about 70 per cent of the surface geostrophic wind speed in the warm sector. Records for the United States indicate that the average speed of depressions is 32 km hr^{-1} in summer and 48 km hr^{-1} in winter. The higher speed in winter reflects the stronger westerly flow in response to a greater meridional temperature gradient. Shallow depressions are mainly steered by the direction of the thermal wind in the warm sector and hence their path closely follows that of the upper jet stream (see Chapter 6A.3). Deep depressions may greatly distort the thermal pattern, however, as a result of the northward transport of warm air and the southward transport of cold air. In such cases, the depression usually becomes slow-moving. The movement of a depression may be additionally guided by energy sources such as a warm sea surface, which generates cyclonic vorticity, or by mountain barriers. The depression may cross obstacles, such as the Rocky Mountains or the Greenland Ice Sheet, as an upper low or trough and subsequently redevelop, aided by the lee effects of the barrier or by fresh injections or contrasting air masses.

The location and intensity of storm tracks appears to be critically influenced by ocean-surface temperatures. Investigations for the North Pacific and the North Atlantic demonstrate interactions with the atmospheric circulation on a near-hemispheric scale. Figure 7.27B, for example, suggests that an extensive relatively warm surface in the north-central Pacific in the winter of 1971–2 caused a northward displacement of the westerly jet stream together with a compensating southward displacement over the western United States, bringing in cold air there. This pattern contrasts with that observed during the 1960s (see Figure 7.27A), when a persistent cold anomaly in the central Pacific, with warmer water to the east, led to frequent storm development in the intervening zone of strong temperature gradient. The associated upper airflow produced a ridge over western North America with warm winters in California and Oregon. Models of the global atmospheric circulation support the view that persistent anomalies of sea-surface temperature exert an important control on local and large-scale weather conditions.

H NON-FRONTAL DEPRESSIONS

Not all depressions originate as frontal waves. Tropical depressions are indeed mainly non-frontal and these are considered in Chapter 9. In middle and high latitudes, four types that develop in distinctly different situations are of particular importance and interest: the lee depression, the thermal low, the polar air depression, and the cold low.

1 The lee depression

Westerly airflow that is forced over a north–south mountain barrier undergoes vertical contraction over the ridge and expansion on the lee side. This vertical movement creates compensating lateral expansion and contraction, respectively. Hence there is a tendency for divergence and anticyclonic curvature over the crest, and convergence and cyclonic curvature in the lee of the barrier. Wave troughs may be set up in this way on the lee side of low hills (see Figure 5.11) as well as major mountain chains like the Rocky Mountains. The airflow characteristics and the size of the barrier determine whether or not a closed low-pressure system actually develops. Such depressions, which at least initially tend to remain 'anchored' by the barrier, are frequent in winter to the south of the Alps, when the mountains block the low-level flow of north-westerly airstreams. Fronts may occur in these depressions, but it is important to recognize that the low does not form as a wave along a frontal zone.

2 The thermal low

These lows occur almost exclusively in summer, resulting from intense daytime heating of continental areas. Figure 6.1C illustrates their vertical structure. The most impressive examples are the summer low-pressure cells over Arabia, the northern part of the Indian subcontinent and Arizona. The Iberian peninsula is another region commonly affected by such lows. They occur over south-western Spain on 40–60 per cent of days in July and August. Typically, their intensity is only 2–4 mb and they extend to about 750 mb, less than in other subtropical areas. The weather accompanying them is usually hot and dry, but if sufficient moisture is present the instability caused by heating may lead to showers and thunderstorms. Thermal lows normally disappear at night, when the heat source is cut off, but in fact those of India and Arizona persist.

3 Polar air depressions

Polar air depressions are a loosely defined class of mesoscale to subsynoptic-scale systems (a few

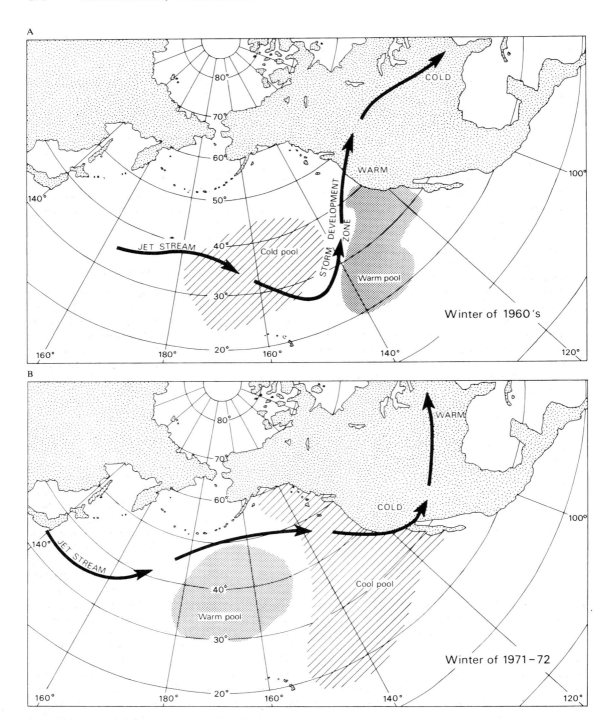

Figure 7.27 Generalized relationships between ocean-surface temperatures, jet-stream tracks, storm-development zones and land temperatures over the North Pacific and North America during (A) average winter conditions in the 1960s, and (B) the winter of 1971–2, as determined by J. Namias.

Source: After Wick 1973.

hundred kilometres across) with a lifetime of 1–2 days. On satellite imagery, they appear as a cloud spiral with one or several cloud bands, as a comma cloud (see Figure 7.18 and Plate 23), or as a swirl in cumulus cloud streets. They develop mainly in the winter months, when unstable mP or mA air currents stream equatorwards along the eastern side of a north–south ridge of high pressure, commonly in the rear of an occluding primary depression (Plate 24). They usually form within a baroclinic zone, e.g. near sea-ice margins, where there are strong sea-surface temperature gradients, and their development may be stimulated by an initial upper-level disturbance.

In the northern hemisphere, the comma cloud type (which is mainly a cold core disturbance of the middle troposphere) is more common over the North Pacific, while the spiral-form polar low occurs more often in the Norwegian Sea. The latter is a low-level warm core disturbance that may have a closed cyclonic circulation up to about 800 mb or may consist simply of one or more troughs embedded in the polar airflow. A key feature is the presence of an ascending, moist, south-westerly flow *relative* to the low centre. This organization accentuates the general instability of the cold airstream to give considerable precipitation, often as snow. Heat input to the cold air from the sea continues by night and day, so in exposed coastal districts showers may occur at any time.

In the southern hemisphere, polar low meso-cyclones appear to be most frequent in the transition seasons, as these are the times of strongest meridional temperature and pressure gradients. Also, over the Southern Ocean the patterns of occurrence and movement are more zonally distributed than in the northern hemisphere.

4 The cold low

The cold low (or *cold pool*) is usually most evident in the circulation and temperature fields of the middle troposphere. Characteristically, it displays symmetrical isotherms about the low centre. Surface charts may show little or no sign of these persistent systems, which are frequent over north-eastern North America and north-eastern Siberia. They probably form as the result of strong vertical motion and adiabatic cooling in occluding baroclinic lows along the Arctic coastal margins. Such lows are especially important during the Arctic winter in that they bring large amounts of medium and high cloud, which offsets radiational cooling of the surface.

Otherwise, they usually cause no 'weather' in the Arctic during this season. It is important to emphasize that tropospheric cold lows may be linked with either low- or high-pressure cells at the surface.

In middle latitudes, cold lows may also form during periods of low-index circulation pattern (see Figure 6.27) by the cutting-off of polar air from the main body of cold air to the north (these are sometimes referred to as *cut-off lows*). This gives rise to weather of polar air-mass type, although rather weak fronts may also be present. Such lows are commonly slow-moving and give persistent unsettled weather with thunder in summer. Heavy precipitation over Colorado in spring and autumn is often associated with cold lows.

I METEOROLOGICAL FORECASTS

National meteorological services perform a variety of activities in order to provide weather forecasts. The principal ones are data collection, the preparation of basic analyses and prognostic charts of atmosphere conditions for use by local weather offices, the preparation of short- and long-term forecasts for the public, special services for aviation, shipping, agricultural and other commercial and industrial users, and the issuance of severe weather warnings.

1 Data sources

The data required for forecasting and other services are provided by worldwide standard synoptic reports at 00, 06, 12, and 18 UTM (see Appendix 3), similar observations made hourly, particularly in support of national aviation requirements, upper-air soundings (at 00 and 12 UTM), satellite data, and other specialized networks such as radar stations for severe weather. Under the World Weather Watch programme, synoptic reports are made at some 4,000 land stations and by 7,000 ships (Figure 7.28). There are about 700 stations making upper-air soundings (temperature, pressure, humidity and wind) (Figure 7.29). These data are transmitted in code via teletype and radio links to regional or national centres and into the high-speed Global Telecommunications System (GTS) connecting World Weather Centres in Melbourne, Moscow and Washington and eleven Regional Meteorological Centres for redistribution. Some 184 member nations co-operate in this activity under the aegis of the World Meteorological Organization.

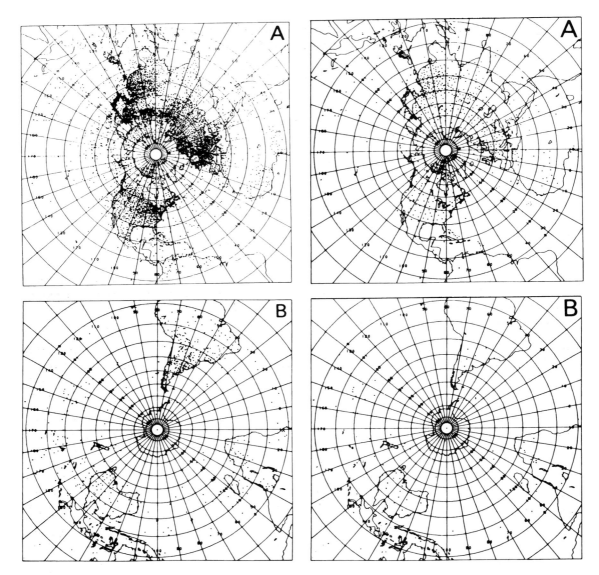

Figure 7.28 Synoptic reports from surface land stations and ships available over the Global Telecommunications System at the National Meteorological Center, Washington, DC.

Figure 7.29 Synoptic reports from upper-air sounding stations available over the Global Telecommunications System at the National Meteorological Center, Washington, DC.

Source: From A. D. Hecht, *Paleoclimate Analysis and Modeling*, Wiley, New York, 1985 (copyright © 1985 by John Wiley and Sons, Inc. Reproduced by permission).

Meteorological information has been collected operationally by satellites of the United States and Russia since 1965 and, more recently, by the European Space Agency, India and Japan. There are two general categories of weather satellite: polar orbiters providing global coverage twice per 24 hours in orbital strips over the poles (such as the United States' NOAA and TIROS series (see Plates 2 and 3) and the former USSR's Meteor); and geosynchronous satellites (such as the Geostationary Operational Environmental Satellites (GOES) and Meteosat), giving repetitive (30-minute) coverage of almost one-third of the earth's surface in low middle latitudes (Figure 7.30). Information on the atmos-

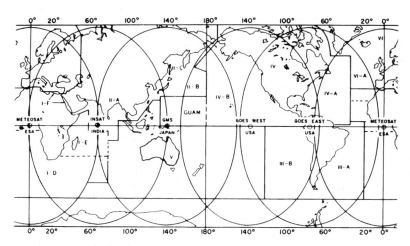

Figure 7.30 Coverage of geostationary satellites and WMO data collection areas (rectangular areas and numbers).

Source: Reproduced by courtesy of NOAA.

phere is collected as digital data or direct readout visible and infrared images of cloud cover and sea-surface temperature, but it also includes global temperature and moisture profiles through the atmosphere obtained from multi-channel infrared and microwave sensors, which receive radiation emitted from particular levels in the atmosphere. Additionally, satellites have a data collection system (DCS) that relays data on numerous environmental variables from ground platforms or ocean buoys to processing centres; GOES can also transmit processed satellite images in facsimile, and the NOAA polar orbiters have an automatic picture transmission (APT) system that is utilized at about 1,000 stations worldwide.

2 Forecasting

Modern forecasting did not become possible until weather information could be collected, assembled and processed rapidly. The first development came in the middle of the last century with the invention of telegraphy, which permitted immediate analysis of weather data by the drawing of synoptic charts. These were first displayed in Britain at the Great Exhibition of 1851. Sequences of weather change were correlated with barometric pressure patterns in both space and time by such workers as Fitzroy and Abercromby, but it was not until later that theoretical models of weather systems were devised – notably the Bjerknes' depression model described earlier (see Figure 7.7).

Forecasts are usually referred to as short-range (1–2 days), extended-range (3–5 days), medium-range (up to 6–10 days) and long-range (monthly or seasonal) outlooks. The first two can for present purposes be considered together.

a Short- and extended-range forecasting

Forecasting procedures developed up to the 1950s were based on synoptic principles but, since the 1960s, practices have been revolutionized by numerical forecasting methods and the adoption of 'nowcasting' techniques.

During the first half of the century, short-range forecasts were based on synoptic principles, empirical rules and extrapolation of pressure changes. The Bjerknes' model of cyclone development for middle latitudes and simple concepts of tropical weather (see Chapter 9) served as the basic tools of the forecaster. The relationship between the development of surface lows and highs and the upper-air circulation was worked out during the 1940s and 1950s by C.-G. Rossby, R. C. Sutcliffe and others, providing the theoretical basis of synoptic forecasting. In this way, the position and intensities of low- and high-pressure cells and frontal systems were predicted.

Since 1955 in the United States – and 1965 in the United Kingdom – routine forecasts have been based on numerical models. These predict the evolution of physical processes in the atmosphere by determinations of the conservation of mass, energy and momentum. The basic principle is that the rise or fall of surface pressure is related to mass convergence or divergence, respectively, in the overlying air column. This prediction method was proposed by L. F. Richardson, who, in 1922, made a laborious test calculation that gave very unsatisfactory results. The major reason for this lack of success was that the net convergence or divergence in an air column is a small residual term compared with the large values of convergence and divergence at different levels in the atmosphere (see Figure 5.7). Small errors arising from observational limitations may, therefore, have a considerable effect on the correctness of the analysis.

Numerical weather prediction (NWP) methods developed in the 1950s use a less direct approach. The first developments assumed a one-level barotropic atmosphere with geostrophic winds and hence no convergence or divergence. The movement of systems could be predicted, but not changes in intensity. Despite the great simplifications involved in the barotropic model, it has been used for forecasting 500 mb contour patterns. The latest techniques employ multi-level baroclinic models and include frictional and other effects; hence the basic mechanisms of cyclogenesis are provided for. It is noteworthy that *fields* of continuous variables, such as pressure, wind and temperature, are handled and that fronts are regarded as secondary, derived features. The vast increase in the number of calculations that these models necessitate has required a new generation of supercomputers to allow the preparation of a forecast map to keep sufficiently ahead of the weather changes!

Forecast practices in the major national centres are basically similar. However, the following specific information relates to the National Centers for Environmental Prediction (NCEP) in Washington, DC. Two types of global forecast are prepared each day; these are 3-day aviation forecasts and 6 to 10-day medium-range forecasts (MRF). A Global Data Assimilation System is used to prepare an analysis of the 00 GMT observations, incorporating as a 'first guess' the prior 6-hour forecast from the MRF model. Different interpolation methods are used to obtain smoothed, gridded data on temperature, moisture, wind and geopotential height for the surface at standard pressure levels (850, 700, 500, 400, 300, 250, 200 and 100 mb) over the globe. The NCEP currently has two basic prediction models: a spectral model with eighteen levels (from the boundary layer into the upper stratosphere) and 160 km horizontal resolution, which is integrated for up to 10 days, and a regionally applicable nested grid model with finer horizontal resolution. It should be noted that typically the computer time required increases several-fold when the grid spacing is halved. The essential forecast products are MSL pressure, temperature and wind velocity for standard pressure levels, 1,000–500 mb thickness, vertical motion and moisture content in the lower troposphere, and precipitation amounts.

Actual weather conditions are now commonly predicted using the Model Output Statistics (MOS) technique developed by the US National Weather Service. Rather than relating weather variables to the predicted pressure/height patterns and taking account of frontal models for example, a series of regression equations are developed for specific locations between the variable of interest and up to ten predictors calculated by the numerical models. Weather elements so predicted for numerous locations include daily maximum/minimum temperature, 12-hour probability of precipitation occurrences and precipitation amount, probability of frozen precipitation, thunderstorm occurrence, cloud cover, and surface winds. These forecasts are distributed as facsimile maps and tables to weather offices for local use.

In the United States, there are also separate centres with responsibilities for forecasting tropical storms (the National Hurricane Center in Miami for the Atlantic area, with warning centres in San Francisco for the eastern North Pacific and Honolulu for the central North Pacific) and also the National Severe Storms Forecasting Center in Kansas City, Missouri. The latter depends greatly on satellite and radar data to issue severe weather alerts.

The United Kingdom Meteorological Office (UKMO) has developed a new Unified Model, which is used in global and limited-area versions. The global model has a 0.855° latitude by 1.25° longitude horizontal grid and nineteen levels between the surface and 5 mb. All essential physical processes (diurnal radiation cycle, surface energy exchanges, cloud development and precipitation) are represented. Certain surface conditions are fixed, such as snow cover and sea ice extent, albedo, soil moisture and thermal capacity, while sea-surface temperatures are entered from daily observations. By repeated calculations for time steps of a few minutes, atmospheric motion, temperature, cloud cover, precipitation, etc. are forecast up to six days ahead. For a more limited area (80°W–40°E, 30–80°N), a 'fine mesh' model (–0.44° latitude, longitude grid) is used to forecast up to 36 hours ahead.

Errors in numerical forecasts arise from several sources. One of the most serious is the limited accuracy of the initial analyses due to data deficiencies.

Coverage over the oceans is sparse, and only a quarter of the possible ship reports may be received within 12 hours; even over land more than one-third of the synoptic reports may be delayed beyond 6 hours. However, satellite-derived information and aircraft reports can help to fill some gaps for the upper air. Another limitation is imposed by the horizontal and vertical resolution of the models and the need to parameterize subgrid processes such as cumulus convection. The small-scale nature of the turbulent motion of the atmosphere means that some weather phenomena are basically unpredictable, for example, the specific locations of shower cells in an unstable air mass. Greater precision than the 'showers and bright periods' or 'scattered showers' is impossible for next-day forecasts. The procedure for preparing a forecast is becoming much less subjective, although in complex weather situations the skill of the experienced forecaster still makes the technique almost as much an art as a science. Detailed regional or local predictions can only be made within the framework of the general forecast situation for the country and demand thorough knowledge of possible topographic or other local effects by the forecaster.

b 'Nowcasting'

Severe weather is typically short-lived (<2 hours) and, due to its mesoscale character (<100 km), it affects local/regional areas, necessitating site-specific forecasts. Included in this category are thunderstorms, flash floods, gust fronts, tornadoes, high winds especially along coasts, over lakes and mountains, heavy snow, and freezing precipitation. Many mesoscale models with grid resolutions of 15–30 km are now in use to study such phenomena. However, the development of radar networks (Plate D), new instruments and high-speed communication links has provided a means of issuing warnings of severe weather within the next hour. Several countries have recently developed integrated satellite and radar systems to provide information on the horizontal and vertical extent of thunderstorms, for example. Such data are supplemented by networks of automatic weather stations (including buoys) that measure wind, temperature and humidity. In addition, for detailed boundary layer and lower troposphere data, there is now an array of vertical sounders – acoustic sounders (measuring wind speed and direction from echoes created by thermal eddies), specialized (Doppler) radar measuring winds in clear air by returns either from insects

(3.5 cm wavelength radar) or from variations in the air's refractive index (10 cm wavelength radar). Nowcasting techniques use highly automated computers and image-analysis systems to integrate data from a variety of sources rapidly. Interpretation of the data displays requires skilled personnel and/or extensive software to provide appropriate information. The prompt warning of wind shear and downburst hazards at airports is one example of the importance of nowcasting procedures.

Overall, the greatest benefits from improved forecasting can be expected in aviation and the electric power industry for forecasts less than 6 hours ahead, in transportation, construction and manufacturing for 12–24-hour forecasts and in agriculture for 2–5-day forecasts. In terms of economic losses, the last category could benefit the most from more reliable and more precise forecasts.

c Medium-range forecasting

The current 6–10-day forecasts prepared by the Climate Analysis Center (CAC) in Washington, DC, are a 'blend' of the results of two numerical models: those produced by NCEP's Medium-Range Forecast (MRF) model and by the European Centre for Medium-Range Weather Forecasts (ECMWF) model. First, a composite mean 500 mb height and height anomaly prediction is produced, because the MRF model has a cold bias with heights in the middle and upper troposphere that are too low, whereas the ECMWF model has an opposite warm bias. Teleconnection patterns of 700 mb are examined to ensure internal consistency. Second, temperature and precipitation are predicted statistically using relationships of the kind illustrated in Figure 7.31. Temperature patterns over the contiguous United States are forecast from correlations with the predicted 6–10-day 700-mb height anomalies. Precipitation amounts derived from the MRF model are shown in three probability categories (above normal, near normal, below normal) (Figure 7.32).

In the United Kingdom Meteorological Office, 12-hourly forecasts are generated using a global model for two to five days ahead; these depict surface isobars, fronts and 1,000–500 mb thickness contours. Model output from several forecast centres is usually compared and, if wide differences are present, the forecaster has to use personal judgement and emphasize the low confidence in the forecast. Currently, forecast errors are large beyond Day 4. Numerical models incorporate the non-linear behaviour of atmospheric dynamics, and this feature is now being

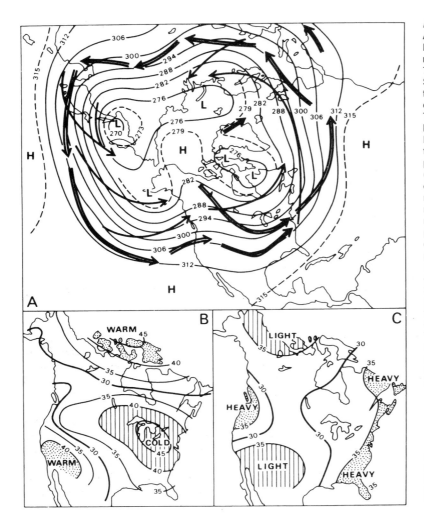

Figure 7.31 Forecast of North American weather for December 1985 made one month ahead. A: predicted 700 mb contours (g.p. dam). Solid arrows indicate main tracks of cyclones, open arrows of anticyclones, at sea level. The forecasting of such tracks has recently been discontinued. B and C: forecast average temperature (B) and average precipitation (C) probabilities (in per cent). There are three classes of temperature, above normal, normal and below normal, and similarly heavy, near normal and light for precipitation. Each of these classes is defined to occur 30 per cent of the time in the long run; near-normal temperature or moderate precipitation occur 30 per cent of the time in the long run; near-normal temperature or moderate precipitation occur 40 per cent of the time. The 30 per cent heavy lines indicate indifference (for any departure from average), but near-normal values are most likely in unshaded areas.

Source: From *Monthly and Seasonal Weather Outlook*, 39(23), 28 November 1985, Climate Analysis Center, NOAA, Washington, DC.

exploited in 'ensemble forecasts'. The numerical model is run in multiple predictions, each with minor differences in the initial conditions. The ensemble solutions can be grouped into clusters and compared with the ensemble mean so as to provide a guide to the likely probability of specific aspects of the forecast. A large degree of similarity (difference) suggests greater (lesser) confidence. Beyond five days, this approach seems particularly appropriate, although for some purposes a deterministic forecast may need to be chosen from the range of probabilities provided.

The methods of medium-range forecasting discussed above are unsuitable for predicting the probable trend of the weather for periods of a month or more, because they are concerned with individual synoptic disturbances with a life-cycle of about three to seven days. Theoretical considerations indicate that the limit of predictability using numerical techniques is less than 15 days. However, for large-scale features of the circulation, and for certain regimes such as blocking patterns, deterministic predictability may extend to 3–4 weeks. It is known, for example, that extended spells of particular types of weather are characteristic of mid-latitudes generally (see Chapter 8A.4).

d Long-range outlooks

There are three approaches to long-range forecasting: dynamical, analogue statistical, and current statistical methods. These are now outlined.

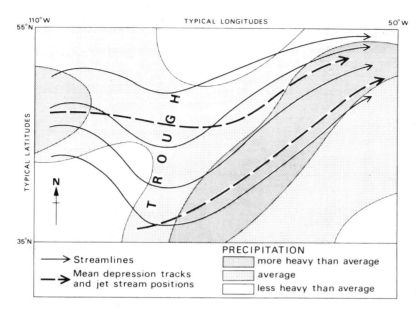

Figure 7.32 Generalized mean relationships between the westerly airflow pattern, depression tracks, jet-stream positions and precipitation zones in mid-latitudes in the northern hemisphere.

Source: After W. Klein; from Harman 1971.

1 *Dynamical methods*. Dynamical 30-day forecasts are being attempted at various centres using longer integrations of the medium-range models. For example, 108 contiguous 30-day forecasts for 14 December 1986 to 31 March 1987 were prepared by the United States National Meteorological Center. It was found that the model generally performed better than a persistence forecast (of existing conditions), but the 30-day mean 500 mb circulation of the northern hemisphere is estimated best from the mean of the first seven to ten days. In particular, the unexpected development of blocking situations reduced the skill on several occasions. The sensitivity of model forecasts to external forcings such as anomalies of sea-surface temperature, soil moisture, snow cover and sea ice are being investigated.

2 *Analogue methods*. A statistical approach to monthly forecasting developed in Germany and Britain is based on the principle that sequences of weather events may tend to follow a similar course if the initial conditions are almost identical. The problem is then to find a period with weather conditions as closely analogous as possible to the present one and to use the past sequence of events as a guide to the future. The analogues are matched from a record of anomaly patterns for temperature and air pressure in a given month, and of sequences of *weather types*.

The latter are actually types of pressure pattern or airflow over the country (see Chapter 8A.3). Each category tends to be associated with a particular type of weather. Composite sea-level pressure and 500 mb maps are prepared for two sets of ten cases (months) that experienced extreme states (i.e. wet/dry, warm/cold) in the subsequent month. These composites and their difference are compared with the current month and used as the basis for predicting the month ahead. The difficulty in analogue prediction arises from the fact that no two patterns or sequences of weather are ever identical. In the preparation of the forecast, therefore, many factors that can affect weather trends, such as sea temperatures and the extent of snow cover, have to be taken into account.

3 *Current statistical methods*. The United States' NCEP currently issues 30-day forecasts twice monthly using a method that has two principal steps. First, individual mean 500 mb height and anomaly forecasts for Days 1–5 and 6–10 are prepared by NCEP and ECMWF models. The likely utility of these forecasts is assessed from statistical studies; for example, in winter and summer the best correlations for the following 30-day mean period are given by the previous 10–28 days, whereas in the autumn and spring shorter intervals of a week or less give the highest correlations to the following month. Thus the

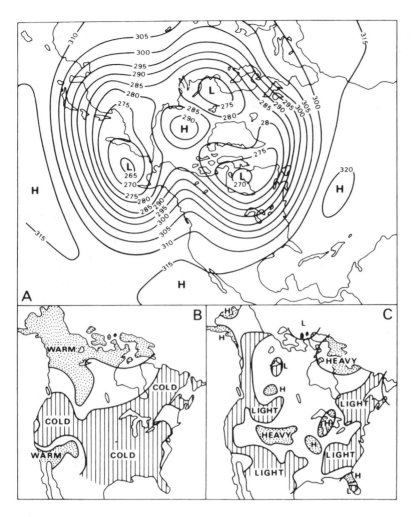

Figure 7.33 Actual North American weather for December 1985 (cf. Figure 7.31). Observed 700 mb contours (5 g.p. dam) corresponding to Figure 7.31A. B and C: observed temperature (B) and precipitation (C), corresponding to Figure 7.31B and C.

Source: From *Monthly and Seasonal Weather Outlook*, 40(1), 1986, Climate Analysis Center, NOAA, Washington, DC.

forecast takes account of the degree of persistence in the circulation pattern expected from one month to the next, the recognition of probable changes of regime in the large-scale atmospheric circulation, the possible effects of features such as snow cover and sea-surface temperature anomalies, and statistical records of the typical locations of troughs and ridges at that season. The second step involves deriving predicted surface temperature and precipitation values from the predicted 700 mb height anomalies for the month. Statistical analyses for the United States by W. H. Klein are the cornerstone of this procedure. These are based partly on analogues where the weather experienced during past cases

with similar 700 mb patterns occurring in the same season serves as a guide to the conditions expected to recur. Figure 7.32 shows that more than average precipitation is probable ahead of the trough where there is maximum vorticity advection, especially in association with an active jet stream, whereas less than average precipitation is usual to the rear of the trough axis. With a low zonal index in the westerlies, long waves may persist in more or less stationary positions for periods of several weeks at a time, or even longer, and their mean positions are, therefore, reflected in surface patterns of temperature and precipitation (see also Chapter 8B.1).

The 30-day outlook for the United States is presented as maps of the probability of three equally likely categories of temperature (near-normal, above/below normal) and precipitation (near-average, above/below the median) (see Figure 7.31), together with tables for many cities.

The 90-day seasonal outlooks prepared by the US Climate Analysis Center are based primarily on persistence statistics that take account of trends and tendencies linking the circulation and weather patterns of the preceding several seasons. Overall, winter is more easily predictable and spring is the most variable season. Recurrent, large-scale abnormalities in climate, such as the El Niño phenomenon (see Chapter 9G.4), may prove useful in seasonal forecasting in view of their worldwide *teleconnections* (i.e. distant areas of the globe where the climatic anomalies show either the same, or almost opposite, tendencies).

Figure 7.33A illustrates the observed height field corresponding to Figure 7.31A for December 1985, showing that the pattern is well represented on the forecast chart. Figure 7.33B and C show that in this case, as is usual, the temperature forecasts are more reliable than those for precipitation.

Long-range forecasts prepared at the UK Meteorological Office have not been published since 1981. However, newer statistical procedures have been developed in the 1980s. They involve a series of steps. The sea-level pressure anomalies over the North Atlantic and Europe have been classified for 1899–1978 into six circulation types for each two-month interval (January–February, etc.) throughout the year. From these data, the probability of the different circulation types succeeding one another has been calculated. The likelihood of each outcome is determined according to an optimum combination of climatic states described by sea-surface temperature anomalies in fifteen tropical and mid-latitude areas, and patterns of sea-level pressure and 1,000–500 mb thickness anomalies over mid-latitudes of the northern hemisphere. On these bases, monthly forecasts of temperature and precipitation are made for ten areas of the British Isles. Statistical techniques are a useful complement to dynamical methods, since they are constrained to be realistic by past data. More recently, the use of an ensemble of extended-range numerical forecasts has been found to have more skill than statistical predictions, and it appears likely that efforts will be concentrated in this direction in the future.

The problems facing the long-range forecasters need to be recognized before criticisms are levelled at them. The complexity of the atmosphere's behaviour makes tentative wording of their predictions necessary and occasional failures inevitable at present.

SUMMARY

Ideal air masses are defined in terms of barotropic conditions, where isobars and isotherms are assumed to be parallel to each other and to the surface. The character of an air mass is determined by the nature of the source area, changes due to air-mass movement, and its age. On a regional scale, energy exchanges and vertical mixing lead to a measure of equilibrium between surface conditions and those of the overlying air, particularly in quasi-stationary high-pressure systems. Air masses are conventionally identified in terms of temperature characteristics (arctic, polar, tropical) and source region (maritime, continental). Primary air masses originate in regions of semi-permanent anticyclonic subsidence over extensive surfaces of like properties. Cold air masses originate either in winter continental anticyclones (Siberia and Canada), where snow cover promotes low temperatures and stable stratification, or over high-latitude sea ice. Some sources are seasonal, like Siberia; others are permanent, such as Antarctica. Warm air masses originate either as shallow tropical continental sources in summer or as deep, moist layers over tropical oceans. Air-mass movement causes stability changes by thermodynamic processes (heating/cooling from below and moisture exchanges) and by dynamic processes (mixing, lifting/subsidence), producing secondary air masses (e.g. mP air). The age of an air mass determines the degree to which it has lost its identity as the result of mixing with other air masses and of vertical exchanges with the underlying surface.

Air-mass boundaries give rise to fronts or baroclinic zones a few hundred kilometres wide. The classical (Norwegian) theory of mid-latitude cyclones considers that fronts are a key feature of their formation and life-cycle. Newer models show that rather than the classical frontal occlusion process, the warm front may become bent-back with warm air seclusion within the polar airstream. Depressions tend to form along major
continued

frontal zones – the polar fronts of the North Atlantic and North Pacific regions and of the southern oceans. Less well-defined arctic fronts lie poleward and there are other seasonal frontal zones, as in the Mediterranean. Air masses and frontal zones move polewards/equatorwards in summer/winter.

Newer cyclone theories regard fronts as rather incidental. Cloud bands and precipitation areas are associated primarily with conveyor belts of warm air. Divergence of air in the upper troposphere is essential for large-scale uplift and low-level convergence. Surface cyclogenesis is therefore favoured on the eastern limb of an upper wave trough. 'Explosive' cyclogenesis appears to be associated with strong wintertime gradients of sea-surface temperature. Cyclones are basically steered by the quasi-stationary long (Rossby) waves in the hemispheric westerlies, the positions of which are strongly influenced by surface features (major mountain barriers and land/sea-surface temperature contrasts). Upper baroclinic zones are associated with jet streams at 300–200 mb, which also follow the long-wave pattern.

The idealized weather sequence in an eastward-moving frontal depression involves increasing cloudiness and precipitation with an approaching warm front; the degree of activity depends on whether the warm-sector air is rising or not (ana- or kata-fronts, respectively). The following cold front is often marked by a narrow band of convective precipitation, but rain both ahead of the warm front and in the warm sector may also be organized into locally intense mesoscale cells and bands due to the 'conveyor belt' of air in the warm sector.

Some low-pressure systems are essentially non-frontal. These include the lee depression encountered in the lee of mountain ranges; thermal lows due to summer heating; the polar air depression commonly formed in an outbreak of maritime arctic air over oceans; and the upper cold low, which is often a cut-off system in upper wave development or an occluded mid-latitude cyclone in the Arctic.

Analysis and forecasting of surface and upper-air weather maps is now highly automated in many national weather services. Short- and extended-range forecasting is based primarily on numerical prediction models. Automated integration of satellite and radar data is being widely developed for 'nowcasting' of severe weather events. Long-range outlooks are primarily statistical, taking account of possible effects of surface conditions on the large-scale circulation structure.

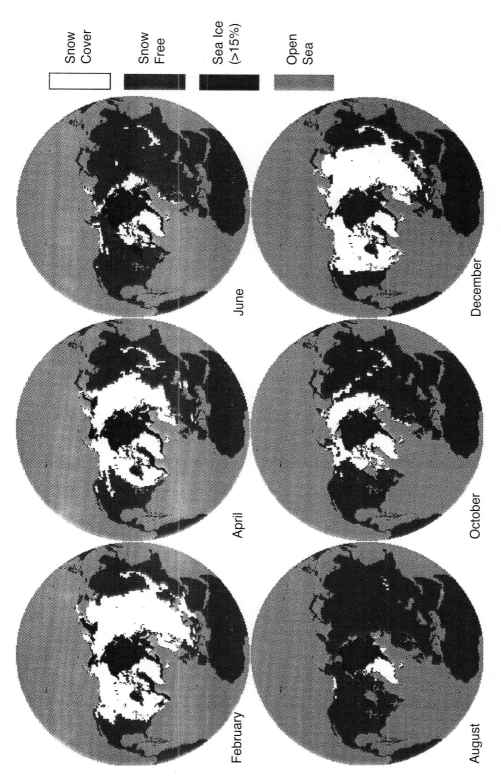

Plate A Average snow cover and sea ice extent in the northern hemisphere for the months of February, April, June, August, October and December derived from weekly data for the period 1978–95. *Source: National Snow and Ice Data Center (NSIDC), Boulder, Colorado. Courtesy National Snow and Ice Data Center.*

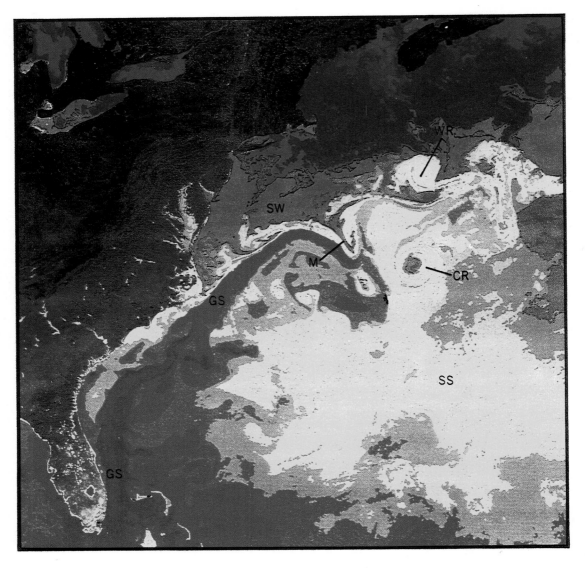

Plate B False-colour satellite image of the western North Atlantic indicating surface water temperatures from cold to warm (blue, green, yellow, red). Features of interest include the Gulf Stream (GS), a Gulf Stream meander (M), a cyclonic cold-core ring (CR) and an anticyclonic warm-core eddy (WR). *Source: Data of NOAA/NESDIS/NCDC/SDSD. Courtesy of Otis B. Brown, Robert Evans and M. Carle, University of Miami; Rosenstiel School of Marine and Atmospheric Science, Florida.*

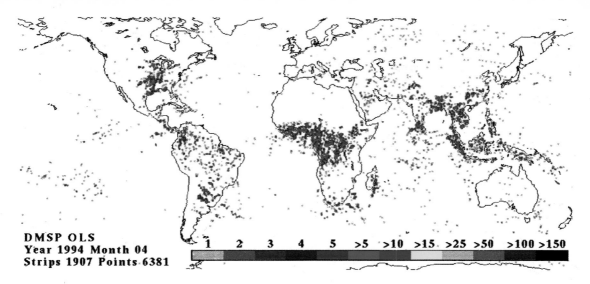

DMSP OLS
Year 1994 Month 04
Strips 1907 Points 6381

1 2 3 4 5 >5 >10 >15 >25 >50 >100 >150

Plate C Global distribution of lightning flashes recorded near local midnight by the visible-band Optical Line Scanner of the US Defense Meteorological Satellite Program during April 1994. *Source: Courtesy of Dr S. Goodman, NASA, Marshall Space Flight Center, Huntsville, Alabama.*

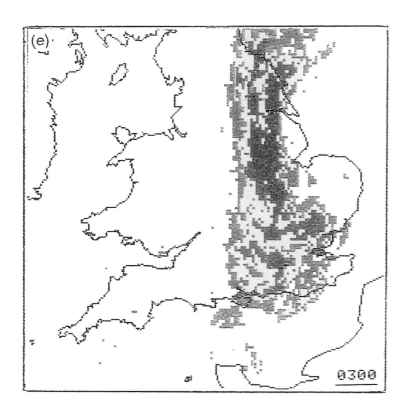

Plate D Instantaneous rainfall intensities (mm/hr), measured by radar, associated with a complex double cold front at 0300 on 23 September 1993 over eastern England. This rain belt had reached its maximum activity some 3–6 hours earlier, at which time local rainfall intensities of more than 32 mm/hr were observed. Green represents less than 2, yellow 2–4, red 4–16 and purple 16–32 mm/hr. *Source: UK weather radar network. Pike 1993, Fig. 3e, p. 204. Reprinted from the Meteorological Magazine by permission of the Meteorological Office. Crown copyright ©.*

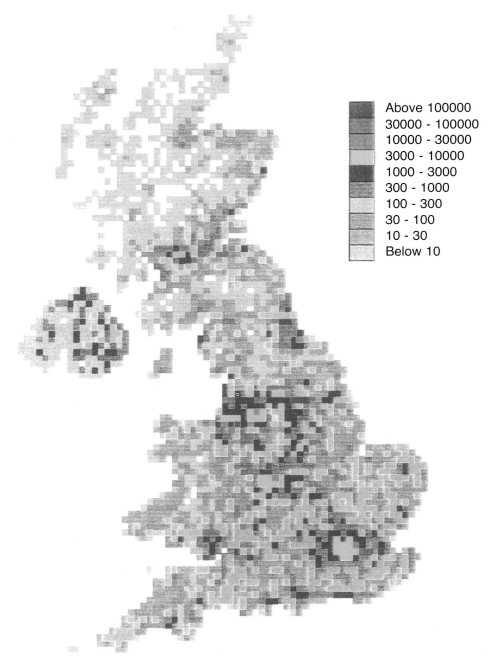

Plate E Total sulphur dioxide emissions (metric tonnes) over the United Kingdom for 1987. Data are shown on 10 x 10 km squares and, particularly, the areas of high emissions from coal-fired power stations in the East Midlands and South Yorkshire should be noted.
Source: Warren Spring Laboratory, Stevenage. Reprinted from the Meteorological Magazine by permission of the Meteorological Office. Crown copyright ©.

8

Weather and climate in temperate and high latitudes

In the two preceding chapters, the general structure of the circulation and air-mass characteristics in middle latitudes have been outlined and the behaviour and origin of extratropical depressions examined. The direct contribution of pressure systems to the daily and seasonal variability of weather in the westerly wind belt is quite apparent to inhabitants of the temperate lands. Nevertheless, there are equally prominent contrasts of regional climate in mid-latitudes that reflect the interaction of geographical and meteorological factors. This chapter gives a selective synthesis of weather and climate in several extratropical regions, drawing mainly on the principles already presented. The climatic conditions of the polar and subtropical margins of the westerly wind belt and the polar regions themselves are examined in the final sections of the chapter. As far as possible, different themes are used to illustrate some of the more significant aspects of the climate in each area.

A EUROPE

1 Pressure and wind conditions

The principal features of the mean North Atlantic pressure pattern are the Icelandic Low and the Azores High. These are present at all seasons (see Figure 6.10), although their location and relative intensity change considerably. The upper flow in this sector undergoes little seasonal change in pattern, but the westerlies decrease in strength by over half from winter to summer. The other major pressure system influencing European climates is the Siberian winter anticyclone, the occurrence of which is intensified by the extensive winter snow cover and the marked continentality of Eurasia. Atlantic depressions frequently move towards the Norwegian or Mediterranean Seas in winter, but if they travel due east they occlude and fill long before they can penetrate into the heart of Siberia. Thus the Siberian high pressure is quasi-permanent at this season, and when it extends westwards severe conditions affect much of Europe. In summer, pressure is low over all of Asia and depressions from the Atlantic tend to follow a more zonal path. Although the depression tracks over Europe do not shift polewards in summer (as a result of the local *southward* displacement of the Atlantic Arctic Front), the depressions at this season are rather less intense and the diminished air-mass contrasts produce weaker fronts.

Wind velocities over Western Europe bear strong relationships to the occurrence and movement of depressions. The strongest winds occur on coasts exposed to the north-west airflow, which follows the passage of frontal systems, or at constricted topographic locations that guide the movement of depressions or funnel airflow into them (Figure 8.1). In the latter regard, the Carcassonne Gap in south-west France provides a preferred southern route for depressions moving eastwards from the Atlantic, and the Rhône and Ebro valleys are funnels for strong airflow into the rear of depressions lying in the western Mediterranean, generating the mistral and cierzo winds, respectively, in winter (see C.1, this chapter). In all locations of Western Europe, the mean velocity of winds on hilltops is at least 100 per cent greater than that in more sheltered locations: winds in open terrain are 25–30 per cent more intense on average than those in sheltered locations; and coastal wind velocities are at least 10–20 per cent less than those over the adjacent sea areas (see Figure 8.1). Although mean wind velocity figures are of some use, for example in making estimates of potential

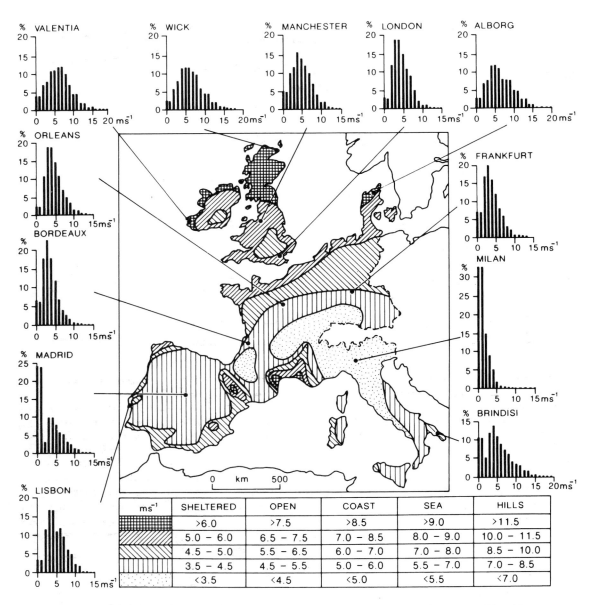

ms⁻¹	SHELTERED	OPEN	COAST	SEA	HILLS
>6.0	>7.5	>8.5	>9.0	>11.5
5.0 – 6.0	6.5 – 7.5	7.0 – 8.5	8.0 – 9.0	10.0 – 11.5
4.5 – 5.0	5.5 – 6.5	6.0 – 7.0	7.0 – 8.0	8.5 – 10.0
3.5 – 4.5	4.5 – 5.5	5.0 – 6.0	5.5 – 7.0	7.0 – 8.5
<3.5	<4.5	<5.0	<5.5	<7.0

Figure 8.1 Average wind velocities (m s⁻¹) over Western Europe, measured 50 m above ground level for sheltered terrain, open plains, sea coast, open sea and hilltops. Frequencies (per cent) of wind velocities for twelve locations are shown.

Source: From Troen and Petersen 1989.

wind power generation (on the assumption that a wind turbine can take advantage of 20–30 per cent of available wind power), for many human activities it is the occurrence of infrequent strong winds, which may occur with very differing frequencies at different locations, that is of especial significance.

2 Oceanicity and continentality

Winter temperatures in north-west Europe are some 11°C (20°F) or more above the latitudinal average (see Figure 2.18), a fact usually attributed solely to the presence of the North Atlantic Current. There

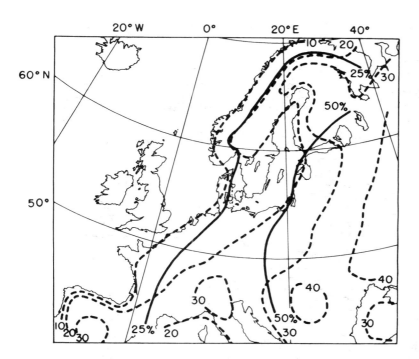

Figure 8.2 Continentality in Europe. The indices of Gorczynski (dashed) and Berg (solid) are explained in the text.

Source: Partly after Blüthgen 1966.

is, however, a complex interaction between the ocean and the atmosphere. The current, which originates from the Gulf Stream off Florida strengthened by the Antilles Current, is primarily a wind-driven current initiated by the prevailing south-westerlies. It flows at a velocity of 16 to 32 km per day and thus, from Florida, the water takes about eight or nine months to reach Ireland and about a year to reach Norway (see Chapter 6D.1). The south-westerly winds transport both sensible and latent heat acquired over the western Atlantic towards Europe, and although they continue to gain heat supplies over the north-eastern Atlantic, this local warming arises in the first place through the drag effect of the winds on the warm surface waters. Warming of air masses over the north-eastern Atlantic is mainly of significance when polar or arctic air flows south-eastwards from Iceland. For example, the temperature in such airstreams in winter may rise by 9°C between Iceland and northern Scotland. By contrast, maritime tropical air cools on average by about 4°C between the Azores and south-west England in winter and summer. One very evident effect of the North Atlantic Current is the absence of ice around the Norwegian coastline. However, as far as the climate of north-western Europe is concerned the primary

factor is the occurrence of prevailing *onshore* winds transferring heat into the area.

The influence of maritime air masses can extend deep into Europe because there are few major topographic barriers to airflow and because of the presence of the Mediterranean Sea. Hence the change to a more continental climatic regime is relatively gradual except in Scandinavia, where the mountain spine produces a sharp contrast between western Norway and Sweden. There are numerous indices expressing this continentality, but most are based on the annual range of temperature. Gorczynski's continentality index (K) is:

$$K = 1.7 \frac{A}{\sin \theta} - 20.4$$

where A is the annual temperature range (°C) and θ is the latitude angle. (The expression assumes that the annual range in solar radiation increases with latitude, but in fact the range is a maximum around 55°N). K is scaled to range from 0 at extreme oceanic stations to 100 at extreme continental stations. However, station values occasionally fall outside these limits. Some values in Europe are London 10, Berlin 21, and Moscow 42. Figure 8.2 shows the variation of this index over Europe.

A very different approach, by Berg, relates the frequency of continental air masses (C) to that of all air masses (N) as an index of continentality, i.e. K = C/N (per cent). Figure 8.2 shows that non-continental air occurs at least half the time over Europe west of 15°E as well as over Sweden and most of Finland.

A further illustration of maritime and continental regimes is provided by a comparison of Valentia (Eire), Bergen and Berlin (Figure 8.3). Valentia has a winter rainfall maximum and equable temperatures as a result of its oceanic situation, whereas Berlin has a considerable temperature range and a summer maximum of rainfall. A theoretically ideal 'equable' climate has been defined as one with a mean temperature of 14°C in all months of the year. Bergen receives larger rainfall totals due to orographic intensification and has a maximum in autumn and winter, its temperature range being intermediate between the other two. Such averages convey only a very general impression of climatic characteristics, and therefore British weather patterns will now be examined in more detail.

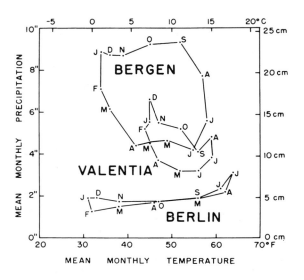

Figure 8.3 Hythergraphs for Valentia (Eire), Bergen and Berlin. Mean temperature and precipitation totals for each month are plotted.

3 British airflow patterns and their climatic characteristics

The daily weather maps for the British Isles sector (50–60°N, 2°E–10°W) from 1873 to the present day have been classified by H. H. Lamb according to the airflow direction or isobaric pattern. He identifies seven major categories: westerly (W), north-westerly (NW), northerly (N), easterly (E) and southerly (S) types – referring to the compass directions from which the airflow and weather systems

Table 8.1 General weather characteristics and air masses associated with Lamb's 'Airflow Types' over the British Isles.

Type	Weather conditions
Westerly	Unsettled weather with variable wind directions as depressions cross the country. Mild and stormy in winter, generally cool and cloudy in summer (mP, mPw, mT).
North-westerly	Cool, changeable conditions. Strong winds and showers affect windward coasts especially, but the southern part of Britain may have dry, bright weather (mP, mA).
Northerly	Cold weather at all seasons, often associated with polar lows. Snow and sleet showers in winter, especially in the north and east (mA).
Easterly	Cold in the winter half-year, sometimes very severe weather in the south and east with snow or sleet. Warm in summer with dry weather in the west. Occasionally thundery (cA, cP).
Southerly	Warm and thundery in summer. In winter, it may be associated with a low in the Atlantic, giving mild damp weather especially in the south-west, or with a high over central Europe, in which case it is cold and dry (mT, or cT, summer; mT or cP, winter).
Cyclonic	Rainy, unsettled conditions, often accompanied by gales and thunderstorms. This type may refer either to the rapid passage of depressions across the country or to the persistence of a deep depression (mP, mPw, mT).
Anticyclonic	Warm and dry in summer, occasional thunderstorms (mT, cT). Cold and frosty in winter with fog, especially in autumn (cP).

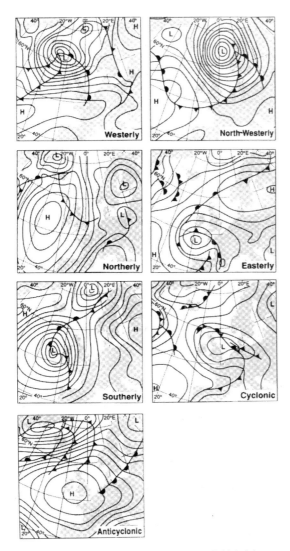

Figure 8.4 Synoptic situations over the British Isles classified according to the primary airflow types of H. H. Lamb.

Source: Lamb; O'Hare and Sweeney 1993. From *Geography* **78**(1). Copyright © The Geographical Association and G. O'Hare.

are moving – and cyclonic (C) or anticyclonic (A) types when depressions or a high-pressure cell, respectively, dominate the weather map (Figure 8.4).

In theory, each category should produce a characteristic type of weather, depending on the season, and the term *weather type* is sometimes used to convey this idea. A few studies have been made of the *actual* weather conditions occurring in different localities

with specific isobaric patterns – a field of study known as *synoptic climatology* – but the term *airflow type* seems preferable for Lamb's categories. The general weather conditions and air masses that are likely to be associated with a particular airflow type over the British Isles are summarized in Table 8.1.

On an annual basis, the most frequent airflow type is westerly; including cyclonic and anticyclonic sub-types, it has a 35 per cent frequency in December–January and is almost as frequent in July–September (Figure 8.5). The minimum occurs in May (15 per cent), when northerly and easterly types reach their maxima (about 10 per cent each). Pure cyclonic patterns are most frequent (13–17 per cent) in July–August and anticyclonic patterns in June and September (20 per cent); cyclonic patterns have ≥10 per cent frequency in all months and anticyclonic patterns ≥13 per cent. Figure 8.5 illustrates the mean daily temperature in central England and the mean daily precipitation over England and Wales for each type in the mid-season months for 1861–1979.

The monthly frequency of the different air mass types over Great Britain was analysed by J. Belasco for 1938–49. There is a clear predominance of north-westerly to westerly polar maritime (mP and mPw) air, which has a frequency of 30 per cent or more over south-east England in all months except March. The maximum frequency of mP air at Kew (London) is 33 per cent (with a further 10 per cent mPw) in July. The proportion is even greater in western coastal districts, with mP and mPw occurring in the Hebrides, for example, on at least 38 per cent of days throughout the year.

North-westerly mP airstreams produce cool, showery weather at all seasons, especially on windward coasts. The air is unstable with cumuliform cloud, although inland in winter and at night the clouds disperse, giving low night temperatures. Over the sea, however, heating of the lower air continues by day and night in winter months, so showers and squalls can occur at any time, and these may affect windward coastal areas. The average daily mean temperatures with mP air are within about ±1°C of the seasonal means in winter and summer, depending on the precise track of the air. More extreme conditions occur with mA air, the temperature departures at Kew being approximately –4°C in summer and winter. The visibility in mA air is usually very good. The contribution of mP and mA air masses to the mean annual rainfall over a five-year period at three stations in northern England and North Wales is given in Table 8.2, although it

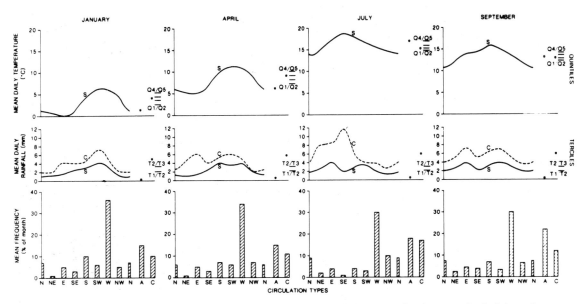

Figure 8.5 Average climatic conditions associated with Lamb's circulation types for January, April, July and September, 1861–1979. Top: mean daily temperature (°C) in central England for the straight (S) airflow types; at the right side are the quintiles of mean monthly temperature (i.e. Q1/Q2 = 20 per cent, Q4/Q5 = 80 per cent). Middle: mean daily rainfall (in millimetres) over England and Wales for the straight (S) and cyclonic (C) subdivisions of each type and terciles of the mean values (i.e. T1/T2 = 33 per cent, T2/T3 = 67 per cent). Bottom: mean frequency (per cent) for each circulation type, including anticyclonic (A) and cyclonic (C).

Source: After Storey 1982 (by permission of the Royal Meteorological Society).

should be noted that both air masses may also be involved in non-frontal polar lows. Over much of southern England, and in areas to the lee of high ground, northerly and north-westerly airstreams usually give clear sunny weather with few showers. There is some indication of this in Table 8.2, for at Rotherham, in the lee of the Pennines, the percentage of the rainfall occurring with mP air is much lower than at Squires Gate on the west coast. Maritime polar air that has followed a more

southerly cyclonic track over the Atlantic, approaching Britain from the south-west, or air that is moving northwards ahead of a depression, has surface properties intermediate with mT air.

Maritime tropical air commonly forms the warm sector of depressions moving from between west and south towards Britain. The weather is unseasonably mild and damp with mT air in winter. There is generally a complete cover of stratus or stratocumulus cloud and drizzle or light rain

Table 8.2 Percentage of the annual rainfall (1956–60) occurring with different synoptic situations.

Station	Synoptic categories								
	Warm front	Warm sector	Cold front	Occlusion	Polar low	mP	cP	Arctic	Thunderstorm
Cwm Dyli (99 m)*	18	30	13	10	5	22	0.1	0.8	0.8
Squires Gate (10 m)†	23	16	14	15	7	22	0.2	0.7	3
Rotherham (21 m)‡	26	9	11	20	14	15	1.5	1.1	3

Source: After Shaw 1962, and R. P. Mathews (unpublished).
Notes: *Snowdonia.
     †On the Lancashire coast (Blackpool).
     ‡In the Don Valley, Yorkshire.

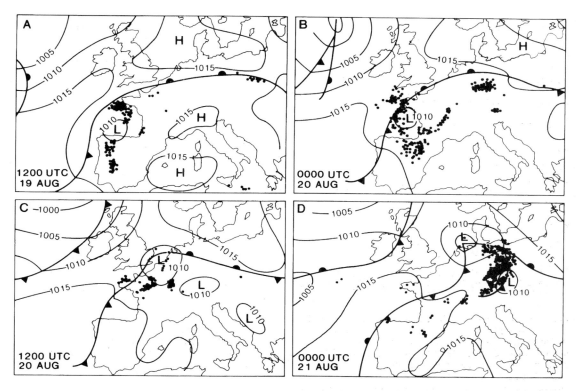

Figure 8.6 Distribution of thunderstorms over Western Europe during the period 19–21 August 1992 (storms shown for the 4-hour period preceding the times given). A small depression formed over the Bay of Biscay and moved eastwards along the boundary of the warm air, developing a strong squall line.

Source: Blackall and Taylor 1993. Reprinted from the *Meteorological Magazine* (Crown copyright ©) by permission of the Controller of Her Majesty's Stationery Office.

(formed by coalescence) may occur, especially over high ground, where low cloud produces hill fog. The clearance of cloud on nights with light winds readily cools the moist air to its dew point, forming mist and fog. Table 8.2 shows that a large proportion of the annual rainfall is associated with warm-front and warm-sector situations and therefore is largely attributable to convergence and frontal uplift within mT air. In summer the cloud cover with this air mass keeps temperatures closer to average than in winter; night temperatures tend to be high, but daytime maxima remain rather low.

During the summer, 'plumes' of warm, moist mT air may spread northwards from the vicinity of Spain into Western Europe. This air is very unstable, having a significant vertical wind shear and a wet-bulb potential temperature that may exceed 18°C. Instability may be increased if cooler Atlantic air advects under the plume from the west. The leading (i.e. northern) edge of the plume is commonly the location for organized systems of thunderstorms to develop in Britain and other parts of Western Europe. Occasionally, depressions develop on the front and move eastwards, bringing widespread storms to the region (Figure 8.6).

True continental polar air affects the British Isles only between December and February and even then it is relatively infrequent. Mean daily temperatures are well below average and maxima rise to only a degree or so above freezing point. The air mass is basically very dry and stable (see easterly type in January, Figure 8.4) but a track over the central part of the North Sea supplies sufficient heat and moisture to cause showers, often in the form of snow, over eastern England and Scotland. In total this provides only a very small contribution to the annual precipitation, as Table 8.2 shows, and on the west coast the weather is generally clear. A transitional cP–cT type of air mass reaches Britain from south-east Europe in all seasons, although less frequently in summer. Such airstreams are dry and stable.

Continental tropical air occurs on average about one day per month in summer, which accounts for the rarity of summer heatwaves, since these south or south-east winds bring hot, settled weather. The lower layers are stable and the air is commonly hazy, but the upper layers tend to be unstable and occasionally intense surface heating may trigger off a thunderstorm (see southerly cyclonic type in July, Figure 8.4). In winter, such modified cT air sometimes reaches Britain from the Mediterranean, bringing fine, hazy, mild weather.

4 Singularities and natural seasons

Most popular weather lore expresses the belief that each season has its own weather (for example, in England, 'February fill-dyke', 'April showers'), and ancient adages suggest that even the sequence of weather may be determined by the conditions established on a given date (e.g. 40 days of wet or fine weather following St Swithin's day, 15 July, in England). Some of these ideas are quite fallacious, but others contain more than a grain of truth if properly interpreted.

The tendency for a certain type of weather to recur with reasonable regularity around the same date is termed a *singularity*. Many calendars of singularities have been compiled, particularly in Europe, but the early ones (which concentrated upon anomalies of temperature or rainfall) did not prove very reliable. Greater success has been achieved by studying singularities of circulation pattern, and catalogues have been prepared for the British Isles by Lamb and for Central Europe by Flohn and Hess and Brezowsky. Lamb's results are based on calculations of the daily frequency of the airflow categories between 1898 and 1947, some examples of which are shown in Figure 8.7. A noticeable feature is the infrequency of the westerly type in spring, the driest season of the year in the British Isles and also in northern France, northern Germany and in the countries bordering the North Sea. The European catalogue is based on a classification of large-scale patterns of airflow in the lower troposphere (*Grosswetterlage*) over Central Europe originally proposed by F. Baur.

Some of the European singularities that occur most regularly are as follows:

1 A sharp increase in the frequency of westerly and north-westerly type over Britain takes place

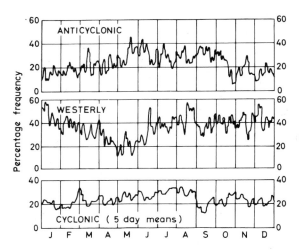

Figure 8.7 The percentage frequency of anticyclonic, westerly and cyclonic conditions over Britain, 1898–1947.
Source: After Lamb 1950.

about the middle of June. These invasions of maritime air also affect Central Europe, and this period is sometimes referred to as the beginning of the European 'summer monsoon'.

2 About the second week in September, Europe and Britain are affected by a spell of anticyclonic weather. This may be interrupted by Atlantic depressions, giving stormy weather over Britain in late September, although anticyclonic conditions again affect Central Europe at the end of the month and Britain during early October.

3 A marked period of wet weather often affects Western Europe and also the western half of the Mediterranean at the end of October, whereas the weather in Eastern Europe generally remains fine.

4 Anticyclonic conditions return to Britain and affect much of Europe about mid-November, giving rise to fog and frost.

5 In early December, Atlantic depressions push eastwards to give mild, wet weather over most of Europe.

In addition to these singularities, major seasonal trends are recognizable and for the British Isles Lamb identified five *natural seasons* on the basis of spells of a particular type lasting for 25 days or more during the period 1898–1947 (Figure 8.8). Insofar as is possible to think in terms of a 'normal year', the seasons are:

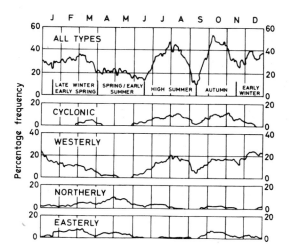

Figure 8.8 The frequency of long spells (25 days or more) of a given airflow type over Britain, 1898–1947. The diagram showing all long spells also indicates a division of the year into 'natural seasons'.

Source: After Lamb 1950.

1 *Spring–early summer* (the beginning of April to mid-June). This is a period of variable weather conditions during which long spells are least likely. Northerly spells in the first half of May are the most significant feature, although there is a marked tendency for anticyclones to occur in late May–early June.
2 *High summer* (mid-June to early September). Long spells of various types may occur in different years. Westerly and north-westerly types are the most common and they may be combined with either cyclonic or anticyclonic types. Persistent sequences of cyclonic type occur more frequently than anticyclonic ones.
3 *Autumn* (the second week in September to mid-November). Long spells are again present in most years. Anticyclonic ones mainly in the first half, cyclonic and other stormy ones generally in October–November.
4 *Early winter* (from about the third week in November to mid-January). Long spells are less frequent than in summer and autumn. They are usually of westerly type, giving mild, stormy weather.
5 *Late winter and early spring* (from about the third week in January to the end of March). The long spells at this time of year can be of very different types, so that in some years it is mid-

winter weather, while in other years there is an early spring from about late February.

5 Synoptic anomalies

The mean climatic features of pressure, wind and the seasonal airflow regime provide an incomplete picture of climatic conditions. Some patterns of circulation occur irregularly and yet, because of their tendency to persist for weeks or even months, form an essential element of the climate.

Blocking patterns are an important example. It was noted in Chapter 6 that the zonal circulation in middle latitudes sometimes breaks down into a cellular pattern. This is commonly associated with a split of the jet stream into two branches over higher and lower middle latitudes and the formation of a cut-off low (see Chapter 7H.4) south of a high-pressure cell. The latter is referred to as a *blocking anticyclone* since it prevents the normal eastward motion of depressions in the zonal flow. Figure 8.9 gives numerical values expressing the frequency of occurrence of blocking for part of the northern hemisphere with five major blocking centres shown (H). A major area of blocking is Scandinavia, particularly in spring. Depressions are diverted north-eastwards towards the Norwegian Sea or south-eastwards into southern Europe. This pattern, with easterly flow around the southern margins of the anticyclone, produces severe winter weather over much of northern Europe. In January–February 1947, for example, easterly flow across Britain as a result of blocking over Scandinavia led to extreme cold and frequent snowfall. Winds were almost continuously from the east between 22 January and 22 February and even daytime temperatures rose little above freezing point. Snow fell in some part of Britain every day from 22 January to 17 March 1947, and major snowstorms occurred as occluded Atlantic depressions moved slowly across the country. Other notably severe winter months – January 1881, February 1895, January 1940 and February 1986 – were the result of similar pressure anomalies with pressure much above average to the north of the British Isles and below average to the south, giving persistent easterly winds.

The average effects of a number of winter blocking situations over north-west Europe are shown in Figures 8.10 and 8.11. Precipitation amounts are above normal, mainly over Iceland and the western Mediterranean, as depressions are steered around the blocking high following the path

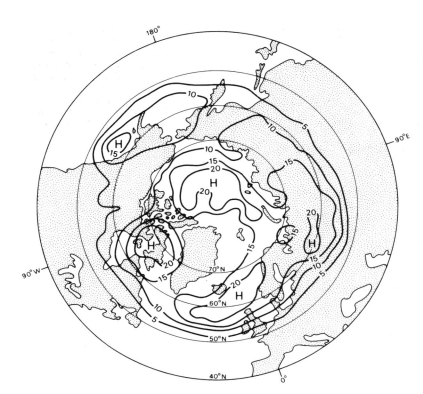

Figure 8.9 Frequency of occurrence of blocking conditions for the 500 mb level for all seasons. Values were calculated as five-day means for 381 × 381 km squares for the period 1946–78.

Source: From Knox and Hay 1985 (by permission of the Royal Meteorological Society).

of the upper jet streams. Over most of Europe, precipitation remains below average and this pattern is repeated with summer blocking. Winter temperatures are above average over the north-eastern Atlantic and adjoining land areas, but below average over Central and Eastern Europe and the Mediterranean due to the northerly outbreaks of cP air (Figure 8.11). The negative temperature anomalies associated with cool northerly airflow in summer cover most of Europe, and only northern Scandinavia and the north-eastern Atlantic have above-average values.

Despite these generalizations, the location of the block is of the utmost importance. For instance, in the summer of 1954 a blocking anticyclone across Eastern Europe and Scandinavia allowed depressions to stagnate over the British Isles, giving a dull, wet August, whereas in 1955 the blocking was located over the North Sea and a fine, warm summer resulted. The 1975–6 drought in Britain and the continent was caused by persistent blocking over north-western Europe. Another less common location of blocking is Iceland. A notable example

was the 1962–3 winter, when persistent high pressure south-east of Iceland led to northerly and north-easterly airflow over Britain. Temperatures in central England at that time were the lowest since 1740, with a mean of 0°C for December 1962 to February 1963. Central Europe was affected by easterly airstreams, and mean January temperatures there were 6°C below average.

6 Topographic effects

In various parts of Europe, topography has a marked effect on the climate, not only of the uplands themselves but also of adjacent areas. Apart from the more obvious effects upon temperatures, precipitation amounts and winds, the major mountain masses also affect the movement of frontal systems. Surface friction over mountain barriers tends to steepen the slope of cold fronts and decrease the slope of warm fronts so that the latter are slowed down and the former accelerated. The cyclogenetic effect of mountain barriers in producing lee depressions has already been discussed (see Chapter 7H.1).

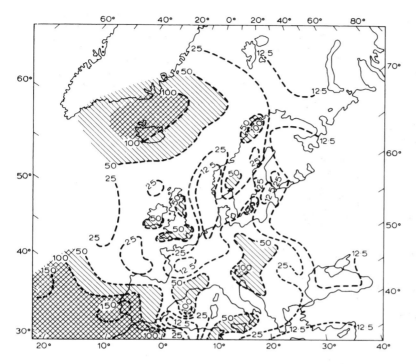

Figure 8.10 The mean precipitation anomaly, as a percentage of the average, during anticyclonic blocking in winter over Scandinavia. Areas above normal are cross-hatched, areas recording precipitation between 50 and 100 per cent of normal have oblique hatching.

Source: After Rex 1950.

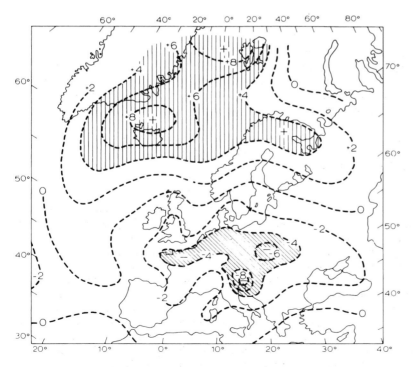

Figure 8.11 The mean surface temperature anomaly (°C) during anticyclonic blocking in winter over Scandinavia. Areas more than 4 °C above normal have vertical hatching, those more than 4 °C below normal have oblique hatching.

Source: After Rex 1950.

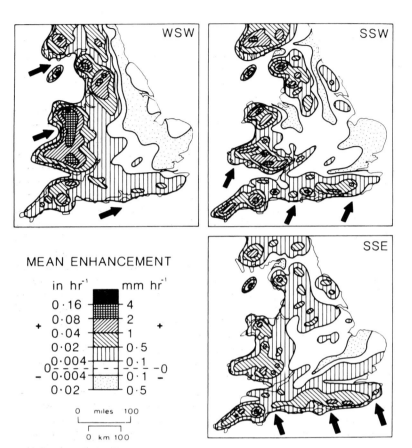

Figure 8.12 Mean orographic enhancement of precipitation over England and Wales, averaged for several days of fairly constant wind direction of about 20 m s⁻¹ and nearly saturated low-level airflow.

Source: After Browning and Hill 1981 (by permission of the Royal Meteorological Society).

The Scandinavian mountains are one of the most significant climatic barriers in Europe as a result of their orientation with regard to westerly airflow. Maritime air masses are forced to rise over the highland zone, giving annual precipitation totals of over 250 cm on the mountains of western Norway, whereas descent in their lee produces a sharp decrease in the amounts. The upper Gudbrandsdalen and Osterdalen in the lee of the Jotunheim and Dovre Mountains receive an average of less than 50 cm, and similar low values are recorded in the Jämtland province of central Sweden around Östersund.

The mountains can function equally in the opposite sense. For example, Arctic air from the Barents Sea may move southwards in winter over the Gulf of Bothnia, usually when there is a depression over northern Russia, giving very low temperatures in Sweden and Finland. Western Norway is rarely affected, since the cold wave is contained to the east of the mountains. In consequence, there is a sharp climatic gradient across the Scandinavian highlands in the winter months.

The Alps provide a quite different illustration of topographic effects. Together with the Pyrenees and the mountains of the Balkans, the Alps effectively separate the Mediterranean climatic region from that of Europe. The penetration of warm air masses north of these barriers is comparatively rare and short-lived. However, with certain pressure patterns, air from the Mediterranean and northern Italy is forced to cross the Alps, losing its moisture through precipitation on the southern slopes. Dry adiabatic warming on the northern side of the mountains can readily raise temperatures by 5–6°C in the upper valleys of the Aar, Rhine and Inn. At Innsbruck, there are approximately 50 days per year with föhn winds, with a maximum in spring, and these occurrences are often responsible for rapid melting of the snow, creating a risk of avalanches. When the airflow across the Alps has a northerly component föhns may occur in northern Italy, but their effects are generally less pronounced.

Features of upland climates in Britain will serve to illustrate some of the diverse effects of altitude. The mean annual rainfall on the west coasts near sea level is about 114 cm, but on the western mountains of Scotland, the Lake District and Wales averages exceed 380 cm per year. The annual record is 653 cm in 1954 at Sprinkling Tarn, Cumbria, and 145 cm fell in a single month (October 1909) just east of the summit of Snowdon. The annual number of rain-days (days with at least 0.25 mm of precipitation) increases from about 165 in south-eastern England and the south coast to over 230 days in north-west Britain. There is little additional increase in the frequency of rainfall with height on the mountains of the north-west, so that the mean rainfall per rain-day rises sharply from 0.5 cm near sea level in the west and north-west to over 1.3 cm in the Western Highlands, the Lake District and Snowdonia. This demonstrates that 'orographic rainfall' here is primarily due to an intensification of the normal precipitation processes and is not a special type. It is more appropriate, therefore, to recognize an orographic component that increases the amounts of rain associated with frontal depressions and unstable airstreams (see Chapter 4E.3).

Even quite low hills such as the Chilterns and South Downs cause a rise in rainfall, receiving about 12–13 cm per year more than the surrounding lowlands. In South Wales, mean annual precipitation increases from 120 cm at the coast to 250 cm on the 500 m high Glamorgan Hills, 20 km inland. Studies using radar and a dense network of rain gauges indicate that orographic intensification is pronounced during strong low-level south-westerly airflows in frontal situations. Most of the enhancement of precipitation rate occurs in the lowest 1,500 m. Figure 8.12 shows the mean enhancement according to wind direction over England and Wales, averaged for several days with fairly constant wind velocities of about 20 m s⁻¹ and nearly saturated low-level flow, attributable to a single frontal system on each day. Differences are apparent in Wales and southern England between winds from the SSW and from the WSW, whereas for SSE airflows the mountains of North Wales and the Pennines have little effect. Note also the areas of negative enhancement on the lee side of mountains.

The sheltering effects of the uplands produce low annual totals on the lee side (with respect to the prevailing winds). Thus, the lower Dee valley in the lee of the mountains of North Wales receives less than 75 cm per year, compared with over 250 cm in Snowdonia.

The complexity of the various factors affecting rainfall in Britain is shown by the fact that a close correlation exists between annual totals in north-west Scotland, the Lake District and western Norway, which are directly affected by Atlantic depressions. At the same time, there is an inverse relationship between annual amounts in the Western Highlands and lowland Aberdeenshire, less than 240 km away. Annual precipitation in the latter area is more closely correlated with that in lowland eastern England. Essentially, the British Isles comprise two major climatic units for rainfall – first, an 'Atlantic' one with a winter season maximum, and, second, those central and eastern districts with 'continental' affinities in the form of a weak summer maximum in most years. Other areas (eastern Ireland, eastern Scotland, northeast England and most of the English Midlands and the Welsh border counties) generally have a wet second half of the year.

The occurrence of snow is another measure of altitude effects. Near sea level, there are on average approximately 5 days per year with snow falling in south-west England, 15 days in the south-east and 35 days in northern Scotland. Between 60 and 300 m, the frequency increases by about 1 day per 15 m of elevation and even more rapidly on higher ground. Approximate figures for northern Britain are 60 days at 600 m and 90 days at 900 m. The number of mornings with snow lying on the ground (more than half the ground covered) is closely related to mean temperature and hence altitude. Average figures range from about 5 days per year or less in much of southern England and Ireland, to between 30 and 50 days on the Pennines and over 100 days on the Grampian Mountains. In the last area (on the Cairngorms) and on Ben Nevis there are several semi-permanent snowbeds at about 1,160 m, and it is estimated that the theoretical climatic snowline – above which there would be *net* accumulation of snow – is at 1,620 m over Scotland.

The seasonal variability of lapse rates in mountain areas was mentioned in Chapter 2B.7. Marked geographical variations also exist even within the British Isles. One measure of these variations is the length of the 'growing season'. Meteorological data can be used to determine an index of growth opportunity by counting the number of days on which the mean daily temperature exceeds an arbitrary threshold value – commonly 6°C. Along the south-west coasts of England the 'growing season', as calculated on this basis, is nearly 365 days per year and in this area it decreases by about 9 days per 30 m of elevation, but in northern England and Scotland the decrease is only about 5 days per 30 m from between 250–270 days near sea level. In continental climates, the altitudinal decrease may be even more gradual; in Central Europe and New England, for example, it is about 2 days per 30 m.

B NORTH AMERICA

The North American continent spans nearly 60° of latitude and, not surprisingly, exhibits a wide range of climatic conditions. Unlike Europe, the west coast is backed by the Pacific Coast Ranges rising to over 2,750 m, which lie across the path of depressions in the mid-latitude westerlies and prevent the extension of maritime influences inland. In the interior of the continent, there are no significant obstructions to air movement, and the absence of any east–west barrier allows air masses from the Arctic or the Gulf of Mexico to sweep across the interior lowlands, causing wide extremes of weather and climate. Maritime influences in eastern North America are greatly limited by the fact that the prevailing winds are westerly, so that the temperature regime is continental. Nevertheless, the Gulf of Mexico is a major source of moisture supply for precipitation over the eastern half of the United States and, as a result, the precipitation regimes are different from those found in East Asia.

We will look first at the broad characteristics of the atmospheric circulation over the continent.

1 Pressure systems

The mean pressure pattern for the middle troposphere displays a prominent trough over eastern North America in both summer and winter (see Figure 6.4). One theory is that this is a lee trough caused by the effect of the western mountain ranges on the upper westerlies, but at least in winter the strong baroclinic zone along the east coast of the continent is undoubtedly a major contributory factor. The implications of this mean wave pattern are that cyclones tend to move south-eastwards over the Mid-west, carrying continental polar air southwards, while the cyclone paths are north-eastward along the Atlantic coast.

In individual months, there may of course be considerable deviations from this average pattern, with important consequences for the weather in different parts of the continent, and, in fact, this relationship provides the basis for the monthly forecasts of the United States Weather Service. For example, if the trough is more pronounced than usual temperature may be much below average in the central, southern and eastern United States, whereas if the trough is weak the westerly flow is stronger with correspondingly less opportunity for cold outbreaks of polar air masses. Sometimes, the trough is displaced to the western half of the continent, causing a reversal of the usual weather pattern, since upper north-westerly airflow can bring cold, dry weather to the west while in the east there are very mild conditions associated with upper south-westerly flow. Precipitation amounts also depend on the depression tracks; if the upper trough is far to the west, depressions form ahead of it (see Chapter 7F) over the south central United States and move north-eastwards towards the lower St Lawrence, giving more precipitation than usual in these areas and less along the Atlantic coast.

The major features of the surface pressure map in January (see Figure 6.10A) are the extension of the subtropical high over the south-western United States (called the Great Basin high) and the separate polar anticyclone of the Mackenzie district of Canada. Mean pressure is low off both the east and west coasts of higher middle latitudes, where oceanic heat sources indirectly give rise to the (mean) Icelandic and Aleutian lows. It is interesting to note that, on average, in December, of any region in the northern hemisphere for any month of the year, the Great Basin region has the most frequent occurrence of highs, whereas the Gulf of Alaska has the maximum frequency of lows. The Pacific coast as a whole has its most frequent cyclonic activity in winter, as does the Great Lakes area, whereas over the Great Plains the maximum is in spring and early summer. Remarkably, the Great Basin in June has the most frequent cyclogenesis of any part of the northern hemisphere in any month of the year. Heating over this area in summer helps to maintain a shallow, quasi-permanent low-pressure cell, in marked contrast with the almost continuous subtropical high-pressure belt in the middle troposphere (see Figure 6.4). Continental heating also indirectly assists in the splitting of the Icelandic low to create a secondary centre over north-eastern Canada. The west-coast summer circulation is dominated by the Pacific anticyclone, while the south-eastern United States is affected by the Atlantic subtropical anticyclone cell (see Figure 6.10).

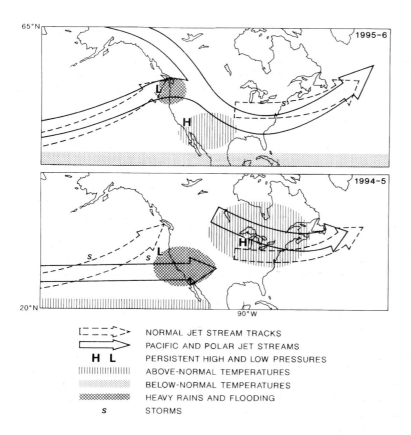

Figure 8.13 Jet streams, pressure distribution and climate for North America during the winters of 1995–6 and 1994–5.

[- - - ->	NORMAL JET STREAM TRACKS
⇒	PACIFIC AND POLAR JET STREAMS
H L	PERSISTENT HIGH AND LOW PRESSURES
‖‖‖‖‖‖‖‖‖‖	ABOVE-NORMAL TEMPERATURES
░░░░░░░░	BELOW-NORMAL TEMPERATURES
✕✕✕✕✕✕✕	HEAVY RAINS AND FLOODING
s	STORMS

Source: US Department of Commerce, Climate Prediction Center. Courtesy US Department of Commerce.

Broadly, there are three prominent depression tracks across the continent in winter (see Figure 7.22). One group moves from the west along a more or less zonal path about 45–50°N, whereas a second loops southwards over the central United States and then turns north-eastwards towards New England and the Gulf of St Lawrence. Some of these depressions originate over the Pacific, cross the western ranges as an upper trough and redevelop in the lee of the mountains. Alberta is a noted area for this process and also for primary cyclogenesis, since the arctic frontal zone is over north-west Canada in winter. This frontal zone involves much-modified mA air from the Gulf of Alaska and cold dry cA (or cP) air. Depressions of the third group form along the main polar frontal zone, which in winter is off the east coast of the United States, and move north-eastwards towards Newfoundland. Sometimes, this frontal zone is present over the continent at about 35°N with mT air from the Gulf and cP air from the north or modified mP air from the Pacific. Polar front depressions forming over

Colorado move north-eastwards towards the Great Lakes and others developing over Texas follow a more or less parallel path, further to the south and east, towards New England. Anomalies in winter climate over North America are strongly influenced by the positions of the jet streams and the movement of associated storm systems. Figure 8.13 illustrates their role in displacing regions of heavy rain and flooding and positive/negative temperature departures in the winters of 1994–5 and 1995–6.

Between the Arctic and Polar Fronts, a third frontal zone is distinguished by Canadian meteorologists. This maritime (arctic) frontal zone is present when mA and mP (or mPc and mPw) air masses interact along their common boundary. The three-front (i.e. four air-mass) model allows a detailed analysis to be made of the baroclinic structure of depressions over the North American continent using synoptic weather maps and cross-sections of the atmosphere. Figure 8.14 illustrates the three frontal zones and associated depressions on 29 May 1963. Along 95°W, from 60 to 40°N,

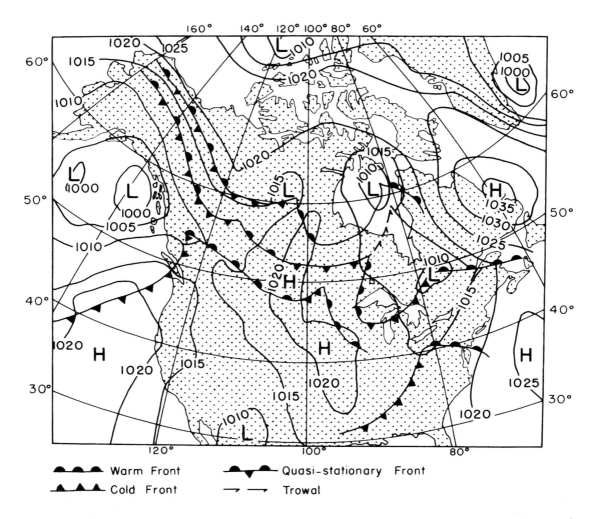

Figure 8.14 A synoptic example of depressions associated with three frontal zones on 29 May 1963 over North America.

Source: Based on charts of the Edmonton Analysis Office and the Daily Weather Report.

the following dew-point temperatures were reported in the four air masses: –8°C, 1°C, 4°C and 13 °C.

In summer, the east-coast depressions are less frequent and the tracks across the continent are displaced northwards, with the main ones moving over Hudson Bay and Labrador–Ungava, or along the line of the St Lawrence. These are associated mainly with a rather poorly defined maritime frontal zone. The Arctic Front is usually located along the north coast of Alaska, where there is a strong temperature gradient between the bare land and the cold Arctic Ocean and pack-ice. East from here, the front is very variable in location from day to day and year to year. Broadly, it occurs most often in the vicinity of northern Keewatin and Hudson Strait, although one study of air-mass temperatures and airstream confluence regions suggests that an arctic frontal zone occurs further south over Keewatin in July and that its mean position (Figure 8.15) is closely related to the boreal forest–tundra boundary. This relationship undoubtedly reflects the importance of arctic air-mass dominance for summer temperatures and consequently for tree-growth possibilities, but the interrelationships between atmospheric systems and vegetation boundaries involve complex feedbacks.

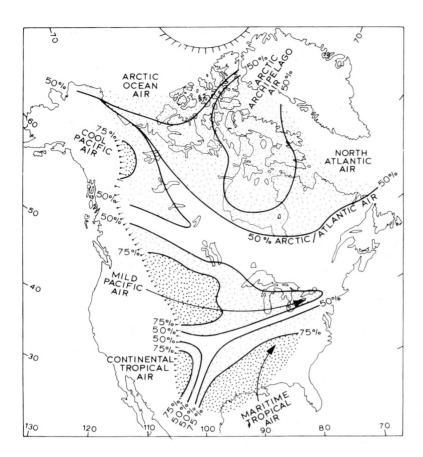

Figure 8.15 Regions in North America east of the Rocky Mountains dominated by the various air-mass types in July for more than 50 per cent and 75 per cent of the time. The 50 per cent frequency lines correspond to mean frontal positions.

Source: After Bryson 1966.

Several circulation singularities have been recognized in North America, as in Europe (see A.4, this chapter). Three that have received considerable attention in view of their prominence are (1) the advent of spring in late March; (2) the midsummer high-pressure jump at the end of June; and (3) the Indian summer in late September (and late October).

The arrival of spring is marked by different climatic responses in different parts of the continent. For example, there is a sharp decrease in March to April precipitation in California, due to the extension of the Pacific high, whereas precipitation intensity increases in the Mid-west (Figure 8.16) as a result of more frequent cyclogenesis in Alberta and Colorado and a northward extension of maritime tropical air over the Mid-west from the Gulf of Mexico. These changes are part of a hemispheric readjustment of the circulation, since at the beginning of April the Aleutian low-pressure cell, which from September to March is located about 55°N,

165°W, splits into two, with one centre in the Gulf of Alaska and the other over northern Manchuria. This represents a decrease in the zonal index (see Chapter 6C.2).

In late June, there is a rapid northward displacement of the Bermuda and North Pacific subtropical high-pressure cells. In North America, this also pushes the depression tracks northwards with the result that precipitation decreases from June to July over the northern Great Plains, part of Idaho and eastern Oregon (see Figure 8.16). Conversely, the south-westerly anticyclonic flow that affects Arizona in June is replaced by air from the Gulf of California, and this causes the onset of the summer rains (see B.3, this chapter). It has been suggested by Bryson and Lahey that these circulation changes at the end of June may be connected with the disappearance of snow cover from the arctic tundra. This leads to a sudden decrease of surface albedo from about 75 to 15 per cent, with consequent changes

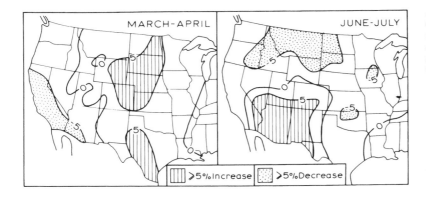

Figure 8.16 The precipitation changes between March–April (left) and June–July (right), as a percentage of the mean annual total, for the central and western United States.

Source: After Bryson and Lahey 1958.

in the heat budget components and hence in the atmospheric circulation.

Frontal wave activity makes the first half of September a rainy period in the northern Mid-west states of Iowa, Minnesota and Wisconsin, but after about the twentieth of the month anticyclonic conditions return with warm airflow from the dry south-west, giving fine weather – the so-called Indian summer. Significantly, the hemispheric zonal index value rises in late September. This anticyclonic weather type has a second phase in the latter half of October, but at this time there are polar outbreaks. The weather is generally cold and dry, although if precipitation does occur there is a high probability of it being in the form of snow.

2 The temperate west coast and cordillera

The oceanic circulation of the North Pacific closely resembles that of the North Atlantic. The drift from the Kuroshio current off Japan is propelled by the westerlies towards the western coast of North America and acts as a warm current between 40 and 60°N. Sea-surface temperatures are several degrees lower than in comparable latitudes off Western Europe, however, due to the smaller volume of warm water involved. Also, in contrast to the Norwegian Sea, the shape of the Alaskan coastline prevents the extension of the drift to high latitudes (see Figure 6.31).

The Pacific coast ranges greatly restrict the inland extent of oceanic influences, and hence there is no extensive maritime temperate climate such as we have described for Western Europe. The major climatic features duplicate those of the coastal mountains of Norway and those of New Zealand and southern Chile in the belt of southern wester-

lies. Topographic factors make the weather and climate of such areas very variable over short distances, both vertically and horizontally, and, therefore, only a few salient characteristics are selected for consideration.

There is a regular pattern of rainy windward and drier lee slopes across the successive north-west to south-east ranges, with a more general decrease towards the interior. The Coast Range in British Columbia has mean annual totals of precipitation exceeding 250 cm, with 500 cm in the wettest places, compared with 125 cm or less on the summits of the Rockies, yet even on the leeward side of Vancouver Island the average figure at Victoria is only 70 cm. Analogous to the 'westerlies–oceanic' regime of north-west Europe, there is a winter precipitation maximum along the littoral, which also extends beyond the Cascades (in Washington) and the Coast Range (in British Columbia), but summers are drier due to the strong North Pacific anticyclone. The regime in the interior of British Columbia is transitional between that of the coastal region and the distinct summer maximum of central North America (Figure 8.17), although at Kamloops in the Thompson valley (annual average 25 cm) there is a slight summer maximum associated with thunderstorm-type rainfall. In general, the sheltered interior valleys receive less than 50 cm per year, and in the driest years certain localities have recorded only 15 cm. Above 1,000 m, much of the precipitation falls as snow (see Figure 8.17) and some of the greatest snow depths in the world are reported from British Columbia, Washington and Oregon. For example, between 1,000 and 1,500 cm falls on the Cascade Range at heights of about 1,500 m, and even as far inland as the Selkirk Mountains the totals

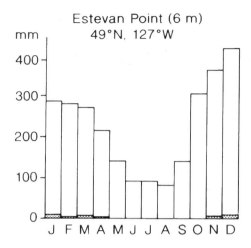

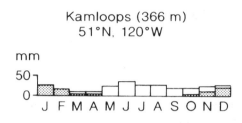

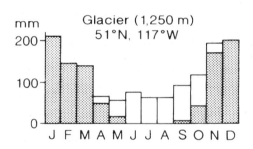

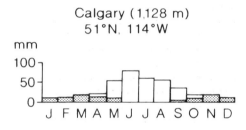

Figure 8.17 Precipitation graphs for stations in western Canada. The shaded portions represent snowfall, expressed as water equivalent.

are considerable. The mean snowfall is 990 cm at Glacier, British Columbia (elevation 1,250 m), and this accounts for almost 70 per cent of the annual precipitation (see Figure 8.17). Near sea level on the outer coast, in contrast, very little precipitation falls as snow (for example, Estevan Point). It is estimated that the climatic snowline rises from about 1,600 m on the west side of Vancouver Island to 2,900 m in the eastern Coast Range. Inland, its elevation increases from 2,300 m on the west slopes of the Columbia Mountains to 3,100 m on the east side of the Rockies. This trend reflects the precipitation pattern referred to above.

Finally, mention must be made of the large diurnal variations that affect the cordilleran valleys. Strong diurnal rhythms of temperature (especially in summer) and wind direction are a feature of mountain climates and their effect is superimposed upon the general climatic characteristics of the area. Cold air drainage produces many remarkably low minima in

the mountain valleys and basins. At Princeton, British Columbia (elevation 695 m), where the mean daily minimum in January is −14°C, there is on record an absolute low of −45°C, for example. This leads in some cases to reversal of the normal lapse rate. Golden in the Rocky Mountain Trench has a January mean of −12°C, whereas 460 m higher at Glacier (1,250 m) it is −10°C.

3 Interior and eastern North America

Central North America has the typical climate of a continental interior in middle latitudes, with hot summers and cold winters (Figure 8.18), yet the weather in winter is subject to marked variability. This is determined by the steep temperature gradient between the Gulf of Mexico and the snow-covered northern plains; also by shifts of the upper wave patterns and jet stream. Cyclonic activity in winter is much more pronounced over central and eastern

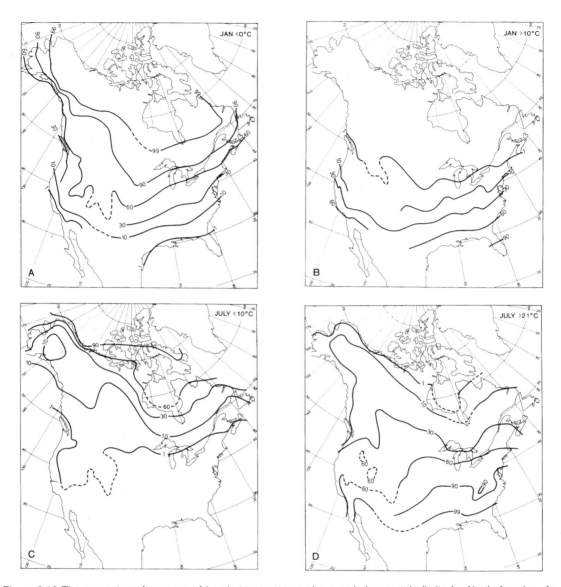

Figure 8.18 The percentage frequency of hourly temperatures above or below certain limits for North America. A: January temperatures <0 °C; B: January temperatures >10 °C; C: July temperatures <10 °C; D: July temperatures >21 °C.

Source: After Rayner 1961.

North America than in Asia, which is dominated by the Siberian anticyclone (see Figure 7.22), and consequently there is no climatic type with a winter minimum of precipitation in eastern North America.

The general temperature conditions in winter and summer are illustrated in Figure 8.18, showing the frequency with which hourly temperature readings exceed or fall below certain limits. The two chief features of all four maps are (1) the dominance of the meridional temperature gradient, away from coasts, and (2) the continentality of the interior and east compared with the 'maritimeness' of the west coast. On the July maps, additional influences are evident and these are referred to below.

a Continental and oceanic influences

The continentality of North America is illustrated in Figure 8.19. This map identifies regions with a continental (positive) or maritime (negative) climate using a new index which, unlike that of Gorczynski described earlier (p. 189), is not based directly on the annual temperature range scaled by latitude (see Note 1). The pattern demonstrates the key role of the distance from the ocean in the direction of the prevailing (westerly) winds, the small moderating influence of inland water bodies such as Hudson Bay and the Great Lakes, which have ice cover in winter, and the effect of topographic barriers in limiting the inland penetration of maritime airstreams.

The Labrador coast is fringed by the waters of a cold current, analogous to the Oyashio off East Asia, but in both cases the prevailing westerlies greatly limit their climatic significance. The Labrador Current maintains drift ice off Labrador and Newfoundland until June and gives very low summer temperatures along the Labrador coast (see Figure 8.18C). The lower incidence of freezing temperatures in this area in January is related to the movement of some depressions into the Davis Strait, carrying Atlantic air northwards. A major role of the Labrador Current is in the formation of fog. Advection fogs are very frequent between May and August off Newfoundland, where the Gulf Stream and Labrador Current meet. Warm, moist, southerly airstreams are cooled rapidly over the cold waters of the Labrador Current and with steady, light winds such fogs may persist for several days, creating hazardous conditions for shipping. Southward-facing coasts are particularly affected and at Cape Race (Newfoundland), for example, there are on average 158 days per year with fog (visibility less than 1 km) at some time of day. The summer concentration is shown by the figures for Cape Race during individual months: May – 18 (days), June – 18, July – 24, August – 21, and September – 18.

Oceanic influence along the Atlantic coasts of the United States is very limited, and although there is some moderating effect of minimum temperatures at coastal stations this is scarcely evident on generalized maps such as Figure 8.18. More significant climatic effects are in fact found in the neighbourhood of Hudson Bay and the Great Lakes. Hudson Bay remains very cool in summer, with water temperatures of about 7–9°C, and this depresses temperatures along its shore, especially in the east

(see Figure 8.18C and D). Mean July temperatures are 12°C at Churchill (59°N) and 8°C at Inukjuak (58°N), on the west and east shores respectively, compared for instance with 13 °C at Aklavik (68 °N) on the Mackenzie delta. The influence of Hudson Bay is even more striking in early winter, when the land is snow-covered. Westerly airstreams crossing the open water are warmed by 11°C on average in November, and moisture added to the air leads to considerable snowfall in western Ungava (see the graph for Inukjuak, Figure 8.22). By the beginning of January, the Bay is frozen over almost entirely and no effects are evident. The Great Lakes influence their surroundings in much the same way (Figure 8.18). Heavy winter snowfalls are a notable feature of the southern and eastern shores of the Great Lakes. In addition to contributing moisture to north-westerly streams of cold cA and cP air, the heat source of the open water in early winter produces a low-pressure trough, which increases the snowfall as a result of convergence. Yet a further factor is frictional convergence and orographic uplift at the shoreline. Mean annual snowfall exceeds 250 cm along much of the eastern shore of Lake Huron and Georgian Bay, the south-eastern shore of Lake Ontario, the north-eastern shore of Lake Superior and its southern shore east of about 90° 30 ′W. Extremes include 114 cm in one day at Watertown, New York, and 894 cm during the 1946–7 winter season at nearby Bennetts Bridge, both of which are close to the eastern end of Lake Ontario.

Transport in cities in these snow belts is quite frequently disrupted during winter snowstorms. The Great Lakes also provide an important tempering influence during winter months by raising average daily minimum temperatures at lakeshore stations by some 2–4°C above those at inland locations. In mid-December, the upper 60 m of Lake Erie has a uniform temperature of 5°C.

b Warm and cold spells

Two types of synoptic condition are of particular significance for temperatures in the interior of North America. One is the 'cold wave' caused by a northerly outbreak of cP air, which in winter regularly penetrates deep into the central and eastern United States and occasionally affects even Florida and the Gulf coast, injuring frost-sensitive crops. Cold waves are arbitrarily defined as a temperature drop of at least 11°C in 24 hours over most of the United States, and at least 9°C in California, Florida

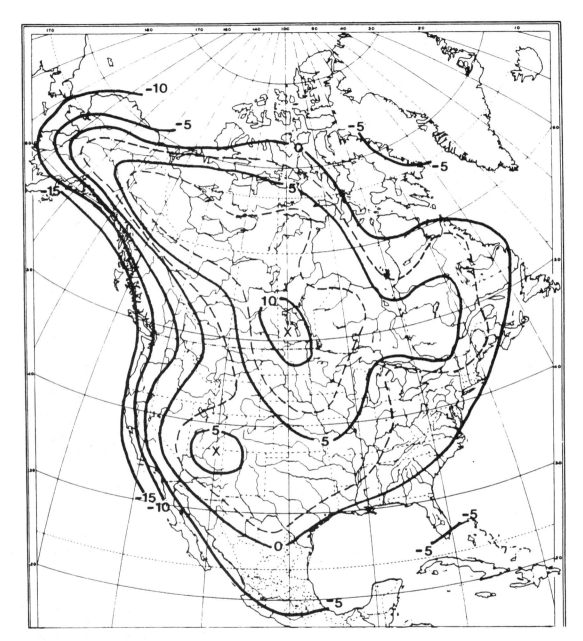

Figure 8.19 Relative continentality in North America.

Source: Driscoll and Yee Fong 1992. From *International Journal of Climatology*, copyright © John Wiley & Sons Ltd. Reproduced with permission.

and the Gulf Coast, to below a specified minimum depending on location and season. The winter criterion decreases from 0°C in California, Florida and the Gulf Coast to –18°C over the northern Great Plains and the north-eastern states. The cold spells commonly occur with the build-up of a north–south anticyclone in the rear of a cold front. The polar air gives clearer, dry weather with strong, cold winds, although if the winds follow snowfall, fine, powdery snow may be whipped up by the wind, creating blizzard conditions. These are quite common over the northern plains.

Another type of temperature fluctuation is associated with the *chinook* winds in the lee of the Rockies (see Chapter 5C.2). The chinook is particularly warm and dry as air of Pacific origin, after losing its moisture over the mountains, descends the eastern slopes and warms at the dry adiabatic lapse rate. The onset of the chinook produces temperatures well above the seasonal normals so that snow is often thawed rapidly, and in fact the Indian word 'Chinook' means snow-eater. Temperature rises of as much as 22°C have been observed in five minutes, and the occurrence of such warm spells is reflected by the high extreme maxima in winter months at Medicine Hat (Figure 8.20). In Canada, the chinook effect may be observed a considerable distance from the Rockies into south-west Saskatchewan, but in Colorado its influence is rarely felt more than about 50 km from the foothills. In south-eastern Alberta, the belt of strong westerly chinook winds and elevated temperatures extends 150–200 km eastward of the Rocky Mountains. Temperature anomalies average 5–9°C above winter normals, and a triangular sector south-east of Calgary, towards Medicine Hat, experiences maximum anomalies of up to 15–25°C, relative to mean daily maximum temperature values. Chinook events with westerly winds >35m s^{-1} occur on 45–50 days between November and February in this area as a result of the relatively low and narrow ridge line of the Rocky Mountains between 49 and 50 °N, compared with the mountains around Banff and further north.

Chinook conditions commonly develop in a Pacific airstream that is replacing a winter high-pressure cell over the high western plains. Sometimes the cold, stagnant cP air of the anticyclone is not dislodged by the descending chinook and a marked inversion is formed, but on other occasions the boundary between the two air masses may reach ground level locally and, for example, the western

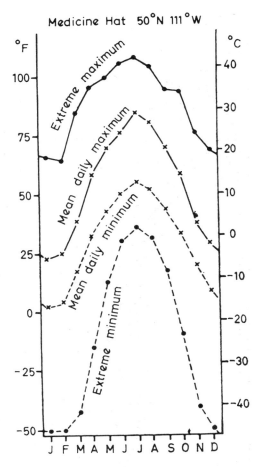

Figure 8.20 Mean and extreme temperatures at Medicine Hat, Alberta.

suburbs of Calgary may record temperatures above 0°C while those to the east of the city remain below –15°C.

c Precipitation and the moisture balance

Longitudinal influences are apparent in the distribution of annual precipitation, although this is in large measure a reflection of the topography. The 60 cm annual isohyet in the United States approximately follows the 100°W meridian (Figure 8.21), and westwards to the Rockies is an extensive dry belt in the rain shadow of the western mountain ranges. In the south-east, totals exceed 125 cm, and 100 cm or more is received along the Atlantic coast as far north as New Brunswick and Newfoundland.

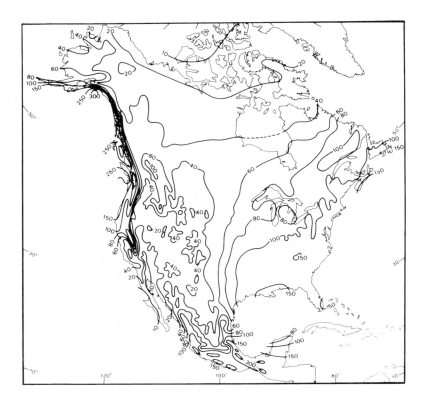

Figure 8.21 Mean annual precipitation (in centimetres) over North America for the period 1931–60. Isohyets in the Arctic underestimate the true totals by 30–50 per cent due to problems in recording snowfall accurately with precipitation gauges.

Source: Based on Hare and Hay, Court, and Mosiño Alemán and Garćia.

The major sources of moisture for precipitation over North America are the Pacific Ocean and the Gulf of Mexico. The former need not concern us here, since comparatively little of the precipitation falling over the interior appears to be derived from that source. The Gulf source is extremely important in providing moisture for precipitation over central and eastern North America, but the predominance of south-westerly airflow means that little precipitation falls over the western Great Plains (see Figure 8.21). Over the south-eastern United States, there is considerable evapotranspiration and this helps to maintain moderate annual totals northwards and eastwards from the Gulf by providing additional water vapour for the atmosphere. Along the east coast, the Atlantic Ocean is an additional significant source of moisture for winter precipitation.

There are at least eight major types of seasonal precipitation regime in North America (Figure 8.22); the winter maximum of the west coast and the transition type of the intermontane region in middle latitudes have already been mentioned, and the subtropical types are discussed in the next section. Four primarily mid-latitude regimes are distinguished east of the Rocky Mountains:

1 A warm season maximum is found over much of the continental interior (e.g. Rapid City). In an extensive belt from New Mexico to the prairie provinces more than 40 per cent of the annual precipitation falls in summer. In New Mexico, the rain occurs mainly with late summer thunderstorms, but May–June is the wettest time over the central and northern Great Plains due to more frequent cyclonic activity. Winters are quite dry over the plains, but the mechanism of the occasional heavy snowfalls is of interest. They occur over the north-western plains during easterly upslope flow, usually in a ridge of high pressure. Further north in Canada, the maximum is commonly in late summer or autumn, when depression tracks are in higher middle latitudes. There is a local maximum in autumn on the eastern shores of Hudson Bay (e.g. Inukjuak) due to the effect of open water.

2 Eastward and southward of the first zone there is a double maximum in May and September. In

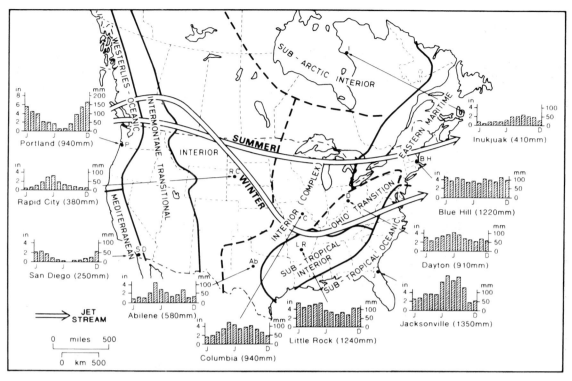

Figure 8.22 North American rainfall regime regions and histograms showing mean monthly precipitations for each region (January, June and December are indicated). Note that the jet stream is anchored by the Rockies in more or less the same position at all seasons.

Source: Mostly after Trewartha 1981; additions by Henderson-Sellers and Robinson 1986.

the upper Mississippi region (e.g. Columbia), there is a secondary minimum, paradoxically in July–August when the air is especially warm and moist, and a similar profile occurs in northern Texas (e.g. Abilene). An upper-level ridge of high pressure over the Mississippi valley seems to be responsible for reduced thunderstorm rainfall in midsummer, and a tongue of subsiding dry air extends southwards from this ridge towards Texas. However, during the period June–August 1993 massive flooding occurred in the Midwestern parts of the Mississippi and Missouri Rivers as the result of up to twice the January– July average precipitation being received, with many point rainfalls exceeding amounts appropriate for recurrence intervals exceeding 100 years (Figure 8.23). The three summer months saw excesses of 50 cm above the average rainfall with totals of 90 cm or more. Strong moist south-westerly airflows recurred throughout

the summer with a quasi-stationary cold front oriented from south-west to north-east across the region. The flooding resulted in 48 deaths, destroyed 50,000 homes and caused damage losses of $10 billion. In September, renewed cyclonic activity associated with the seasonal southward shift of the polar front, at a time when mT air from the Gulf is still warm and moist, typically causes a resumption of rainfall. Subsequently, however, drier westerly airstreams affect the continental interior as the general airflow becomes more zonal.

The diurnal occurrence of precipitation in the central United States is rather unusual for a continental interior. Sixty per cent or more of the summer precipitation falls during nocturnal thunderstorms (2000–0800 hours True Solar Time) in central Kansas, parts of Nebraska, Oklahoma and Texas. Hypotheses suggest that the nocturnal thunderstorm rainfall that occurs, especially

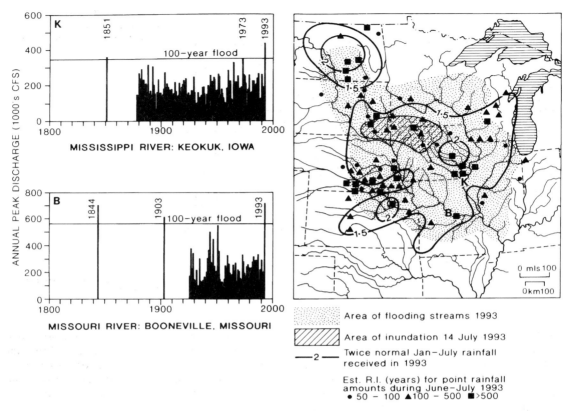

Figure 8.23 Distribution of flooding streams and inundation in the US Mid-west during the period June–August 1993. Peak discharges for the Mississippi River at Keokuk, Iowa, (K) and the Missouri River at Booneville, Missouri, (B) are shown, together with the historic annual peak discharge record. The isopleths indicate the multiples of the 30-year average January–July precipitation that fell in the first seven months of 1993, and the symbols the estimated recurrence intervals (R.I. years) for point rainfall amounts received during June–July 1993.
Sources: Parrett *et al*. 1993 and Lott 1994. Courtesy US Geological Survey.

with extensive mesoscale convective systems (see p. 98), may be linked to a tendency for nocturnal convergence and rising air over the plains east of the Rocky Mountains. The terrain profile appears to play a role here, as a large-scale inversion layer forms at night over the mountains, setting up a low-level jet east of the mountains just above the boundary layer. This southerly flow, at 500–1,000 m above the surface, can supply the necessary low-level moisture influx and convergence for the storms (cf. Figure 4.27).

3   East of the upper Mississippi, in the Ohio valley and south of the lower Great Lakes, there is a transitional regime between that of the interior and the east-coast type. Precipitation is reasonably abundant in all seasons, but the summer maximum is still in evidence (e.g. Dayton).

4   In eastern North America (New England, the Maritimes, Quebec and south-east Ontario), precipitation is fairly evenly distributed throughout the year (e.g. Blue Hill). In Nova Scotia and locally around Georgian Bay there is a winter maximum, due in the latter case to the influence of open water. In the Maritimes it is related to winter (and also autumn) storm tracks.

It is worth comparing the eastern regime with the summer maximum that is found over East Asia. There the Siberian anticyclone excludes cyclonic precipitation in winter and monsoonal influences are felt in the summer months.

The seasonal distribution of precipitation is of vital interest for agricultural purposes. Rain falling in summer, for instance, when evaporation losses

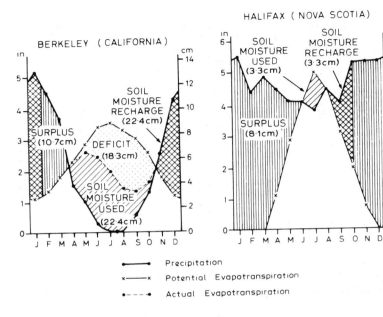

Figure 8.24 The moisture balances at Berkeley, California, and Halifax, Nova Scotia.

Precipitation
x———x Potential Evapotranspiration
•----• Actual Evapotranspiration

Source: After Thornthwaite and Mather 1955.

are high, is less effective than an equal amount in the cool season. Figure 8.24 illustrates the effect of different regimes in terms of the moisture balance, calculated according to Thornthwaite's method (see Appendix 1B). At Halifax (Nova Scotia), sufficient moisture is stored in the soil to maintain evaporation at its maximum rate (i.e. actual evaporation = potential evaporation), whereas at Berkeley (California) there is a computed moisture deficit of nearly 5 cm in August. This is a guide to the amount of irrigation water that may be required by crops, although in dry regimes the Thornthwaite method generally underestimates the real moisture deficit.

The mean monthly potential evaporation (PE) is calculated in the method developed by C. W. Thornthwaite from tables based on a complex equation relating PE to air temperature (Appendix 1B). Figure 8.25 illustrates the distribution of moisture regions over North America based on the moisture index (I_m):

$$I_m = \frac{100 \ (\text{water surplus} - \text{water deficit})}{\text{PE}}$$

Table 8.3 summarizes the climate types related to the moisture index. The boundary separating the moist climates of the east from the dry climates of the west (apart from the west coast) follows the 95th meridian. The major humid areas are along the Appalachians, in the north-east and along the

Pacific coast, while the most extensive arid areas are in the intermontane basins, the High Plains, the American south-west and parts of northern Mexico. Some aspects of the precipitation climatology of this arid area are examined in C.3, this chapter.

C THE SUBTROPICAL MARGINS

1 The Mediterranean

The characteristic west-coast climate of the subtropics is the Mediterranean type with hot, dry summers and mild, relatively wet winters. It is interposed between the temperate maritime type and the arid subtropical desert climate. As will be seen later, the boundary between the temperate maritime climate of Western Europe and that of the Mediterranean can be delimited on the basis of the seasonality of

Table 8.3 The Thornthwaite moisture index.

I_m	climate	symbol
>100	perhumid	A
20–100	humid	B (with 4 subdivisions)
0–20	moist subhumid	C_2
−33 to 0	dry subhumid	C_1
−67 to −33	semi-arid	D
−100 to −67	arid	E

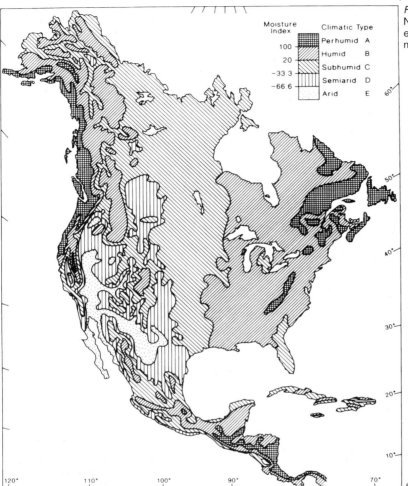

Figure 8.25 Moisture regions of North and Central America employing Thornthwaite's moisture index.

Moisture Index   Climatic Type

100	Perhumid	A
20	Humid	B
-33.3	Subhumid	C
-66.6	Semiarid	D
	Arid	E

Source: From Mather 1988.

rainfall. However, another diagnostic feature is the relatively sharp increase in solar radiation across a zone running along northern Spain, south-east France, northern Italy and to the east of the Adriatic (Figure 8.26). The Mediterranean regime is transitional in a special way, because it is controlled by the westerlies in winter and by the subtropical anticyclone in summer. The seasonal change in position of the subtropical high and the associated subtropical westerly jet stream in the upper troposphere is evident in Figure 8.27. The type region is peculiarly distinctive, extending more than 3,000 km into the Eurasian continent. Additionally, the configuration of seas and peninsulas produces great regional variety of weather and climate. The Californian region,

with similar conditions (see Figure 8.22), is of very limited extent, and attention is therefore concentrated on the Mediterranean basin itself.

The winter season sets in quite suddenly in the Mediterranean as the summer eastward extension of the Azores high-pressure cell collapses. This phenomenon can be observed on barographs throughout the region, but particularly in the western Mediterranean, where a sudden drop in pressure occurs on about 20 October and is accompanied by a marked increase in the probability of precipitation. The probability of receiving rain in any five-day period increases dramatically from 50–70 per cent in early October to 90 per cent in late October. This change is associated with the first invasions by cold

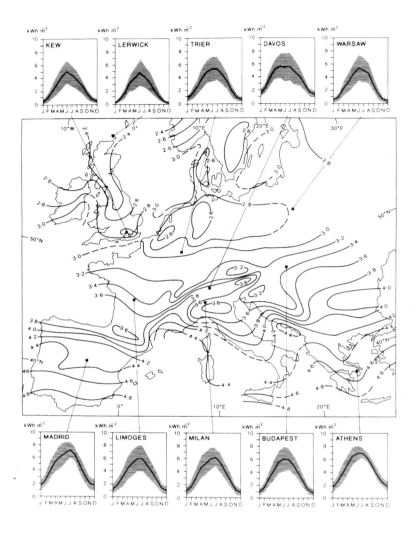

Figure 8.26 Average annual means of daily global irradiation on a horizontal surface (kWh/m⁻²) for Western and Central Europe calculated for the period 1966–75. 10-year means of monthly means of daily sums, together with standard deviations (shaded band), are also shown for selected stations.

Source: Palz 1984. Reproduced by permission of the Directorate-General, Science, Research and Development, European Commission, Brussels, and W. Palz.

fronts, although thunder-shower rain has been common since August. The pronounced winter precipitation over the Mediterranean largely results from the relatively high sea-surface temperatures at that season, the sea temperature in January being about 2°C higher than the mean air temperature. Incursions of colder air into the region lead to convective instability along the cold front, producing frontal and orographic rain. Incursions of arctic air are relatively infrequent (there being, on average, six to nine invasions by cA and mA air each year), but penetration by unstable mP air is much more common. It typically gives rise to deep cumulus development and is critical in the formation of Mediterranean depressions. The initiation and movement of these depressions (Figure 8.28) is asso-

ciated with a branch of the Polar Front Jet Stream located at about 35°N. This jet develops during low index phases, when the westerlies over the eastern Atlantic are distorted by a blocking anticyclone at about 20°W, leading to a deep stream of arctic air flowing southwards over the British Isles and France.

Atlantic depressions entering the western Mediterranean as surface lows make up 9 per cent of those affecting the region (see Figure 8.28); 17 per cent form as baroclinic waves south of the Atlas Mountains (the so-called Saharan depressions, which are most important sources of rainfall in late winter and spring) and fully 74 per cent develop in the western Mediterranean to the lee of the Alps and Pyrenees (see Chapter 7H.1). The combination of the lee effect and that of unstable surface air over

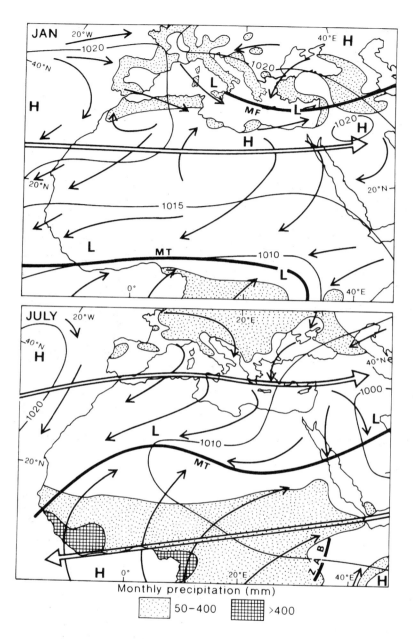

Figure 8.27 The distribution of surface pressure, winds and precipitation for the Mediterranean and North Africa during January and July. The average positions of the Subtropical Westerly and Tropical Easterly Jet Streams, together with the Monsoon Trough (MT), the Mediterranean Front (MF) and the Zaire Air Boundary (ZAB), are also shown.

Monthly precipitation (mm)

50–400   >400

Source: Partly after *Weather in the Mediterranean*, HMSO, 1962 (Crown Copyright Reserved).

the western Mediterranean explains the frequent formation of these *Genoa-type depressions* when conditionally unstable mP air invades the region. These depressions are exceptional in that the instability of the local air in the warm sector gives unusually intense precipitation along the warm front, and the unstable mP air produces heavy showers and thunderstorm rainfall to the rear of the cold front,

especially between 5 and 25°E. This warming of mP (or mA) air is so characteristic as to produce air designated as *mediterranean*. The mean boundary between this mediterranean air mass and cT air flowing north-eastwards from the Sahara is referred to as the Mediterranean front (see Figure 8.27). There may be a temperature discontinuity as great as 12–16°C across it in late winter. Saharan

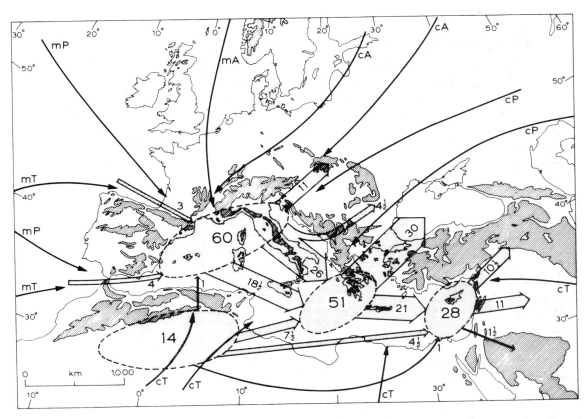

Figure 8.28 Tracks of Mediterranean depressions, showing average annual frequencies, together with air-mass sources.
Source: After *Weather in the Mediterranean*, HMSO, 1962 (Crown Copyright Reserved).

depressions and those from the western Mediterranean move eastwards forming a belt of low pressure associated with this frontal zone and frequently drawing cT northwards ahead of the cold front as the warm, dust-laden *scirocco* (especially in spring and autumn when Saharan air may spread into Europe). The movement of Mediterranean depressions is greatly complicated both by relief effects and by their regeneration in the eastern Mediterranean by fresh cP air from Russia or south-east Europe. Although many depressions travel eastwards over Asia, there is a strong tendency for low-pressure centres to move north-eastward over the Black Sea and Balkans, especially as spring advances. Winter weather in the Mediterranean presents considerable variation, however, particularly as the Subtropical Westerly Jet Stream is highly mobile and may occasionally even coalesce with the southwards-displaced Polar Front Jet Stream.

With high index zonal circulation over the Atlantic and Europe, depressions may pass far enough to the north that their cold-sector air does not reach the Mediterranean, and then the weather there is generally settled and fine. Between October and April, anticyclones are the dominant circulation type for at least 25 per cent of the time over the whole Mediterranean area and in the western basin for 48 per cent of the time. This is reflected in the high mean pressure over the latter area in January (see Figure 8.27). Consequently, although the winter half-year is the rainy period, there are rather few rain-days. On average, rain falls on only six days per month during winter in northern Libya and south-east Spain, yet there are twelve rain-days per month in western Italy, the western Balkan peninsula and the Cyprus area. The higher frequencies (and totals) are related to the areas of cyclogenesis and to the windward sides of peninsulas.

Table 8.4 Number of days with a strong mistral in the south of France.

Speed	J	F	M	A	M	J	J	A	S	O	N	D	Year
≥11 m s⁻¹ (21 kt)	10	9	13	11	8	9	9	7	5	5	7	10	103
≥17 m s⁻¹ (33 kt)	4	4	6	5	3	2	0.6	1	0.6	0	0	4	30

Source: After *Weather in the Mediterranean*, HMSO, 1962.

Regional winds are also related to the meteorological and topographic factors. The familiar cold, northerly winds of the Gulf of Lions (the *mistral*), which are associated with northly mP airflows, are best developed when a depression is forming in the Gulf of Genoa east of a high-pressure ridge from the Azores anticyclone. Katabatic and funnelling effects strengthen the flow in the Rhône valley and similar localities, so that violent winds are sometimes recorded. The mistral may last for several days until the outbreak of polar or continental air ceases. The frequency of these winds depends on their definition. The average frequency of strong mistrals in the south of France is shown in Table 8.4 (based on occurrence at one or more stations from Perpignan to the Rhône in 1924–7). Similar winds may occur along the Catalan coast of Spain (the *tramontana*, see Figure 8.29) and also in the northern Adriatic (the *bora*) and northern Aegean Seas when polar air flows southwards in the rear of an eastward-moving depression and is forced over the mountains (cf. Chapter 5C.2). In Spain, cold, dry, northerly winds occur in several different regions. Figure 8.30 shows the *galerna* of the north coast and the *cierzo* of the Ebro valley.

The generally wet, windy and mild winter season in the Mediterranean is succeeded by a long indecisive spring lasting from March to May, with many false starts of summer weather.

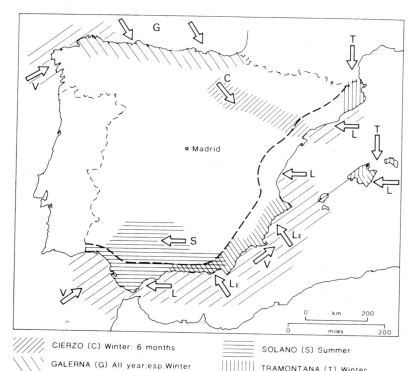

Figure 8.29 Areas affected by the major regional winds in Spain as a function of season.

//// CIERZO (C) Winter: 6 months

\\\\ GALERNA (G) All year: esp. Winter

\\\\ LEBECHE (Le) Spring and Summer

—— LEVANTE (L) All year: N and W limit

≡≡≡ SOLANO (S) Summer

|||| TRAMONTANA (T) Winter

//// VENDAVAL (V) Winter: 6 months

Source: From Tout and Kemp 1985 (by permission of the Royal Meteorological Society).

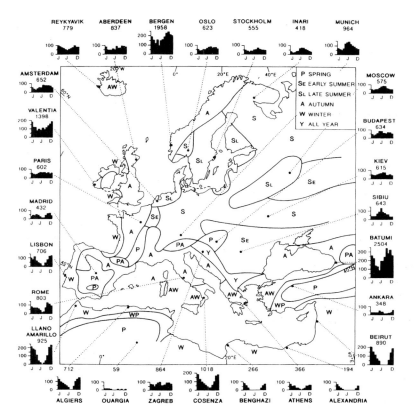

Figure 8.30 Seasons of maximum precipitation for Europe and North Africa, together with average monthly and annual figures (mm) for 28 stations.

Sources: Thorn 1965 and Huttary 1950. Reprinted from D. Martyn (1992) *Climates of the World*, with kind permission from Elsevier Science NL, Sara Burgerhartstraat 25, 1055 KV Amsterdam, the Netherlands.

The spring period, like that of early autumn, is especially unpredictable. In March 1966, a trough moving across the eastern Mediterranean, preceded by a warm southerly *khamsin* and followed by a northerly airstream, brought up to 70 mm of rain in only four hours to an area of the southern Negev Desert. Although April is normally a dry month in the eastern Mediterranean, Cyprus having an average of only three days with 1 mm of rainfall or more, high rainfalls can occur, as in April 1971 when four depressions affected the region. Two of these were Saharan depressions moving eastwards beneath the zone of diffluence on the cold side of a westerly jet and the other two were intensified in the lee of Cyprus. The rather rapid collapse of the Eurasian high-pressure cell in April, together with the discontinuous northward and eastward extension of the Azores anticyclone, encourages the northward displacement of depressions, and, even if higher latitude air does penetrate south to the Mediterranean, the sea surface there is relatively cooler and the air is more stable than during the winter.

By mid-June, the Mediterranean basin is dominated by the expanded Azores anticyclone to the west, while to the south the mean pressure field shows a low-pressure trough extending across the Sahara from southern Asia (see Figure 8.27). The winds are predominantly northerly (e.g. the *etesians* of the Aegean) and represent an eastward continuation of the north-easterly trades. Locally, sea breezes reinforce these winds, but on the Levant coast they cause surface south-westerlies. The day-to-day weather of many parts of the North African coast is largely conditioned by land and sea breezes, involving air up to 1,500 m deep. Depressions are by no means absent in the summer months, but they are usually weak since the anticyclonic character of the large-scale circulation encourages subsidence, and air-mass contrasts are much reduced compared with winter. Thermal lows form from time to time over Iberia and Anatolia, although thundery outbreaks are infrequent due to the low relative humidity.

The most important regional winds in summer are of continental tropical origin. There are a variety

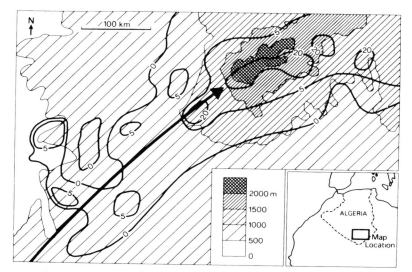

Figure 8.31 Track of a storm and the associated 3-hour rainfall (mm) during September 1950 around Tamanrasset in the vicinity of the Ahaggar Mountains, southern Algeria.

Source: Partly after Goudie and Wilkinson 1977.

of local names for these usually hot, dry and dusty airstreams – *scirocco* (Algeria and the Levant), *lebeche* (south-east Spain) and *khamsin* (Egypt) – which move northwards ahead of eastward-moving depressions. In the Negev, the onset of an easterly *khamsin* may cause the relative humidity to fall suddenly to less than 10 per cent and temperatures to rise to as much as 48°C. In southern Spain, the easterly *solano* brings hot, humid weather to Andalucia in the summer half-year, whereas the coastal *levante* – which has a long fetch over the Mediterranean – is moist and somewhat cooler (see Figure 8.29). Such regional winds occur when the Azores high extends over Western Europe with a low-pressure system to the south.

Many stations in the Mediterranean receive only a few millimetres of rainfall in at least one summer month, yet it is important to realize that the seasonal distribution does not conform to the pattern of simple winter maximum over the whole of the Mediterranean basin. Figure 8.30 shows that this is found in the eastern and central Mediterranean, whereas Spain, southern France, northern Italy and the northern Balkans have more complicated profiles with a maximum in autumn or peaks in both spring and autumn. This double maximum can be interpreted generally as a transition between the continental interior type with summer maximum and the Mediterranean type with winter maximum. A similar transition region occurs in the south-western United States (see Figure 8.22), but local topography in this intermontane zone introduces further irregularities into the regimes.

2 North Africa

The dominance of high-pressure conditions in the Sahara is marked by the low average precipitation figures for this region. Over most of the central Sahara, the mean annual precipitation is less than 25 mm, except for the high plateaus of the Ahaggar and Tibesti, which receive more than 100 mm. Parts of western Algeria have gone at least two years without more than 0.1 mm of rain in any 24-hour period, and most of south-west Egypt as much as five years. However, 24-hour storm rainfalls approaching 50 mm (more than 75 mm over the high plateaus) may be expected in scattered localities and, during a 35-year period of record, excessive short-period rainfall intensities occurred in the vicinity of west-facing slopes in Algeria, such as at Tamanrasset (46 mm in 63 minutes) (Figure 8.31), El Golea (8.7 mm in 3 minutes) and Beni Abes (38.5 mm in 25 minutes). During the summer, rainfall variability is introduced into the southern Sahara by the variable northward penetration of the Monsoon Trough (see Figure 9.2B), which allows on occasion tongues of moist south-westerly air to penetrate far north and produce short-lived low-pressure centres. Study of these Saharan depressions has permitted a clearer picture to emerge of the region. In the upper troposphere at about 200 mb

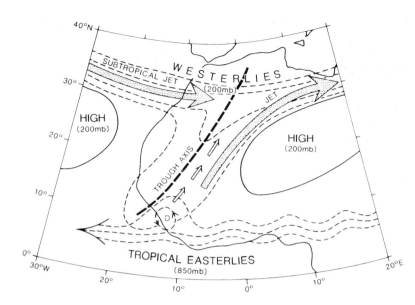

Figure 8.32 Interaction between the westerlies and the tropical easterlies leading to the production of Saharan depressions (D), which move northeastwards along a trough axis.

Source: After Nicholson and Flohn 1980 (copyright © 1980/1982 by D. Reidel Publishing Company. Reprinted by permission).

(12 km), the westerlies overlie the poleward flanks of the subtropical high-pressure belt. Occasionally, the individual high-pressure cells contract away from one another as meanders develop in the westerlies between them, which may extend equatorwards to interact with the low-level underlying tropical easterlies (Figure 8.32). This interaction may lead to the development of lows, which then move north-east along the meander trough associated with rain and thunder. By the time they reach the central Sahara, they are frequently 'rained out' and give rise to duststorms, but they may be reactivated further north by the entrainment of moist mediterranean air. The conditions of subtropical cell separation and the interaction of westerly and easterly circulations are most likely to occur around the equinoxes, or sometimes in winter if the otherwise dominant Azores high-pressure cell contracts westwards. A less extreme case of the equatorward extension of the westerlies is the occasional penetration of cold fronts south from the Mediterranean, bringing heavy rain to restricted desert areas (Figure 8.33). In December 1976, such a depression produced up to 40 mm of rain during two days in southern Mauretania.

3 The semi-arid south-western United States

Both the mechanisms and patterns of the climates of areas dominated by the subtropical high-pressure

cells are rather obscure at present. The inhospitable nature of these arid regions inhibits data collection, and yet the proper interpretation of infrequent meteorological events requires a close network of stations maintaining continuous records over long periods. This difficulty is especially apparent in the

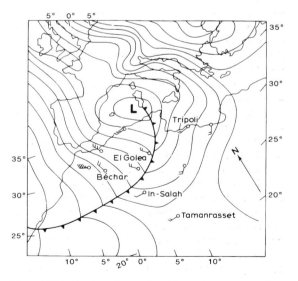

Figure 8.33 Southern extension of a cold front from a Mediterranean depression into the Sahara at 1200 hours on 17 January 1972.

Source: From Breed et al. 1979.

interpretation of desert precipitation data, because much of the rain falls in very local storms irregularly scattered in both space and time. It is convenient to treat aspects of this climatic type here, in that most of the reliable data relate to the less arid regions marginal to the subtropical cell centres, and in particular to the south-western United States.

A series of observations at Tucson (730 m), Arizona, between 1895 and 1957 showed a mean annual precipitation of 27.7 cm falling on an average of about 45 days per year, with extreme annual figures of 61.4 cm and 14.5 cm. Two moister periods in late November to March (receiving 30 per cent of the mean annual precipitation) and late June to September (50 per cent) are separated by more arid seasons from April to June (8 per cent) and October to November (12 per cent). The winter rains are generally prolonged and of low intensity (more than half the falls have an intensity of less than 0.5 cm per hour), falling from altostratus clouds associated with the cold fronts of depressions that are forced to take southerly routes by strong blocking to the north. This occurs during phases of equatorial displacement of the Pacific subtropical high-pressure cell. The re-establishment of the cell in spring, before the main period of intense surface heating and convectional showers, is associated with the most persistent drought periods. Dry westerly to south-westerly flow from the eastern edge of the Pacific subtropical anticyclone is responsible for the low rainfall at this season. During one 29-year period in Tucson, there were eight spells of more than 100 consecutive days of complete drought and twenty-four periods of more than 70 days. The dry conditions occasionally lead to dust storms. Yuma records nine per year, on average, associated with winds averaging 10–15 m s^{-1}. They occur both with cyclonic systems in the cool season and with summer convective activity. Phoenix experiences six to seven per year, mainly in summer, with visibility reduced below 1 km in nearly half of these events.

The period of summer precipitation (known as the summer 'monsoon') is quite sharply defined, beginning in the last week in June and lasting until the middle of September. Precipitation mainly occurs from convective cells initiated by surface heating, convergence or, less commonly, orographic lifting when the atmosphere is destabilized by upper-level troughs in the westerlies. These summer convective storms form in mesoscale clusters, the individual storm cells together covering less than 3 per cent of the surface area at any one time, and

persisting for less than an hour on average. The storm clusters move across the country in the direction of the upper-air motion, and often seem to be controlled in movement by the existence of low-level jet streams at elevations of about 2,500 m. The airflow associated with these storms is generally southerly along the southern and western margins of the Atlantic (or Bermudan) subtropical high, so that in contrast to the winter months the moisture is derived mainly from the Gulf of California during 'surges' associated with the south-south-westerly low-level Sonoran jet (850–700 mb), and secondarily from the Gulf of Mexico during south-easterly flows. The southerly airflow regime at the surface and 700 mb (see Figures 6.4 and 6.10) over the south-west often sets in abruptly around 1 July and it is therefore recognized as a singularity (see A.4, this chapter, and Figure 8.16).

Precipitation from these cells is extremely local (see Plate 10), and commonly concentrated in the mid-afternoon and evening. Intensities are much higher than in winter, half the summer rain falling at more than 10 mm per hour. A particularly well-documented storm occurred over Phoenix, Arizona, at 0600 hours on 22 June 1972. Moving north-east, it produced up to 127 mm of rainfall in the following six hours, including hailstones measuring 19 mm in diameter (Figure 8.34). During the 29-year period, about one-quarter of the mean annual precipitation fell in storms giving 2.5 cm or more per day. These intensities are much less than those associated with rainstorms in the humid tropics, but the sparsity of vegetation in the drier regions allows the rain to produce considerable surface erosion. Thus the highest measurements of surface erosion in the United States are from areas having 30–40 cm of rain per year.

4 The interior and east coast of the United States

The climate of the subtropical south-eastern part of the United States has no exact counterpart in Asia, which is affected by the summer and winter monsoon systems. These are discussed in the next chapter and only the distinctive features of the North American subtropics are examined here. Seasonal wind changes are experienced in Florida, which is within the westerlies in winter and lies on the northern margin of the tropical easterlies in summer, but this is not comparable with the regime in South and South-east Asia. Nevertheless, the summer season rainfall maximum (see Figure 8.22 for Jacksonville)

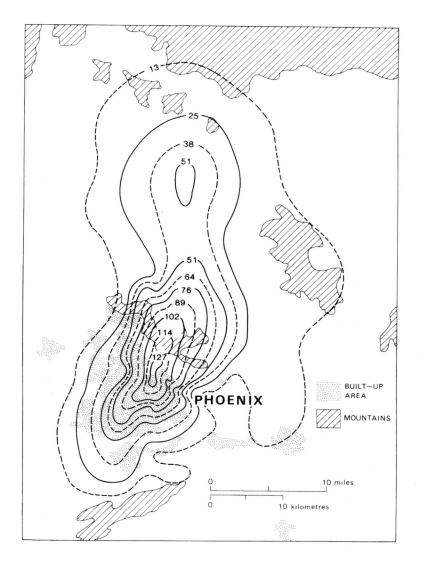

Figure 8.34 A rainstorm over Phoenix, Arizona, on 22 June 1972. The rainfall between 0600 and 1200 hours is shown in millimetres.

BUILT-UP AREA

MOUNTAINS

0   10 miles

0   10 kilometres

Source: After Durrenberger and Ingram 1978 (courtesy State Climatologist, Arizona).

is a result of this changeover. In June, the upper flow over the Florida peninsula changes from north-westerly to southerly as a trough moves westwards and becomes established in the Gulf of Mexico. This deep, moist southerly airflow provides appropriate conditions for convection, and indeed Florida probably ranks as the area with the highest annual number of days with thunderstorms – ninety or more, on average, in the vicinity of Tampa. These often occur in late afternoon, although two factors apart from diurnal heating are thought to be important. One is the effect of sea breezes converging from both sides of the peninsula, and the other is the

northward penetration of disturbances in the easterlies (see Chapter 9). The latter may of course affect the area at any time of day. The westerlies resume control in September–October, although Florida remains under the easterlies during September, when Caribbean tropical cyclones are most frequent and consequently the rainy season is prolonged.

Tropical cyclones contribute an average of 10–15 per cent of the annual rainfall near the Gulf coast and in Florida (Plate 25). According to *Storm Data* reports for 1975–94, hurricanes striking the southern and eastern USA account for over 40 per cent of the total property damage and 20 per cent

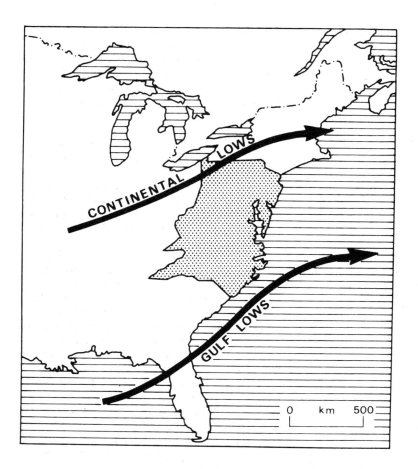

Figure 8.35 The two major depression tracks that affect the winter climate of the mid-Atlantic region of the USA.

Source: Knappenberger and Michaels 1993. From *International Journal of Climatology*, copyright © John Wiley & Sons Ltd. Reproduced with permission.

of the crop damage attributed to extreme weather events in the country. The single most costly natural disaster up to 1989 was Hurricane Hugo ($9 billion), but this was far surpassed by the $27 billion losses caused by Hurricane Andrew over southern Florida and Louisiana in August 1992. In contrast, injuries caused by hurricanes average only 250 per year, as a result of storm warnings and the evacuation of endangered communities.

Winter precipitation along much of the eastern seaboard of the United States is dominated by an apparent oscillation between depression tracks following the Ohio valley (continental lows) and the south-east Atlantic coast (Gulf lows), only one of which is normally dominant during a single winter. The fomer bring below-average winter rainfall and snowfall, but above-average temperatures, to the mid-Atlantic region, whereas the reverse conditions are associated with depressions following the south-east Atlantic coast track (Figure 8.35).

The region of the Mississippi lowlands and the southern Appalachians to the west and north is not simply transitional to the 'interior type', at least in terms of rainfall regime (see Figure 8.22). The profile shows a winter–spring maximum and a secondary summer maximum. The cool season peak is related to westerly depressions moving north-eastwards from the Gulf coast area, and it is significant that the wettest month is commonly March, when the mean jet stream is farthest south. The summer rains are associated with convection in humid air from the Gulf, although this convection becomes less effective inland as a result of the subsidence created by the anticyclonic circulation in the middle troposphere referred to previously (see B.3c, this chapter).

5 Australasia

The South Atlantic and Indian Ocean subtropical high-pressure cells 'bud off' anticyclones, which

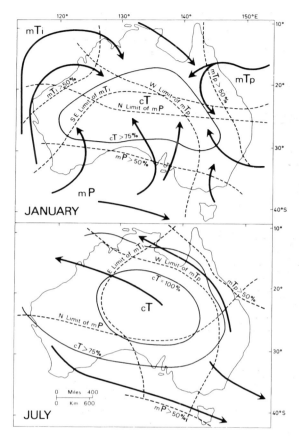

Figure 8.36 Air-mass frequencies, source areas, wind directions and dominance of the cT high-pressure cell over Australia in summer (above) and winter (below).
Source: After Gentilli 1971.

fronts (sometimes termed 'polar') (Figure 8.37). Within these troughs, the subtropical jet stream meanders equatorwards, accelerates (particularly in winter, when it reaches an average velocity of 60 m s^{-1} – its mean annual value being 39 m s^{-1}) and generates upper-air depressions, which move south-eastwards along the fronts (see Figure 8.32 for similar features in North Africa). The variation in strength of the continental anticyclonic tendency and the passage of inter-anticyclonic fronts cause periodic inflows of surrounding maritime tropical air masses from the Pacific (mTp) and the Indian (mTi) Oceans, incursions of maritime polar air (mP) from the south, and variations in strength of the local source of continental tropical (cT) air masses (see Figure 8.36).

The high-pressure conditions over Australia promote especially high temperatures over central and western parts of the continent, towards which there is a major heat transport in summer. These pressures keep average rainfall amounts low, and these normally total less than 250 mm annually over 37 per cent of Australia. In winter, upper-air depressions along the inter-anticyclonic fronts bring rain to south-eastern regions and also, in conjunction with mTi incursions, to south-west Australia. In summer, the southward movement of the Inter-tropical Convergence Zone and its transformation into a Monsoon Trough brings on the wetter season in northern Australia (see Chapter 9D), and the south-east trades blow directly onshore to give rain along the eastern seaboard.

New Zealand is subject to a similar set of climatic controls as the southern part of Australia (Figure 8.38). Anticyclones, separated by troughs associated with cold fronts often deformed into wave depressions, cross the region on average once a week. Their most southerly track (38.5°S) is taken in February. The eastward rate of anticyclonic movement averages about 570 km/day in May to July and 780 km/day in October to December. Anticyclones, occurring some 7 per cent of the time, are associated with settled weather, light winds, sea breezes and some fog. On the eastern (leading) edge of the high-pressure cell the airflow is usually cool, maritime and south-westerly, interspersed with south or south-easterly flow producing drizzle. On the western side of the cell, the airflow is commonly north or north-westerly, bringing mild and humid conditions. In autumn, high-pressure conditions increase from their normal 7 per cent frequency to as much as 22 per cent, giving a drier period.

move eastwards intensifying south-east of South Africa and west of Australia. These warm-core anti-cyclones, extending through the troposphere, are formed by descending air. The continental intensification of the constant eastward progression of these cells causes pressure maps to give the impression of the existence of a stable anticyclone over Australia (Figure 8.36). Some forty anticyclones traverse Australia annually, being somewhat more numerous in spring and summer than in autumn and winter. As in the Indian Ocean, the frequency of anticyclonic centres is greatest in a latitudinal belt around 30°S in winter and 35–40°S in summer; they occur rarely south of 45°S.

Between successive anticyclones are troughs of lower pressure, which contain inter-anticyclonic

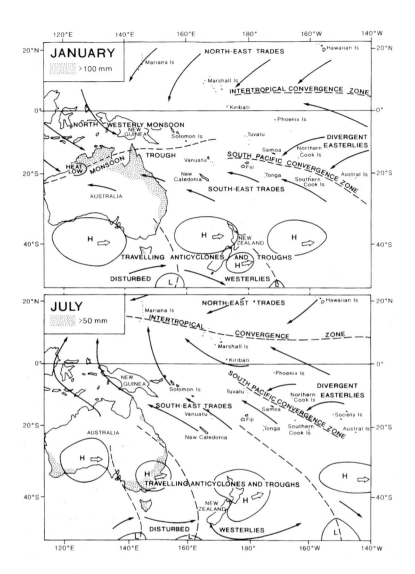

Figure 8.37 Main climatological features of Australasia and the south-west Pacific. Areas with >100 mm (January) and >50 mm (July) mean monthly precipitation for Australia are also shown.

Source: After Steiner, from Salinger, Basher et al. 1995. From International Journal of Climatology, copyright © John Wiley & Sons Ltd. Reproduced with permission.

Simple troughs with undeformed cold fronts and relatively simple interactions between the trailing and leading edge conditions of the anticyclones persist in about 44 per cent of the time during winter, spring and summer, compared with only 34 per cent in autumn. About the same frequency is occupied by frontal deformation into wave depression conditions. If a wave depression forms on the cold front to the west of New Zealand, it usually moves south-east along the front, passing to the south of the country, whereas such a depression forming over New Zealand may take 36–48 hours

to clear the country, bringing prolonged rain in the process. Rainfall in general is predominantly controlled by relief, especially by the Southern Alps. West- or north-west-facing mountains receive an average annual precipitation in excess of 2,500 mm, with some parts of South Island exceeding 8,000 mm (see Figure 8.38 and also Figure 4.16). The highest daily rainfall was recorded in Milford Sound on 17 April 1939 at 559 mm. The eastern lee areas have much lower amounts, with less than 500 mm in some parts. North Island has a winter precipitation maximum, but South Island, under greater influence

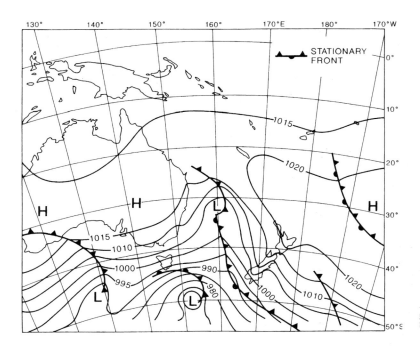

Figure 8.38 The synoptic situation at 0000 hours on 1 September 1982, resulting in heavy rainfall in the Southern Alps of New Zealand.

Sources: After Hessell; from Wratt et al. 1996. From *Bulletin of the American Meteorological Society* by permission of the American Meteorological Society.

of depressions moving in the belt of the southern westerlies, has a more variable seasonal maximum.

D HIGH LATITUDES

1 The southern westerlies

The strong zonal airflow in the belt of the southern westerlies, which is apparent only on averaged maps, is driven by a belt of major meridional heat exchange that produces a major frontal zone characterized by the continual passage of depressions and ridges of higher pressure. This belt extends southwards from about 30°S in July and 40°S in January (see Figures 7.19 and 8.37) throughout the Southern Ocean to the Antarctic Trough of low pressure, which fluctuates between 60 and 72°S. The Antarctic Trough is a region of cyclonic stagnation and decay which tends to be located furthest south during equinoctial periods. In the vicinity of New Zealand, the westerly airflow persists throughout the year at an elevation of 3–15 km in the belt 20–50°S, reaching its maximum strength as a jet stream at 150 mb (13.5 km) (25–30°S) flowing at 60 m s^{-1} in May–August and decreasing to 26 m s^{-1} in February. In the Pacific, the strength of the westerlies depends on the meridional pressure differences between 40 and 60°S, being on average greatest all

the year south of Western Australia and west of southern Chile.

Many depressions form as waves on the interanticyclonic fronts, which move south-eastwards into the belt of the westerlies. Others form in the belt itself at preferred locations such as south of Cape Horn, and at around 45°S in the Indian Ocean in summer and in the South Atlantic off the South American coast and around 50°S in the Indian Ocean in winter. The Polar Front (see Figure 7.21) is most closely associated with the sea-surface temperature gradient across the Antarctic convergence, whereas the sea ice boundaries further south are surrounded by equally cold surface water (Figure 8.39B).

In the South Atlantic, depressions travel at about 1,300 km/day near the northern edge of the belt, slowing to 450–850 km/day within 5–10° latitude of the Antarctic trough. In the Indian Ocean, eastward velocities range from 1,000 to 1,300 km/day in the belt 40–60°S, reaching the maximum in a core at 45–50°S. Pacific depressions tend to be similarly located and generally form, travel and decay within a period of about a week. As in the northern hemisphere, a high zonal index results from a strong meridional pressure gradient and is associated with wave disturbances propagated eastwards at high speed with irregular and often violent winds and

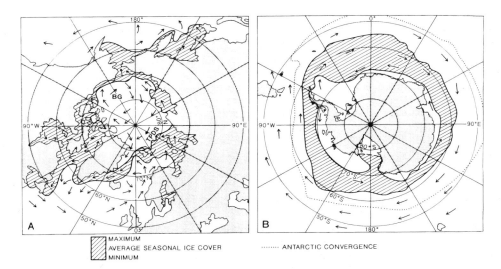

MAXIMUM
AVERAGE SEASONAL ICE COVER
MINIMUM

········· ANTARCTIC CONVERGENCE

Figure 8.39A Surface currents in the Arctic, together with average autumn minimum and spring maximum sea ice extent.
Sources: Maythum 1993, and Barry 1983.

Figure 8.39B Southern Ocean surface circulation, convergence zones and seasonal ice limits in March and September.
Source: After Barry 1986. Copyright © Plenum Publishing Corp., New York. Published by permission.

zonally oriented fronts. A low zonal index results in high-pressure ridges extending further south and low-pressure centres located further north. However, break-up of the flow, leading to blocking, is less common and less persistent in the southern hemisphere than in the northern.

The southern westerlies are linked to the belt of travelling anticyclones and troughs by cold fronts, which connect the inter-anticyclonic troughs of the latter with the wave depressions of the former. Although storm tracks of the westerlies are usually well to the south of Australia, fronts may extend north into the continent, particularly from May, when the first rains occur in the south-west, and in the July midwinter month an average of three depression centres skirt the south-west coast. When a deep depression moves to the south of New Zealand, the passage of the cold front causes that country to be covered first by a warm, moist westerly or northerly airflow and then by cooler southerly air. Often a series of such depressions may follow at intervals of 12–36 hours, each cold front being followed by progressively colder air. Further east over the South Pacific, the northern fringe of the belt of the southern westerlies is influenced by north-westerly winds, changing to west or south-

west as depressions move to the south. This weather pattern is interrupted by periods of easterly winds if depression systems track along lower latitudes than usual.

2 The sub-Arctic

The longitudinal differences in mid-latitude climates persist into the northern polar margins, giving rise to maritime and continental sub-types, modified by the extreme radiation conditions in winter and summer. For example, radiation receipts in summer along the Arctic coast of Siberia compare favourably, by virtue of the long daylight, with those in lower middle latitudes. The maritime type is found in coastal Alaska, Iceland, northern Norway and adjoining parts of the former USSR. Winters are cold and stormy, with very short days. Summers are cloudy but mild with mean temperatures of about 10°C. For example, Vardø in northern Norway (70°N, 31°E) has monthly mean temperatures of –6°C in January and 9°C in July, while Anchorage, Alaska (61°N, 150°W) records –11°C and 14°C, respectively. Annual precipitation is generally between 60 and 125 cm, with a cool season maximum and about six months of snow cover.

The weather is mainly controlled by depressions, which are weakly developed in summer. In winter, the Alaskan area is north of the main depression tracks and occluded fronts and upper troughs are prominent, whereas northern Norway is affected by frontal depressions moving into the Barents Sea. Iceland is similar to Alaska, although depressions often move slowly over the area and occlude, whereas others moving north-eastwards along the Denmark Strait bring mild, rainy weather.

The interior, cold-continental climates have much more severe winters, although precipitation amounts are smaller. At Yellowknife (62°N, 114°W), for instance, the mean January temperature is only –28°C. In these regions, *permafrost* (permanently frozen ground) is widespread and often of great depth. In summer, only the top 1–2 m of ground thaw and as the water cannot readily drain away this 'active layer' often remains waterlogged. Although frost may occur in any month, the long summer days usually give three months with mean temperatures above 10°C, and at many stations extreme maxima are 32°C or more (see Figure 8.18). The Barren Grounds of Keewatin, however, are much cooler in summer due to the extensive areas of lake and muskeg, and only July has a mean daily temperature of 10°C. Labrador–Ungava to the east, between 52° and 62°N, is rather similar with very high cloud amounts and maximum precipitation in June–September (Figure 8.40). In winter, conditions fluctuate between periods of very cold, dry, high-pressure weather and spells of dull, bleak, snowy weather as depressions move eastwards or occasionally northwards over the area. In spite of the very low mean temperatures in winter, there have been occasions when maxima have exceeded 4°C during incursions of maritime Atlantic air. Such variability is not found in eastern Siberia, which is intensely continental, apart from the Kamchatka Peninsula, with the northern hemisphere's *cold pole* located in the remote north-east (see Figure 2.11A). Verkhoyansk and Oimyakon have a January mean of –50°C, and both have recorded an absolute minimum of –67.7°C. Stations located in the valleys of Northern Siberia record, on average, strong to extreme frosts 50 per cent of the time during six months of the year, but very warm summers (Figure 8.41).

3 The polar regions

Common to both polar regions is the bi-annual alternation between polar night and polar day, and

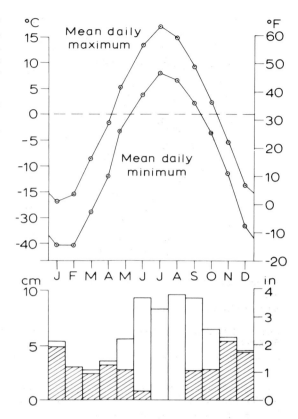

Figure 8.40 Selected climatological data for McGill Sub-Arctic Research Laboratory, Schefferville, 1955–62. The shaded portions of the precipitation represent snowfall, expressed as water equivalent.
Source: Data from J. B. Shaw and D. G. Tout.

the prevalence of snow and ice surfaces. These factors control the surface energy budget regimes and low annual temperatures of polar regions (see Chapter 10B). The polar regions are also energy sinks for the global atmospheric circulation (see Chapter 6C.1) and in both cases they are overlain by large-scale circulation vortices in the middle troposphere and above (see Figures 6.3 and 6.4). In many other respects, the two polar regions differ markedly because of geographical factors. The north polar region comprises the Arctic Ocean, with its almost year-round sea ice cover (see Plate A), surrounding tundra land areas, the Greenland Ice Sheet and numerous smaller ice caps in Arctic Canada, Svalbard and the Siberian Arctic Islands. In contrast, the south polar region is occupied by the Antarctic continent, with an ice plateau 3 to 4

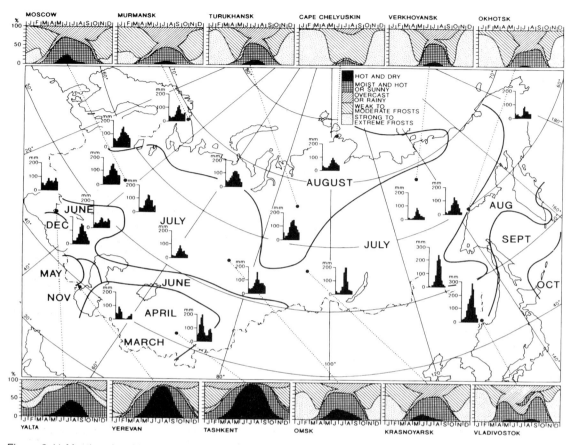

Figure 8.41 Months of maximum precipitation, annual regimes of mean monthly precipitation and annual regimes of mean monthly frequencies of five main weather types in the former USSR.

Source: Reprinted from P. E. Lydolph (1977) *Climates of the Soviet Union*, with kind permission from Elsevier Science NL, Sara Burgerhartstraat 25, 1055 KV Amsterdam, the Netherlands.

km high, floating ice shelves in the Ross Sea and Weddell Sea embayments, and surrounded by a seasonally ice-covered ocean. Accordingly, the Arctic and Antarctic are treated separately.

a The Arctic

At 75°N, the sun is below the horizon for about ninety days, from early November until early February. Winter air temperatures over the Arctic Ocean average about –32°C, but they are usually 10–12°C higher some 1,000 m above the surface as a result of the strong radiative temperature inversion. The winter season is generally stormy in the Eurasian sector, where low-pressure systems enter the Arctic Basin from the North Atlantic, whereas anticyclonic conditions predominate north of Alaska over the Beaufort and Chukchi Seas. In spring, high

pressure prevails, centred over the Canadian Arctic Archipelago–Beaufort Sea.

The average 3 to 4 m thickness of sea ice in the Arctic Ocean permits little heat loss to the atmosphere and largely decouples the ocean and atmosphere systems in winter and spring. The winter snow accumulation on the ice averages 25–30 cm depth. Only when the ice fractures, forming a *lead*, or where persistent offshore winds and/or upwelling warm ocean water form an area of open water and new ice (called a *polynya*), is the insulating effect of sea ice disrupted. The ice in the western Arctic circulates clockwise in a gyre driven by the mean anticyclonic pressure field. Ice from the northern margin of this gyre, and ice from the Eurasian sector, moves across the North Pole in the Transpolar Drift Stream and exits the Arctic via Fram

Strait and the East Greenland Current (see Figure 8.39A). This export largely balances the annual thermodynamic ice growth in the Arctic Basin. In late summer, the Eurasian shelf seas and the coastal section of the Beaufort Sea are mostly ice-free.

In summer, the Arctic Ocean has mostly overcast conditions with low stratus and fog (see Figure 4.13). Snowmelt and extensive meltwater puddles on the ice keep air temperatures around freezing. Low-pressure systems tend to predominate, entering the basin from either the North Atlantic or Eurasia. Precipitation may fall as rain or snow, with the largest monthly totals in late summer–early autumn. However, the mean annual net precipitation minus evaporation over the Arctic, based on atmospheric moisture transport calculations, is only about 16 cm.

On Arctic land areas, there is a stable snow cover from mid-September until early June, when melt occurs within 10–15 days. As a result of the large decrease in surface albedo, the surface energy budget undergoes a dramatic change to large positive values (Figure 8.42). The tundra is generally wet and boggy as a result of the *permafrost table* only 0.5–1.0 m below the surface, which prevents drainage. Thus the net radiation is expended primarily for evapotranspiration. Permanently frozen ground is over 500 m thick in parts of Arctic North America and Siberia and extends under the adjacent Arctic coastal shelf areas. Much of the Queen Elizabeth Islands, the Northwest Territories of Canada and the Siberian Arctic Islands is cold, dry polar desert, with gravel or rock surfaces, or ice caps and glaciers. Nevertheless, 10–20 km inland from the Arctic coasts in summer, daytime heating disperses the stratiform cloud and afternoon temperatures may rise to 15–20°C.

The Greenland ice sheet, 3 km thick and covering an area of 1.7 million km², contains enough water to raise global sea level by over 7 m if it were all melted. However, there is no melting above the equilibrium line altitude (where accumulation balances ablation), which is at 2,000 m (1,000 m) elevation in the south (north) of Greenland. The ice sheet largely creates its own climate. It deflects cyclones moving from Newfoundland, either northwards into Baffin Bay or north-eastwards towards Iceland. These storms give heavy snowfall in the south and on the western slope of the ice sheet. A persistent shallow inversion overlays the ice sheet with downslope katabatic winds averaging 10 m s⁻¹, except when storm systems cross the area.

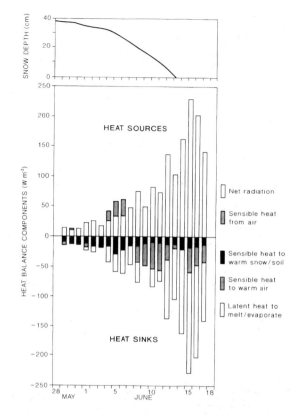

Figure 8.42 The effect of tundra snow cover on the surface energy budget at Barrow, Alaska, during the spring melt. The lower graph shows the daily net radiation and energy terms.

Source: Weller and Holmgren 1974. From *Journal of Applied Meteorology*, by permission of the American Meteorological Society.

b Antarctica

Except for protruding peaks in the Transantarctic Mountains and Antarctic Peninsula, and the Dry Valleys of Victoria Land (77°S, 160°E) over 97 per cent of Antarctica is covered by a vast continental ice sheet. The ice plateau averages 1,800 m elevation in West Antarctica and 2,600 m in East Antarctica, where it rises above 4,000 m (82°S, 75°E). In September, sea ice averaging 0.7–1.0 m in thickness covers 20 million km² of the Southern Ocean, but 80 per cent of this melts each summer.

Over the ice sheet, temperatures are almost always well below freezing. The South Pole (2,800 m elevation) has a mean summer temperature of −28°C and a winter temperature of −58°C. Vostok (3,500 m) recorded −89°C in July 1983, a world record

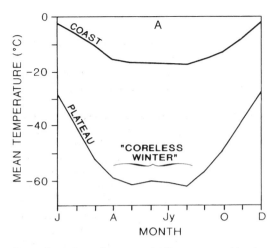

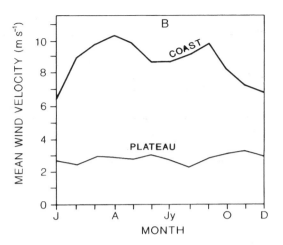

Figure 8.43 Annual course of (A) mean monthly air temperature (°C) and (B) wind speed (m s⁻¹) for 1980–89 at Dome C (3,280 m), 74.5 °S, 123.0 °E (plateau) and D-10, an automatic weather station at 240 m, 66.7 °S, 139.8 °E (coast).

Source: Stearns et al. 1993. From D. H. Bromwich and C. R. Stearns (eds) Antarctic Meteorology and Climatology, 1993, American Geophysical Union.

minimum. Mean monthly temperatures are consistently close to their winter value for the six months between equinoxes, creating a so-called 'coreless winter' (Figure 8.43). The radiative loss of energy is balanced by atmospheric poleward energy transfer. Nevertheless, there are considerable day-to-day temperature changes associated with cloud cover increasing downward long-wave radiation, or winds mixing warmer air from above the inversion down to the surface. Over the plateau, the inversion strength is about 20–25°C. Precipitation is almost impossible to measure, as a result of blowing and drifting snow, but snow pit studies indicate an annual accumulation varying from less than 50 mm over the high plateaus above 3,000 m elevation to 500–800 mm in some coastal areas of the Bellingshausen Sea and parts of East Antarctica.

Lows in the southern westerlies have a tendency to spiral clockwise towards Antarctica, especially from south of Australia towards the Ross Sea, from the South Pacific towards the Weddell Sea, and from the western South Atlantic towards Kerguelen Island and East Antarctica (Figure 8.44). Over the adjacent Southern Ocean, cloudiness exceeds 80 per cent year-round at 60–65°S (see Figures 2.8 and 4.13) due to the frequent cyclones, but coastal Antarctica has more synoptic variability, associated with alternating lows and highs. Over the interior, cloud cover is generally less than 40–50 per cent and half of this amount in winter.

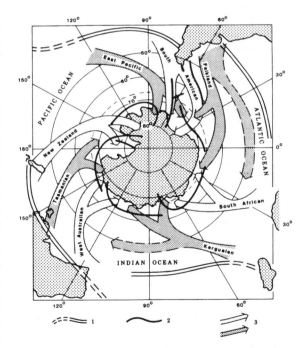

Figure 8.44 Southern hemisphere cyclone paths affecting Antarctica and major frontal zones in winter.
1. Polar Front; 2. Antarctic Front; 3. Cyclone trajectories.

Source: Carleton 1987. From J. Oliver and R. Fairbridge (eds) Encyclopedia of Climatology, Van Nostrand Reinhold, NY, 1987. Copyright © Chapman & Hall, New York. Reproduced by permission.

The poleward air circulation in the tropospheric polar vortex (see Figure 6.3) leads to subsiding air over the Antarctic Plateau and outward flow over the ice sheet surface. The winds represent a balance between gravitational acceleration, Coriolis force (acting to the left), friction and inversion strength. On the slopes of the ice sheet, there are stronger downslope katabatic flows, and extreme speeds are observed in some coastal locations. Cape Denison (67°S, 143°E), Adelie Land, recorded average daily wind speeds of >18 m s^{-1} on over 60 per cent of days in 1912–13.

SUMMARY

Seasonal changes in the Icelandic low and the Azores high, together with variations in cyclone activity, control the climate of Western Europe. The eastward penetration of maritime influences related to these atmospheric processes and to the warm waters of the North Atlantic Current is illustrated by mild winter temperatures, the seasonality of precipitation regimes and indices of continentality in Western Europe. Topographic effects on precipitation, snowfall, length of growing seasons and local winds are particularly marked over the Scandinavian Mountains, the Scottish Highlands and the Alps. Weather types in the British Isles can be described in terms of seven basic airflow patterns, the frequency and effects of which vary considerably with season. Recurrent weather spells about a particular date (singularities), such as the tendency for anticyclonic weather in mid-September, have been recognized in Britain and major seasonal trends in occurrence of airflow regimes can be used to define five natural seasons. Abnormal weather conditions (synoptic anomalies) are associated particularly with blocking anticyclones, which are especially prevalent over Scandinavia and may give rise to cold, dry winters and warm, dry summers.

The climate of North America is similarly affected by pressure systems that generate air masses of varying seasonal frequency. In winter, the subtropical high-pressure cell extends north over the Great Basin with anticyclonic cP air to the north over Hudson Bay. Major depression belts occur at about 45–50°N, from the central USA to the St Lawrence, and along the east coast of Newfoundland. The Arctic Front is located over north-west Canada, the Polar Front lies along the north-east coast of the United States, and between the two a maritime (arctic) front may occur over Canada. In summer, the frontal zones move north, the Arctic Front lying along the north coast of Alaska, Hudson Bay and the St Lawrence being the main locations of depression tracks. Three major North American singularities concern the advent of spring in early March, the midsummer northward displacement of the subtropical high-pressure cell, and the Indian summer of September–October. In western North America, the Coast Ranges inhibit the eastward spread of precipitation, which may vary greatly locally (e.g. in British Columbia), especially as regards snowfall. The strongly continental interior and east of the continent experiences some moderating effects of Hudson Bay and the Great Lakes in early winter, but with locally significant snow belts. The climate of the east coast is dominated by continental pressure influences. Cold spells are produced by winter outbreaks of high-latitude cA/cP air in the rear of cold fronts. Zonal westerly airflows give rise to chinook winds in the lee of the Rockies. The major moisture sources of the Gulf of Mexico and the North Pacific produce regions of differing seasonal regime: the winter maximum of the west coast is separated by a transitional intermontane region from the interior, with a general warm season maximum; the north-east has a relatively even seasonal distribution. Moisture gradients, which strongly influence vegetation and soil types, are predominantly east–west in central North America, in contrast to the north–south isotherm pattern.

The semi-arid south-western United States comes under the complex influence of the Pacific and Bermudas high-pressure cells, having extreme rainfall variations, with winter and summer maxima mainly due to depression and local thunderstorms, respectively. The interior and east coast of the United States is dominated by westerlies in winter and southerly thundery airflows in summer.

The subtropical margin of Europe consists of the Mediterranean region, lying between the belts dominated by the westerlies and the

continued

Saharan–Azores high-pressure cells. The collapse of the Azores high-pressure cell in October allows depressions to move and form over the relatively warm Mediterranean Sea, giving well-marked orographic winds (e.g. mistral) and stormy, rainy winters. Spring is an unpredictable season marked by the collapse of the Eurasian high-pressure cell to the north and the strengthening of the Saharan–Azores anticyclone. In summer, the latter gives dry, hot conditions with strong local southerly airstreams (e.g. scirocco). The simple winter rainfall maximum is most characteristic of the eastern and southern Mediterranean, whereas in the north and west, autumn and spring rains become more important. North Africa is dominated by high-pressure conditions. Infrequent rainfalls may occur in the north with extratropical systems and to the south with Saharan depressions.

Australian weather is determined largely by travelling anticyclone cells from the southern Indian Ocean and intervening low-pressure troughs and fronts. In the winter months, such frontal troughs give rains in the south-east. The climatic controls in New Zealand are similar to those in southern Australia, but South Island is greatly influenced by depressions in the southern westerlies. Rainfall amounts vary strongly with the relief.

The southern westerlies (30–40° to 60–70°S) dominate the weather of the Southern Ocean. The strong, mean zonal flow conceals great day-to-day synoptic variability and frequent frontal passages. The persistent low-pressure systems in the Antarctic trough produce the highest year-round zonally averaged global cloudiness.

The Arctic margins have six to nine months of snow cover and extensive areas of permanently frozen ground (permafrost) in the continental interiors, whereas the maritime regions of northern Europe and northern Canada–Alaska have cold, stormy winters and cloudy milder summers influenced by the passage of depressions. Northeast Siberia has an extreme continental climate.

The Arctic and Antarctic differ markedly because of the types of surface – a perennially ice-covered Arctic Ocean surrounded by land areas and a high Antarctic ice plateau surrounded by the Southern Ocean and thin seasonal sea ice. The Arctic is affected by mid-latitude cyclones from the North Atlantic and in summer from Northern Asia. A surface inversion dominates Arctic conditions in winter and year-round over Antarctica. In summer, stratiform cloud blankets the Arctic and temperatures are near 0°C. Sub-zero temperatures persist year round on the Antarctic continent and katabatic winds dominate the surface climate. Precipitation amounts are low, except in a few coastal areas, in both polar regions.

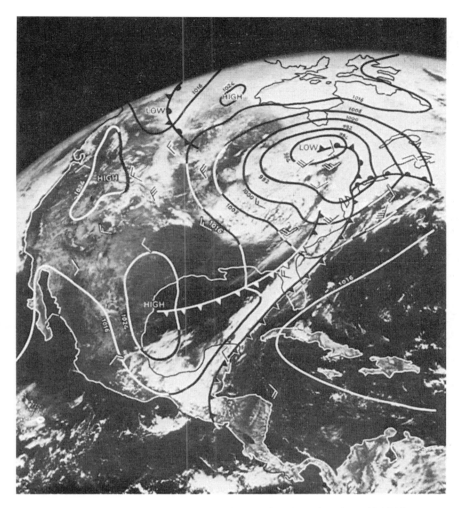

Plate 20 Satellite photograph of North and Central America taken from 37,000 km on an April day. A depression lies to the south-east of Hudson Bay and a belt of rain or drizzle marks the cold front from Yucatan to New England. This cloud band suggests high-level tropical/mid-latitude interaction. The Great Lakes are snowy and windy but elsewhere conditions are mostly clear with light winds (*NOAA photograph; courtesy National Geographic Society*).

Plate 21 Infra-red photograph from a NOAA satellite taken at 1502 hours GMT on 9 September 1983 showing a convective cloud band associated with a polar trough conveyor belt (low-level jet) wrapped around the leading edge of a cold pool lying to the north-west of a stratiform cloud band marking a polar front conveyor belt (upper-level jet) (*from Browning 1985, fig. 21b, p. 314; see also Figure 7.18, this volume*). (*Permission from Dept of Electrical Engineering and Electronics, University of Dundee.*) (*Reproduced by permission of the Controller of Her Majesty's Stationery Office.*)

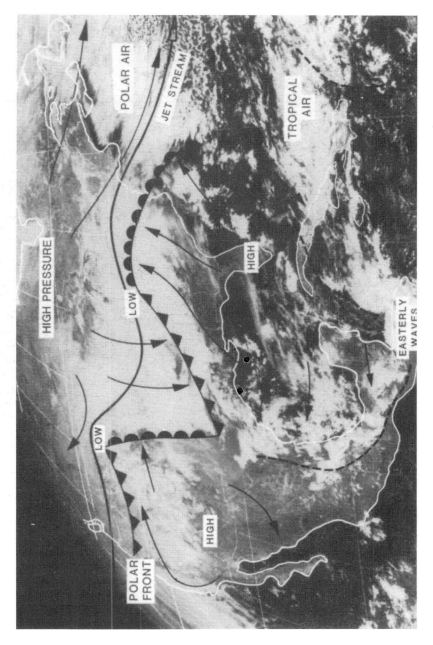

Plate 22 Satellite photograph of North America taken at 1700 hours GMT on 12 February 1979 from a GOES weather satellite located some 35,800 km above the equator. Two depressions are located below an upper jet stream located between polar and subtropical high-pressure cells. The more westerly depression is forming in the lee of the Rocky Mountains; cold polar air streaming off New England is becoming cloudy over the warmer sea; weak easterly waves (dashed) have developed equatorward of the subtropical high-pressure cells over the Caribbean and Central America (*courtesy NOAA*).

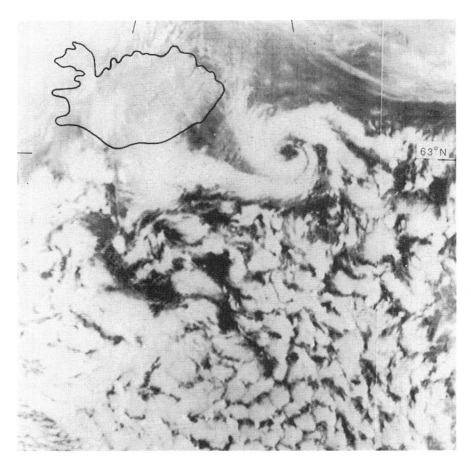

Plate 23 A polar low near Iceland, 14 January 1984, as seen on a visible band DMSP satellite image. This mesoscale low and the closed cellular cloud patterns to the south developed in a northerly airflow behind an occluded depression situated over the coast of Norway (*courtesy National Snow and Ice Data Center, University of Colorado, Boulder*).

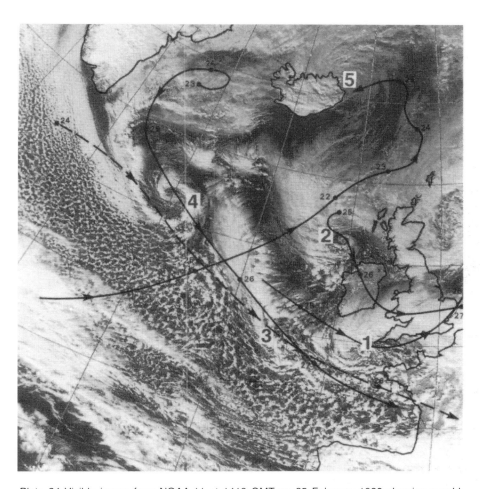

Plate 24 Visible image from NOAA-11 at 1418 GMT on 25 February 1989 showing a cold, north-westerly airstream over the North Atlantic with open cellular convection, bounded on the south by a weak surface front. To the north-east are several low pressure systems: 5 is the filling remains of a frontal depression which crossed the area in a north-easterly direction between the 21 and 25 February, behind which the cold north-westerly outbreak occurred. The passage of this depression assisted the formation of four polar low vortices (1-4) which formed between 24 and 25 February and moved in a south-easterly direction, suppressing convection in their vicinity; 3 moved the fastest but 1 had a central pressure of only 950 mb and brought strong winds and damage to the south of France. The dates marked on the trajectories of the low-pressure centres show their positions in the early morning of those days (*photograph courtesy of NERC Satellite Station, University of Dundee; description from the* Meteorological Magazine, *1989, p. 116, by permission of the Controller of Her Majesty's Stationery Office*).

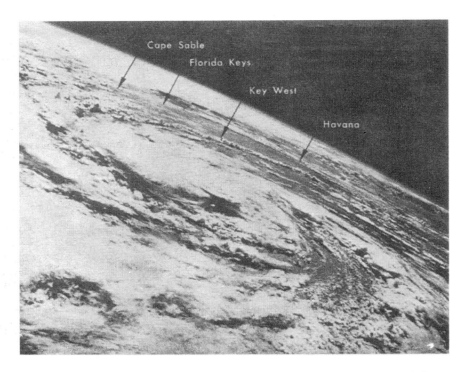

Plate 25 Hurricane 'Gladys' just west of Florida, photographed towards the south from Apollo 7 manned spacecraft at an elevation of 179 km on 17 October 1968. At this time, wind speeds up to 40 m s^{-1} were reported (*NASA photograph*).

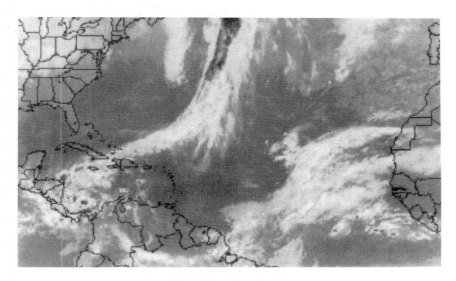

Plate 26 Satellite image showing a major North Atlantic front extending south-west into the Caribbean where it links with a small tropical disturbance lying between Jamaica and Central America. Time 0300 hrs GMT on the 27 November 1995 (*reproduced from the Internet by permission from INTELLicast Corp. Copyright WSI Corp.*)

Plate 27 A satellite infra-red mosaic of eastern Asia and the western North Pacific showing two mid-latitude depression systems and typhoons 'Wendy' (28°N, 126°E) and 'Virginia' (22°N, 147°N) on 29 July 1978, about 0900 local time (Tokyo). The typhoons had maximum winds of about 36 ms^{-1} and sea-level pressure minima of about 965 mb ('Wendy') and 975 mb ('Virginia'). A subtropical high-pressure ridge about 35°N separates the tropical and mid-latitude storms minimizing any interaction (*Defense Meteorological Satellite Program imagery, National Snow and Ice Data Center, University of Colorado, Boulder*).

Plate 28 Visible image taken by METEOSAT-2, geostationary some 35,900 km above the equator, at 1130 hours GMT on 25 September 1983. A broad ITCZ lies across West and East Africa exhibiting centres of varied convective activity including cumulus towers over Central and East Africa. The subtropical high-pressure belt is particularly well developed in the northern hemisphere, as are depressions in the belts of the northern and southern westerlies. A broad wedge of cloud extends west over the South Atlantic from Angola and Gabon in the tropical easterly flow and there is a narrow belt of low cloud in the zone of coastal up-welling off Namibia (*METEOSAT image supplied by the European Space Agency*).

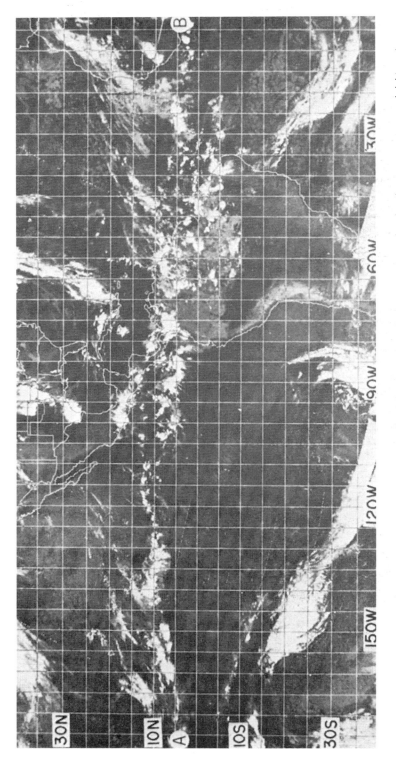

Plate 29 Infra-red photograph showing the Intertropical Convergence Zone on 1 June 1971, along which active thunderstorms appear as bright spots. There is also a cyclonic cloud band in the western South Pacific representing the South Pacific Convergence Zone (*NOAA-1 photograph, World Meteorological Organization 1973*).

Plate 30 View south over Florida from the Gemini V manned spacecraft at an elevation of 180 km on 22 August 1965, with Cape Kennedy launching site in the foreground. Cumulus clouds have formed over the warmer land, with a tendency to align in east-west 'streets', and are notably absent over Lake Okeechobee. In the south, thunder-head anvils can be seen (*NASA photograph*).

Plate 31 An air view looking south-eastwards towards the line of high cumulus towers marking the convergence zone near the Wake Island wave trough shown in Figure 9.7 (*from Malkus and Riehl 1964*).

Plate 32 Computer-enhanced satellite image showing hurricanes (Humberto H, Iris I, and Louis L), a tropical storm (Karen K) and thunderstorms (eastern Gulf of Mexico and Panama/Colombia) over the western Atlantic region at the end of August 1995. On only 10 occasions since the start of records in 1893 have three hurricanes and a tropical storm been reported simultaneously in this region.

Plate 33 Two hurricanes with opposite rotary circulations formed simultaneously on either side of the equator just east of New Guinea. Cyclone Namn is in the southern hemisphere. The GMS-3 image was taken at 2100 hours LST (1000 hours GMT) on 18 May 1986. The arrows show the directions and distances of the movement of the hurricane eyes during the subsequent 15 hours (*courtesy National Space Development Agency of Japan; description courtesy of Mr S. Jones*).

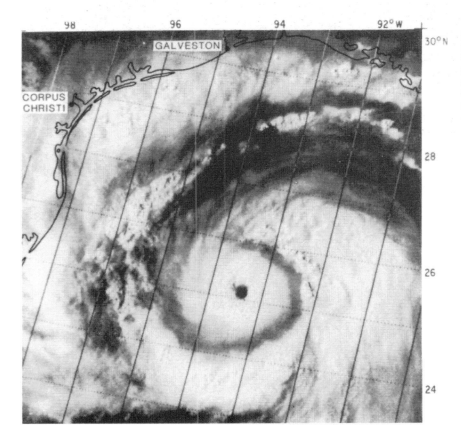

Plate 34 Visible satellite image of hurricane Allen, located south of the Texas coast, taken at 2123 hours GMT on 8 August 1980 showing the clear eye and spiral cloud bands (*photograph courtesy National Hurricane Center, Miami, Florida, USA*).

Plate 35 Visible satellite image showing five large tropical cloud clusters topped by cirrus shields situated between latitudes 5° and 10°N in the vicinity of West Africa, together with one squall-line cloud cluster at 15°N having a well-defined arc cloud squall line on its leading (south-west) edge. Taken by SMS-1 satellite at 1130 hours GMT on 5 September 1974 (*courtesy NOAA*).

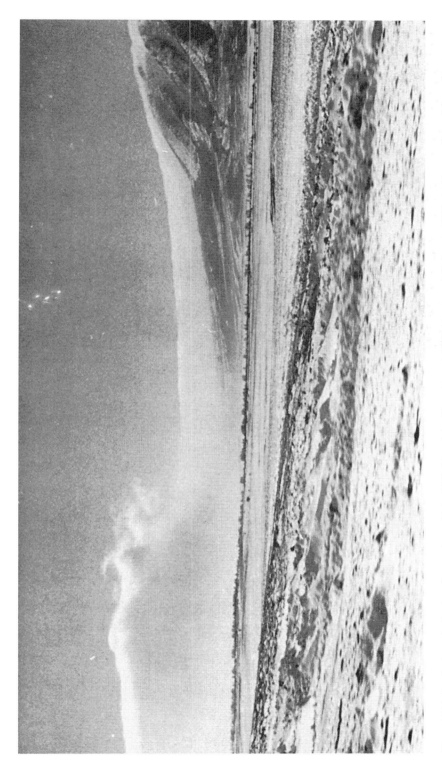

Plate 36 Hydraulic jump in the lee (i.e. eastern side) of Elk Mountain, Wyoming, at 1345 hours on 1 February 1973, looking south-west from the north-eastern edge of the Elk Mountain (*photograph courtesy Dr August H. Auer, Jr, University of Wyoming*).

Plate 37 View westwards over Edinburgh showing Arthur's Seat in the foreground and the Castle rising above urban smog in the middle distance. The smog thickens to the north (*right*) over the industrial suburb of Leith, merging with sea mist over the Firth of Forth at the far right. This photograph was taken some considerable time ago, before pollution controls were enforced (*courtesy Aerofilms Ltd*).

9

Tropical weather and climate

Fifty per cent of the surface of the globe lies between latitudes 30°N and 30°S, and over 75 per cent of the world's population inhabits climatically tropical lands. Tropical climates are, therefore, of especial geographical interest.

The latitudinal limits of tropical climates vary greatly with longitude and season, and tropical weather conditions may reach well beyond the Tropics of Cancer and Capricorn. For example, the summer monsoon extends to 30°N in South Asia, but only to 20°N in West Africa, while in late summer and autumn tropical hurricanes may affect 'extra-tropical' areas of East Asia and eastern North America. Not only do the tropical margins extend seasonally polewards, but in the zone between the major subtropical high-pressure cells there is frequent interaction between temperate and tropical disturbances. Plate 26 illustrates a situation where there is such interaction between low and middle latitudes, whereas Plate 27, in contrast, shows distinct tropical and mid-latitude storms. In general, the tropical atmosphere is far from being a discrete entity and any meteorological or climatological boundaries must be arbitrary. There are, nevertheless, a number of distinctive features of tropical weather and these are discussed below.

Several basic factors help to shape tropical weather processes and also affect their analysis and interpretation. First, the Coriolis parameter approaches zero at the equator, so that winds may depart considerably from geostrophic balance. Pressure gradients are also generally weak, except for tropical storm systems. For these reasons, tropical weather maps usually depict streamlines, not isobars or geopotential heights. Second, temperature gradients are characteristically weak. Spatial and temporal variations in moisture content are much more significant diagnostic characteristics of climate. Third, diurnal land/sea breeze regimes play a major role in

coastal climates, in part as a result of the almost constant day length and strong solar heating. There are also semi-diurnal pressure oscillations of 2–3 mb, with minima around 0400 and 1600 hours and maxima around 1000 and 2200 hours. Fourth, the annual regime of incoming solar radiation, with the sun overhead at the equator in March and September and over the Tropics at the respective summer solstices, is reflected in the seasonal variations of rainfall at some stations. However, as discussed below, dynamic factors greatly modify this classical picture.

A THE INTERTROPICAL CONVERGENCE

The tendency for the trade wind systems of the two hemispheres to converge in the Equatorial (low-pressure) Trough has already been noted (see Chapter 6B). Views on the exact nature of this feature have been subject to continual revision. From the 1920s to the 1940s, the frontal concepts developed in mid-latitudes were applied in the tropics, and the streamline confluence of the north-east and south-east trades was identified as the Intertropical Front (ITF). Over continental areas such as West Africa and South Asia, where in summer hot, dry continental tropical air meets cooler, humid equatorial air, this term has some limited applicability (Figure 9.1). Sharp temperature and moisture gradients can occur, but the front is seldom a weather-producing mechanism of the mid-latitude type. Elsewhere in low latitudes, true fronts (with a marked density contrast) are rare.

Recognition in the 1940s and 1950s of the significance of wind field convergence in tropical weather production led to the designation of the trade wind convergence as the Intertropical Convergence Zone (ITCZ). This feature is apparent on a *mean* streamline map, but areas of convergence grow and decay,

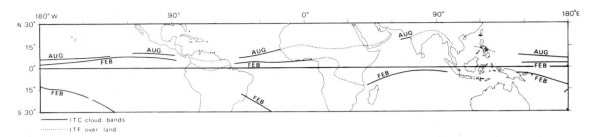

Figure 9.1 The position of the Equatorial Trough (Intertropical Convergence Zone or Intertropical Front in some sectors) in February and August. The cloud band in the south-west Pacific in February is known as the South Pacific Convergence Zone; over South Asia and West Africa the term Monsoon Trough is used.
Sources: After Saha 1973, Riehl 1954 and Yoshino 1969.

either *in situ* or within disturbances moving west-wards (see Plate 28), over periods of a few days. Moreover, convergence is infrequent even as a climatic feature in the doldrum zones (see Figure 6.14). Satellite photography has shown that over the oceans the position and intensity of the ITCZ varies greatly, even from day to day.

The ITCZ is predominantly an oceanic feature where its position changes so as to lie always over the warmest surface waters, such that small differences of sea-surface temperature may cause considerable changes in the location of the ITCZ. A sea-surface temperature of at least 27.5°C seems to provide a threshold for organized convective activity, and above this temperature organized convection is essentially competitive between different regions potentially available to form part of a continuous ITCZ. The convective rainfall belt of the ITCZ has very sharply defined latitudinal limits. For example, along the West African coast the following mean annual rainfalls are recorded:

12°N	1,939 mm
15°N	542 mm
18°N	123 mm

In other words, moving southwards into the ITCZ, precipitation increases by 440 per cent in a meridional distance of only 330 km.

As climatic features, the Equatorial Trough and the ITCZ are asymmetric about the equator, lying on average to the north. They also move seasonally away from the equator (see Figure 9.1) in association with the *thermal equator* (zone of seasonal maximum temperature). The location of the thermal equator is directly related to solar heating (see Figures 9.2 and 2.11), and there is an obvious link between this and the Equatorial Trough in terms of

thermal lows. However, if the ITC were to coincide with the Equatorial Trough then this zone of cloudiness would decrease incoming solar radiation, reducing the surface heating needed to maintain the low-pressure trough. In fact, this does not happen. Solar energy is available to heat the surface because the maximum surface wind convergence, uplift and cloud cover is commonly located several degrees equatorward of the trough. In the Atlantic (Figure 9.2B), for example, the cloudiness maximum is distinct from the Equatorial Trough in August. Figure 9.2 illustrates regional differences in the Equatorial Trough and ITCZ. Convergence of two trade wind systems occurs over the central North Atlantic in August and the eastern North Pacific in February. In contrast, the Equatorial Trough is defined by easterlies on its poleward side and westerlies on its equatorward side over West Africa in August and New Guinea in February.

The dynamics of low-latitude atmosphere–ocean circulations are also involved. The convergence zone in the central equatorial Pacific moves seasonally between about 4°N in March–April and 8°N in September, giving a single pronounced rainfall maximum in March–April. This appears to be a response to the relative strengths of the north-east and south-east trades. The ratio of South Pacific/North Pacific trade wind strength exceeds 2 in September but falls to 0.6 in April. Interestingly, the ratio varies in phase with the ratio of Antarctic–Arctic sea ice areas; Antarctic ice is at a maximum in September when Arctic ice is at its minimum. The convergence axis is often aligned close to the zone of maximum sea-surface temperatures, but is not anchored to it. Indeed, the SST maximum located within the Equatorial Counter-current (see Figure 6.31) is a result of the interactions between the trade winds

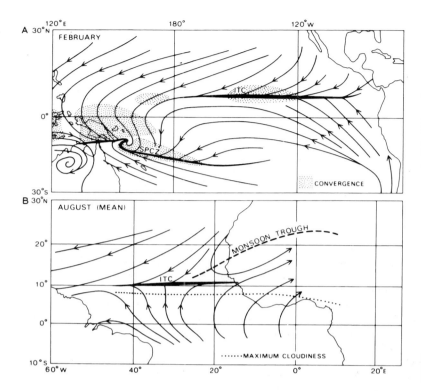

Figure 9.2 Illustrations of (A) streamline convergence forming an Intertropical Convergence (ITC) and South Pacific Convergence Zone (SPCZ) in February, and (B) the contrasting patterns of Monsoon Trough over West Africa, streamline convergence over the central tropical North Atlantic, and axis of maximum cloudiness to the south for August.

Sources: A: C. S. Ramage, personal communication, 1986. B: From Sadler 1975a.

and horizontal and vertical motions in the ocean surface layer.

Aircraft studies have revealed the complex structure of the central Pacific ITCZ. When moderately strong trades provide horizontal moisture convergence, convective cloud bands form, but the convergent lifting may be insufficient for rainfall in the absence of upper-level divergence. Moreover, although the south-east trades cross the equator, the mean monthly resultant winds between 115° and 180 ° W have, throughout the year, a more southerly component north of the equator and a more northerly one south of it, giving a zone of divergence (due to the sign change in the Coriolis parameter) along the equator.**

In the south-western sectors of the Pacific and Atlantic Oceans, satellite cloudiness studies indicate the presence of two semi-permanent confluence zones (see Figure 9.1). These do not occur in the eastern South Atlantic and South Pacific, where there are cold ocean currents. The convergence shown in the western South Pacific in February (summer) is now recognized as an important discontinuity and zone of maximum cloudiness (see Plate

29) termed the South Pacific Convergence Zone (SPCZ); it extends from the eastern tip of Papua New Guinea to about 30°S, 120°W. At sea-level, moist north-easterlies, west of the South Pacific subtropical anticyclone, converge with south-easterlies ahead of high-pressure systems moving eastwards from Australia/New Zealand. The low-latitude section west of 180° longitude is part of the ITCZ system, related to warm surface waters. However, the maximum precipitation is south of the axis of maximum sea-surface temperature, and the surface convergence is south of the precipitation maximum in the central South Pacific. The south-eastward orientation of the SPCZ is caused by interactions with the mid-latitude westerlies. Its south-eastern end is associated with wave disturbances and jet stream clouds on the South Pacific polar front. The link across the subtropics appears to reflect upper-level tropical mid-latitude transfers of moisture and energy, especially during subtropical storm situations. Hence, the SPCZ shows substantial short-term and interannual variability in its location and development. The interannual variability is strongly associated with the phase of the

Southern Oscillation (see p.133). During the northern summer, the SPCZ is poorly developed, whereas the ITCZ is strong all across the Pacific. During the southern summer, the SPCZ is well developed, with a weak ITCZ over the western tropical Pacific. After April, the ITCZ strengthens over the western Pacific and the SPCZ weakens as it moves westwards and equatorwards. In the Atlantic, the ITCZ normally begins its northward movement in April–May, when South Atlantic sea-surface temperatures start to fall and the subtropical high-pressure cell there and the south-east trades intensify. In cold, dry years this movement can begin as early as February and in warm, wet years as late as June.

B TROPICAL DISTURBANCES

It was not until the 1940s that detailed accounts were given of types of tropical disturbances other than the long-recognized tropical cyclone. However, our view of tropical weather systems has been radically revised following the advent of operational meteorological satellites in the 1960s. Special programmes of meteorological measurements at the surface and in the upper air together with aircraft and ship observations have been carried out in the Pacific and Indian Oceans, the Caribbean sea and the tropical eastern Atlantic.

It appears that five categories of weather system can be distinguished according to their space and time scales (see Figure 9.3). The smallest, with a lifespan of a few hours, is the *individual cumulus*, 1–10 km in diameter, which is generated by dynamically induced convergence in the trade wind boundary layer. In fair weather, cumulus clouds are generally aligned in 'cloud streets', more or less parallel to the wind direction (see Plate 30), or form polygonal honeycomb-pattern cells, rather than scattered at random. This seems to be related to the boundary-layer structure and wind speed (see p. 85). There is little interaction between the air layers above and below the cloud base under these conditions, but in disturbed weather conditions updraughts and downdraughts cause interaction between the two layers, which intensifies the convection. Individual cumulus towers, associated with violent thunderstorms, develop particularly in the Intertropical Convergence Zone, sometimes reaching to more than 20,000 m in height and having updraughts of 10–14 m s⁻¹ (see Plate 5). In this way, the smallest scale of system can aid the

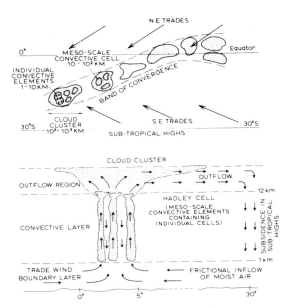

Figure 9.3 The mesoscale and synoptic structure of the equatorial trough zone (ITCZ), showing a model of the spatial distribution (above) and of the vertical structure (below) of convective elements which form the cloud clusters.

Source: From Mason 1970.

development of larger disturbances. Convection is most active over sea surfaces with temperatures exceeding 26°C.

The second category of system develops through cumulus clouds becoming grouped into *mesoscale convective areas* (MCAs) up to 100 km across (see Figure 9.3). In turn, several MCAs may comprise a *cloud cluster*, 100–1,000 km in diameter. These subsynoptic-scale systems are rather arbitrarily defined because they were initially identified as amorphous cloud areas on satellite images. They have been studied primarily from satellite photographs of the tropical Atlantic and Pacific Oceans (Plate 1 and Plate 29). Their definition is rather arbitrary, but they may extend over an area 2° square up to 12° square. It is important to note that the peak convective activity has passed when cloud cover is most extensive through the spreading of cirrus canopies. Clusters in the Atlantic, defined as more than 50 per cent cloud cover extending over an area of 3° square, show maximum frequencies of ten to fifteen clusters per month near the ITC and also at 15–20°N in the western Atlantic over zones of high sea-surface temperature. They consist of a cluster of mesoscale convective cells with the system having a deep layer of convergent airflow (see Figure 9.3). Some persist for only one to two days, but others develop within synoptic-scale waves. Many aspects of their development and role remain to be determined.

The fourth category of tropical weather system includes the synoptic-scale waves and cyclonic vortices (discussed more fully below) and the fifth group is represented by the planetary-scale waves. The planetary waves (with a wavelength from 10,000 to 40,000 km) need not concern us in detail. Two types occur in the equatorial stratosphere and another in the equatorial upper troposphere. While they may interact with lower tropospheric systems, they do not appear to be direct weather mechanisms. The synoptic-scale systems that determine much of the 'disturbed weather' of the tropics are sufficiently important and varied to be discussed under the headings of wave disturbances and cyclonic storms.

1 Wave disturbances

Several types of wave travel westwards in the equatorial and tropical tropospheric easterlies; the differences between them probably result from regional and seasonal variations in the structure of the tropical atmosphere. Their wavelength is about 2,000–4,000 km, and they have a lifespan of one to two weeks, travelling some 6–7° longitude per day.

The first wave type to be described in the tropics was the *easterly wave* of the Caribbean area. This system is quite unlike a mid-latitude depression. There is a weak pressure trough, which usually slopes eastwards with height (Figure 9.4), and typically the main development of cumulonimbus cloud and thundery showers is *behind* the trough line. This pattern is associated with horizontal and vertical motion in the easterlies. Behind the trough, low-level air undergoes convergence, while ahead of it there is divergence (see Chapter 5B.1). This follows from the equation for the conservation of potential vorticity (cf. Chapter 7G), which assumes that the air travelling at a given level does not change its potential temperature (i.e. dry adiabatic motion; see Chapter 4A):

$$\frac{f + \zeta}{\Delta p} = k$$

where f = the Coriolis parameter, ζ = relative vorticity (cyclonic positive) and Δp = the depth of the tropospheric air column. Air overtaking the trough line is moving both polewards (f increasing)

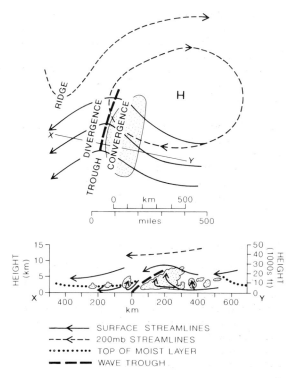

Figure 9.4 A model of the areal (above) and vertical (below) structure of an easterly wave. Cloud is stippled and the precipitation area is shown in the vertical section. The streamline symbols refer to the areal structure, and the arrows on the vertical section indicate the horizontal and vertical motions.

Source: Partly after Riehl and Malkus.

and towards a zone of cyclonic curvature (ζ increasing), so that if the left-hand side of the equation is to remain constant Δp must increase. This vertical expansion of the air column necessitates horizontal contraction (convergence). Conversely, there is divergence in the air moving southwards ahead of the trough and curving anticyclonically. The true divergent zone is characterized by descending, drying air with only a shallow moist layer near the surface, while in the vicinity of the trough and behind it the moist layer may be 4,500 m or more deep. When the easterly airflow is *slower* than the speed of the wave, the reverse pattern of low-level convergence ahead of the trough and divergence behind it is observed as a consequence of the potential vorticity equation. Often this is the case in the middle troposphere, so that the pattern of vertical motion shown in Figure 9.4 is augmented.

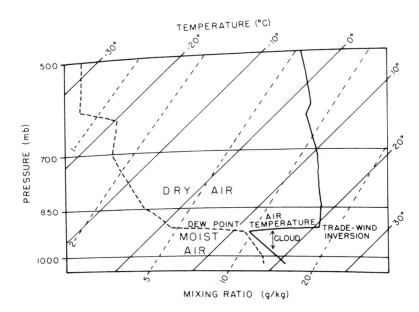

Figure 9.5 The vertical structure of trade wind air at 30°N, 140°W at 0300 GMT on 10 July 1949. The mixing ratio is the saturation value.

Source: Based on Riehl 1954.

The passage of such a transverse wave in the trades commonly produces the following weather sequence:

1 In the ridge ahead of the trough: fine weather, scattered cumulus cloud, some haze.
2 Close to the trough line: well-developed cumulus, occasional showers, improving visibility.
3 Behind the trough: veer of wind direction, heavy cumulus and cumulonimbus, moderate or heavy thundery showers and a decrease of temperature.

Satellite photography indicates that the simple easterly wave is rather less common than has been supposed. Many Atlantic disturbances show an 'inverted V' waveform in the low-level wind field and associated cloud, or a 'comma' cloud related to a vortex. They are often apparently linked with a wave pattern on the ITC further south. West African disturbances that move out over the eastern tropical Atlantic usually exhibit low-level confluence and upper-level diffluence *ahead* of the trough, giving maximum precipitation rates in this same sector. Many disturbances in the easterlies have a closed cyclonic wind circulation at about the 600 mb level.

It is obviously difficult to trace the growth of wave disturbances over the oceans and in continental areas with sparse data coverage. However, some generalizations can be made. At least eight out of ten disturbances develop some 2–4° latitude pole-ward of the Equatorial Trough. Convection is probably set off by convergence of moisture in the airflow, accentuated by friction, and then maintained by entrainment into the thermal convective plumes (see Figure 9.3). About ninety tropical disturbances develop during the June–November hurricane season in the tropical Atlantic, apparently one system every three to five days. More than half of those disturbances originate over Africa. According to N. Frank, a high ratio of African depressions in the storm total in a given season indicates tropical characteristics, whereas a low ratio suggests storms originating from cold lows and the baroclinic zone between Saharan air and cooler, moist monsoon air. Many of them can be tracked westwards into the eastern North Pacific. A quarter of the disturbances intensify into tropical depressions and 10 per cent become 'named' storms.

Developments in the Atlantic are closely related to the structure of the trades. In the eastern sectors of subtropical anticyclones, active subsidence maintains a pronounced inversion at 450 to 600 m (Figure 9.5). Thus, the cool eastern tropical oceans are characterized by extensive, but shallow, marine stratocumulus, which gives little rainfall. Downstream the height of the inversion base rises (Figure 9.6) because the subsidence decreases away from the eastern part of the anticyclone and cumulus towers penetrate through the inversion from time

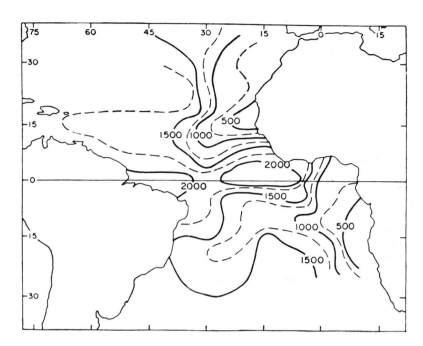

Figure 9.6 The height (in metres) of the base of the trade wind inversion over the tropical Atlantic.

Source: From Riehl 1954.

to time, spreading moisture into the dry air above. Easterly waves tend to develop in the Caribbean when the trade wind inversion is weak or even absent during summer and autumn, whereas in winter and spring accentuated subsidence aloft inhibits their growth, although disturbances may move westwards above the inversion. Another factor that may initiate waves in the easterlies is the penetration of cold fronts into low latitudes. This is common in the sector between two subtropical high-pressure cells, where the equatorward portion of the front tends to fracture, generating a westward-moving wave.

The influence of these features on regional climate is illustrated by the rainfall regime. For example, there is a late summer maximum at Martinique in the Windward Islands (15°N) when subsidence is weak, although some of the autumn rainfall is associated with tropical storms. In many trade wind areas, the rainfall occurs in a few rainstorms associated with some form of disturbance. Over a ten-year period, Oahu (Hawaii) had an average of twenty-four rainstorms per year, ten of which accounted for more than two-thirds of the annual precipitation. There is quite high variability of rainfall from year to year in such areas, since a small reduction in the frequency of disturbances can have a large effect on rainfall totals.

In the central equatorial Pacific, the trade wind systems of the two hemispheres converge in the Equatorial Trough, and wave disturbances may be generated if the trough is sufficiently removed from the equator (usually to the north) to provide a small Coriolis force to begin cyclone motion. These disturbances quite often become unstable, forming a cyclonic vortex as they travel westwards towards the Philippines, but the winds do not necessarily attain hurricane strength. The synoptic chart for part of the north-west Pacific on 17 August 1957 (Figure 9.7) shows three developmental stages of tropical low-pressure systems. An incipient easterly wave has formed west of Hawaii, which, however, filled and dissipated during the next 24 hours. A well-developed wave is evident near Wake Island, having spectacular cumulus towers extending up to over 9,100 m along the convergence zone some 480 km east of it (see Plate 31). This wave developed within 48 hours into a circular tropical storm with winds up to 20 m s^{-1}, but not into a full hurricane. A strong, closed, north-west moving circulation is situated east of the Philippines. Equatorial waves may form on both sides of the equator in an easterly current located between about 5°N and S. In such cases, divergence ahead of a trough in the northern hemisphere is paired with convergence behind a trough line located further to the west in

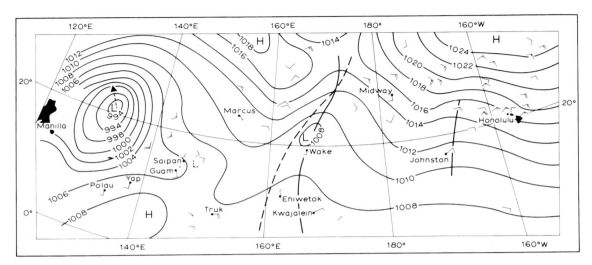

Figure 9.7 The surface synoptic chart for part of the north-west Pacific on 17 August 1957. The movements of the central wave trough and of the closed circulation during the following 24 hours are shown by the dashed line and arrow, respectively. The dashed L just east of Saipan indicates the location in which another low-pressure system subsequently developed. Plate 31 shows the cloud formation along the convergence zone just east of Wake Island.

Source: From Malkus and Riehl 1964.

the southern hemisphere. The reader may confirm that this should be so by applying the equation for the conservation of potential vorticity, remembering that both f and ζ operate in the reverse sense in the southern hemisphere.

2 Cyclones

a Hurricanes

The most notorious type of cyclone is the hurricane (or typhoon). Some eighty or so cyclones each year are responsible, on average, for 20,000 fatalities, as well as causing immense damage to property and a serious shipping hazard, due to the combined effects of high winds, high seas, flooding from the heavy rainfall and coastal storm surges. As a result, considerable attention has been given to forecasting their development and movement, so their origin and structure are beginning to be understood. Naturally, the catastrophic force of a hurricane makes it a very difficult phenomenon to investigate, but some assistance is now obtained from aircraft reconnaissance flights sent out during the 'hurricane season', from radar observations of cloud and precipitation structure, and from satellite photography (see Plate 32).

The typical hurricane system has a diameter of about 650 km, less than half that of a mid-latitude depression, although typhoons in the western Pacific are often much larger. The central pressure at sea level is commonly 950 mb and exceptionally falls below 900 mb. Named tropical storms are those defined as having one-minute average wind velocities of at least 18 m s^{-1} at the surface. If these winds intensify to at least 33 m s^{-1}, the named storm becomes a hurricane. Five intensity classes are distinguished: category 1, weak (winds of 33–42 m s^{-1}); 2, moderate (43–49 m s^{-1}); 3, strong (50–58 m s^{-1}); 4, very strong (59–69 m s^{-1}) and 5, devastating (70 m s^{-1} or more). Hurricane Camille, which struck coastal Mississippi in August 1969, was a category 5 storm, while Hurricane Andrew, which devastated southern Florida in August 1992, was category 4 at landfall. The great vertical development of cumulonimbus clouds, with tops at over 12,000 m, reflects the immense convective activity concentrated in such systems. Radar and satellite studies show that the convective cells are normally organized in bands that spiral inwards towards the centre.

Although the largest hurricanes are characteristic of the Pacific, the record is held by the Caribbean hurricane 'Gilbert', which was generated 320 km east of Barbados on 9 September 1988 and moved

Table 9.1 Annual frequencies and usual seasonal occurrence of tropical cyclones (maximum sustained winds exceeding 25 m s^{-1}), 1958–77.

Location	Annual frequency	Main occurrence
Western North Pacific	26.3	July–October
Eastern North Pacific	13.4	August–September
Western North Atlantic	8.8	August–October
Northern Indian Ocean	6.4	May–June; October–November
Northern hemisphere total	54.6	
South-west Indian Ocean	8.4	January–March
South-east Indian Ocean	10.3	January–March
Western South Pacific	5.9	January–March
Southern hemisphere total	24.5	
Global total	79.1	

Source: After Gray 1979.
Note: Area totals are rounded.

westwards at an average speed of 24–27 km hr^{-1}, dissipating off the east coast of Mexico. Aided by an upper tropospheric high-pressure cell north of Cuba, Hurricane Gilbert intensified very rapidly, the pressure at its centre dropped to 888 mb (the lowest ever recorded in the western hemisphere), and maximum windspeeds near the core were in excess of 55 m s^{-1}. More than 500 mm of rain fell on the highest parts of Jamaica in only nine hours. However, the most striking feature of this record storm was its size, being some three times that of average Caribbean hurricanes. At its maximum extent, the hurricane had a diameter of 3,500 km, disrupting the ITCZ along more than one-sixth of the earth's equatorial circumference and drawing in air from as far away as Florida and the Galapagos Islands.

The main tropical cyclone activity in both hemispheres is in late summer and autumn during times of maximum northward and southward displacements of the Equatorial Trough (see Table 9.1). A small number of storms affect both the western North Atlantic and North Pacific areas as early as May and as late as December, and have occurred in every month in the latter area. In the Bay of Bengal, there is also a secondary early summer maximum. A tropical cyclone that struck coastal Bangladesh on 24–30 April 1991 caused over 130,000 deaths from drowning, due to flooding, and left over ten million people homeless. The annual frequency of cyclones shown in Table 9.1 is only approximate, since in some cases it may be uncertain as to whether or not the winds actually

exceeded hurricane force. In addition, storms in the more remote parts of the South Pacific and Indian Oceans frequently escaped detection prior to the use of weather satellites.

A number of conditions are necessary, even if not always sufficient, for hurricane formation. One requirement as shown by Figure 9.8 is an extensive ocean area with a surface temperature greater than 27°C. Cyclones rarely form near the equator, where the Coriolis parameter is close to zero, or in zones of strong vertical wind shear (i.e. beneath a jet stream), as both factors inhibit the development of an organized vortex. There is also a definite connection between the seasonal position of the Equatorial Trough and zones of hurricane formation, which is borne out by the fact that no hurricanes occur in the South Atlantic (where the trough never lies south of 5°S) or in the south-east Pacific (where the trough remains north of the equator). On the other hand, satellite photographs over the north-east Pacific show an unexpected number of cyclonic vortices in summer, many of which move westwards near the trough line at about 10–15°N. About 60 per cent of tropical cyclones seem to originate 5–10° latitude poleward of the Equatorial Trough in the doldrum sectors, where the trough is at least 5° latitude from the equator. The development regions of hurricanes lie mainly over the western sections of the Atlantic, Pacific and Indian Oceans, where the subtropical high-pressure cells do not cause subsidence and stability and the upper flow is divergent. About twice per season, in the western equatorial Pacific, tropical cyclones form almost

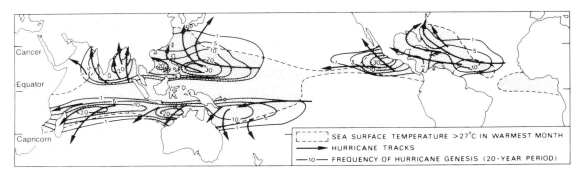

Figure 9.8 Frequency of hurricane genesis (numbered isopleths) for a 20-year period. The principal hurricane tracks and the areas of sea surface having water temperatures greater than 27°C in the warmest month are also shown.

Source: After Palmén 1948 and Gray 1979.

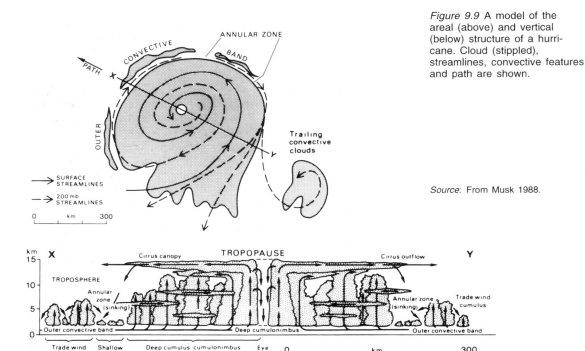

Figure 9.9 A model of the areal (above) and vertical (below) structure of a hurricane. Cloud (stippled), streamlines, convective features and path are shown.

Source: From Musk 1988.

simultaneously in each hemisphere near 5° latitude and along the same longitude (Plate 33). The cloud and wind patterns in these cyclone 'twins' are roughly symmetrical with respect to the equator.

Early theories of hurricane development held that convection cells generated a sudden and massive release of latent heat to provide energy for the storm. Although convection cells were regarded as an integral part of the hurricane system, their scale was thought to be too small for them to account for the growth of a storm hundreds of kilometres in diameter. However, recent research is modifying this picture considerably. Energy is apparently transferred from the cumulus-scale to the large-scale circulation of the storm through the organization of the clouds into spiral bands (see Figure 9.9 and Plate 34), although the nature of the process is still being investigated. There is now ample evidence to show

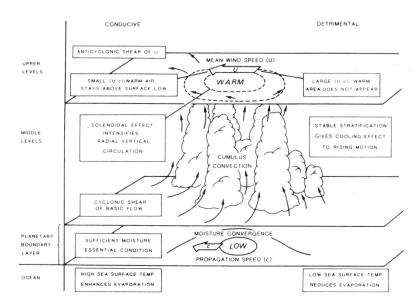

Figure 9.10 A schematic model of the conditions conducive (left) or detrimental (right) to the growth of a tropical storm in an easterly wave; *U* is the mean upper-level wind speed and *c* is the rate of propagation of the system. The warm vortex creates a thermal gradient that intensifies both the radial motion around it and the ascending air currents, termed the solenoidal effect.

Source: From Kurihara 1985 (copyright © Academic Press. Reproduced by permission).

that hurricanes form from pre-existing disturbances, but while many of these disturbances develop as closed low-pressure cells, few attain full hurricane intensity. The key to this problem is high-level outflow (Figure 9.10). This does not require an upper tropospheric anticyclone but can occur on the eastern limb of an upper trough in the westerlies. This outflow in turn allows the development of very low pressure and high wind speeds near the surface. A distinctive feature of the hurricane is the warm vortex, since other tropical depressions and incipient storms have a cold core area of shower activity. The warm core develops through the action of 100–200 cumulonimbus towers releasing latent heat of condensation; about 15 per cent of the area of cloud bands is giving rain at any one time. Observations show that although these 'hot towers' form less than 1 per cent of the storm area within a radius of about 400 km, their effect is sufficient to change the environment. The warm core is vital to hurricane growth because it intensifies the upper anticyclone, leading to a 'feedback' effect by stimulating the low-level influx of heat and moisture, which further intensifies convective activity, latent heat release and therefore the upper-level high pressure. This enhancement of a storm system by cumulus convection is termed conditional instability of the second kind, or CISK (cf. the basic parcel instability described on p. 79). The thermally direct circulation converts the heat increment into

potential energy and a small fraction of this – about 3 per cent – is transformed into kinetic energy. The remainder is exported by the anticyclonic circulation at about the 12 km (200 mb) level.

In the *eye*, or innermost region of the storm (see Figure 9.9 and Plate 34), adiabatic warming of descending air accentuates the high temperatures, although since high temperatures are also observed in the eye-wall cloud masses, subsiding air can only be one contributory factor. Without this sinking air in the eye, the central pressure could not fall below about 1,000 mb. The eye has a diameter of some 30–50 km, within which the air is virtually calm and the cloud cover may be broken. The mechanics of the eye's inception are still largely unknown. If the rotating air conserved absolute angular momentum, wind speeds would become infinite at the centre and clearly this is not the case. The strong winds surrounding the eye are more or less in cyclostrophic balance, with the small radial distance providing a large centripetal acceleration (see p. 107). The air rises when the pressure gradient can no longer force it further inwards. It is possible that the cumulonimbus anvils play a vital role in the complex link between the horizontal and vertical circulations around the eye by redistributing angular momentum in such a way as to set up a concentration of rotation near the centre.

The supply of heat and moisture combined with low frictional drag at the sea surface, the release of latent heat through condensation and the removal of the air aloft are essential conditions for the maintenance of hurricane intensity. As soon as one of these ingredients diminishes the storm decays. This can occur quite rapidly if the track (determined by the general upper tropospheric flow) takes the vortex over a cool sea surface or over land. In the latter case, the increased friction causes greater cross-isobar air motion, temporarily increasing the convergence and ascent. At this stage, increased vertical wind shear in thunderstorm cells may generate tornadoes, especially in the north-east quadrant of the storm (in the northern hemisphere). However, the most important effect of a land track is that cutting off of the moisture supply removes one of the major sources of heat. Rapid decay also occurs when cold air is drawn into the circulation or when the upper-level divergence pattern moves away from the storm.

Hurricanes usually move at 16–24 km hr^{-1}, controlled primarily by the rate of movement of the upper warm core. Commonly, they recurve polewards around the western margins of the subtropical high-pressure cells, entering the circulation of the westerlies, where they die out or degenerate into extra-tropical depressions (see Plate 18). Some of these systems retain an intense circulation and the high winds and waves can still wreak havoc. This is not uncommon along the Atlantic coast of the United States and occasionally eastern Canada. Similarly, in the western North Pacific, recurved typhoons are a major element in the climate of Japan (see D, this chapter) and may occur in any month. There is an average frequency of twelve typhoons per year over southern Japan and neighbouring sea areas.

To sum up: the hurricane develops from an initial disturbance, which, under favourable environmental conditions, grows first into a tropical depression and then a tropical storm. The tropical storm stage may persist for four to five days, whereas the hurricane stage usually lasts for only two to three days (four to five days in the western Pacific). The main energy source is latent heat derived from condensed water vapour, and for this reason hurricanes are generated and continue to gather strength only within the confines of warm oceans. The cold-cored tropical storm is transformed into a warm-cored hurricane in association with the release of latent heat in cumulonimbus towers, and this establishes or intensifies an upper tropospheric anticyclonic cell. Thus high-level outflow maintains the ascent and low-level inflow in order to provide a continuous generation of potential energy (from latent heat) and the transformation of this into kinetic energy. The inner eye that forms by sinking air is an essential element in the life-cycle.

The science of hurricane forecasting is a complex one, not least because of its increasing involvement with the subject of teleconnections (see G.2, this chapter). Recent studies of North Atlantic/Caribbean annual hurricane frequencies suggest that an above-average hurricane frequency is related to three major factors:

1   The west phase of the Atlantic Quasi-biennial Oscillation (QBO). The QBO involves periodic changes in the velocities of, and vertical shear between, the zonal upper tropospheric (50 mb) winds and the lower stratospheric (30 mb) winds. The onset of such an oscillation can be predicted with some confidence almost a year in advance. The east phase of the QBO is associated with strong easterly winds in the lower stratosphere between latitudes 10°N and 15°N, producing a large vertical wind shear. This phase usually persists for 12 to 15 months and inhibits hurricane formation. The west QBO phase exhibits weak easterly winds in the lower stratosphere and small vertical wind shear. This phase, typically lasting 13 to 16 months, is associated with 50 per cent more named storms, 60 per cent more hurricanes and 200 per cent more major hurricanes than is the east phase.

2   West African precipitation during the previous year along the Gulf of Guinea (August–November) and in the western Sahel (August–September). The former moisture source appears to account for some 40 per cent of major hurricane activity, the latter for only 5 per cent. Between the late 1960s and 1980s, the Sahel drought was associated with a marked decrease in Atlantic tropical cyclones and hurricanes, mainly through strong upper-level shearing winds over the tropical North Atlantic and a decrease in the propagation of easterly waves over Africa in August and September.

3   ENSO predictions for the following year (see G, this chapter). There is an inverse correlation between the frequency of El Niños and that of Atlantic hurricanes.

b Other tropical disturbances

Not all low-pressure systems in the tropics are of the intense tropical cyclone variety. There are two other major types of cyclonic vortex. One is the *monsoon depression* that affects South Asia during the summer. This disturbance is somewhat unusual in that the flow is westerly at low levels and easterly in the upper troposphere (see Figure 9.27). It is more fully described in C.4, this chapter.

The second type of system is usually relatively weak near the surface, but well-developed in the middle troposphere. In the eastern North Pacific and Indian Ocean, such lows are referred to as *subtropical cyclones*. Some develop from the cutting-off in low latitudes of a cold upper-level wave in the westerlies (cf. Chapter 7H.4). They possess a broad eye some 150 km in radius with little cloud, surrounded by a belt of cloud and precipitation about 300 km wide. In late winter and spring, a few such storms make a major contribution to the rainfall of the Hawaiian Islands. These cyclones are very persistent and tend eventually to be reabsorbed by a trough in the upper westerlies. Other subtropical cyclones occur over the Arabian Sea in summer and make a major contribution to summer ('monsoon') rains in north-west India. These systems show upward motion mainly in the upper troposphere. Their development may be linked to air export at upper levels of cyclonic vorticity from the persistent heat low over South Asia, especially Arabia.

An infrequent and distinctly different weather system, known as a *temporal*, occurs along the Pacific coasts of Central America in autumn and early summer. Its main feature is an extensive layer of altostratus, fed by individual convective cells, producing sustained moderate rainfall. These systems originate in the ITCZ over the eastern tropical North Pacific Ocean and are maintained by large-scale lower tropospheric convergence, localized convection and orographic uplift.

3 Tropical cloud clusters

Mesoscale convective systems (MCSs) are widespread in tropical and subtropical latitudes. The mid-latitude mesoscale convective complexes discussed in Chapter 4.G are an especially severe category of MCS. Satellite studies of cold (high) cloud-top signatures show that tropical systems typically extend over a 3,000–6,000 km² area. Maximum frequencies are found over tropical South America and the maritime continent of Indonesia–Malaysia and adjacent western Pacific Ocean warm pool. Other land areas include Australia, India and Central America in their respective summer seasons.

As a result of the diurnal regimes of convective activity, MCSs are more frequent at sunset compared with sunrise by 60 per cent over the continents and 35 per cent more frequent at sunrise than sunset over the oceans. Most of the intense systems (MCCs) occur over land, particularly where there is abundant moisture and usually downwind of orographic features that favour the formation of low-level jets.

Mesoscale convective systems fall into two categories: non-squall and squall-line. The former contain one or more mesoscale precipitation areas. They occur diurnally, for example, off the north coast of Borneo in winter, where they are initiated by convergence of a nocturnal land breeze and the north-east monsoon flow (Figure 9.11). By morning (0800 LST), cumulonimbus cells give precipitation. The cells are linked by an upper-level cloud shield, which persists when the convection dies about around noon as a sea-breeze system replaces the nocturnal convergent flow.

Squall-line systems in the tropics (Figure 9.12) form the leading edge of a line of cumulonimbus cells (Plate 35). The squall line and gust front move forwards in the low-level flow and by the formation of new cells. These mature and eventually dissipate to the rear of the main line. The process is analogous to that of mid-latitude squall lines (see Figure 7.26) but the tropical cells are weaker. Squall-line systems cross Malaya from the west during the south-west monsoon season. They appear to be initiated by the convergent effects of land breezes in the Malacca Straits. These particular linear systems, known as *sumatras*, give heavy rain and often thunder.

In West Africa, systems known as *disturbance lines* are an important feature of the climate in the summer half-year, when low-level south-westerly monsoon air is overrun by dry, warm Saharan air. The meridional air-mass contrast helps to set up the lower-tropospheric African easterly jet shown in Figure 9.38, which illustrates the zonation of climate and associated weather systems. Disturbance lines tend to form when there is divergence in the upper troposphere north of the Tropical Easterly Jet (see also Figure 9.40). They are several hundred kilometres long and travel westward associated with upper easterlies at about 50 km hr⁻¹ giving squalls and thundery showers before dissipating over cold-water areas of the North Atlantic. Spring and autumn rainfall in West Africa is derived in large part from these disturbances. Figure 9.13 for

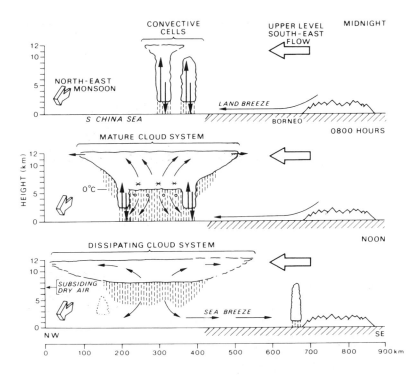

Figure 9.11 Schematic development of a non-squall cloud cluster off the north coast of Borneo: large arrows indicate the major circulation; small arrows, the local circulation; vertical shading, the zones of rain; stars, ice crystals; and circles, melted raindrops.

Source: After Houze et al. 1981.

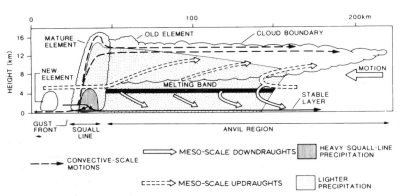

Figure 9.12 Cross-section of a tropical squall-line cloud cluster showing locations of precipitation and ice particle melting. Dashed arrows show the air motion generated by the squall-line convection and the broad arrows the mesoscale circulation.

Source: After Houze; from Houze and Hobbs 1982.

Kortright (Freetown, Sierra Leone) illustrates the daily rainfall amounts in 1960–1 associated with disturbance lines at 8°N. The summer monsoon rains make up the greater part of the total here, but their contribution diminishes northwards.

C THE ASIAN MONSOON

The name *monsoon* is derived from the Arabic word '*mausim*', which means season, so explaining its application to large-scale seasonal reversals of the wind regime. The Asiatic seasonal wind reversal is notable for its immense extent and the penetration of its influence beyond tropical latitudes (Figure 9.14). However, such seasonal wind shifts at the surface are quite widespread and occur in many regions that would not traditionally be considered as monsoonal. Although there is a rough accordance between these traditional regions and those experiencing over 60 per cent frequency of the prevailing octant, it is obvious that a variety of unconnected mechanisms can produce significant seasonal wind

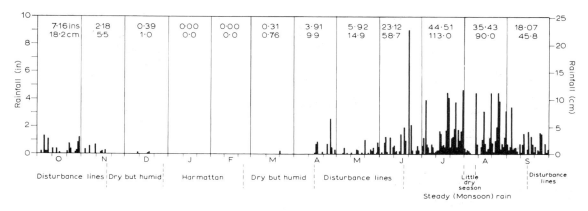

Figure 9.13 Daily rainfall at Kortwright, Freetown, Sierra Leone, October 1960–September 1961.
Source: After Gregory 1965.

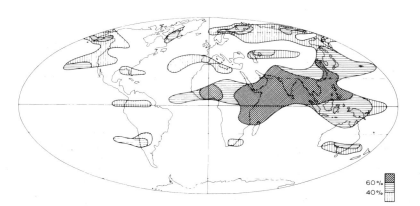

Figure 9.14 Regions experiencing a seasonal surface wind shift of at least 120°, showing the frequency of the prevailing octant.

60%
40%

Source: After Khromov.

shifts. Nor is it possible to establish a simple relationship between seasonality of rainfall (Figure 9.15) and seasonal wind shift. Areas traditionally designated as 'monsoonal' include some of the tropical and near-tropical regions experiencing a summer rainfall maximum and most of those having a double rainfall maximum. It is clear that a combination of criteria (for example, from Figures 9.14 and 9.15) is necessary for an adequate definition of monsoonal areas.

In summer, the Equatorial Trough and the subtropical anticyclones are everywhere displaced northwards in response to the changing pattern of solar heating of the earth, and in South Asia this movement is magnified by the effects of the land mass. However, the attractive simplicity of the traditional explanation, which envisages a monsoonal 'sea breeze' directed towards a summer thermal low

pressure over the continent, unfortunately provides an inadequate basis for understanding the workings of the system. The Asiatic monsoon regime is a consequence of the interaction of planetary and regional factors, both at the surface and in the upper troposphere. It is convenient to look at each season in turn, and Figure 9.16 shows the generalized meridional circulations at 90°E over India and the Indian Ocean in winter (December–February), spring (April) and autumn (September), together with those associated with active and break periods during the June–August summer monsoon.

1 Winter

Near the surface, this is the season of the outblowing 'winter monsoon', but aloft westerly airflow dominates. This, as we have seen, reflects

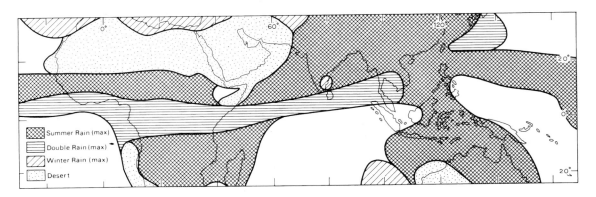

Figure 9.15 The annual distribution of tropical rainfall. The shaded areas refer to periods during which more than 75 per cent of the mean annual rainfall occurs. Areas with less than 250 mm y⁻¹ (10 in y⁻¹) are classed as deserts, and the unshaded areas are those needing at least seven months to accumulate 75 per cent of the annual rainfall and are thus considered to exhibit no seasonal maximum.

Source: After Ramage 1971.

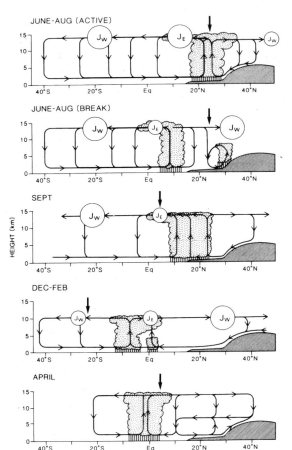

the hemispheric pressure distribution. A shallow layer of cold, high-pressure air is centred over the continental interior, but this has disappeared even at 700 mb (see Figure 6.4) where there is a trough over East Asia and zonal circulation over the continent. The upper westerlies split into two currents to the north and south of the high Tibetan (Qinghai–Xizang) Plateau (Figure 9.17), to reunite again off the east coast of China (Figure 9.18). The plateau, which exceeds 4,000 m over a vast area, is a tropospheric cold source in winter, particularly over its western part, although the strength of this source depends on the extent and duration of snow cover (snow-free ground acts as a heat source for the atmosphere in *all* months). Below 600 mb, the tropospheric heat sink gives rise to a shallow, cold plateau anticyclone, which is best developed in December and January. The two jet-stream branches have been attributed to the disruptive effect of the

Figure 9.16 Schematic representation of the meridional circulation over India at 90°E at five characteristic periods of the year; winter monsoon (December–February); approach of the monsoon season (April); the active summer monsoon (June–August); a break in the summer monsoon (June–August); and the retreat of the summer monsoon (September). Easterly (J_E) and westerly (J_W) jet streams are shown at sizes depending on their strength; the arrows mark the positions of the overhead sun; and zones of maximum precipitation are indicated.

Source: After Webster 1987a. From J. S. Fein and P. L. Stephens (eds) *Monsoons*. Copyright © 1987. Reproduced by kind permission of John Wiley and Sons, Inc.

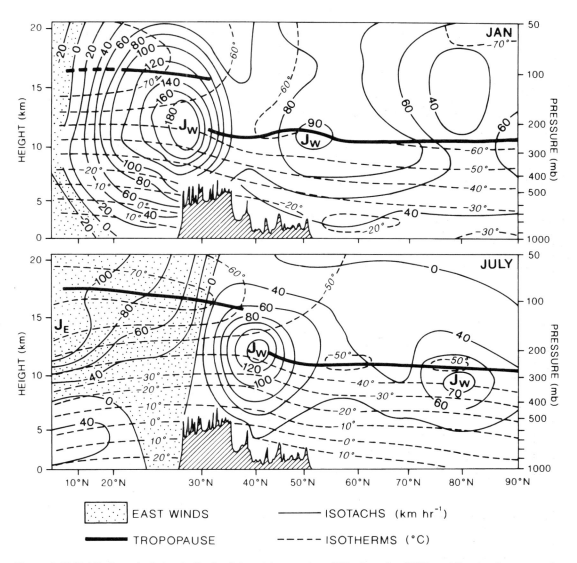

Figure 9.17 Distribution of wind velocity (km/hr) and temperature (°C) along the 90°E meridian for January and July, showing the westerly jet streams (J$_W$) and the tropopause. Note the variable intervals in the height and latitudinal scales.

Source: After Pogosyan and Ugarova 1959. From *Meteorologiya Gidrologiya*.

topographic barrier on the airflow, but this is limited to altitudes below about 4 km. In fact, the northern jet is highly mobile and may be located far from the Tibetan Plateau. Two currents are also found to occur farther west, where there is no obstacle to the flow. The branch over northern India corresponds to a strong latitudinal thermal gradient (from November to April) and it is probable that this factor, combined with the thermal effect of the

barrier to the north, is responsible for the anchoring of the southerly jet. This southern branch is the stronger, with an average speed of more than 40 m s⁻¹ at 200 mb, compared with about 20–25 m s⁻¹ in the northern one. Where the two unite over north China and south Japan the average velocity exceeds 66 m s⁻¹ (Figure 9.19).

Air subsiding beneath this upper westerly current gives dry outblowing northerly winds from the

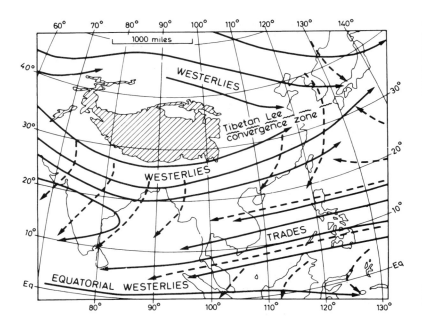

Figure 9.18 The characteristic air circulation over South and East Asia in winter. Solid lines indicate airflow at about 3,000 m, and dashed lines at about 600 m. The names refer to the wind systems aloft.

Sources: After Thompson 1851, Flohn 1968, Frost and Stephenson 1965, and others.

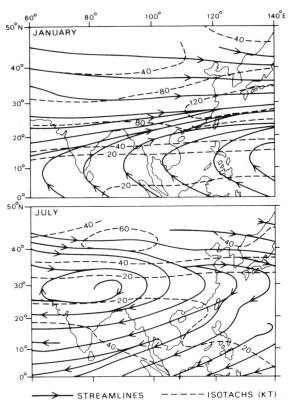

— STREAMLINES   ---- ISOTACHS (KT)

subtropical anticyclone over north-west India and Pakistan. The surface wind direction is north-westerly over most of northern India, becoming north-easterly over Burma and Bangladesh and easterly over peninsular India. Equally important is the steering of winter depressions over northern India by the upper jet. The lows, which are not usually frontal, appear to penetrate across the Middle East from the Mediterranean and are important sources of rainfall for northern India and Pakistan (e.g. Kalat, Figure 9.20), especially as it falls when evaporation is at a minimum. The equatorial trough of convergence and precipitation lies between the equator and about latitude 15°S (see Figure 9.16).

Some of these westerly depressions continue eastwards, redeveloping in the zone of jet stream confluence about 30°N, 105°E over China, beyond the area of subsidence in the immediate lee of Tibet (see Figure 9.18), and it is significant that the mean axis of the winter jet stream over China shows a close correlation with the distribution of winter rainfall (Figure 9.21). Other depressions affecting central and north China travel within the westerlies north of Tibet or are initiated by outbreaks of fresh cP

Figure 9.19 Mean 200 mb streamlines and isotachs in knots over South-east Asia for January and July, based on aircraft reports and sounding data.

Source: From Sadler 1975b (courtesy Dr J. C. Sadler, University of Hawaii).

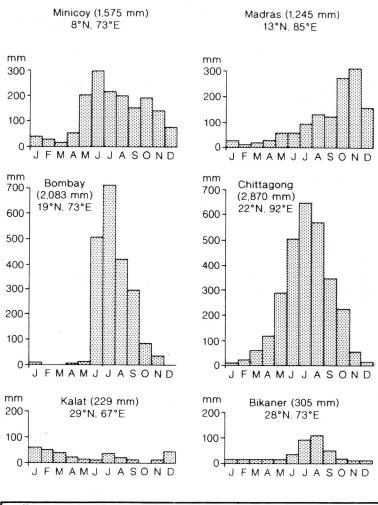

Minicoy (1,575 mm)
8°N, 73°E
Madras (1,245 mm)
13°N, 85°E
Bombay
(2,083 mm)
19°N, 73°E
Chittagong
(2,870 mm)
22°N, 92°E
Kalat (229 mm)
29°N, 67°E
Bikaner (305 mm)
28°N, 73°E
mm
J F M A M J J A S O N D

Figure 9.20 Average monthly rainfall (mm) at six stations in the Indian region. The annual total is given after the station name.

Source: Based on 'CLIMAT' normals of the World Meteorological Organization for 1931–60.

AXIS OF JET STREAM
WIND SPEED (METERS PER SEC.)
RAINFALL
20 - 40 CM.
40 - 60 CM.
OVER 60 CM.
JET AXIS

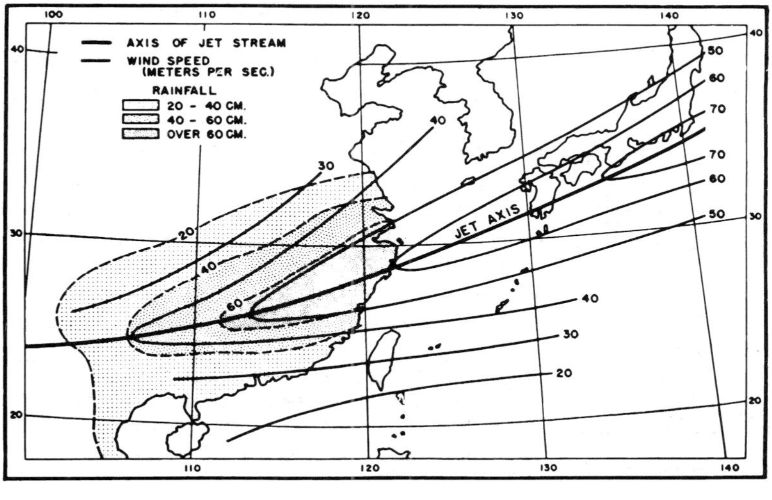

Figure 9.21 The mean winter jet stream axis at 12 km over the Far East and the mean winter precipitation over China in cm.

Source: After Mohri and Yeh; from Trewartha 1958.

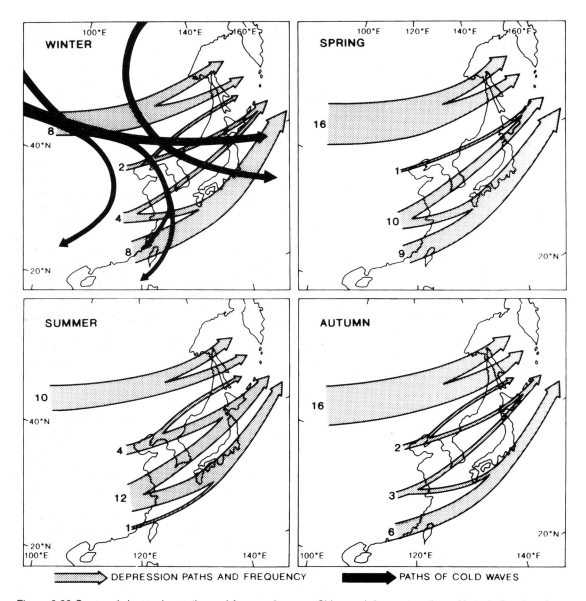

Figure 9.22 Seasonal depression paths and frequencies over China and Japan, together with typical paths of winter cold waves.

Sources: Compiled from various sources, including Tao, S. 1984, Zhang, J. and Lin, Z. 1985, Sheng, C. *et al.* 1986 and Domrös, M. and Gongbing, P. 1988. Reproduced by permission of Springer-Verlag, Berlin.

air. In the rear of these depressions are invasions of very cold air (e.g. the *buran* blizzards of Mongolia and Manchuria). The effect of such cold waves, comparable with the *northers* in the central and southern United States, is greatly to reduce mean temperatures (Figure 9.22). Winter mean

temperatures in less-protected southern China are considerably below those at equivalent latitudes in India; for example, temperatures at Calcutta and Hong Kong (both at approximately 22½°N) are 19°C and 16°C in January and 22°C and 15°C in February, respectively.

2 Spring

The key to change during this transition season is again found in the pattern of the upper airflow. In March, the upper westerlies begin their seasonal migration northwards, but whereas the northerly jet strengthens and begins to extend across central China and into Japan, the southerly branch remains positioned south of Tibet, although weakening in intensity.

In April, there is weak convection over India, where the circulation is dominated by subsiding air originating along the convective ITCZ trough centred over the equator and following the overhead sun northwards over the warm Indian Ocean (see Figure 9.16). The weather over northern India becomes hot, dry and squally in response to the greater solar radiation heating. Mean temperatures at Delhi rise from 23°C in March to 33°C in May. The thermal low-pressure cell (see Chapter 7H.2) reaches its maximum intensity at this time, but although onshore coastal winds develop, the onset of the monsoon is still a month away and other mechanisms cause only limited precipitation. Some precipitation occurs in the north with 'westerly disturbances', particularly towards the Ganges delta, where the low-level inflow of warm, humid air is overrun by dry, potentially cold air, triggering squall lines known as *nor'westers*. In the north-west, where less moisture is available, the convection generates violent squalls and dust-storms termed *andhis*. The mechanism of these storms is not yet fully known, although high-level divergence in the waves of the subtropical westerly jet stream appears to be essential. The early onset of summer rains in Bengal, Bangladesh, Assam and Burma (e.g. Chittagong, Figure 9.20) is favoured by an orographically produced trough in the westerlies at 300 mb, which is located at about 85–90°E in May. Low-level convergence of maritime air from the Bay of Bengal, combined with the upper-level divergence ahead of the 300 mb trough, generates thunder squalls. Tropical disturbances in the Bay of Bengal are another source of these early rains. Rain also falls during this season over Sri Lanka and south India (e.g. Minicoy, Figure 9.20) in response to the northward movement of the Equatorial Trough.

3 Early summer

Generally, during the last week in May the southern branch of the high-level jet begins to break down, becoming intermittent and then gradually shifting northwards over the Tibetan Plateau. At 500 mb and below, however, the plateau exerts a blocking effect on the flow and the jet axis there jumps from the south to the north side of the plateau from May to June. Over India, the Equatorial Trough pushes northwards with each weakening of the upper westerlies south of Tibet, but the final *burst* of the monsoon, with the arrival of the humid, low-level south-westerlies, is not accomplished until the upper-air circulation has switched to its summer pattern (see Figures 9.19 and 9.23). Increased continental convection overcomes the spring subsidence and the return upper-level flow to the south is deflected by the Coriolis force to produce a strengthening easterly jet located at about 10–15°N and a westerly jet to the south of the equator (see Figure 9.16). One theory suggests that this takes place in June as the col between the subtropical anticyclone cells of the west Pacific and the Arabian Sea at the 300 mb level is displaced north-westwards from a position about 15°N, 95°E in May towards central India. The north-westward movement of the monsoon (see Figure 9.24) is apparently related to the extension over India of the upper tropospheric easterlies.

The recognition of the upper airflow has widespread effects in southern Asia. It is directly linked with the *Mai-yu* rains of China (which reach a peak about 10–15 June), the onset of the south-west Indian monsoon and the northerly retreat of the upper westerlies over the whole of the Middle East.

It must nevertheless be emphasized that it is still uncertain how far these changes are caused by events in the upper air or indeed whether the onset of the monsoon initiates a readjustment in the upper-air circulation. The presence of the Tibetan Plateau is certainly of importance even if there is no significant barrier effect on the upper airflow. The plateau surface is strongly heated in spring and early summer (R_n is about 180 W m^{-2} in May) and nearly all of this is transferred via sensible heat to the atmosphere. This results in the formation of a shallow heat low on the plateau, overlain, at about 450 mb, by a warm anticyclone (see Figure 6.1). The plateau atmospheric boundary layer now extends over an area about twice the size of the plateau surface itself. Easterly airflow on the southern side of the upper anticyclone undoubtedly assists in the northward shift of the subtropical westerly jet stream. At the same time, the pre-monsoonal convective activity over the south-eastern rim of the plateau provides a

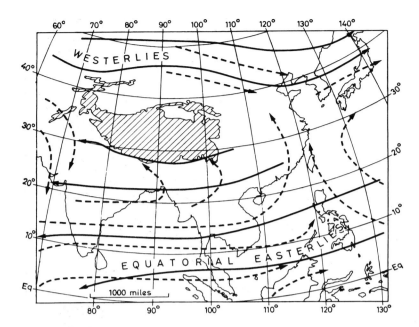

Figure 9.23 The characteristic air circulation over South and East Asia in summer. Solid lines indicate air flow at about 6,000 m and dashed lines at about 600 m. Note that the low-level flow is very uniform between about 600 and 3,000 m.

Sources: After Thompson 1951, Flohn 1968, Frost and Stephenson 1965, and others.

further heat source, by latent heat release, for the upper anticyclone. The seasonal wind reversals over and around the Tibetan Plateau have led Chinese meteorologists to distinguish a 'Plateau Monsoon' system, distinct from that over India.

4 Summer

By mid-July, monsoon air covers most of South and South-east Asia (see Figure 9.23), and in India the Equatorial Trough is located at about 25°N. North of the Tibetan Plateau there is a rather weak upper westerly current with a (subtropical) high-pressure cell over the plateau. The south-west monsoon in South Asia is overlain by strong upper easterlies (see Figure 9.19) with a pronounced jet at 150 mb (about 15 km), which extends westwards across South Arabia and Africa (Figure 9.25). No easterly jets have so far been observed over the tropical Atlantic or Pacific. The jet is related to a steep lateral temperature gradient, with the upper air getting progressively colder to the south.

An important characteristic of the tropical easterly jet is the location of the main belt of summer rainfall on the right (i.e. north) side of the axis upstream of the wind maximum and on the left side downstream, except for areas where the orographic effect is predominant (see Figure 9.25).

The mean jet maximum is located at about 15°N, 50–80°E.

The monsoon current does not give rise to a simple pattern of weather over India, despite the fact that much of the country receives 80 per cent or more of its annual precipitation during the monsoon season (Figure 9.26). In the north-west, a thin wedge of monsoon air is overlain by subsiding continental air. The inversion prevents convection and consequently little or no rain falls in the summer months in the arid north-west of the sub-continent (e.g. Bikaner and Kalat, Figure 9.20). This is similar to the Sahel zone in West Africa, discussed below.

Around the head of the Bay of Bengal and along the Ganges valley the main weather mechanisms in summer are the 'monsoon depressions' (Figure 9.27), which usually move westwards or north-westwards across India, steered by the upper easterlies (Figure 9.28), mainly in July and August. On average, they occur about twice a month, apparently when an upper trough becomes superimposed over a surface disturbance in the Bay of Bengal. Monsoon depressions have cold cores, are generally without fronts and are some 1,000–1,250 km across, with a cyclonic circulation up to about 8 km, and a typical lifetime of two to five days. They produce average daily rainfalls of 120–200 cm, occurring mainly

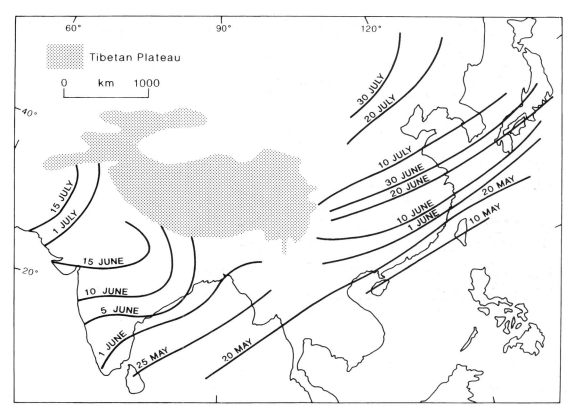

Figure 9.24 Mean onset date of the summer monsoon over South and East Asia.

Source: After Tao Shi-yan and Chen Longxun. From Domrös and Gongbing 1988. Reproduced by permission of Professor Tao Shi-yan and the Chinese Geographical Society.

as convective rains in the south-west quadrant of the depression. The main rain areas typically lie south of the Equatorial or Monsoon Trough (Figure 9.29) (in the south-west quadrant of the monsoon depressions, resembling an inverted mid-latitude depression). Figure 9.30 shows the extent and magnitude of a particularly severe monsoon depression. Such storms occur mainly in two zones: (1) the Ganges valley east of 76°E; (2) a belt across central India at around 21°N, at its widest covering 6° of latitude. Monsoon depressions also tend to occur on the windward coasts and mountains of India, Burma and Malaya. Without such disturbances, the distribution of monsoon rains would be controlled to a much larger degree by orography.

It is known that part of the south-west monsoonal flow occurs in the form of a 15–45 m s^{-1} jet stream at a level of only 1,000–1,500 m. This jet, strongest during active periods of the Indian monsoon, flows north-westwards from Madagascar (Figure 9.31) and crosses the equator from the south over East Africa, where its core is often marked by a streak of cloud (similar to that in Plate 14) and where it may bring excessive local rainfall. The jet is displaced northwards and strengthens from February to July; by May, it has become constricted against the Abyssinian Highlands, it accelerates still more and is deflected eastwards across the Arabian Sea towards the west coast of the Indian peninsula. This low-level jet, unique in the trade wind belt, flows offshore from the Horn of Africa, bringing up cool waters and contributing to a temperature inversion that is also produced by dry upper air originating over Arabia or East Africa and by subsidence due to the convergent upper easterlies. The flow from the south-west over the Indian Ocean is relatively dry near the equator and near shore, apart from a shallow moist layer near the base. Downwind towards India, however, there is a strong temperature and moisture interaction between the

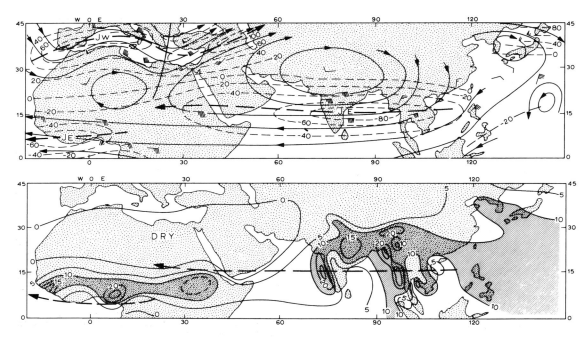

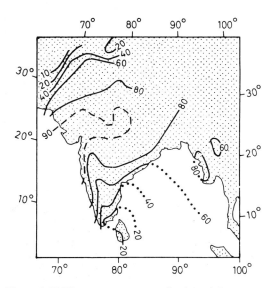

Figure 9.25 The easterly tropical jet stream. Above: The location of the easterly jet streams at 200 mb on 25 July 1955. Streamlines are shown in solid lines and isotachs (wind speed) dashed. Wind speeds are given in knots (westerly components positive, easterly negative). Below: The average July rainfall (shaded areas receive more than 25 cm) in relation to the location of the easterly jet streams.

Source: From Koteswaram 1958

Figure 9.26 The percentage contribution of the monsoon rainfall (June to September) to the annual total.

Sources: After Rao and Ramamoorthy, in Indian Meteorological Department 1960; and Ananthakrishnan and Rajagopalachari, in Hutchings 1964.

ocean surface and the low-level jet flow, such that deep convection builds up and convective instability is released, especially as the airflow slows down and converges near the west coast of India and as it is forced up over the Western Ghats. A portion of this south-west monsoon airflow is deflected by the Western Ghats to form 100 km diameter offshore vortices lasting two to three days and capable of bringing 100 mm of rain in 24 hours along the western coastal belt of the peninsula. At Mangalore (13°N), there are on average 25 rain-days per month in June, 28 in July and 25 in August. The monthly rainfall averages are 98 cm, 106 cm and 58 cm, respectively, accounting for 75 per cent of the annual total. In the lee of the Ghats, amounts are much reduced and there are semi-arid areas receiving less than 64 cm per year.

In southern India, excluding the south-east, there is a marked tendency for less rainfall when the Equatorial Trough is farthest north. Figure 9.20 shows a maximum at Minicoy in June, with a secondary peak in October as the Equatorial Trough and its associated disturbances withdraw southwards. This double peak occurs in much of interior

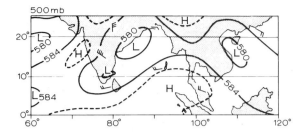

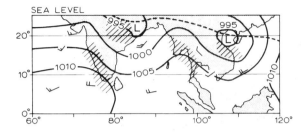

Figure 9.27 Monsoon depressions of 1200 GMT, 4 July 1957. The upper diagram shows the height (in tens of metres) of the 500 mb surface, the lower one the sea-level isobars. The broken line in the lower diagram represents the Equatorial Trough, and precipitation areas are shown by the oblique shading.

Source: Based on the IGY charts of the Deutscher Wetterdienst.

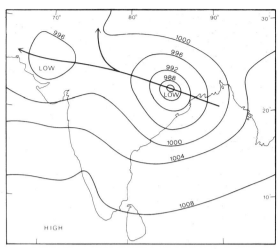

Figure 9.28 The normal track of monsoon depressions, together with a typical depression pressure distribution (mb).

Source: After Das 1987. From J. S. Fein and P. L. Stephens (eds) Monsoons. Copyright © 1987. Reproduced by permission of John Wiley and Sons, Inc.

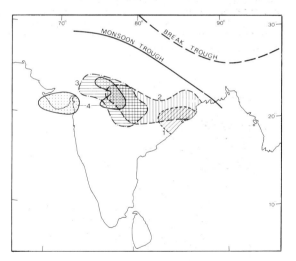

Figure 9.29 The location of the Monsoon Trough in its normal position during an active summer monsoon phase (solid) and during breaks in the monsoon (dashed).* Areas 1–4 indicate four successive daily areas of heavy rain (>50 mm/day) during the period 7–10 July 1973 as a monsoon depression moved west along the Ganges valley. Areas of lighter rainfall were much more extensive.

Source: *After Das 1987. From J. S. Fein and P. L. Stephens (eds) Monsoons. Copyright © 1987. Reproduced by permission of John Wiley and Sons, Inc.

peninsular India south of about 20°N and in western Sri Lanka, although autumn is the wettest period.

There is a variable pulse alternating between active and *break* periods in the May–September summer monsoon flow (see Figure 9.16) which, particularly at times of its strongest expression (e.g. 1971), produces periodic rainfall (Figure 9.32). During active periods, the convective Monsoon Trough is located in a southerly position, giving heavy rain over north and central India and the west coast (see Figure 9.16). Consequently, there is a strong upper-level outflow to the south, which strengthens both the easterly jet north of the equator and the westerly jet to the south over the Indian Ocean. The other upper-air outflow to the north fuels the weaker westerly jet there. Convective activity moves east from the Indian Ocean to the cooler eastern Pacific with an irregular periodicity (on average 40–50 days for the most marked waves), finding maximum expression at the 850 mb level and clearly being connected with the Walker circulation. After the passage of an active convective wave there is a more stable break in the summer monsoon when the ITCZ shifts to the south, the

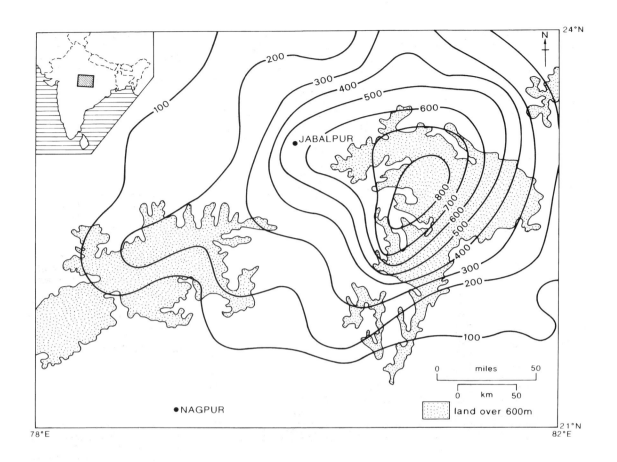

Figure 9.30 Rainfall (mm) produced in three days over a 50,000 km² area of central India north-east of Nagpur by a severe, westward-moving monsoon depression, during September 1926.

Source: Dhar and Nandargi 1993. From *International Journal of Climatology*, copyright © John Wiley & Sons Ltd. Reproduced with permission.

easterly jet weakens and subsiding air is forced to rise by the Himalayas along a *break trough* located above the foothills (see Figure 9.16), which replaces the Monsoon Trough during break periods. This circulation brings rain to the foothills of the Himalayas and the Brahmaputra valley at a time of generally low rainfall elsewhere. The shift of the ITCZ to the south of the subcontinent is associated with a similar movement and strengthening of the westerly jet to the north, weakening the Tibetan anticyclone or displacing it north-eastwards. The lack of rain over much of the subcontinent during break periods may be due in part to the eastward extension across India of the subtropical high-pressure cell centred over Arabia at this time.

It is important to realize that the monsoon rains are highly variable from year to year, emphasizing the role played by disturbances in generating rainfall within the generally moist south-westerly airflow. Droughts occur with some regularity in the Indian subcontinent: between 1890 and 1975 there were nine years of extreme drought (Figure 9.33) and at least five others of significant drought. These droughts are brought about by a combination of a late burst of the summer monsoon and an increase in the number and length of the break periods. Although breaks are most common in August–September, and last on average five days, they may occur at any time during the summer and can last as long as three weeks.

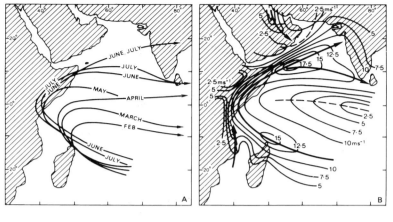

Figure 9.31 The mean monthly positions (A) and the mean July velocity (m s⁻¹) (B) of the low-level (1 km) Somali jet stream over the Indian Ocean.

Source: After Findlater 1971 (reproduced by permission of the Controller of Her Majesty's Stationery Office).

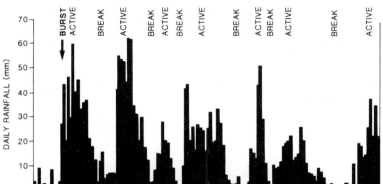

Figure 9.32 Mean daily rainfall (mm) along the west coast of India during the period 16 May to 30 September 1971, showing a pronounced burst of the monsoon followed by active periods and breaks of a periodic nature. All years do not exhibit these features as clearly.

Source: After Webster 1987b. From J. S. Fein and P. L. Stephens (eds) *Monsoons*. Copyright © 1987. Reproduced by permission of John Wiley and Sons, Inc.

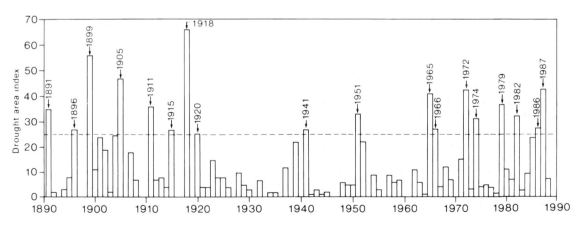

Figure 9.33 The yearly drought area index for the Indian subcontinent for the period 1891–1988, based upon the percentage of the total area experiencing moderate, extreme or severe drought. Years of extreme drought are dated. The dashed line indicates the lower limit of major droughts.

Source: After Bhalme and Mooley 1980. Updated by courtesy of H. M. Bhalme (reproduced by permission of the American Meteorological Society).

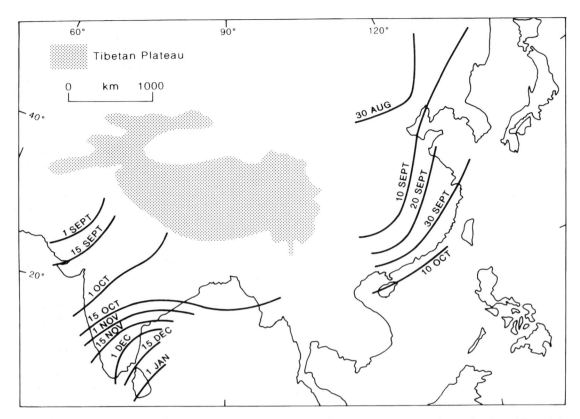

Figure 9.34 Mean onset date of the winter monsoon (i.e. retreat of the summer monsoon) over South and East Asia.
Sources: After Tao Shi-yan and Chen Longxun. Reproduced by permission of Professor Tao Shi-yan and the Chinese Geographical Society.

The strong surface heat source over the Tibetan Plateau, which is most effective during the day, gives rise to a 50–85 per cent frequency of deep cumulonimbus clouds over central and eastern Tibet in July. Late afternoon rain or hail showers are generally accompanied by thunder, but half or more of the precipitation falls at night, accounting for 70–80 per cent of the total in south-central and south-eastern Tibet. This may be related to large-scale plateau-induced local wind systems. However, the central and eastern plateau also has a frequency maximum of shear lines and associated weak lows at 500 mb during May to September. These plateau systems are shallow (2–2.5 km) and only 400–1,000 km in diameter, but they are associated with cloud clusters on satellite imagery in summer.

5 Autumn

Autumn sees the southward swing of the Equatorial Trough and the zone of maximum convection, which lies just to the north of the weakening easterly jet (see Figure 9.16). The break-up of the summer circulation systems is associated with the withdrawal of the monsoon rains, which is much less clearly defined than their onset (Figure 9.34). By October, the easterly trades of the Pacific affect the Bay of Bengal at the 500 mb level and generate disturbances at their confluence with the equatorial westerlies. This is the major season for Bay of Bengal cyclones and it is these disturbances, rather than the onshore north-easterly monsoon, that cause the October/November maximum of rainfall in south-east India (e.g. Madras, Figure 9.20).

During October, the westerly jet re-establishes itself south of the Tibetan Plateau, often within a few days, and cool season conditions are restored over most of South and East Asia.

D EAST ASIAN AND AUSTRALIAN SUMMER MONSOONS

China has no equivalent to India's hot, pre-monsoon season. The low-level, north-easterly winter monsoon (reinforced by subsiding air from the upper westerlies) persists in north China, and even in the south it begins to be replaced by maritime tropical air only in April–May. Thus, at Guangzhou (Canton) mean temperatures rise from only 17°C in March to 27°C in May, some 6°C lower than the mean values over northern India.

Westerly depressions are most frequent over China in spring (see Figure 9.22). They form more readily over Central Asia at this season as the continental anticyclone begins to weaken; also, many develop in the jet stream confluence zone in the lee of the plateau. The average number crossing China per month during 1921–31 was as follows:

J	F	M	A	M	J	J	A	S	O	N	D	Year
7	8	9	11	10	8	5	3	3	6	7	7	86

The zonal westerlies retreat northwards over China in May–June and the westerly flow becomes concentrated north of the Tibetan Plateau. The equatorial westerlies spread across South-east Asia from the Indian Ocean, giving a warm, humid air mass at least 3,000 m deep, but the summer monsoon over southern China is apparently influenced less by the westerly flow over India than by southerly airflow over Indonesia near 100°E. Also, contrary to earlier views, the Pacific is only a moisture source when tropical south-easterlies extend westwards to affect the east coast.

The Mai-yu 'front' involves both the monsoon trough and the East Asian–West Pacific Polar Front, with weak disturbances moving eastwards along the Yangtze valley and occasional cold fronts from the north-west. Its location shifts northwards in three stages, from south of the Yangtze River in early May to north of it by the end of the month and into northern China in mid-July (see Figure 9.24), where it remains until late September.

The surface airflow over China in summer is southerly (Table 9.2) and the upper winds are weak, with only a diffuse easterly current over southern China. According to traditional views, the monsoon current reaches northern China by July. The annual rainfall regime shows a distinct summer maximum with, for example, 64 per cent of the annual total occurring at Tianjin (Tientsin) (39°N) in July and August. Nevertheless, much of the rain falls during thunderstorms associated with shallow lows, and the existence of the ITCZ in this region is doubtful (see Figure 9.1). The southerly winds, which predominate over northern China in summer, are not necessarily linked to the monsoon current farther south. Indeed, this idea is the result of incorrect interpretation of streamline maps (of instantaneous airflow direction) as ones showing air trajectories (or the actual paths followed by air parcels). The depiction of the monsoon over China in Figure 9.24 is, in fact, based on a wet-bulb temperature value of 24°C. Cyclonic activity in northern China is attributable to the West Pacific Polar Front, forming between cP air and much-modified mT air (Figure 9.35).

In central and southern China, the three summer months account for about 40–50 per cent of the annual average precipitation, with another 30 per cent or so being received in spring. In south-east China, there is a rainfall singularity in the first half of July; a secondary minimum in the profile seems to result from the westward extension of the Pacific subtropical anticyclone over the coast of China.

A similar pattern of rainfall maxima occurs over southern and central Japan (Figure 9.36), comprising two of the six natural seasons that have been recognized there. The main rains occur during the *Bai-u* season of the south-east monsoon resulting from waves, convergence zones and closed circulations moving mainly in the tropical airstream around the Pacific subtropical anticyclone, but partly originating in a south-westerly stream that is the extension of the monsoon circulation of South Asia (Figure 9.23). The south-east circulation is displaced westwards from Japan by a zonal expansion of the subtropical anticyclone during late July

Table 9.2 Surface circulation over China.

	January	July
North China	60 % of winds from W, NW and N	57 % of winds from SE, S and SW
South-east China	88 % of winds from N, NE and E	56 % of winds from SE, S and SW

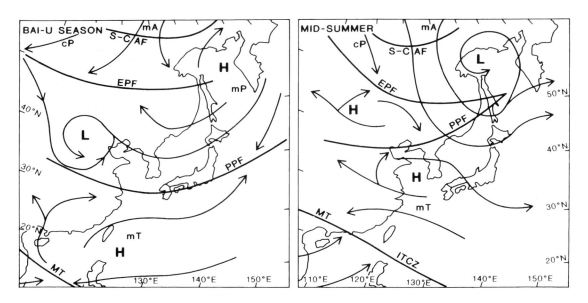

Figure 9.35 Schematic surface circulation pattern and frontal locations (Siberian–Canadian Arctic S–C AF, Eurasian Polar EPF, Pacific Polar PPF and Monsoon Trough MT/ Inter-tropical Convergence Zone MT/ITCZ) over East Asia during the Bai-u (i.e. July–August) season.

Source: Matsumoto 1985. Reproduced by permission, University of Tokyo.

and August, giving a period of more settled sunny weather. The secondary precipitation maximum of the *Shurin* season during September and early October coincides with an eastward contraction of the Pacific subtropical anticyclone, allowing low-pressure systems and typhoons from the Pacific to swing north towards Japan. Although much of the Shurin rainfall is believed to be of typhoon origin (see Figure 9.36), some is undoubtedly associated with the southern sides of depressions moving along the southward-migrating Pacific Polar Front to the north (see Figures 9.22 and 9.35), because there is a marked tendency for the autumn rains to begin first in the north of Japan and to spread south-wards. The manner in which the location of the western margin of the North Pacific subtropical high-pressure cell affects the climates of China and Japan is well illustrated by the changing seasonal trajectories of typhoon paths over East Asia (Figure 9.37). The northward and southward migrations of the cell zonal axis through 15° of latitude, the north-western high-pressure cell extensions over eastern China and the Sea of Japan in August, and its south-eastern contraction in October are especially marked.

Northern Australia experiences a monsoon regime during the austral summer. Low-level westerlies

develop in late December associated with a thermal low over northern Australia. Analogous to the vertical wind structure over Asia in July, there are easterlies in the upper troposphere. Various wind and rainfall criteria have been used to define monsoon onset. Based on the occurrence of (weighted) surface to 500 mb westerly winds, over-lain by 300–100 mb easterlies at Darwin (12.5°S, 131°E), the main onset date is 28 December and retreat date 13 March. Despite an average duration of 75 days, monsoon conditions lasted only 10 days in January 1961 and 1986 but 123–125 days in 1985 and 1974. Active phases with deep westerlies and rainfall each occur on just over half the days in a season, although there is little overlap between them. However, summer rainfall may also occur during deep easterlies associated with tropical squall lines and tropical cyclones. Active monsoon conditions typically persist for four to fourteen days, with breaks lasting about twenty to forty days.

E CENTRAL AND SOUTHERN AFRICA

1 The African monsoon

The annual climatic regime over West Africa has many similarities to that over South Asia, the surface

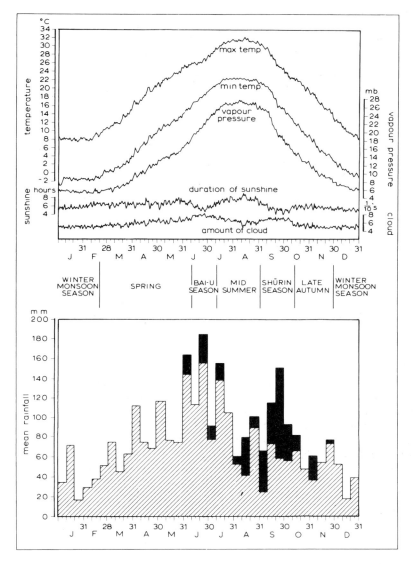

Figure 9.36 Above: Seasonal variation of daily normals at Nagoya, southern Japan, suggesting six natural seasons.* Below: Average 10-day precipitation amounts for a station in southern Japan, indicating in black the proportion of rainfall produced by typhoon circulations. The latter reaches a maximum during the Shurin season.†

Sources: *From Maejima 1967.
†After Saito 1959, from Trewartha 1981.

airflow being determined by the position of the leading edge of a Monsoon Trough (see Figure 9.2). This airflow is south-westerly to the south of the trough and easterly to north-easterly to its north (Figure 9.38). The major difference between the circulations of the two regions is due largely to the differing geography of the land–sea distribution and, particularly, to the topography. The lack of a large mountain range to the north of West Africa allows the Monsoon Trough to migrate regularly and continuously with the seasons without the pronounced northward surge

associated with the early summer burst of the Indian monsoon. In general, the West African Monsoon Trough oscillates between annual extreme locations of about 2°N and 25°N (Figure 9.39). In 1956, for example, these extreme positions were 5°N on 1 January and 23°N in August. The leading edge of the Monsoon Trough is complex in structure (see Figure 9.40B) and its position may oscillate greatly from day to day through several degrees of latitude.

In winter, the south-westerly monsoon airflow over the coasts of West Africa is very shallow (i.e.

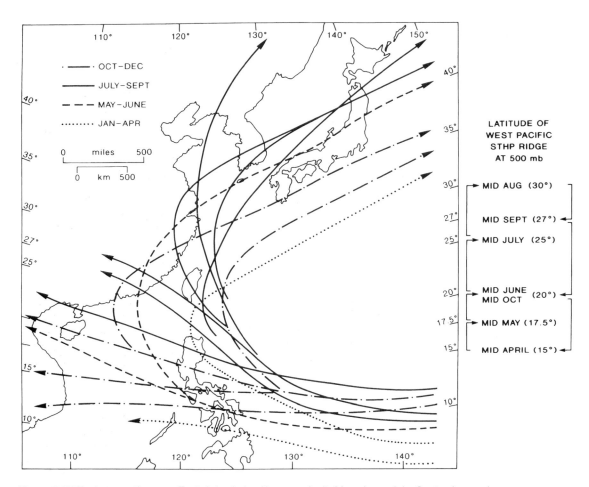

Figure 9.37 Typhoon paths over East Asia during January–April, May–June, July–September and October–December related to the mean latitude of the central ridge axis line of the subtropical high-pressure cell at 500 mb over the western Pacific.

Sources: Compiled from various sources, including Lin, Z. 1982 and Tao, S. 1984. Reproduced by permission of the Chinese Geographical Society.

1,000 m) with 3,000 m of overriding easterly winds, which are themselves overlain by strong (>20 m s⁻¹) winds (see Figure 9.41). North of the Monsoon Trough, the surface north-easterlies (i.e. the 2,000 m deep Harmattan flow) blow clockwise outwards from the subtropical high-pressure centre, being compensated above 5,000 m by an anticlockwise westerly airflow that, at about 12,000 m and 20–30°N, is concentrated into a subtropical westerly jet stream of average speed 45 m s⁻¹. Mean January surface temperatures decrease from about 26°C along the southern coast to 14°C in southern Algeria.

With the approach of the northern summer, the strengthening of the South Atlantic subtropical high-

pressure cell, combined with the increased continental temperatures, establishes a strong south-westerly airflow at the surface that spreads northwards behind the Monsoon Trough, lagging about six weeks behind the progress of the overhead sun. The northward migration of the trough oscillates diurnally with a northward progress of up to 200 km in the afternoons following a smaller southward retreat in the mornings. The northward spread of moist, unstable and relatively cool south-westerly airflow from the Gulf of Guinea brings rain in differing amounts to extensive areas of West Africa. Aloft, easterly winds spiral clockwise outwards from the subtropical high-pressure centre (see Figure 9.41) and are

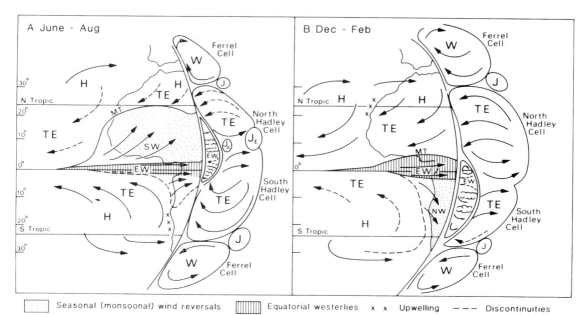

Seasonal (monsoonal) wind reversals   ▥ Equatorial westerlies   x x Upwelling   — — — Discontinuities

Figure 9.38 The major circulation in Africa in (A) June–August and (B) December–February. H: subtropical high-pressure cells; EW: equatorial westerlies (moist, unstable but containing the Congo high-pressure ridge); NW: the north-westerlies (summer extension of EW in the southern hemisphere); TE: tropical easterlies (trades); SW: south-westerly monsoonal flow in the northern hemisphere; W: extratropical westerlies; J: subtropical westerly jet stream; J_A and J_E: the (easterly) African jet streams; and MT: Monsoon Trough.

Source: From Rossignol-Strick 1985 (by permission Elsevier Science Publishers B.V., Amsterdam).

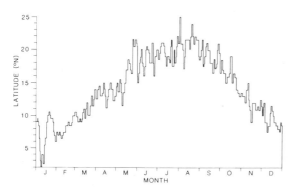

Figure 9.39 The daily position of the Monsoon Trough at longitude 3°E during 1957. This year experienced an exceptionally wide swing over West Africa, with the trough reaching 2°N in January and 25°N on 1 August. Within a few days after the latter date, the strongly oscillating trough had swung southwards through 8° of latitude.

Source: After Clackson 1957, from Hayward and Oguntoyinho 1987.

concentrated between June and August into two tropical easterly jet streams, the stronger (>20 m $^{-1}$) at about 15,000–20,000 m and the weaker (>10 m s^{-1}) at about 4,000–5,000 m (see Figure 9.40B). The lower jet occupies a broad band from 13°N to 20 °N, on the underside of which oscillations produce easterly waves, which may develop into squall lines. By July, the south-westerly monsoon airflow has spread far to the north, the leading trough reaching its extreme northern location, about 20°N, in August. At this time, four major climatic belts can be identified over West Africa (see Figure 9.40A):

1   A coastal belt of cloud and light rain related to frictional convergence within the monsoon flow, overlain by subsiding easterlies.
2   A quasi-stationary zone of disturbances associated with deep stratiform cloud yielding prolonged light rains. Low-level convergence south of the easterly jet axes, apparently associated with easterly wave disturbances from east central Africa, causes instability in the monsoon air.
3   A broad zone underlying the easterly jet streams, which help to activate disturbance lines and thunderstorms. North–south lines of deep

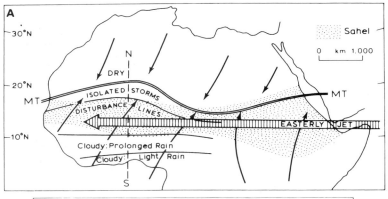

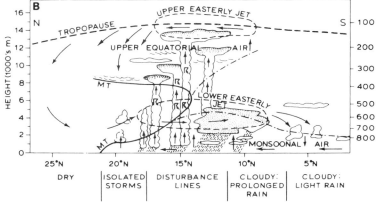

Figure 9.40 The structure of the circulation over North Africa in August. A: Surface airflow and Easterly Tropical Jet. B: Vertical structure and resulting precipitation zones over West Africa.

Sources: A: Reproduction from the *Geographical Magazine*, London. B: From Morley 1982 (copyright © Academic Press; reproduced by permission), and Musk 1983.

Notes : ℞ = thunderstorm activity; MT = Monsoon Trough.

cumulonimbus cells may move westwards steered by the jets. The southern, wetter part of this zone is termed the Soudan, the northern part the Sahel, but popular usage assigns the name Sahel to the whole belt.

4 Just south of the Monsoon Trough, the shallow tongue of humid air is overlain by drier subsiding air. Here there are only isolated storms, scattered showers, and occasional thunderstorms.

In contrast to winter conditions, August temperatures are lowest (i.e. 24–25°C) along the cloudy southern coasts and increase towards the north, where they average 30°C in southern Algeria.

Both the summer airflows, the south-westerlies below and the easterlies aloft, are subject to perturbations, which contribute significantly to the rainfall during this season. Three types of perturbation are particularly prevalent:

1 Waves in the south-westerlies. These are northward surges of the humid airflow, having periodicities of four to six days, producing bands of summer monsoon rain some 160 km broad and 50–80 km in north–south extent, which have the most marked effect 1,100–1,400 km south of the surface Monsoon Trough, the position of which oscillates with the surges.

2 Waves in the easterlies. These develop on the interface between the lower south-westerly and the upper easterly airflows. These waves are from 1,500 to 4,000 km long from north to south and move across West Africa towards the west between mid-June and October with a periodicity of three to five days at speeds of about 5–10° of longitude per day (i.e. 18–35 km hr⁻¹), sometimes developing closed cyclonic circulations. At the height of the summer monsoon, they produce most rainfall at around latitude 14°N, between 300 and 1,100 km south of the Monsoon Trough. On average, some fifty easterly waves per year cross Dakar. Some of these carry on in the general circulation across the

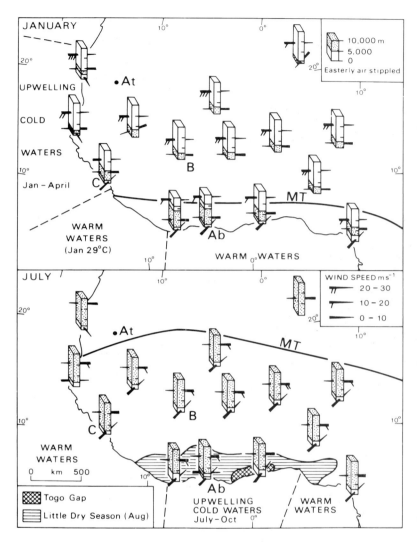

Figure 9.41 Mean wind speeds (m s⁻¹) and directions in January and July over West Africa up to about 15,000 m. Ocean water temperatures and the positions of the Monsoon Trough are also shown, as are the area affected by the August Little Dry Season and the location of the anomalous Togo Gap. The locations of Abidjan (Ab), Atar (At), Bamako (B) and Conakry (C) are given (see precipitation graphs in Figure 9.42).

Source: From Hayward and Oguntoyinbo 1987.

Atlantic, and it has been estimated that 60 per cent of West Indian hurricanes originate in West Africa as easterly waves.

3 Squall lines. Easterly waves vary greatly in intensity. Some give rise to little cloud and rain, whereas others develop as squall lines when the wave extends down to the surface, producing updraughts, heavy rain and thunder. Squall line formation is assisted where surface topographic convergence of the easterly flow occurs (e.g. near Lake Chad and north-west of the Niger delta). These disturbance lines travel at up to 60 km hr⁻¹ from east to west across southern West Africa for distances of up to 3,000 km (but aver-

aging 600 km) between June and September, yielding 40–90 mm of rain per day. Some coastal locations suffer about forty squall lines per year, which account for more than 50 per cent of the annual rainfall (see Plate 35).

Annual rainfall decreases from 2,000–3,000 mm in the coastal belt (e.g. Conakry, Guinea) to about 1,000 m at latitude 20°N (Figure 9.42). Near the coast, more than 300 mm per day of rain may fall during the rainy season but further north the variability increases due to the irregular extension and movement of the Monsoon Trough. Squall lines and other disturbances give a zone of maximum rainfall

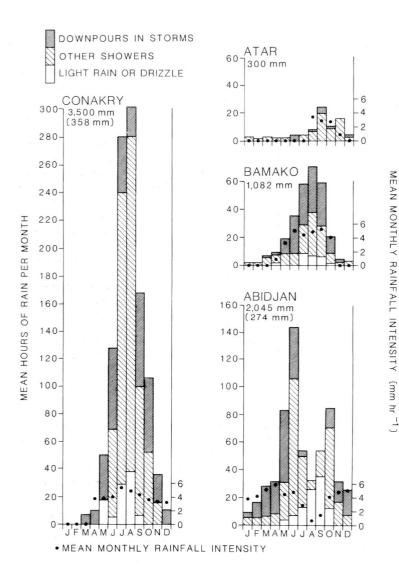

Figure 9.42 Mean number of hours of rain per month for four West African stations. Also shown are types of rainfall, mean annual totals (mm) and, in parentheses, maximum recorded daily rainfalls (m) for Conakry (August) and Abidjan (June). Dots show the mean monthly rainfall intensities (mm hr⁻¹). Note the pronounced Little Dry Season at Abidjan. Station locations are marked on Figure 9.41.

Source: From Hayward and Oguntoyinbo 1987.

located 800–1,000 km south of the surface position of the Monsoon Trough (see Figure 9.40B). Monsoon rains in the coastal zone of Nigeria (4°N) contribute 28 per cent of the annual total (about 2,000 mm), thunderstorms 51 per cent and disturbance lines 21 per cent. At 10°N, 52 per cent of the total (about 1,000 mm) is due to disturbance lines, 40 per cent to thunderstorms and only 9 per cent to the monsoon. Over most of the country, rainfall from disturbance lines has a double frequency maximum, thunderstorms a single one in summer (see Figure 9.43 for Minna, 9.5°N). In the

northern parts of Nigeria and Ghana, rain falls in the summer months, mostly from isolated storms or disturbance lines. The high variability of these rains from year to year characterizes the drought-prone Sahel environment.

The summer rainfall in the northern Soudana–Sahelian belts is determined partly by the northward penetration of the Monsoon Trough, which may range up to 500–800 km beyond its average position (Figure 9.44), and by the strength of the easterly jet streams. The latter affects the frequency of disturbance lines.

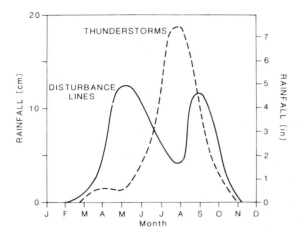

Figure 9.43 The contributions of disturbance lines and thunderstorms to the average monthly precipitation at Minna, Nigeria (9.5°N).

Source: After Omotosho 1985 (by permission of the Royal Meteorological Society).

Anomalous climatic effects occur in a number of distinct West African localities at different times of the year. Although the temperatures of coastal waters always exceed 26°C and may reach 29°C in January, there are two areas of locally upwelling cold waters (see Figure 9.41). One lies north of Conakry along the coasts of Senegal and Mauretania, where dominant offshore north-easterly winds in January–April skim off the surface waters, causing cooler (20°C) water to rise, dramatically lowering the temperature of the afternoon onshore

breezes (see Figure 5.15). The second area of cool ocean (19–22°C) is located along the central southern coast west of Lagos during the period July–October, for a reason that is as yet unclear. From July to September, an anomalously dry land area is located along the southern coastal belt (see Figure 9.41) during what is termed the *Little Dry Season*. The reason is that at this time the Monsoon Trough is in its most northerly position and this coastal zone, lying 1,200–1,500 km to the south of it and, more important, 400–500 km to the south of its major rain belt, has relatively stable air (see Figure 9.40B), a condition assisted by the relatively cool offshore coastal waters. Embedded within this relatively cloudy but dry belt is the smaller Togo Gap, lying between 0° and 3°E and having during the summer above-average sunshine, subdued convection, relatively low rainfall (i.e. less than 1,000 mm) and low thunderstorm activity. It is considered that the trend of the coast here parallels the dominant low-level south-westerly winds, so limiting surface frictionally induced convergence in an area where temperatures and convection are in any case inhibited by low coastal water temperatures.

2 Southern Africa

Southern Africa lies between the South Atlantic and Indian Ocean subtropical high-pressure cells in a region subject to the interaction of tropical easterly and extra-tropical westerly airflows. Both these high-pressure cells shift west and intensify (see Figure 6.11) in the southern winter but, because the

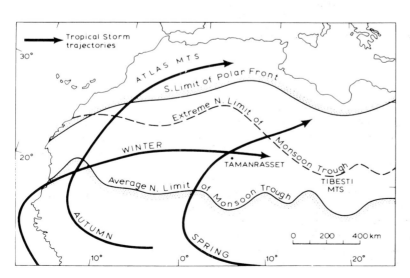

Figure 9.44 Extent of precipitation systems affecting western and central North Africa and typical tracks of Soudano–Sahelian depressions.

Source: After Dubief and Yacono; from Barry 1991.

January

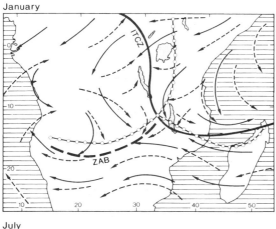

July

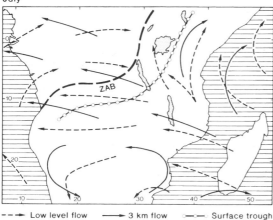

--- Low level flow   ⟶ 3 km flow   ⊶⊶ Surface trough

Figure 9.45 Airflow over southern Africa during January and July, together with the locations of the Intertropical Convergence Zone (ITCZ), the Zaire Air Boundary (ZAB) and the major surface low-pressure troughs.

Source: After Taljaard; from Tyson 1986, *Climatic Change and Variability in Southern Africa*, copyright © Oxford University Press, Southern Africa.

South Atlantic cell always extends 3° latitude further north than the Indian Ocean cell, it brings low-level westerlies to Angola and Zaire at all seasons and high-level westerlies to central Angola in the southern summer. The seasonal longitudinal shifts of the subtropical high-pressure cells are especially significant to the climate of southern Africa in respect of the Indian Ocean cell. Whereas the 7–13° longitudinal shift of the South Atlantic cell has relatively little effect, the westward movement of 24–30° during the southern winter by the Indian Ocean cell brings an easterly flow at all levels to

most of southern Africa. The seasonal airflows and convergence zones are shown in Figure 9.45.

In summer (i.e. January) low-level westerlies over Angola and Zaire meet the north-east monsoon of East Africa along the Intertropical Convergence Zone (ITCZ), which extends east as the boundary between the recurved (westerly) winds from the Indian Ocean and the deep tropical easterlies further south. To the west, these easterlies impinge on the Atlantic westerlies along the Zaire Air Boundary (ZAB). The ZAB is subject to daily fluctuations and low-pressure systems form along it, either being stationary or moving slowly westward. When these are deep and associated with southward-extending troughs they may produce significant rainfall. It should be noted that the complex structure of the ITCZ and ZAB means that the major surface troughs and centres of low pressure do not coincide with them but are situated some distance upwind in the low-level airflow (see Figure 9.45), particularly in the easterlies. This low-level summer circulation is dominated by a combination of these frontal lows and convectional heat lows. By March, a unified high-pressure system has been established, giving a northerly flow of moist air, which produces autumn rains in western regions. In winter (i.e. July), the ZAB separates the low-level westerly and easterly airflows from the Atlantic and Indian Oceans, although both are overlain by a high-level easterly flow. At this time, the northerly displacement of the general circulation brings low- and high-level westerlies with rain to the southern Cape.

Thus tropical easterly airflows affect much of southern Africa throughout the year. A deep easterly flow dominates south of about 10°S in winter and south of 15–18°S in summer. Over East Africa, a north-easterly monsoonal flow occurs in summer, replaced by a south-easterly flow in winter. Easterly waves form in these airflows, similar to, but less mobile than, those in other tropical easterlies. These waves form at the 850–700 mb level (i.e. 200–3,000 m) in flows associated with easterly jets, often producing squall lines, belts of summer thunder cells and heavy rainfall. These waves are most common between December and February, when they may produce at least 40 mm of rain per day, but are rare between April and October. Tropical cyclones in the South Indian Ocean occur particularly around February (see Figure 9.8 and Table 9.1), when the ITCZ lies at its extreme southerly position. These storms recurve south along the east coast of

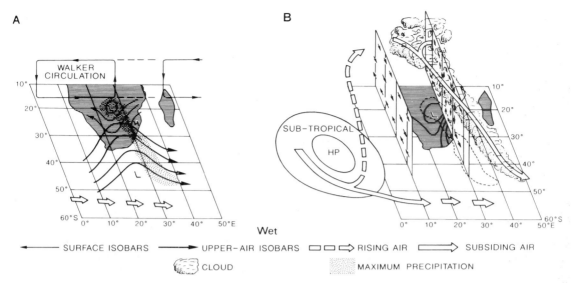

A WALKER CIRCULATION

B SUB-TROPICAL HP

Wet

◄──── SURFACE ISOBARS ────► ─────► UPPER–AIR ISOBARS ▭▭⇨ RISING AIR ▭⇨ SUBSIDING AIR

☁ CLOUD   ⬡ MAXIMUM PRECIPITATION

Figure 9.46 Combination of A, a high-phase Walker circulation, continental low pressure and an upper westerly wave, and B, a strengthened South Atlantic subtropical high-pressure cell, resulting in above-normal rainfall over southern Africa.

Source: After M. S. J. Harrison; from Tyson 1986, *Climatic Change and Variability in Southern Africa*, copyright © Oxford University Press, Southern Africa.

Tanzania and Mozambique, but their influence is limited mainly to the coastal belt.

With few exceptions, deep westerly airflows are limited to the most southerly locations of southern Africa, especially in winter. As in northern mid-latitudes, disturbances in the westerlies involve:

1 Quasi-stationary Rossby waves.
2 Travelling waves, particularly marked at and above the 500 mb level, with axes tilted west-wards with height, divergence ahead and conver-gence in the rear, moving eastwards at a speed of some 550 km/day, having a periodicity of two to eight days and with associated cold fronts.
3 Cut-off low-pressure centres. These are intense, cold-cored depressions, most frequent during March–May and September–November, and rare during December–February.

A feature of the climate of southern Africa is the prevalence of wet and dry spells, associated with broader features of the global circulation. Above-normal rainfall, occurring as a north–south belt over the region, is associated with a high-phase Walker circulation (see p. 133) having an ascending limb over southern Africa; a strengthening of the ITCZ; an intensification of tropical lows and easterly waves, often in conjunction with a westerly wave

aloft to the south; and a strengthening of the South Atlantic subtropical high-pressure cell (see Figure 9.46). Such a wet spell may occur particularly during the spring to autumn period. Below-normal rainfall is associated with a low-phase Walker circu-lation having a descending limb over southern Africa; a weakening of the ITCZ; a tendency to high pressure with a diminished occurrence of tropical lows and easterly waves; and weakening of the South Atlantic subtropical high-pressure cell. At the same time, there is a belt of cloud and rain lying to the east in the western Indian Ocean associated with a rising Walker limb and enhanced easterly disturbances in conjunction with a westerly wave aloft south of Madagascar (Figure 9.47).

F AMAZONIA

Amazonia lies athwart the equator (Figure 9.48) and contains some 30 per cent of the total global biomass. The continuously high temperatures (24–28°C) combine with the high transpiration to cause the region to behave at times as if it were a source of maritime equatorial air.

Important influences over the climate of Amazonia are the North and South Atlantic subtropical high-pressure cells. From these, stable

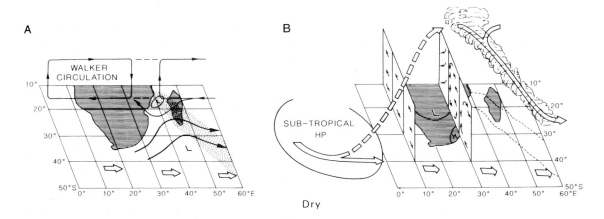

Figure 9.47 Combination of A, a low-phase Walker circulation, continental high pressure, increased frequency of easterly waves and tropical lows over the western Indian Ocean, an upper westerly wave south of Madagascar, and B, a weakened South Atlantic subtropical high-pressure cell, resulting in below-normal rainfall over southern African and above-normal to the east.

Source: After M. S. J. Harrison; from Tyson 1986, *Climatic Change and Variability in Southern Africa*, copyright © Oxford University Press, Southern Africa.

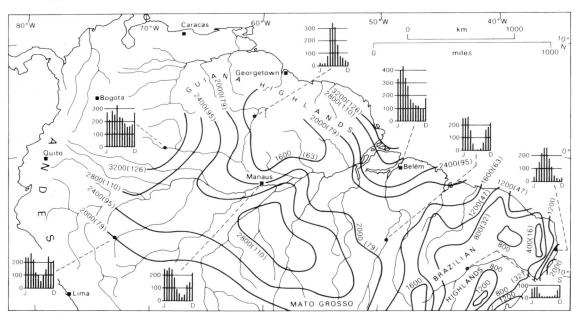

Figure 9.48 Mean annual precipitation (mm) over the Amazon basin, together with mean monthly precipitation amounts for eight stations.

Source: From Ratisbona 1976.

easterly mT air invades Amazonia in a shallow (1,000–2,000 m), relatively cool and humid layer, overlain by warmer and drier air from which it is separated by a strong temperature inversion and humidity discontinuity. This shallow airflow gives some precipitation in coastal locations but produces drier conditions inland unless it is subjected to strong convection when a heat low is established over the continental interior. At such times, the inversion rises to 3,000–4,000 m and may break down altogether associated with heavy precipitation, particularly in late afternoon or evening. The South Atlantic subtropical high-pressure cell expands westwards over Amazonia in July, producing drier conditions as shown by the rainfall at inland stations such as Manaus (see Figure 9.48), but in September it begins to contract and the build-up of the continental heat low ushers in the October–April rainy season in central and southern Amazonia. The North Atlantic subtropical high-pressure cell is less mobile than its southern counterpart but varies in a more complex manner, having maximum westward extensions in July and February and minima in November and April. In northern Amazonia, the rainy season is May to September. Rainfall over the region as a whole is mainly due to a low-level convergence associated with convective activity, a poorly defined Equatorial Trough, instability lines, occasional incursions of cold fronts from the southern hemisphere, and relief effects.

Strong thermal convection over Amazonia can commonly produce more than 40 mm/day of rainfall over a period of a week and much higher average intensities over shorter periods. When it is recognized that 40 mm of rainfall in one day releases sufficient latent heat to warm the troposphere by 10°C, it is clear that sustained convection at this intensity is capable of fuelling the Walker circulation (see Figure 9.52). During high phases of ENSO, air rises over Amazonia, whereas during the low phases the drought over north-east Brazil is intensified. In addition, convective air moving polewards may strengthen the Hadley circulation. This air tends to accelerate, due to the conservation of angular momentum, and strengthen the westerly jet streams such that correlations have been found between Amazonian convective activity and North American jet stream intensity and location.

The Intertropical Convergence Zone (ITCZ) does not exist in its characteristic form over the interior of South America, and its passage affects rainfall only near the east coast. The intensity of this zone varies, being least when both the North and South Atlantic subtropical high-pressure cells are strongest (i.e. in July), giving a pressure increase that causes the equatorial trough to fill. The ITCZ swings to its most northerly position during July–October, when invasions of more stable South Atlantic air are associated with drier conditions over central Amazonia, and to its most southerly in March–April

(Figure 9.49). At Manaus, surface winds are predominantly south-easterly from May to August and north-easterly from September to April, whereas the upper tropospheric winds are north-westerly or westerly from May to September and southerly or south-easterly from December to April. This reflects the development in the austral summer of an upper tropospheric anticyclone that is located over the Peru–Bolivia Altiplano. This upper high is a result of sensible heating of the elevated plateau and the release of latent heat in frequent thunderstorms over the Altiplano, analogous to the situation over Tibet. Outflow from this high subsides in a broad area extending from eastern Brazil to West Africa. The drought-prone region of eastern Brazil is particularly moisture-deficient during periods when the ITCZ remains in a northerly position and relatively stable mT air from a cool South Atlantic surface is dominant (see Chapter 3E.5). Dry conditions may occur during January–May during strong ENSO events (see p. 277), when the descending branch of the Walker circulation covers most of Amazonia.

Significant Amazonian rainfall, particularly in the east, originates along mesoscale lines of instability, which form near the coast due to converging trade winds and afternoon sea breezes, or to the interaction of nocturnal land breezes with onshore trade winds. These lines of instability move westwards in the general airflow at speeds of about 50 km hr⁻¹, moving faster in January than in July and exhibiting a complex process of convective-cell growth, decay, migration and regeneration. Many of these instability lines reach only 100 km or so inland, decaying after sunset (Figure 9.50). However, the more persistent instabilities may produce a rainfall maximum about 500 km inland, and some remain active for up to 48 hours such that their precipitation effects reach as far west as the Andes. Other meso- to synoptic-scale disturbances form within Amazonia, especially between April and September. Precipitation also occurs with the penetration of cool mP air masses from the south, especially between September and November, which are heated from below and become unstable (see Figure 9.49).

Surges of cold polar air (*friagens*) during the winter months can cause freezing temperatures in southern Brazil, with cooling to 10°C even in Amazonia. In June–July 1994, such events caused devastation to Brazil's coffee production. Typically, an upper-level trough crosses the Andes of central Chile

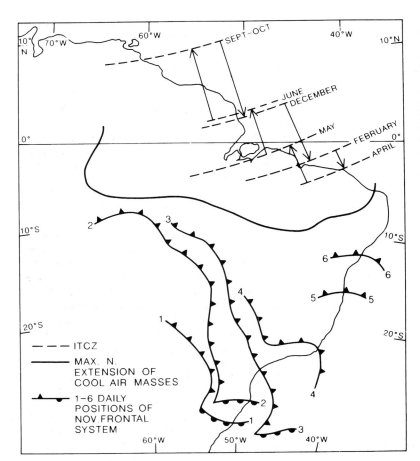

Figure 9.49 The synoptic elements of Brazil. The seasonal positions of the coastal Intertropical Convergence Zone; the maximum northerly extension of cool southerly mP air masses; and the positions of a typical frontal system during six successive days in November as the centre of the low pressure moves south-eastwards into the South Atlantic.

Source: From Ratisbona 1976.

from the eastern South Pacific and an associated southerly airflow transports cold air north-eastwards over southern Brazil. Concurrently, a surface high-pressure cell may move northwards from Argentina, with the associated clear skies producing additional radiative cooling.

The tropical easterlies over the northern and eastern margins of Amazonia are susceptible to the formation of easterly waves and closed vortices, which move westwards generating rain bands. Relief effects are naturally most noteworthy as airflow approaches the eastern slopes of the Andes, where large-scale orographic convergence in a region of high evapotranspiration contributes to the high precipitation all through the year.

G EL NIÑO–SOUTHERN OSCILLATION (ENSO) EVENTS

1 The Pacific Ocean

The Southern Oscillation is an irregular variation, see-saw, or standing wave in atmospheric mass and pressure (having a period of two to ten years) involving exchanges of air between the subtropical high-pressure cell over the eastern South Pacific and a low-pressure region centred on the western Pacific and Indonesia (Figure 9.51). Its complicated mechanism is held by some experts to centre on the control over the strength of the Pacific trade winds exercised by the activity of the subtropical high-pressure cells, particularly the one over the South Pacific. Others, recognizing the ocean as an enormous heat energy source, believe that near-surface temperature variations in the tropical Pacific may

January.

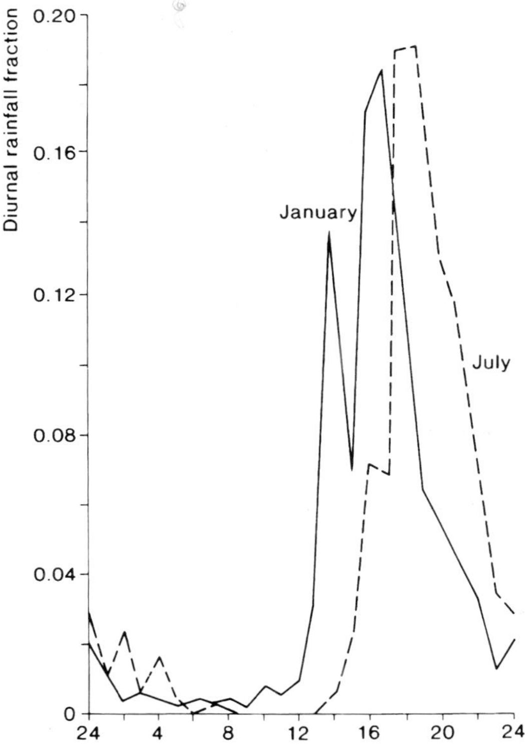

Diurnal rainfall fraction

January

July

Figure 9.50 Hourly rainfall fractions for Belém, Brazil, for January and July. The rain mostly results from convective cloud clusters developing offshore and moving inland, more rapidly in January.

Source: After Kousky 1980.

act somewhat similar to a flywheel to drive the whole ENSO system. It is important to note that a deep (i.e. 100 m+) pool of the world's warmest surface water builds up in the western equatorial Pacific between the surface and the thermocline due to intense insolation, low heat loss from evaporation in this region of light winds, and to the piling up of surface water driven westwards by the easterly trade winds. The pool of heat energy is dissipated periodically during El Niño by the changing ocean currents and by release into the atmosphere.

The Southern Oscillation is associated with the phases of the Walker circulation that have already been introduced in Chapter 6C.1. The high phases of the Walker circulation (i.e. non-ENSO or La Niña events), which occur on average three years out of four, alternate with low phases (i.e. ENSO or El Niño events). Sometimes, however, the Southern Oscillation is not in evidence and neither phase is dominant. The level of activity of the Southern Oscillation in the Pacific is expressed by the Southern Oscillation Index (SOI), which is a complex measure involving sea-surface and air temperatures, pressures at sea level and aloft, and rainfall at selected locations.

During non-ENSO, high phases (Figure 9.52A) strong easterly trade winds in the eastern tropical Pacific produce upwelling along the west coast of South America, resulting in a north-flowing cold current (the Peru or Humboldt), locally termed La Niña – the girl – on account of its richness in plankton and fish. The low sea temperatures produce a

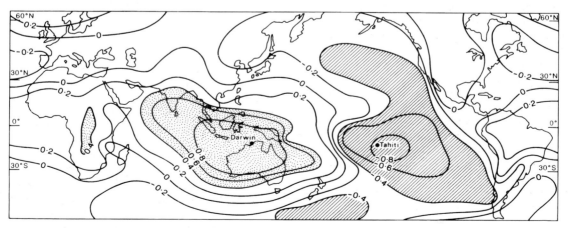

Figure 9.51 The correlation of mean annual sea-level pressures with that at Darwin, Australia, illustrating the two major cells of the Southern Oscillation.

Source: Rasmusson 1985. Copyright © *American Scientist* **73**, 1985.

A. DEC-FEB (NON-ENSO)

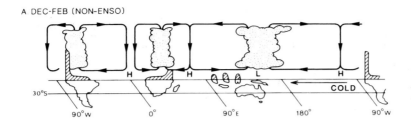

B. DEC-FEB 1982-3 (ENSO)

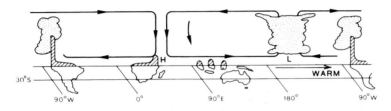

C. DEC-FEB SEA SURFACE TEMPERATURES (NON-ENSO)

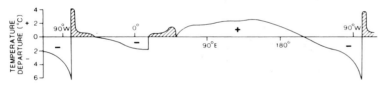

D. STRONG EASTERLY WIND (NON-ENSO)   E. WEAK EASTERLY WIND (ENSO)

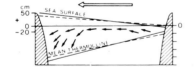 

Figure 9.52 Schematic cross-sections of the Walker circulation along the equator based on computations of Y. Tourre (1984). A: Mean December–February regime (non-ENSO); rising air and heavy rains occur over the Amazon basin, central Africa and Indonesia–western Pacific. B: December–February 1982–3 ENSO pattern; the ascending Pacific branch is shifted east of the Date Line and suppressed convection occurs elsewhere due to subsidence. C: Departure of sea-surface temperature from its equatorial zonal mean, corresponding to non-ENSO case (A). D: Strong trades cause sea level to rise and the thermocline to deepen in the western Pacific for case (A). E: Winds relax, sea level rises in the eastern Pacific as water mass moves back eastwards and the thermocline deepens off South America during ENSO events.

Source: Based on Wyrtki (by permission World Meteorological Organization 1985).

shallow inversion, thereby further strengthening the trade winds (i.e. effecting positive feedback), which skim water off the surface of the Pacific, where warm surface water accumulates (Figure 9.52D). This action also causes the thermocline to lie at shallow depths (about 40 m) in the east, as distinct from 100–200 m in the western Pacific. The strengthening of the easterly trades causes cold water upwelling to spread westwards and the cold tongue of surface water extends in that direction sustained by the South Equatorial Current. This westward-flowing current is wind-driven and is compensated by a deeper surface slope. The westward contraction of warm Pacific water into the central and western

tropical Pacific (Figure 9.52C) produces an area of instability and convection fed by moisture in a convergence zone under the dual influence of both the Intertropical Convergence Zone and the South Pacific Convergence Zone. The rising air over the western Pacific feeds the return airflow in the upper troposphere (i.e. at 200 mb), closing and strengthening the Walker circulation. However, this airflow also strengthens the Hadley circulation, particularly its meridional component northwards in the northern winter and southwards in the southern winter.

Each year, usually starting in December, a weak southward flow of warm water replaces the northward-flowing Peru Current and its associated

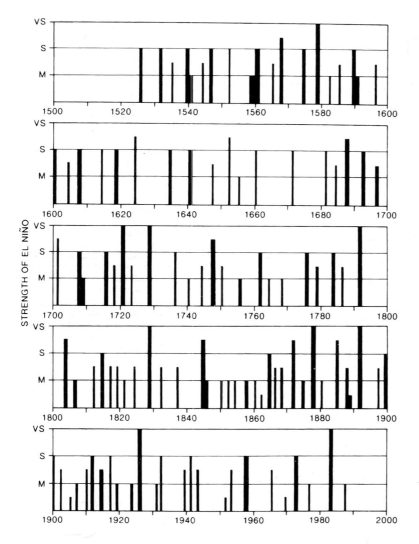

Figure 9.53 El Niño events 1525–1987 classified according to very strong, strong and medium.

Source: Quinn and Neal 1992. Reprinted from Climate since AD 1500, R. S. Bradley and P. D. Jones (eds) 1992. Copyright © Routledge, London.

cold upwelling southward to about 6°S along the coast of Ecuador. This phenomenon, known as El Niño (the child, after the Christ child), strengthens at irregular intervals of two to ten years (its average interval is four years) when warm surface water becomes much more extensive and the coastal upwelling ceases entirely. This has catastrophic ecological and economic consequences to fish and bird life, and to the fishing and guano industries of Ecuador, Peru and northern Chile. Figure 9.53 shows the occurrence of El Niño events between 1525 and 1987 classified according to their intensity.

ENSO events result from a radical reorganization of the Walker circulation in two main respects:

1 Pressure declines and the trades weaken over the eastern tropical Pacific (Figure 9.52B), wind-driven upwelling slackens, allowing the ITCZ to extend southwards to Peru. This increase of sea-surface temperatures by 1–4°C reduces the west-to-east sea-surface temperature gradient across the Pacific and also tends to decrease pressure over the eastern Pacific. The latter causes a further decrease of trade wind activity, a decrease in upwelling of cold water, an advection of warm water and a further increase in sea-surface temperatures – in other words, the onset of El Niño activates a positive feedback loop in the eastern Pacific atmosphere–ocean system.

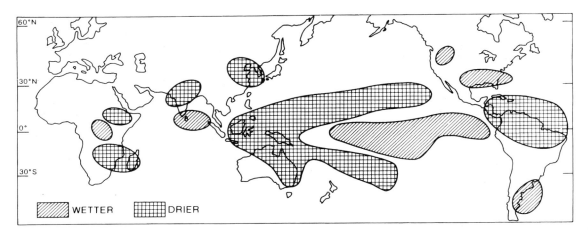

Figure 9.54 The coincidence of ENSO events with regional climates that are wetter or drier than normal.

Sources: After Rasmusson and Ropelowski, also Halpert. From M. H. Glantz *et al.* (eds) *Teleconnections Linking Worldwide Climate Anomalies*, 1990. Composite reproduced by permission of Cambridge University Press.

2 Over the western tropical Pacific, the area of maximum sea temperatures and convection responds to the above weakening of the Walker circulation by moving eastwards into the central Pacific (Figure 9.52B). This is partly due to an increase of pressure in the west but also to a combined movement of the ITCZ southwards and the SPCZ north-eastwards. Under these conditions, bursts of equatorial westerly winds spread a huge tongue of warm water (i.e. warmer than 27.5°C) eastwards over the central Pacific as large-scale, internal oceanic (Kelvin) waves. It has been suggested that this eastward flow may sometimes be triggered off or strengthened by the occurrence of cyclone pairs north and south of the equator (see Plate 33). This eastward flow of warm water depresses the thermocline off South America (Figure 9.52E), further preventing cold water reaching the surface and terminating the El Niño effect.

Thus, whether La Niña or El Niño develops, bringing westward-flowing cold surface water or eastward-flowing warm surface water, respectively, to the central Pacific, depends on the competing processes of upwelling versus advection. The most intense phase of an El Niño event commonly lasts about one year, and the change to El Niño usually occurs about March–April, when the trade winds and the cold tongue are at their weakest. The changes to the Pacific atmosphere–ocean circulation during El Niño are facilitated by the fact that the time taken for ocean surface currents to adjust to

major wind changes decreases markedly with decreasing latitude, as is demonstrated by the seasonal reversal of the south-west and north-east monsoon drift off the Somali coast in the Indian Ocean. Large-scale atmospheric circulation is subject to a negative-feedback constraint involving a negative correlation between the strengths of the Walker and Hadley circulations. Thus the weakening of the Walker circulation during an ENSO event leads to a relative strengthening of the associated Hadley circulation.

2 Teleconnections

Teleconnections are defined as linkages over great distances of atmospheric and oceanic variables; clearly the linkages between climatic conditions in the eastern and western tropical Pacific Ocean represent an archetypal teleconnection. Figure 9.54 illustrates the coincidence of ENSO events with regional climates that are wetter or drier than normal.

In Chapter 6C.1, we have referred to Walker's observed teleconnection between ENSO events and the lower than normal monsoon rainfall over South and South-east Asia (Figure 9.55). This is due to the eastward movement of the zone of maximum convection over the western Pacific. However, it is important to recognize that ENSO mechanisms form only part of the South Asian monsoon phenomenon and that, for example, parts of India can experience droughts in the absence of El Niño and that the onset of the monsoon can also depend on the control

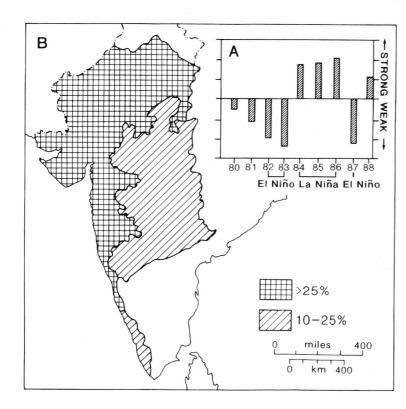

B

A

STRONG WEAK

80 81 82 83 84 85 86 87 88
El Niño La Niña El Niño

▦ >25%

▨ 10–25%

0   miles   400

0   km   400

Figure 9.55 The proposed connection between the Indian summer monsoon and El Niño. A: The observed strength of the Asian summer monsoon (1980–88) showing its weakness during the three strong El Niño years 1982, 1983 and 1987. B: Areas of India where the summer monsoonal rainfall deficits (as a percentage less than the 1901–70 average) were significantly more frequent in the El Niño years.

Sources: A: Browning 1996. B: Gregory 1988. IGU Study Group on Recent Climate Change.

exercised by the amount of Eurasian snow cover on the persistence of the continental high-pressure cell.

The eastward movement of the western Pacific zone of maximum convection in the ENSO phase also decreases summer monsoon rainfall over northern Australia, as well as extra-tropical rainfall over eastern Australia in the winter–spring season. During the latter, a high-pressure cell over Australia brings widespread drought, but this is compensated for by enhanced rainfall over Western Australia associated with northerly winds there.

Over the Indian Ocean, the dominant seasonal weather control is exercised by the monsoon seasonal reversals, but there is still a minor El Niño-like mechanism over south-east Africa and Madagascar, which results in a decrease of rainfall during ENSO events.

It is apparent that ENSO teleconnections affect extra-tropical regions as well as tropical ones. During the most intense phase of El Niño, two high-pressure cells, centred at 20°N and 20°S, develop over the Pacific in the upper troposphere, where anomalous heating of the atmosphere is at a

maximum. These cells strengthen the Hadley circulation, cause upper-level tropical easterlies to develop near the equator, as well as subtropical jet streams to be intensified and displaced equatorwards, especially in the winter hemisphere. During the intense ENSO event of the northern winter of 1982–3, such changes caused floods and high winds in parts of California and the US Gulf states, together with heavy snowfalls in the mountains of the western USA. In the northern hemisphere winter, ENSO events with equatorial heating anomalies are associated with a strong trough and ridge teleconnection pattern, known as the Pacific–North American (PNA) pattern (Figure 9.56), which may bring cloud and rain to the south-west United States and north-west Mexico.

The Atlantic Ocean shows some tendency towards a modest effect resembling El Niño, but the western pool of warm water is much smaller, and the east–west tropical differences much less, than in the Pacific. Nevertheless, ENSO events in the Pacific have some bearing on the behaviour of the Atlantic atmosphere–ocean system, for example

A 200mb

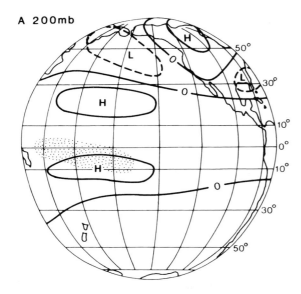

B Sea level

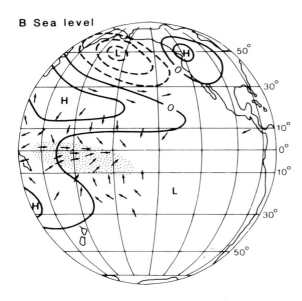

Figure 9.56 Schematic Pacific–North America (PNA) circulation pattern in the upper troposphere during an ENSO event in December–February. The shading indicates a region of enhanced rainfall associated with anomalous westerly surface wind convergence in the equatorial western Pacific.

Source: After Shukla and Wallace 1983.

the establishment of the convective low-pressure centre over the central and eastern Atlantic subtropical high-pressure cell and of the general trade wind flow in the Atlantic. This results in the development of a stronger subsidence inversion layer, as well as subjecting the western tropical Atlantic to greater ocean mixing, giving lower sea-surface temperatures, less evaporation and less convection. This tends to:

1 Increase drought over north-east Brazil. However, ENSO events account for only some 10 per cent of precipitation variations in north-east Brazil.
2 Increase wind shear over the North Atlantic/ Caribbean region such that moderate to strong ENSO events are correlated with the occurrence of some 44 per cent fewer Atlantic hurricanes than occur with non-ENSO events.

A further Pacific influence involves the manner in which the ENSO strengthening of the southern subtropical jet stream may partly explain the heavy rainfall experienced over southern Brazil, Paraguay and northern Argentina during an intense El Niño. Another Atlantic teleconnection may reside in the North Atlantic Oscillation (NAO), a large-scale alternation of atmospheric mass between the Azores high-pressure and the Icelandic low-pressure cells. The relative strength of these two pressure systems appears to affect the rainfall of both north-west Africa and the sub-Saharan zone.

H OTHER SOURCES OF CLIMATIC VARIATIONS IN THE TROPICS

The major systems of tropical weather and climate have now been discussed, yet various other elements help to create contrasts in tropical weather in both space and time.

1 Cool ocean currents

Between the western coasts of the continents and the eastern rims of the subtropical high-pressure cells the ocean surface is relatively cold (see Figure 6.31). This is the result of the importation of water from higher latitudes by the dominant currents and the slow upwelling (sometimes at the rate of about 1 m in 24 hours) of water from intermediate depths due to the Ekman effect (see Chapter 6D.1) and to the coastal divergence (see Figure 6.34). This concentration of cold water gently cools the local air to dew point. As a result, dry warm air degenerates into a relatively cool, clammy, foggy atmosphere with a comparatively low temperature and little range along the west coast of North America off California (see Plate 16), off South America between latitudes 4 and 31°S, and off south-west Africa (8 and 32°S). Thus Callao, on the Peruvian coast, has a mean annual temperature of 19.4°C, whereas Bahia (at the same latitude on the Brazilian coast) has a corresponding figure of 25°C.

The cooling effect of offshore cold currents is not limited to coastal stations, as it is carried inland during the day at all times of the year by a pronounced sea-breeze effect (see Chapter 5C.3). Along the west coasts of South America and south-west Africa the sheltering effect from the dynamically stable easterly trades aloft provided by the nearby Andes and Namib Escarpment, respectively, allows incursions of shallow tongues of cold air to roll in from the south-west. These tongues of air are capped by strong inversions at between 600 and 1,500 m, reinforcing the regionally low trade wind inversions (see Figure 9.6) and thereby precluding the development of strong convective cells, except where there is orographically forced ascent. Thus, although the cool maritime air perpetually bathes the lower western slopes of the Andes in mist and low stratus cloud and Swakopmund (south-west Africa) has an average of 150 foggy days a year, little rain falls on the coastal lowlands. Lima (Peru) has a total mean annual precipitation of only 4.6 cm, although it receives frequent drizzle during the winter months of June to September, and Swakopmund in Namibia has a mean annual rainfall of 1.6 cm. Heavier rain occurs on the rare instances when large-scale pressure changes cause a cessation of the diurnal sea breeze or when modified air from the South Atlantic or South Indian Ocean is able to cross the continents at a time when the normal dynamic stability of the trade winds is disturbed. In south-west Africa, the inversion is most likely to break down during either October or April, allowing convectional storms to form, and Swakopmund recorded 5.1 cm of rain on a single day in 1934. Under normal conditions, however, the occurrence of precipitation is mainly limited to the higher seaward mountain slopes. Further north, tropical west-coast locations in Angola and Gabon show that cold upwelling is a more variable phenomenon in both space and time; coastal rainfall varies strikingly with changing sea-surface temperatures (Figure 9.57). In South America, from

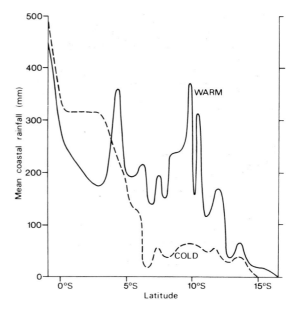

Figure 9.57 March rainfall along the south-western coast of Africa (Gabon and Angola) associated with warm and cold sea-surface conditions.

Source: After Nicholson and Entekhabi; from Nicholson 1989 (by permission of the Royal Meteorological Society, redrawn).

Colombia to northern Peru, the diurnal tide of cold air rolls inland for some 60 km, rising up the seaward slopes of the Western Cordillera and overflowing into the longitudinal Andean valleys like water over a weir (Figure 9.58). Plate 36 illustrates a similar overflow, giving rise to a downstream hydraulic jump where the flow decelerates in the lee of mountains in Wyoming. On the west-facing slopes of the Andes of Colombia, air ascending or banked up against the mountains may under suitable conditions trigger off convectional instability in the overlying trades and produce thunderstorms. In south-west Africa, however, the 'tide' flows inland for some 130 km and rises up the 1,800 m Namib Escarpment without producing much rain because convectional instability is not generated and the adiabatic cooling of the air is more than offset by radiational heating from the warm ground.

2 Topographic effects

Relief and surface configuration have a marked effect on rainfall amounts in tropical regions, where hot, humid air masses are frequent. At the south-western foot of Mount Cameroon, Debundscha

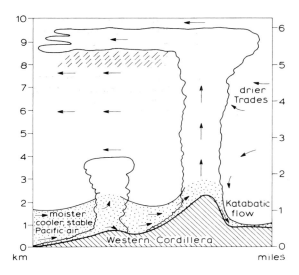

Figure 9.58 The structure of the sea breeze in western Colombia.

Source: After Lopez; from Fairbridge 1967.

(9 m elevation) receives 1,116 cm yr^{-1} on average (1960–80) from the south-westerly monsoon. In the Hawaiian Islands, the mean annual total exceeds 760 cm on the mountains, with one of the world's largest mean annual totals of 1,199 cm at 1,569 m elevation on Mount Waialeale (Kauai), but land on the lee side suffers correspondingly accentuated sheltering effects with less than 50 cm over wide areas. On Hawaii itself, the maximum falls on the eastern slopes at about 900 m, whereas the 4,200 m summits of Mauna Loa and Mauna Kea, which rise above the trade wind inversion, receive only 25–50 cm. On the Hawaiian island of Oahu, the maximum precipitation occurs on the western slopes, just leeward of the 850 m summit with respect to the easterly trade winds. Measurements in the Koolau

Mountains, Oahu, show that the orographic factor is pronounced during summer, when precipitation is associated with the easterlies, but in winter, when precipitation is from cyclonic disturbances, it is more evenly distributed (Table 9.3).

The Khasi Hills in Assam are an exceptional instance of the combined effect of relief and surface configuration. Part of the monsoon current from the head of the Bay of Bengal (see Figure 9.23) is channelled by the topography towards the high ground and the sharp ascent, which follows the convergence of the airstream in the funnel-shaped lowland to the south, results in some of the heaviest annual rainfall totals recorded anywhere. Mawsyuran (1,400 m elevation), 16 km west of the more famous station of Cherrapunji, has a mean annual total (1941–69) of 1,221 cm and can claim to be the wettest spot in the world. Cherrapunji (1,340 m) during the same period averaged 1,102 cm; extremes recorded there include 569 cm in July and 2,440 cm in 1974 (see Figure 3.12). However, throughout the monsoon area, topography plays a secondary role in determining rainfall distribution to the synoptic activity and large-scale dynamics.

Really high relief produces major changes in the main weather characteristics and is best treated as a special climatic type. In equatorial East Africa, the three volcanic peaks of Mount Kilimanjaro (5,800 m), Mount Kenya (5,200 m) and Ruwenzori (5,200 m) nourish permanent glaciers above 4,700–5,100 m. Annual precipitation on the summit of Mount Kenya is about 114 cm, similar to amounts on the plateau to the south, but on the southern slopes between 2,100 and 3,000 m, and on the eastern slopes between about 1,400 and 2,400 m, totals exceed 250 cm. Kabete (at an elevation of 1,800 m near Nairobi) exhibits many of the features of tropical highland climates, having a small annual temperature range (mean monthly

Table 9.3 Precipitation in the Koolau Mountains, Oahu, Hawaii.

Location	Elevation	Source of rainfall		
		Trade winds	Cyclonic disturbances	
		28 May–3 Sept. 1957	2–29 Jan. 1957	5–6 March 1957
Summit	850 m	71.3 cm	49.9 cm	32.9 cm
760 m west of summit	625 m	121.0 cm	54.4 cm	37.0 cm
7,600 m west of summit	350 m	32.9 cm	46.7 cm	33.4 cm

Source: After Mink 1960.

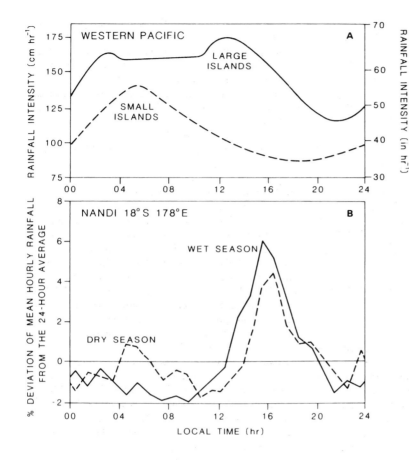

Figure 9.59 Diurnal variation of rainfall intensity for tropical islands in the Pacific. A: Large and small islands in the western Pacific. B: Wet and dry seasons for Nandi (Fiji) in the south-west Pacific (percentage deviation from the daily average).

Sources: A: After Gray and Jacobson 1977. B: After Finkelstein, in Hutchings 1964.

temperatures are 19°C for February and 16°C for July), a high diurnal temperature range (averaging 9.5°C in July and 13°C in February) and a large average cloud cover (mean 7–8/10ths).

3 Diurnal variations

Diurnal weather variations are particularly evident at coastal locations in the trade wind belt and in the Indonesia–Malaysian Archipelago. Land and sea breeze regimes (see Chapter 5C.3) are well developed, as the heating of tropical air over land can be up to five times that over adjacent water surfaces. The sea breeze normally sets in between 0800 and 1100 hours, reaching a maximum velocity of 6–15 m s⁻¹ about 1300 to 1600 and subsiding around 2000. It may be up to 1,000–2,000 m in height, with a maximum velocity at an elevation of 200–400 m, and it normally penetrates some 20–60 km inland.

In northern Australia, sea-breeze phenomenon apparently extends up to 200 km inland from the Gulf of Carpentaria by late evening. During the August–November dry season, this may create suitable conditions for the bore-like 'Morning Glory' – a linear cloud roll and squall line that propagates, usually from the north-east, on the inversion created by the maritime air and nocturnal cooling. Sea breezes are usually associated with a heavy build-up of cumulus cloud and afternoon downpours. On large islands under calm conditions the sea breezes converge towards the centre so that an afternoon maximum of rainfall is observed. Under steady trade winds, the pattern is displaced downwind so that descending air may be located over the centre of the island. A typical case of an afternoon maximum is illustrated by Figure 9.59B for Nandi (Viti Levu, Fiji) in the south-west Pacific. The station has a lee exposure in both wet and dry seasons. This rainfall pattern is commonly believed to be widespread in the tropics, but over the open sea and on small islands a night-time maximum (often with a peak near dawn) seems to occur, and even large islands can display this nocturnal regime when there is little synoptic activity. Figure 9.59A illustrates this nocturnal pattern at four small island locations in the western Pacific. Even large islands may show this effect, as well as the afternoon maximum associated with sea breeze convergence and convection. There are several theories concerning the nocturnal rainfall peak. Recent studies point to a radiative effect, involving more effective nocturnal cooling of cloud-free areas around the mesoscale cloud systems. This favours subsidence, which, in turn, enhances low-level convergence into the cloud systems and strengthens the ascending air currents. Strong cooling of cloud tops, relative to their surroundings, may also produce localized destabilization and encourage droplet growth by mixing of droplets at different temperatures (see Chapter 4D). This effect would be at a maximum near dawn. Another factor is that the sea–air temperature difference, and consequently the oceanic heat supply to the atmosphere, is largest about 0300–0600 hours. Yet a further hypothesis suggests that the semi-diurnal pressure oscillation encourages convergence and therefore convective activity in the early morning and evening, but divergence and suppression of convection around midday.

The Malayan peninsula displays very varied diurnal rainfall regimes in summer. The effects of land and sea breezes, anabatic and katabatic winds and topography greatly complicate the rainfall pattern by their interactions with the low-level south-westerly monsoon current. For example, there is a nocturnal maximum in the Malacca Straits region associated with the convection set off by the convergence of land breezes from Malaya and Sumatra (cf. p. 247), whereas on the east coast of Malaya the maximum occurs in the late afternoon to early evening, when sea breezes extend about 30 km inland against the monsoon south-westerlies, and convective cloud develops in the deeper sea breeze current over the coastal strip. On the interior mountains the summer rains have an afternoon maximum due to the unhindered convection process.

I FORECASTING TROPICAL WEATHER

In the past two decades, significant progress has been achieved in tropical weather forecasting. This has resulted from many of the advances in observing technology and in global numerical modelling discussed in Chapter 7I. Of particular importance in the tropics has been the availability of geostationary satellite data on global cloud conditions, wind vectors, sea-surface temperatures, and vertical profiles of temperature and moisture. Weather radar installations are also available at major centres in India, Central America and the Far East; and at some locations in Africa and the south-west Pacific; but up to now there are few in South America.

1 Short- and extended-range forecasts

The evolution and motion of tropical weather systems are primarily connected with areas of wind speed convergence and horizontal wind shear as identified on low-level kinematic analyses depicting streamlines and isotachs and associated cloud systems, and their changes can be identified from half-hourly geostationary satellite images and weather radars; these are useful for 'nowcasting' and warnings. However, cloud clusters are known to be highly irregular in their persistence beyond 24 hours. They are also subject to strong diurnal variations and orographic influences, which need to be evaluated. Analysis of diurnal variations in temperature with differing cloud states for wet and dry seasons can be a useful aid to local forecasting. Giving equal weight to persistence and climatology produces good results for low-level winds, for example. The forecasting of tropical storm movement also relies mainly on satellite imagery and radar data. For 6–12 hour forecasts, extrapolations can be made from the smoothed track over the preceding 12–24 hours. The accuracy of landfall location forecasts for the storm centre is typically within about 150 km. There are specialized centres for such regional forecasts and warnings in Miami, Guam, Darwin, Hong Kong, New Delhi and Tokyo. Forecasts for periods of two to five days have received limited attention. In the winter months, the tropical margins, especially of the northern hemisphere, may be affected by mid-latitude circulation features. Examples include cold fronts moving southwards into Central America and the Caribbean, or northwards from Argentina into Brazil. The motion of such systems can be anticipated from numerical model forecasts prepared at major centres such as NCEP and ECMWF.

2 Long-range forecasts

Three areas of advance deserve attention. Predictions of the number of Atlantic tropical storms and hurricanes and of the number of days on which these occur have been developed from statistical relations with the El Niño state, mean April–May sea-level pressure over the Caribbean and the easterly or westerly phase of the stratospheric tropical winds at 30 mb (see pp. 17 and 246). Cyclones in the following summer season are more numerous when during the spring season 30 and 50 mb zonal winds are westerly and increasing, ENSO is in the La Niña (cold) mode and there is below-normal pressure in the Caribbean. Wet conditions in the Sahel appear to favour the development of disturbances in the eastern and central Atlantic. An initial forecast is made in November for the following season (based on stratospheric wind phase and August–November rainfall in the western Sahel) and a second one using information on nine predictors through July of the current year.

At least five forecast models have been developed to predict ENSO fluctuations with a lead time of up to twelve months; three involve coupled atmosphere–ocean GCMs, one is statistical and one uses analogue matching. Each of the methods shows a comparable level of moderate skill over three seasons ahead, with a noticeable decrease in skill in the northern spring. The ENSO phase strongly affects seasonal rainfall in north-east Brazil, for example, and other tropical continental areas, as well as modifying the winter climate of parts of North America through the interaction of tropical sea-surface temperature anomalies and convection on mid-latitude planetary waves.

Summer monsoon rainfall in India is related to the ENSO, but the linkages are mostly simultaneous, or the monsoon events even *lead* the ENSO changes. El Niño (La Niña) years are associated with droughts (floods) over India. Numerous predictors of monsoon rainfall over all India have been proposed, including spring temperatures and pressure indicative of the heat low, cross-equatorial airflow in the Indian Ocean, 500 and 200 mb circulation features, ENSO phase, and Eurasian winter snow cover. A key predictor of Indian rainfall is the latitude of the 500 mb ridge along 75°E in April, but the most useful operational approach seems to be a statistical combination of such parameters, with a forecast issued in May for the June–September period. The important question of the spatial pattern of monsoon onset, duration and retreat and this variability has not yet been addressed.

Rainfall over sub-Saharan West Africa is predicted by the UK Meteorological Office using statistical methods. For the Sahel, drier conditions are associated with a decreased inter-hemispheric gradient of sea-surface temperatures in the tropical Atlantic and with an anomalously warm equatorial Pacific. Rainfall over the Guinea coast is increased when the South Atlantic is warmer than normal.

SUMMARY

The tropical atmosphere differs significantly from that in middle latitudes. Temperature gradients are generally weak and weather systems are mainly produced by airstream convergence triggering convection in the moist surface layer. Strong longitudinal differences in climate exist as a result of the zones of subsidence (ascent) on the eastern (western) margins of the subtropical high-pressure cells. In the eastern oceans, there is typically a strong trade wind inversion at about 1 km with dry subsiding air above, giving fine weather. Downstream, this stable lid is raised gradually by the penetration of convective clouds as the trades flow westwards. Cloud masses are frequently organized into amorphous 'clusters' on a subsynoptic scale; some of these have linear squall lines, which are an important source of precipitation in West Africa. The trade wind systems of the two hemispheres converge, but not in a spatially or temporally continuous manner. This Intertropical Convergence Zone also shifts polewards over the land sectors in summer, associated with the monsoon regimes of South Asia, West Africa and northern Australia. There is a further South Pacific Convergence Zone in the southern summer.

Wave disturbances in the tropical easterlies vary regionally in character. The 'classical' easterly wave has maximum cloud build-up and precipitation behind (east of) the trough line. This distribution follows from the conservation of potential vorticity by the air. About 10 per cent of wave disturbances later intensify to become tropical storms or cyclones. This development requires a warm sea surface and low-level convergence to maintain the sensible and latent heat supply and upper-level divergence to maintain ascent. Cumulonimbus 'hot towers' nevertheless account for a small fraction of the spiral cloud bands. Tropical cyclones are most

continued

numerous in the western oceans of the northern hemisphere in the summer–autumn seasons.

The monsoon seasonal wind reversal of South Asia is the product of global and regional influences. The orographic barrier of the Himalayas and Tibetan Plateau plays an important role. In winter, the subtropical westerly jet stream is anchored south of the mountains. Subsidence occurs over northern India, giving north-easterly surface (trade) winds. Occasional depressions from the Mediterranean penetrate to north-western India–Pakistan. The circulation reversal in summer is triggered by the development of an upper-level anticyclone over the elevated Tibetan Plateau with upper-level easterly flow over India. This change is accompanied by the northward extension of low-level south-westerlies in the Indian Ocean, which appear first in southern India and along the Burma coast and then extend north-westwards. The summer 'monsoon' over East Asia also progresses from south-east to north-west, but the Mai-yu rains are mainly a result of depressions moving north-eastwards and thunderstorms. Rainfall is concentrated in spells associated with 'monsoon depressions', which travel westward steered by the upper easterlies. Monsoon rains fluctuate in intensity, giving rise to 'active' and 'break' periods in response to southward and northward displacements of the Monsoon Trough, respectively. There is also considerable year-to-year variability.

The West African monsoon has many similarities to that of India, but its northward advance is unhindered by a mountain barrier to the north. Four zonal climatic belts, related to the location of overlying easterly jet streams and east–west-moving disturbances, are identified. The Sahel zone is reached by the Monsoon Trough, but overlaying subsiding air greatly limits rainfall.

The climate of equatorial Africa is strongly influenced by low-level westerlies from the South Atlantic high (year-round) and easterlies in winter from the South Indian Ocean anticyclone. These flows converge along the Zaire Air Boundary (ZAB) with easterlies aloft. In summer, the ZAB is displaced southwards and north-easterlies over the eastern Pacific meet the westerlies along the ITCZ, oriented north–south from 0° to 12°S. The characteristics of African disturbances are complex and poorly known. Deep easterly flow affects most of Africa south of 10°S (winter) or 15–18°S (summer), although the southern westerlies affect South Africa in winter.

In Amazonia, where there are broad tropical easterlies but no well-defined ITCZ, the subtropical highs of the North and South Atlantic both influence the region. Precipitation is associated with convective activity triggering low-level convergence, with meso- to synoptic-scale disturbances forming *in situ*, and with instability lines generated by coastal winds that move inland.

The equatorial Pacific Ocean sector plays a major role in climate anomalies throughout much of the tropics. At irregular, three- to five-year intervals, the tropical easterly winds over the eastern–central Pacific weaken, upwelling ceases off South America and the usual convection over Indonesia shifts eastwards towards the central Pacific. Such warm ENSO events, which replace the normal La Niña mode, have global repercussions since teleconnection links extend to some extratropical areas, particularly East Asia and North America.

Variability in tropical climates also occurs through diurnal effects, such as land–sea breezes, local topographic and coastal effects on airflow, and the penetration of extra-tropical weather systems and airflow into lower latitudes.

Short-range tropical weather prediction is commonly limited by sparse observations and the poorly understood disturbances involved. Seasonal predictions show some success for the evolution of the ENSO regime, Atlantic hurricane activity and West African rainfall.

10

Boundary layer climates

Meteorological phenomena encompass a wide range of space and time scales, from the instantaneous gusts of wind that swirl up leaves and litter to the global-scale wind systems that shape the annual planetary climate. The weather systems discussed in Chapter 7 are conventionally designated as synoptic-scale systems, whereas tornadoes and thunderstorms (with a spatial scale of 1–50 km and a time scale of a few hours) are referred to as *mesoscale systems*. Other wind systems of comparable scale to the latter, like mountain and valley winds and land–sea breezes, can give rise to distinctive *local climates* (see Chapter 5C). Small-scale turbulence, with wind eddies of a few metres dimension and lasting only a few seconds, represent the domain of *micrometeorology*, or boundary layer climates.

Small-scale climates occur within the planetary boundary layer (see Chapter 5) and have vertical scales of the order of 10^3 m, horizontal scales of some 10^4 m, and time scales of about 10^5 seconds (i.e. 1 day). The boundary layer is typically 1 km thick, but varies between 20 m and several kilometres in different locations and at different times in the same location. Within this layer diffusion processes, both mechanical and convective in character, transport (i.e. flux) mass, momentum and energy, as well as exchanging aerosols and chemicals between the lower atmosphere and the earth's surface. The boundary layer is especially prone to nocturnal cooling and diurnal heating, and within it the wind velocity decreases from the free velocity aloft to lower (frictional) values near the surface, and ultimately to the zero-velocity roughness length (see Chapter 5).

Diffusion processes within the boundary layer are of two types:

1 *Eddy diffusion*. Eddies are parcels of air that transport energy, momentum and moisture from one location to another. Most usually, they can be resolved into upward-spiralling vortices leading to transfers from the earth's surface to the atmosphere or from one vertical layer of air to another. These eddies can be defined by generalized streamlines (i.e. resolved fluctuations). They range in size from a few centimetres (10^{-2} m) in diameter above a heated surface to 1–2 m (10^0 m) resulting from small-scale convection and surface roughness, and grade into dust devils (10^1 m, lasting 10^1–10^2 s) and tornadoes (10^3 m, lasting 10^2–10^3 s).

2 *Turbulent diffusion*. These are apparently random (i.e. unresolved) fluctuations of instantaneous velocities having variations of a second or less.

Atmospheric airflow disturbances in the boundary layer near the earth's surface can be classified into four main types, depending on their altitudinal and horizontal scales (i.e. synoptic and mesoscale), thermal, topographic and surface friction on 'flat' surfaces of varying roughness (see Figure 10.1). Their time and length scales, together with their kinetic energy, are illustrated in Figure 10.2, in comparison with those for a range of human activities. For our purposes, we can consider such phenomena in relation to the climatic processes within an area bare of vegetation, a crop canopy, a forest stand or a cluster of city buildings.

A SURFACE ENERGY BUDGETS

We will first review the process of energy exchange between the atmosphere and an unvegetated surface. In Chapter 2D, it was noted that the surface energy budget equation is usually written:

$$R_n + H + LE + G$$

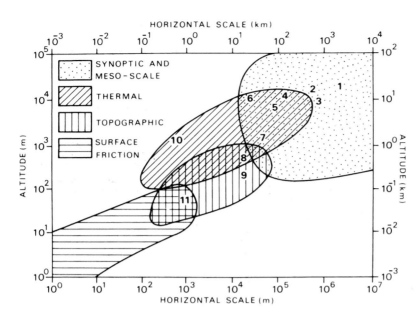

Figure 10.1 Types of airflow disturbances in the near-surface boundary layer of the atmosphere. Examples of specific phenomena are: (1) depression (altitude 12 km; horizontal scale 1,500–2,000 km); (2) hurricane (15 km; 700 km); (3) easterly wave (10 km; 800 km); (4) thunderstorm (13 km; 100 km); (5) mountain lee wave (7 km; 75 km); (6) ITCZ convective cell (15 km; 10–100 km); (7) sea breeze (1 km; 50 km); (8) urban heat island (300 m; 20 km); (9) urban friction hump (of buildings); (10) small cumulus cell (2,000 m; 100–1,000 ml); (11) forest (60 m; 1,500 m).

Source: From Anderson 1971.

where R_n, the net all-wavelength radiation,

= $[S(1 - a)] + L_n$

S = incoming short-wave radiation,

a = fractional albedo of the surface, and

L_n = the net outgoing long-wave radiation.

R_n is usually positive by day, since the absorbed solar radiation exceeds the net outgoing long-wave radiation; at night, when $S = 0$, R_n is determined by the negative magnitude of L_n.

The surface energy flux terms are:

G = ground heat flux,

H = turbulent sensible heat flux to the atmosphere,

LE = turbulent latent heat flux to the atmosphere (E = evaporation; L = latent heat of vaporization).

Positive values denote a flux *away* from the surface interface. By day, the available net radiation is balanced by turbulent fluxes of sensible heat (H) and latent heat (LE) into the atmosphere and by conductive heat flux into the ground (G). At night, the negative R_n caused by net outgoing long-wave radiation is offset by the supply of conductive heat from the soil (G) and turbulent heat from the air (H) (Figure 10.3A). Occasionally, condensation may contribute heat to the surface.

Commonly, there is a small residual heat storage (ΔS) in the soil in spring/summer and a return of heat to the surface in autumn/winter. Where a

vegetation canopy is present there may also be a small additional biochemical heat storage, due to photosynthesis, as well as physical heat storage by leaves and stems (see Figure 10.3B).

An additional energy component to be considered in areas of mixed canopy cover (forest/grassland, desert/oasis), and in water bodies, is the horizontal transfer (*advection*) of heat by wind and currents (ΔA; see Figure 10.3B). The atmosphere transports both sensible and latent heat.

B NON-VEGETATED NATURAL SURFACES

1 Rock and sand

The energy exchanges of dry desert surfaces are relatively simple and straightforward. Figure 10.4 illustrates the instantaneous noon and evening fluxes on a granite surface in July in California and the resulting large temperature range. Surface properties modify the heat penetration, as shown in Figure 10.5 from mid-August measurements in the Sahara. The maximum surface temperatures reached on bare, dark-coloured basalt and light-coloured sandstone are almost identical, but the greater thermal conductivity of the former (3.1 W m^{-1} K^{-1} for basalt, versus 2.4 W m^{-1} K^{-1} for sandstone) gives a greater diurnal range and deeper penetration of the diurnal temperature wave, to about 1 m in the basalt. In

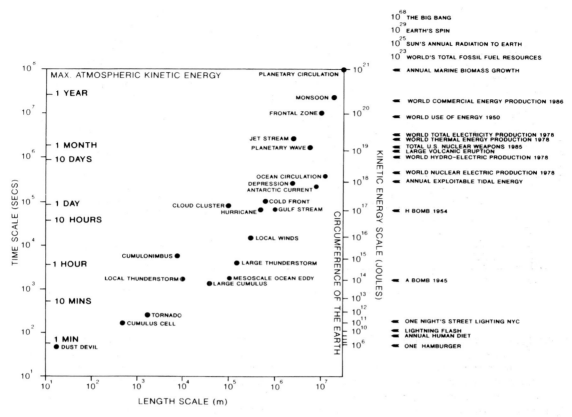

10^{68} THE BIG BANG
10^{29} EARTH'S SPIN
10^{25} SUN'S ANNUAL RADIATION TO EARTH
10^{23} WORLD'S TOTAL FOSSIL FUEL RESOURCES
ANNUAL MARINE BIOMASS GROWTH

WORLD COMMERCIAL ENERGY PRODUCTION 1986

WORLD USE OF ENERGY 1950

WORLD TOTAL ELECTRICITY PRODUCTION 1978
WORLD THERMAL ENERGY PRODUCTION 1978
TOTAL U.S. NUCLEAR WEAPONS 1985
LARGE VOLCANIC ERUPTION
WORLD HYDRO-ELECTRIC PRODUCTION 1978

WORLD NUCLEAR ELECTRIC PRODUCTION 1978
ANNUAL EXPLOITABLE TIDAL ENERGY

H BOMB 1954

A BOMB 1945

ONE NIGHT'S STREET LIGHTING NYC
LIGHTNING FLASH
ANNUAL HUMAN DIET
ONE HAMBURGER

Figure 10.2 The relationship between the time and length scales of a range of meteorological phenomena, together with their kinetic energy equivalents (in joules). The kinetic energy equivalents of a number of other human and natural phenomena are also shown. It is interesting that the Big Bang was equivalent in energy to 10^{62} hamburgers!

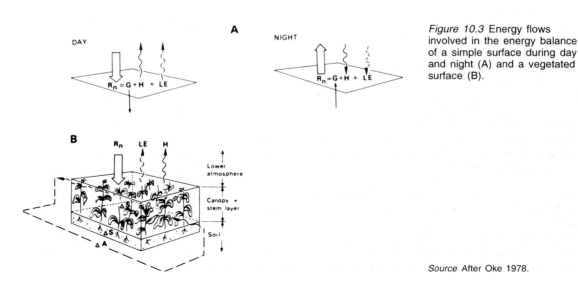

Figure 10.3 Energy flows involved in the energy balance of a simple surface during day and night (A) and a vegetated surface (B).

Source After Oke 1978.

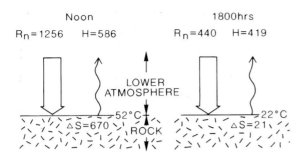

Figure 10.4 Energy balance of a granite surface in California on 18 July at noon (solar altitude 70°) and 1800 hours (solar altitude 10°). Figures in W m⁻².

Source: Based on data from Miller 1965.

sand (see Figure 10.5C), the temperature wave is negligible at 30 cm due to the low conductivity of intergranular air. Note that the surface range of temperature is several times that in the air. Sand also has an albedo of about 0.35, compared with 0.2 for a rock surface.

A representative diurnal pattern of energy exchange over desert surfaces is shown in Figure 10.6. The 2 m air temperature varies between 17 and 29°C, although the surface of the dry lake bed reaches 57°C at midday. R_n reaches a maximum at

about 1300 hours. At this time, most of the heat is transferred to the air by turbulent convection, while in the early morning the heating goes into the ground. At night, this soil heat is returned to the surface, offsetting radiational cooling. Over a 24-hour period, about 90 per cent of the net radiation goes into sensible heat, 10 per cent into ground flux.

2 Water

For a water body, the energy fluxes are very differently apportioned. Figure 10.7 illustrates the seasonal regime for Lake Mead, Arizona, in 1952–3. The incoming short-wave radiation penetrates to about 10 m depth (see Chapter 2B.5) and there is an important horizontal advective term (ΔA) due to the changing density stratification in the lake. Warmer water rises to the surface in winter (ΔA positive) whereas in summer there is a large loss as a result of turbulent mixing of the water. There is a strong annual cycle in the flux into and out of the water body (ΔW), whereas the evaporative loss in excess of 200 cm per year occurs at all seasons. Wind effects in autumn cause LE to exceed the net radiation term. Figure 10.8 shows the average energy balance components for the tropical Atlantic Ocean during a summer day (20 June–2 July 1969).

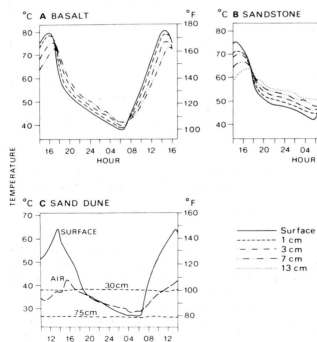

Figure 10.5 Diurnal temperatures near, at, and below the surface in the Tibesti region, central Sahara, in mid-August 1961. A: at the surface and at 1 cm, 3 cm and 7 cm below the surface of a basalt. B: at the surface and at 1 cm, 3 cm, 7 cm and 13 cm below the surface of a light-coloured sandstone. C: in the surface air layer, at the surface, and at 30 cm and 75 cm below the surface of a sand dune.

Source: After Peel 1974.

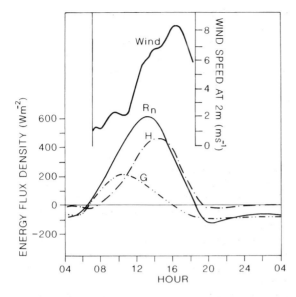

Figure 10.6 Energy flows involved at a dry-lake surface at El Mirage, California (35°N), on 10–11 June 1950. Wind speed due to surface turbulence was measured at a height of 2 m.

Source: After Vehrencamp 1953 and Oke 1978.

The resulting simple thermal economy is based on the assumption that the horizontal advective term (ΔA) is zero and that the total energy input is into the upper 27 m of the ocean. Thus, between 0600 and 1600 hours almost all of the net radiation is absorbed by the water (i.e. ΔW is positive) and at all other times the ocean water is heating the air by the transfer of sensible heat and by the latent heat of evaporation.

3 Snow and ice

Surfaces that consist of snow or ice for large parts of the year present more specialized or complex examples of local energy budgets, and in some respects provide a transition between non-vegetated and vegetated natural surfaces. As a whole, the north polar regions provide a heat sink that is a key element in the global energy balance. In detail, they exhibit a variety of local surfaces (ice-covered ocean almost 60 per cent; tundra 16.5 per cent; boreal forests 13.5 per cent; glaciers 11.2 per cent), all of which are snow-covered during the long winter. These winter months are naturally characterized by rather similar energy balances (Figure

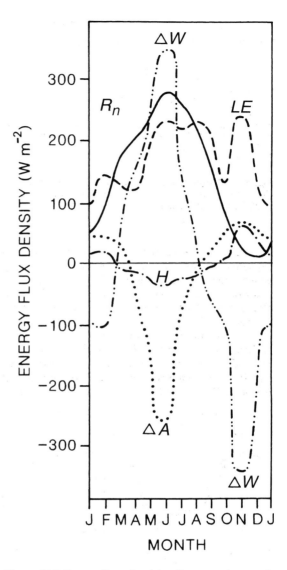

Figure 10.7 Energy flows involving the upper layers of water. Annual figures for Lake Mead, Arizona (36.1°N) during 1952–3.

Source: After Sellers 1965.

10.9) but there are important differences during the summer, when albedo becomes a critical surface parameter. The spring transition on land is very rapid (see Figure 8.42). However, it is important to recognize that, during the winter, areas of ocean covered by thin sea ice and open leads in the ice have 300 W m^{-2} available – more than the net radiation for boreal forests in summer. During the

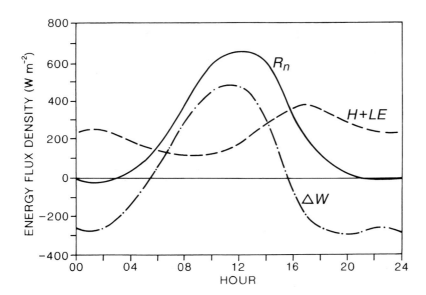

Figure 10.8 Average diurnal variation of the energy balance components in and above the tropical Atlantic Ocean during the period 20 June – 2 July 1969.

Source: After Holland. From Oke 1987. From *Boundary Layer Climates*, 2nd edn 1987. By permission of Routledge and Methuen and Co., London, and T. R. Oke.

summer, the radiation budget for sea ice more than 3 metres thick is quite low and for ablating glaciers lower still. Melting snow involves the additional energy balance component (ΔM), which is the net latent heat storage change (positive) due to melting (Figure 10.10). In this example of snow melt at Bad Lake, Saskatchewan on 10 April 1974, the value of R_n was kept low by the high albedo of the snow (0.65). As the air was always warmer than the melting snow, there was a flow of sensible heat from the air at all times (i.e. H negative). Prior to noon, almost all the net radiation went into snow heat storage, causing melting, which peaked in the afternoon (ΔM maximum). Net radiation was responsible for about 68 per cent of the snow melt and convection ($H + LE$) for 31 per cent. Snow melt takes place earlier in the boreal forests than in the tundra, which lies further north, and as the albedo of the uncovered spruce forest tends to be lower than that of the tundra, the net radiation of the forest tends to be significantly greater than that of the tundra. Thus, south of the treeline in the polar regions the boreal forest acts as the major heat source; north of it the tundra assumes this role.

C VEGETATED SURFACES

From the viewpoint of energetics, as well as of climate within the plant canopy, it is useful to consider short crops and forests separately.

1 Short green crops

Short green crops, up to a metre or so high, supplied with sufficient water and exposed to similar solar radiation conditions, all have a similar receipt of net radiation (R_n). This is largely because of the small range of albedos, 20–30 per cent for short green crops compared with 9–18 per cent for forests. Canopy structure appears to be the primary reason for this difference.

General figures for rates of energy dispersal at noon on a June day in a 20 cm high stand of grass in the higher mid-latitudes are shown in Table 10.1.

Figures 10.11A and B show the diurnal and annual energy balances of a field of short grass near Copenhagen (56°N). For an average 24-hour period in June about 58 per cent of the incoming radiation is involved in evapotranspiration, whereas in December the small amount of net outgoing radiation (i.e. R_n negative) is composed of 55 per cent heat supplied by the soil and 45 per cent sensible heat transfer from the air to the grass.

Table 10.1   Rates of energy dispersal (W m^{-2}) at noon in a 20 cm stand of grass (in higher mid-latitudes on a June day).

Net radiation at the top of the crop	550
Physical heat storage in leaves	6
Biochemical heat storage (i.e. growth processes)	22
Received at soil surface	200

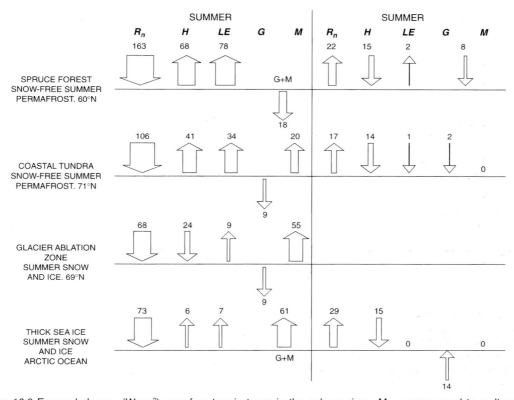

Figure 10.9 Energy balances (W m^{-2}) over four terrain types in the polar regions. M = energy used to melt snow.

Source: Weller and Wendler 1990. Reprinted from *Annals of Glaciology*, with the permission of the International Glaciological Society.

According to T. R. Oke, it is possible to generalize regarding the microclimates of short growing crops (see Figure 10.12):

1 *Temperature*. In early afternoon, there is a temperature maximum just below the vegetation crown, where the maximum energy absorption is occurring; the temperature is lower near the soil surface, where heat flows into the soil. At night, the crop cools mainly by long-wave emission and by some continued transpiration, producing a temperature minimum at about two-thirds the height of the crop. Under calm conditions, a temperature inversion may form just above the crop.

2 *Wind speed*. This is at a minimum in the upper crop canopy, where the foliage is most dense. There is a slight increase below and a marked increase above.

3 *Water vapour*. The maximum diurnal evapotranspiration rate and supply of water vapour

occurs at about two-thirds the crop height, where the canopy is most dense.

4 *Carbon dioxide*. During the day, CO_2 is absorbed by the photosynthesis of growing plants and emitted at night due to respiration. This maximum sink and source of CO_2 is at about two-thirds the crop height.

Finally, it is instructive to look at the conditions accompanying the growth of irrigated crops. Figures 10.13A and B show the energy relationships in a 1 m high stand of irrigated sudan grass at Tempe, Arizona, on 20 July 1962. The air temperature varied between 25 and 45°C. During the day, evapotranspiration in the dry air was near its potential and *LE* (anomalously high due to a local temperature inversion) exceeded R_n, the deficiency being made up by a transfer of sensible heat from the air (*H* negative). Evaporation continued during the night due to a quite high wind speed (7 m s^{-1})

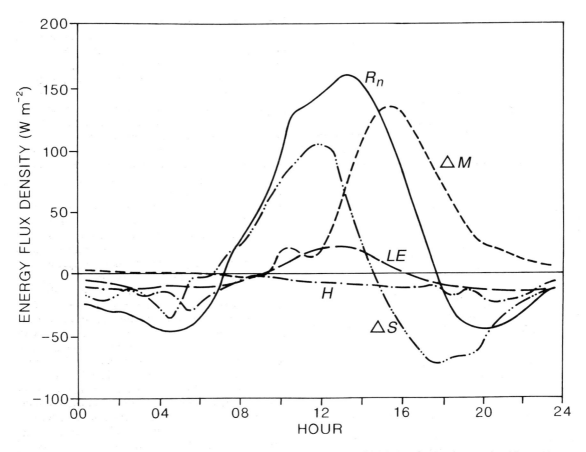

Figure 10.10 Energy balance components for a melting snow cover at Bad Lake, Saskatchewan (51°N) on 10 April 1974.

Source: Granger and Male. Modified by Oke 1987. From *Boundary Layer Climates*, 2nd edn 1987. By permission of Routledge and Methuen and Co., London, and T. R. Oke.

sustained by the continued heat flow from the air. The evapotranspiration thus gives comparatively low diurnal temperatures within irrigated desert crops.

Where irrigation involves inundating the surface with a shallow water layer, as in a rice paddy field, the energy balance components and thus the local climate take on something of the character of water bodies (see B, this chapter). Figure 10.14 gives the energy balance for a Japanese paddy field on a clear September day, showing that the radiation absorption by the water layer and underlying ground (Δ (W + S)) exceeds the sensible heat (H) convected into the air between sunrise and noon. In the afternoon and at night the water becomes the most important heat source, when turbulent losses to the

atmosphere are mainly in the form of the latent heat of evaporation and evapotranspiration (*LE*), rather than by the flow of sensible heat (*H*).

2 Forests

The vertical structure of a forest, which depends on the vegetational species, the ecological associations, the age of the stand and other botanical considerations, largely determines the forest microclimate. Much of the climatic influence of a forest can be explained in terms of the geometry of the forest, including morphological characteristics, size, coverage and stratification. Morphological characteristics include amount of branching (bifurcation), the periodicity of growth (i.e. evergreen or deciduous),

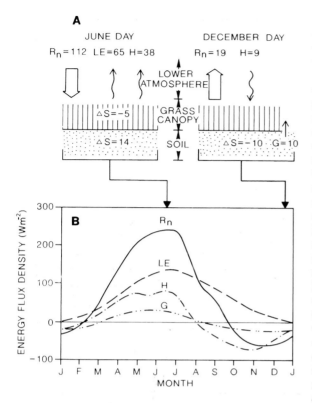

Figure 10.11 Energy fluxes over short grass near Copenhagen (56°N). A: totals for a day in June (17 hours daylight; maximum solar altitude 58°) and December (7 hours daylight; maximum solar altitude 11°). Units are W m^{-2}. B: seasonal curves of net radiation (R_n), latent heat (LE), sensible heat (H) and ground-heat flux (G).

Source: Data from Miller 1965; and after Sellers 1965.

together with the size, density and texture of the leaves. Tree size is obviously important. In temperate forests the sizes may be closely similar, whereas

in tropical forests a great variety of sizes may be present locally. Crown coverage is important in terms of the physical obstruction presented by the canopy to radiation exchange and air movements.

An obvious example of the microclimatic effects of different spatial forest organizations can be gained by comparing the features of tropical rain forests and temperate forests. In tropical forests the average height of the taller trees is of the order of 46–55 m, individuals rising to over 60 m. The dominant height of temperate forests is up to 30 m, so that neither temperate nor most tropical forests can compare in height with the western American redwoods (*Sequoia sempervirens*). Tropical forests commonly possess a great variety of species, seldom less than forty per hectare (100 hectares = 1 km^2) and sometimes over 100, compared with less than twenty-five (occasionally only one) tree species with a trunk diameter greater than 10 cm in Europe and North America. Many British woodlands, for example, have almost continuous canopy stratification from low shrubs to the tops of 36 m beeches, whereas tropical forests are strongly stratified with dense undergrowth, unbranching trunks, and commonly two upper strata of foliage. This stratification, the second or lower of which is usually the more dense, results in rather more complex microclimates in tropical forests than in temperate stands.

It is convenient to describe the climatic effects of forest stands in terms of their modification of energy transfers and of the airflow, their modification of the humidity environment, and their modification of the thermal environment.

a Modification of energy transfers

Forest canopies significantly change the pattern of incoming and outgoing radiation. The short-wave reflectivity of forests depends somewhat on the characteristics of the trees and their density. Coniferous

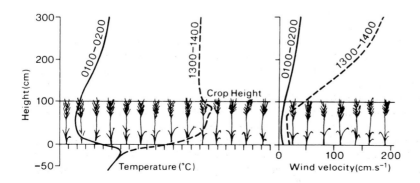

Figure 10.12 Temperature and wind-velocity profiles within and above a metre-high stand of barley at Rothamsted, southern England, on 23 July 1963 at 0100–0200 hours and 1300–1400 hours.

Source: After Long et al. 1964.

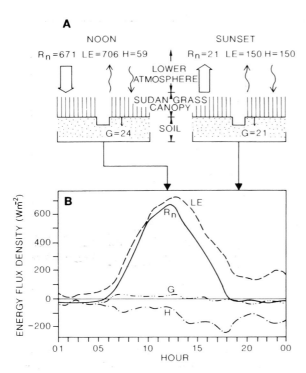

Figure 10.13 Energy flows involved in the energy diurnal balance of irrigated sudan grass at Tempe, Arizona, on 20 July 1962.
Source: After Sellers 1965.

forests have albedos of about 8–14 per cent, and values for deciduous woods range between 12 and 18 per cent, increasing as the canopy becomes more open. Values for semi-arid savannas and scrub woodland are much higher.

Besides reflecting energy, the forest canopy traps energy (Figure 10.15); it has been calculated that for dense red beeches (*Fagus sylvatica*) 80 per cent of the incoming radiation is intercepted by the tree-tops and less than 5 per cent reaches the forest floor. The greatest trapping occurs in sunny conditions, because when the sky is overcast the diffuse incoming radiation has greater possibility of penetration laterally to the trunk space (Figure 10.16). Visible light, however, does not give an altogether accurate picture of total energy penetration, because more ultraviolet than infrared radiation is absorbed in the crowns. For example, only 10.6 per cent of short-wave radiation (less than 0.5 µm) reached a forest floor in Nigeria, as against 45.3 per cent greater than 0.6 µm. As far as light penetration is concerned, there are obviously great variations depending on type of tree, tree spacing, time of year, age, crown density and height. About 50–75 per cent of the outside light intensity may penetrate to the floor of a birch–beech forest, 20–40 per cent for pine, 10–25 per cent for spruce and fir; but for tropical Congo forests the figure may be as low as 0.1 per cent, and one of 0.01 per cent has been recorded for a dense elm stand in Germany. One of the most important effects of this is to reduce the length of daylight. For deciduous trees, more than 70 per cent of the light may penetrate when they are leafless. Tree age is also important in that this controls both crown cover and height. Figure 10.16 shows this rather complicated effect for spruce in the Thuringian Forest, Germany. For a Scots pine (*Pinus sylvestris*) forest in Germany, 50 per cent of the outside light intensity was recorded at 1.3 years, only 7 per cent at 20 years and 35 per cent at 130 years.

b Modification of airflow

Forests impede both the lateral and the vertical movement of air, but it is more convenient to treat the latter in connection with thermal modifications. In general, air movement within forests is slight compared with that in the open, and quite large variations of outside wind velocity have little effect inside woods (Figure 10.17). Measurements for European forests show that 30 m of penetration reduces wind velocities to 60–80 per cent, 60 m to 50 per cent and 120 m to only 7 per cent. A speed of 2.2 m s^{-1} outside a Brazilian evergreen forest was found to be reduced to 0.5 m s^{-1} at about 100 m within, and to be negligible at 1,000 m. In the same location, external storm winds of 28 m s^{-1} were reduced to 2 m s^{-1} some 11 km deep in the forest. Where there is a complex vertical structuring of the forest, wind velocities become more complex. Thus whereas in the crowns (23 m) of a Panama rain-forest the wind velocity was 75 per cent of that outside, it was only 20 per cent in the undergrowth (2 m). Other influences include the density of the stand and the season. For dense pine stands in Idaho, simultaneous recordings showed the wind velocity to be 0.6 m s^{-1} in a cut area, 0.4 m s^{-1} half cut, and 0.1 m s^{-1} in the uncut stand. The effect of season on wind velocities in deciduous forests is shown in Figure 10.17. Observations in a Tennessee mixed-oak forest showed January forest wind velocities to be 12 per cent of those in the open, whereas those in August had dropped to 2 per cent.

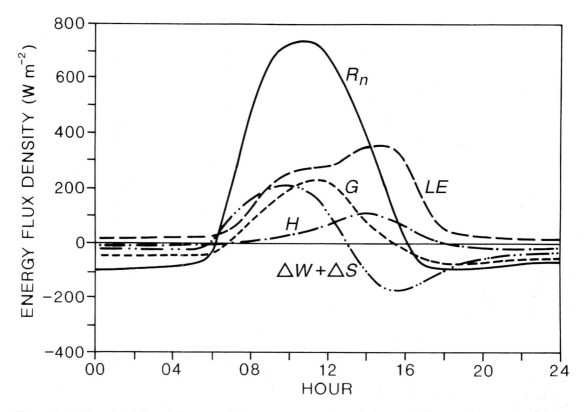

Figure 10.14 Diurnal variation of the energy balance components in and above a shallow, water-covered paddy field in Japan on a clear September day. *W* = net energy storage in the water, *S* = net energy storage in the soil.

Source: After Yabuki. From Oke 1987. From *Boundary Layer Climates*, 2nd edn 1987. By permission of Routledge and Methuen and Co., London, and T. R. Oke.

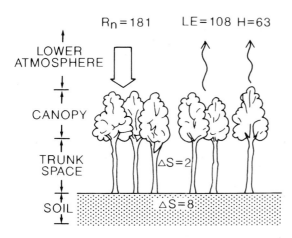

Figure 10.15 Energy flows in a 30-year-old oak stand in the Tellerman Experimental Forest, Voronezh District, USSR, on an average summer day (June to August).

Source: After Sukachev and Dylis 1968.

Knowledge of the effect of forest barriers on winds has been utilized in the construction of wind breaks to protect crops and soil, and, for example, the cypress breaks of the southern Rhône valley and the Lombardy poplars (*Populus nigra*) of the Netherlands form distinctive features of the landscape. It has been found that the denser the obstruction the greater the shelter immediately behind it, although the downwind extent of its effect is reduced by lee turbulence set up by the barrier. The maximum protection is given by the filtering mechanism produced by a break of about 40 per cent penetrability (Figure 10.18). An obstruction begins to have an effect about eighteen times its own height upwind, and the downwind effect can be increased by the *back coupling* of more than one belt (see Figure 10.18).

There are also some less obvious microclimatic effects of forest barriers. One of the most important

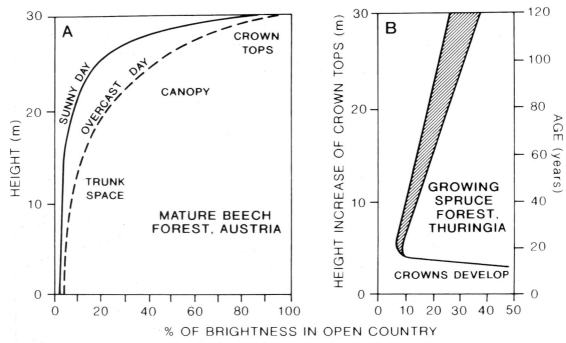

Figure 10.16 The amount of light beneath the forest canopy as a function of cloud cover and crown height: (A) for a thick stand of 120–150-year-old red beeches (*Fagus sylvatica*) at an elevation of 1,000 m on a 20° south-east-facing slope near Lunz, Austria; (B) for a Thuringian spruce forest in Germany during more than 100 years of growth, during which the crown height increased to almost 30 m.

Source: After Geiger 1965.

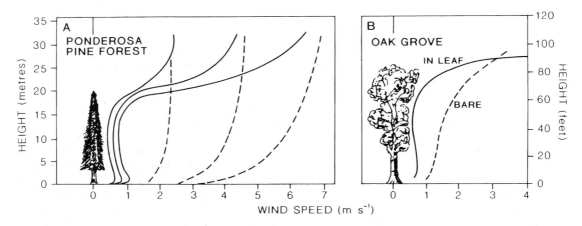

Figure 10.17 Influence on wind velocity profiles exercised by: (A) a dense stand of 20 m high ponderosa pines (*Pinus ponderosa*) in the Shasta Experimental Forest, California. The dashed lines indicate the corresponding wind profiles over open country for general wind speeds of about 2.3, 4.6 and 7.0 m s^{-1}, respectively; (B) a grove of 25 m high oak trees, both bare and in leaf.

Sources: A: After Fons, and Kittredge 1948. B: After R. Geiger and H. Amann, and Geiger 1965.

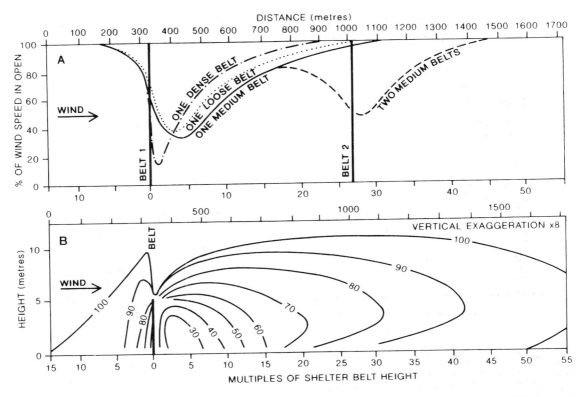

Figure 10.18 The influence of shelter belts on wind velocity distributions (expressed as percentages of the velocity in the open). A: the effects of one shelter belt of three different densities, and of two back-coupled medium-dense shelter belts. B: the detailed effects of one half-solid shelter belt.

Sources: A: After W. Nägeli, and Geiger 1965. B: After Bates and Stoeckeler, and Kittredge 1948.

is that the reduction of horizontal air movement in forest clearings increases the frost hazard on winter nights. Another aspect is the removal of dust and fog droplets from the air by the filtering action of forests. Measurements 1.5 km upwind, on the lee side and 1.5 km downwind of a kilometre-wide German forest gave dust counts (particles per litre) of 9,000, less than 2,000 and more than 4,000, respectively. Measurements in association with a 2 m high and 13 m thick shelter belt on the south-east Hokkaido coast, Japan, in July 1952 showed this filtering effect on advection fog rolling in from the sea, in that 20 m downwind of the obstruction the humidity was only 0.1 g m^{-1} (mean wind velocity 2.55 m s^{-1}), compared with 0.3 g m^{-1} (mean wind velocity 3.4 m s^{-1}) a similar distance upwind. In extreme cases, so much fog can be filtered from laterally moving air that *negative interception* can occur, where there is a higher precipitation catch within a forest than outside. The winter rainfall catch outside a eucalyptus forest near Melbourne, Australia, was 50 cm, whereas inside the forest it was 60 cm.

c Modification of the humidity environment

The humidity conditions within forest stands contrast strikingly with those in the open. Evaporation from the forest floor is usually much less because of the decreased direct sunlight, lower wind velocities, lower maximum temperatures, and generally higher forest air humidity. Evaporation from the bare floors of pine forests is 70 per cent of that in the open for Arizona in summer and only 42 per cent for the Mediterranean region, although such measurements have little real significance in that water losses from vegetated surfaces are controlled by the plant evapo-transpiration.

Unlike many cultivated crops, forest trees exhibit a wide range of physiological resistance to transpiration processes and, as a consequence, the

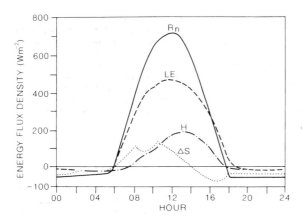

Figure 10.19 A computer simulation of energy flows involved in the diurnal energy balance of a primary tropical broadleaved forest in the Amazon during a high-sun period on the second dry day following a 22 mm daily rainfall.

Source: After a Biosphere Atmosphere Transfer Scheme (BATS) model from Dickinson and Henderson-Sellers 1988 (by permission of the Royal Meteorological Society, redrawn).

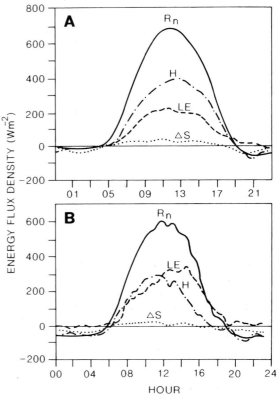

Figure 10.20 Energy components on a July day in two forest stands. A: Scots and Corsican pine at Thetford, England (52°N), on 7 July 1971. Cloud cover was present during the period 0000–0500 hours. B: Douglas fir stand at Haney, British Columbia (49°N), on 10 July 1970. Cloud cover was present during the period 1100–2000 hours.

Sources: A: data from Gay and Stewart 1974; after Oke 1978; B: data from McNaughton and Black 1973; after Oke 1978.

proportions of forest energy flows involved in evapotranspiration (LE) are varied, as is the inversely linked contribution by sensible heat exchange (H). In view of the global climatic significance of the Amazonian tropical broadleaved forest, it is especially interesting that estimates suggest that after rain up to 80 per cent of the net solar radiation (R_n) is involved in evapotranspiration (LE) (Figure 10.19). Figure 10.20 compares diurnal energy flows during July in respect of a pine forest in eastern England and a fir forest in British Columbia. In the former area, only 0.33 R_n is employed for LE due to the high resistance of the pines to transpiration, whereas 0.66 R_n is similarly employed in the British Columbia fir forest, especially during the afternoon. Like short green crops, however, only a very small proportion of R_n is used ultimately for tree growth, an average figure being about 1.3 W m^{-2}, some 60 per cent of which produces wood tissue and 40 per cent forest litter.

During daylight, leaves transpire water through open pores, or *stomata*, so this loss is controlled by the length of day, the leaf temperature (modified by evaporational cooling), surface area, the tree species and its age, as well as by the meteorological factors of available radiant energy, atmospheric vapour pressure and wind speed. Total evaporation figures are therefore extremely varied; the evaporation of water intercepted by the vegetation surfaces

also enters into the totals, in addition to direct transpiration. Calculations made for a catchment covered with Norway spruce (*Picea abies*) in the Harz Mountains of Germany showed an estimated annual evapotranspiration of 34 cm and additional interception losses of 24 cm.

The humidity of forest stands is very much linked to the amount of evapotranspiration and increases with the density of vegetation present (see Figure 10.21A). The increase in forest humidity over that outside averages 3–10 per cent of relative humidity and is especially marked in summer (Figure 10.21B). Mean annual relative humidity forest excesses for Germany and Switzerland are for beech 9.4 per cent,

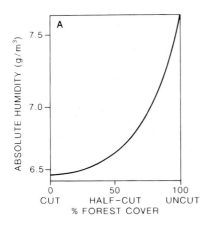

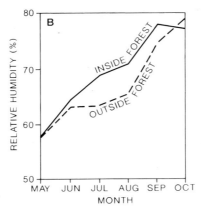

Figure 10.21 The effects (A) of percentage of white pine (*Pinus monticola*) cover on absolute humidity in northern Idaho in July and August 1931–3; and (B) of season on relative humidity in a Michigan birch–beech–maple forest at 1700 hours during May to October 1927–36.

Source: A: After Kittredge 1948. B: After US Dept of Agriculture *Yearbook* 1941.

Norway spruce 8.6 per cent, larch 7.9 per cent and Scots pine 3.9 per cent. However, humidity comparisons in these terms are rather unsatisfactory in that forest temperatures differ strikingly from those in the open. Forest vapour pressures were found to be higher within an oak stand in Tennessee than outside for every month except December. Tropical forests exhibit almost complete night saturation irrespective of elevation in the trunk space, whereas during the day humidity is inversely related to elevation. Measurements in Amazonia show that in dry conditions daytime humidities in the lower trunk space (1.5 m) are near 20 g kg^{-1}, compared with 18 g kg^{-1} at the top of the canopy (36 m); the deficit of absolute humidity (with respect to saturation) is reduced at the 1.5 m level to between one-third and one-half that at the top.

At the opposite forest extreme, recent research in boreal forests has shown that they have low photosynthetic and carbon drawdown rates, and consequently low transpiration rates. Over the year, the uptake of CO_2 by photosynthesis is balanced by its loss through respiration. Over the season, the evapotranspiration rate of boreal (mainly spruce) forests is surprisingly low (less than 2 mm per day). The very low albedos, coupled with low evapotranspiration energy losses, lead to high available energy, high sensible heat fluxes and the development of a deep convective planetary boundary layer. The latter is particularly marked during spring and early summer due to intense mechanical and sensible heat-driven turbulence.

The influence of forest structures on precipitation is still very much an unresolved problem. This is partly due to the difficulties of comparing rain-gauge catches in the open with those near forests, within clearings or beneath trees. For example, on the windward side of a forest the dominance of low-level upcurrents decreases the amount of precipitation actually caught in the rain gauge, whereas the reverse occurs where there are lee side downdraughts. In small clearings, the low wind velocities cause little turbulence around the opening of the gauge and catches are generally greater than outside the forest, although the actual precipitation amounts may be identical. On the other hand, it is sometimes found that the larger the clearing the more prevalent are downdraughts and consequently the precipitation catch increases. In a 25 m high pine and beech forest in Germany, catches in clearings of 12 m diameter were only 87 per cent of that upwind of the forest, but this catch rose to 105 per cent in clearings of 38 m diameter. An analysis of precipitation records for Letzlinger Heath (Germany) before and after afforestation suggested a mean annual increase of 6 per cent, with the greatest excesses occurring during drier years. It is generally agreed, however, that forests have little effect on cyclonic rain, but that they may have a marginal orographic effect in increasing lifting and turbulence, which is of the order of 1–3 per cent in temperate regions.

A far more important obstructional influence of forests on precipitation is in terms of the direct interception of rainfall by the canopy. This obviously varies with crown coverage, with season, and with the rainfall intensity. Measurements in German beech forests indicate that, on average, they intercept 43 per cent of precipitation in summer and 23 per cent in winter. Pine forests may intercept up to

94 per cent of low-intensity precipitation but as little as 15 per cent of high intensities, the average for temperate pines being about 30 per cent. The intercepted precipitation either evaporates on the canopy, runs down the trunk, or drips to the ground. Assessment of the total precipitation reaching the ground (the throughfall) requires very detailed measurements of the stem flow and the contribution of drips from the canopy. Canopy evaporation is not necessarily a total loss of moisture from the woodland, since the solar energy used in the evaporating process is not available to remove soil moisture or transpiration water, but the vegetation does not derive the benefit of the cycling of the water through it via the soil. Evaporation from the canopy is very much a function of net radiation receipts (20 per cent of the total precipitation evaporates from the canopy of Brazilian evergreen forests), and of the type of species. Some Mediterranean oak forests yield virtually no stem flow, and their 35 per cent interception almost all evaporates from the canopy.

Investigations of the water balance of forests provide evidence that evergreen forests may permit 10–50 per cent more evapotranspiration than grass in the same climatic conditions. Grass normally reflects 10–15 per cent more solar radiation than coniferous tree species and hence less energy is available for evaporation. In addition, trees have a greater surface roughness, which increases turbulent air motion and, therefore, the evaporation efficiency. Evergreens also allow transpiration to occur throughout the year. Nevertheless, many more detailed and careful studies are required to check these results and to test the various hypotheses.

d Modification of the thermal environment

From what has been said it is apparent that forest vegetation has an important effect on microscale temperature conditions. Shelter from the sun, blanketing at night, heat loss by evapotranspiration, reduction of wind speed, and the impeding of vertical air movement all influence the temperature environment. The most obvious effect of canopy blanketing is that inside the forest daily maximum temperatures are lower and minima are higher (Figure 10.22). This is particularly apparent during periods of high summer evapotranspiration, which depress daily maximum temperatures and cause mean monthly temperatures in tropical and temperate forests to fall well below those outside. In temperate forests at sea level, the mean annual

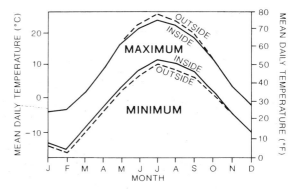

Figure 10.22 Seasonal regimes of mean daily maximum and minimum temperatures inside and outside a birch–beech–maple forest in Michigan.
Source: After US Dept of Agriculture *Yearbook* 1941.

temperature may be about 0.6°C lower than that in surrounding open country, the mean monthly differences may reach 2.2°C in summer but not exceed 0.1°C in winter, and on hot summer days the difference can be more than 2.8°C. Mean monthly temperatures and temperature ranges for temperate beech, spruce and pine forests are given in Figure 10.23, which also shows that when trees do not transpire greatly in the summer (*e.g.* the *forteto* oak maquis of the Mediterranean) the high day temperatures reached in the sheltered woods may cause mean monthly figures to reverse the trend exhibited by temperate forests. Even within individual climatic regions it is difficult to generalize, however, because at elevations of 1,000 m the lowering of temperate forest mean temperatures below those in open country may be double that at sea level.

The complex vertical structure of forest stands is a further factor in complicating forest temperatures. Even in relatively simple stands vertical differences are very apparent. For example, in a ponderosa pine forest (*Pinus ponderosa*) in Arizona the recorded mean June–July maximum was increased by 0.8°C simply by raising the thermometer from 1.5 to 2.4 m above the forest floor. In stratified tropical forests the thermal picture is much more complicated. The dense canopy heats up very much during the day and quickly loses its heat at night, showing a much greater diurnal temperature range than the undergrowth (Figure 10.24A). Whereas daily maximum temperatures of the second storey are intermediate between those of the tree tops and the undergrowth, the nocturnal minima are higher than either tree

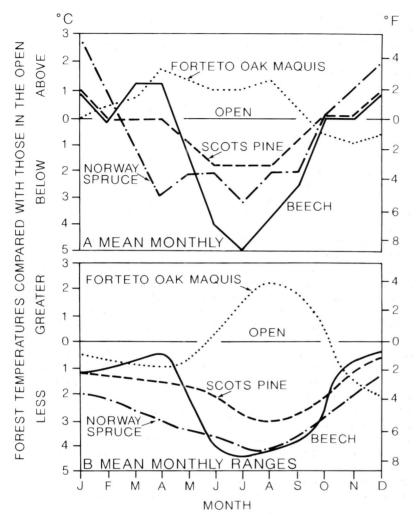

Figure 10.23 Seasonal regimes of (A) mean monthly temperatures and (B) mean monthly temperature ranges, compared with those in the open, for four types of Italian forest. Note the anomalous conditions associated with the *forteto* oak maquis, which transpires little.

Source: Food and Agriculture Organization of the United Nations 1962.

tops or undergrowth because the second storey is insulated by trapped air both above and below (Figure 10.24B). During dry conditions in the Amazonian rainforest, there is a similar decoupling of the air in the lower storey from the upper two-thirds of the canopy, as reflected by the reduced amplitude of diurnal temperature range. At night, the pattern is reversed: temperatures respond to the radiative cooling of the lowest two-thirds of the vegetation canopy. Temperature variations within this deeper layer up to 25 m height are now decoupled from those in the treetops and above.

D URBAN SURFACES

From a total of 5.5 billion in 1993, world population is forecast to increase to 8.2 billion in 2025, with the proportion of urban dwellers rising from 40 to 60 per cent during the same period. Thus, from about the turn of the century the majority of the human race will live and work in association with urban climatic influences.

The construction of every house, road or factory destroys existing microclimates and creates new ones of great complexity that depend on the design, density and function of the building. Despite the great internal variation of urban climatic influences, it is

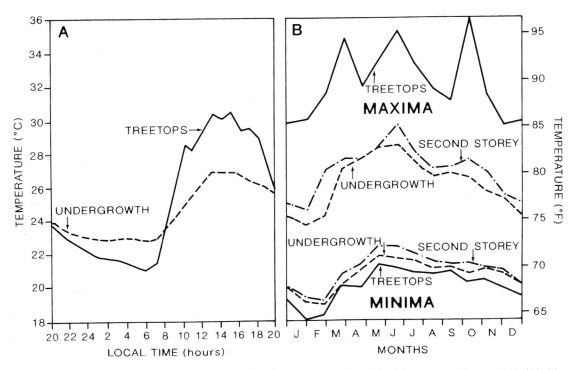

Figure 10.24 The effect of tropical rainforest stratification on temperature.* A: daily march of temperature (10–11 May 1936) in the treetops (24 m) and in the undergrowth (0.7 m) during the wet season in primary rainforest at Shasha Reserve, Nigeria. B: average weekly maximum and minimum temperatures in three layers of primary (Dipterocarp) forest, Mount Maquiling, Philippine Islands.

Sources: *After Richards 1952; A: after Evans; B: after Brown.

possible to make some generalizations regarding the effects of urban structures under three main headings:

1 Modification of atmospheric composition.
2 Modification of the heat budget.
3 Other effects of modifications of surface roughness and composition.

1 Modification of atmospheric composition

Urban pollution has effects that involve changes in the thermal properties of the atmosphere, cutting down the passage of sunlight, and providing abundant condensation nuclei. The urban atmosphere of most modern cities comprises a complex mixture of gases and particulates, including ozone, sulphur dioxide, nitrogen oxides, mineral dust, carbon and complex hydrocarbons; it is convenient to examine its sources under the following headings:

1 *Aerosols.* The production of suspended particular matter (measured in mg m^{-3} or μg m^{-3}), chiefly of carbon, lead and aluminium compounds, and silica. In Los Angeles, aerosol carbon accounts for 40 per cent of the total fine particle mass and is the major cause of severe visibility decreases, yet it is not routinely monitored. Half of this total is from vehicle exhausts and the remainder from industrial and other stationary fuel burning.

2 *Gases.* The production of gases (measured in parts per million – ppm) can be viewed either from the traditional standpoint with its concentration on industrial and domestic coal burning and the production of such gases as sulphur dioxide (SO_2), or from the newer standpoint of petrol and oil combustion and the production of carbon monoxide (CO), hydrocarbons (Hc), nitrogen oxides (NO_x), ozone (O_3) and the like. A three-year survey of 39 urban areas in the United States identified 48 hydrocarbon compounds: 25 paraffins (60 per cent of the total with a median concentration of 266 ppb carbon),

15 aromatics (26 per cent of the total, 116 ppb C) and seven biogenic olefins (11 per cent, 47 ppb C). Biogenic hydrocarbons (olefins) emitted by vegetation are highly reactive, destroying ozone and forming aerosols in rural conditions, but forming ozone under urban conditions.

In dealing with atmospheric pollution it must be remembered, first, that the diffusion or concentration of pollutants is a function both of atmospheric stability (especially the presence of inversions) and of the features of horizontal air motion; second, that aerosols are removed from the atmosphere by settling out and by washing out; and, third, that certain gases are susceptible to complex chains of photochemical changes, which may destroy some gases but produce others.

a Aerosols

As has already been pointed out (see Chapter 1A.2 and A.4), the thermal economy of the globe is affected significantly by the natural production of aerosols that are deflated from deserts, erupted from volcanoes, produced by fires and so on (see Chapter 11D.3). Over the last century, the average dust concentration has increased, particularly in Eurasia, due only in part to such eruptions as those of Mount Agung in Bali (1963) and El Chichón (1982) and Mount Pinatubo (1991). The proportion of atmospheric dust directly or indirectly attributable to human activity has been estimated at 30 per cent (see Chapter 1A.4). As an example of the latter, it is interesting that the North African tank battles of the Second World War disturbed the desert surface to such an extent that the material subsequently deflated was visible in clouds over the Caribbean.

The concentration of small nuclei (0.01–0.1 μm diameter) averages 9,500 cm^{-3} in the British countryside but is typically 150,000 cm^{-3} and can reach 4,000,000 cm^{-3} in cities, as was measured near ground level in the industrial section of Vienna in 1946. Similarly, the concentration of larger particles (0.5–10 μm diameter) has been measured at 25–30 cm^{-3} in the city of Leipzig, as against 1–2 cm^{-3} in the rural environs. The greatest concentrations of smoke generally occur with low wind speeds, low vertical turbulence, temperature inversions, high relative humidities and air moving from the pollution sources of factory districts or areas of high-density housing (see Plate 37). The character of domestic heating and power demands causes city smoke pollution to take on striking seasonal and diurnal cycles, with the greatest concentrations occurring at about 0800 hours in early winter (Figure 10.25). The sudden morning increase is also partly a result of natural processes. Pollution trapped during the night beneath a stable layer a few hundred metres above the surface may be brought back to ground level when thermal convection sets off vertical mixing. Figure 10.26 shows the striking results of the accumulation of air pollution that occurred over the British industrial city of Sheffield in mid-December 1964 during a period of cloudless skies, weak air flow, maximum long-wave radiation and the development of near-surface temperature inversions and radiation fog. These conditions were associated with a smoke concentration of 10 per cent above the monthly average on 14 December, which increased to 100 per cent above the average on 16 December.

The most direct effect of particulate pollution is to reduce incoming radiation and sunshine. Pollution, plus the associated fogs (termed smog), used to cause some British cities to lose 25–55 per cent of incoming solar radiation during the period November–March (see Plate 37). In 1945, it was estimated that the city of Leicester lost 30 per cent of incoming radiation in winter, as against 6 per cent in summer. These losses are naturally greatest when the sun's rays strike the smog layer at a low angle. Compared with the radiation received in the surrounding countryside, Vienna loses 15–21 per cent of radiation when the sun's altitude is 30°, but the loss rises to 29–36 per cent with an altitude of 10°. The effect of smoke pollution is dramatically illustrated by Figure 10.27, which compares conditions before and after the enforcing of the Clean Air Act of 1956 in London. Before 1950, there was a striking difference of sunshine between the surrounding rural areas and the city centre (see Figure 10.27A), which could mean a loss of mean daily sunshine of 16 minutes in the outer suburbs, 25 minutes in the inner suburbs and 44 minutes in the city centre. It must be remembered, however, that smog layers also impeded the reradiation of surface heat at night and that this blanketing effect contributed to higher night-time city temperatures. The use of smokeless fuels and other pollution controls cut London's total smoke emission from 1.4×10^8 kg (141,000 tons) in 1952 to 0.9×10^8 kg (89,000 tons) in 1960, and Figure 10.27B shows the increase in average monthly sunshine figures for the period 1958–67 compared with those of 1931–60.

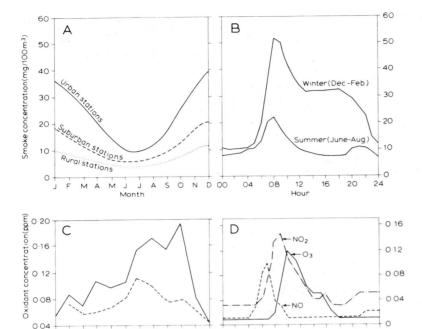

Figure 10.25 Annual and daily pollution cycles. A: annual cycle of smoke pollution in and around Leicester, England, during the period 1937–9, before smoke abatement legislation was introduced. B: diurnal cycle of smoke pollution in Leicester during summer and winter, 1937–9. C: annual cycle of mean daily maximum 1-hour average oxidant concentrations for Los Angeles (1964–5) and Denver (1965) (dashed). D: diurnal cycles of nitric oxide (NO), nitrogen dioxide (NO_2) and ozone (O_3) concentrations in Los Angeles on 19 July 1965.

Sources: A, B: After Meethan 1952 [*et al.* 1980]. C, D: After US DHEW 1970 and Oke 1978.

The abundance of condensation nuclei in city atmospheres, particularly those situated on low-lying land adjacent to a large river, explains the former abundance of city fogs. Between August 1944 and December 1946, for example, suburban Greenwich had a monthly average of more than twenty days with good visibility at 0900 hours, whereas central London had less than fifteen. Occasionally, very stable atmospheric conditions combined with excessive pollution production to give dense smog of a lethal character. During the period 5–9 December 1952, a temperature inversion over London caused a dense fog with visibility of less than 10 m for 48 consecutive hours, resulting in 12,000 more deaths (mainly from chest complaints) during the period December 1952 to February 1953 compared with the same period the previous year. The close association of the incidence of fog with increasing industrialization and urbanization was well shown by the city of Prague, where the mean annual number of days with fog rose from 79 during the period 1860–80 to 217 during 1900–20.

b Gases

Along with the pollution by smoke and other particulate matter produced by the traditional urban and industrial activities involving the combustion of coal and coke has been associated the generation of pollutant gases. Before the Clean Air Act it was estimated that, whereas 80–90 per cent of London's smoke was produced by domestic fires, these were responsible for only 30 per cent of the sulphur dioxide released into the atmosphere – the remainder being contributed by electricity power stations (41 per cent) and factories (29 per cent) (Plate E). Figure 10.26 illustrates the association between pollution by smoke and that by sulphur dioxide in Sheffield 28 years ago; it is significant that on 16 December 1964 the concentration of sulphur dioxide in the city air had risen to three times the monthly average. After the early 1960s, improved technology, the phasing out of coal fires and anti-pollution regulations brought about a striking decline in sulphur dioxide pollution in many European and North American cities (Figure 10.28). The effect of the regulations was not always clear, however, in that the decrease in London's atmospheric pollution was not apparent until eight years after the introduction of the 1956 Clean Air Act, whereas in New York city the observed decrease began in the same year (1964) – *prior* to the air pollution control regulations there.

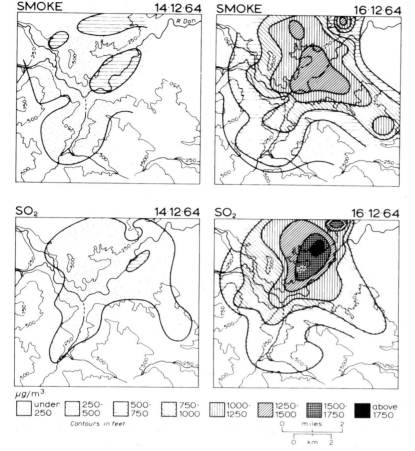

SMOKE   14·12·64

SMOKE   16·12·64

SO₂   14·12·64

SO₂   16·12·64

Figure 10.26 Average values of air pollution by smoke and sulphur dioxide for Sheffield, England, on 14 and 16 December 1964.

μg/m³

| under 250 | 250–500 | 500–750 | 750–1000 | 1000–1250 | 1250–1500 | 1500–1750 | above 1750 |

Contours in feet

0   miles   2

0   km   2

Source: From Garnett 1961.

Urban complexes are increasingly affected by another serious form of pollution resulting from the combustion of petrol and oil by cars, lorries and aircraft, as well as from petrochemical industries. Los Angeles, lying in a topographically constricted basin and often subject to temperature inversions, is the prime example of such pollution, although this affects all modern cities. In Los Angeles, seven million people use four million private cars, consuming 30 million litres of gasoline per day and producing more than 12,000 tons of pollutants. To this is added the results of the consumption of 0.5 million litres per day of diesel fuel by 13,500 lorries and buses and 2.5 million litres of aviation fuel consumed in the vicinity of the city. Even with controls, 7 per cent of the gasoline from private cars is emitted in an unburned or poorly oxidized form, another 3.5 per cent as photochemical smog and 33–40 per cent as carbon monoxide. Smog involves at least four main components: carbon soot, particulate organic matter (POM), sulphate (SO_4^{2-}) and peracyl nitrates (PANS). Half of the aerosol mass is typically POM and sulphate. However, there are important regional differences. For example, the sulphur content of fuels used in California is lower than in the eastern United States and Europe, and NO_2 emissions greatly exceed those of SO_2 in California. The production of the Los Angeles smog, which, unlike traditional city smogs, occurs characteristically during the daytime in summer and autumn (see Figures 10.25C and D), is the result of a very complex chain of chemical reactions termed the disrupted photolytic cycle (Figure 10.29). Ultraviolet radiation (0.37–0.42 μm) dissociates natural NO_2 into NO and O. Monatomic oxygen (O) may then combine with natural oxygen (O_2) to produce ozone (O_3). The ozone in turn reacts

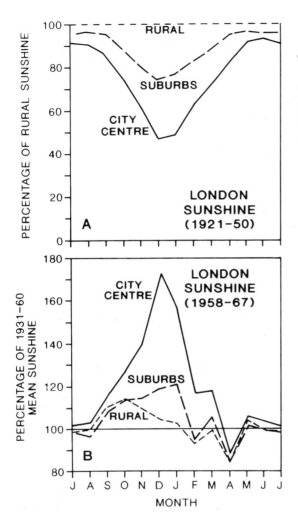

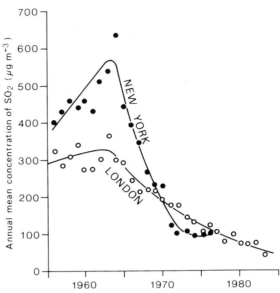

Figure 10.28 Annual mean concentration of sulphur dioxide (μg m^{-3}) measured in New York and London during a 25–30-year period. These show dramatic decreases of urban pollution by SO_2.

Source: From Brimblecombe 1986.

Figure 10.27 Sunshine in and around London. A: mean monthly bright sunshine recorded in the city and suburbs for the years 1921–50, expressed as a percentage of that in adjacent rural areas. This shows clearly the effects of winter atmospheric pollution in the city. B: mean monthly bright sunshine recorded in the city, suburbs and surrounding rural areas during the period 1958–67, expressed as a percentage of the averages for the period 1931–60. This shows the effect of the 1956 Clean Air Act in increasing the receipt of winter sunshine, in particular, in central London.

Sources: A: After Chandler 1965; B: after Jenkins 1969.

with the artificial NO to produce NO_2 (which goes back into the photochemical cycle forming a dangerous positive feedback loop) and oxygen. The hydrocarbons produced by the combustion of petrol combine with oxygen atoms to produce the hydro-carbon free radical HcO*, and these react with the products of the O_3–NO reaction to generate oxygen and photochemical smog. This smog exhibits well-developed annual and diurnal cycles in the Los Angeles basin (see Figures 10.25C and D). Annual levels of photochemical smog pollution in Los Angeles (derived from averages of the daily highest hourly figures) are greatest in late summer and autumn, when clear skies, light winds and temperature inversions combine with high amounts of solar radiation. The diurnal figures of variations in individual components of the disrupted photolytic cycle reflect the complex reactions with, for example, an early morning concentration of NO_2 due to the build-up of traffic, and a peak of O_3 when receipts of incoming radiation are high. The effect of this smog is not only to modify the radiation budget of cities but to produce a human health hazard: in Tokyo, for instance, citizens sometimes wear respiratory masks in the street in self-defence!

c Pollution distribution

Polluted atmospheres commonly assume well-marked physical configurations around urban areas, which are very dependent upon environmental lapse

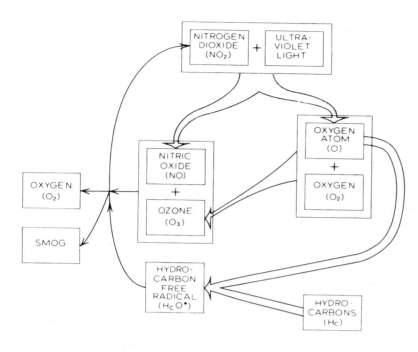

Figure 10.29 The NO_2 photolytic cycle disrupted by hydrocarbons to produce photochemical smog.

Sources: US DHEW 1970 and Oke 1978.

rates, particularly the presence of temperature inversions, and on wind speed. A pollution dome forms as a result of the collection of pollution below an inversion forming the urban boundary layer (Figure 10.30A). A wind speed as low as 2 m s^{-1} is sufficient to displace the Cincinnati pollution dome downwind, and a wind speed of 3.5 m s^{-1} will disperse it into a plume. Figure 10.30B shows a section of an urban plume with the volume above the urban canopy of the building tops filled by buoyant mixing circulations. *Fumigation* is the term used when an inversion lid prevents upward dispersion, but lapse conditions due to morning heating of the surface air allow convective plumes and associated downdraughts to bring pollution back to the surface. Downwind, *lofting* occurs above the temperature inversion at the top of the rural boundary layer, dispersing the pollution upwards. Figure 10.30C illustrates some of the features of a pollution plume up to 160 km downwind of St Louis on 18 July 1975. In view of the complexity of photochemical reactions, it is of note that ozone increases downwind due to photochemical reactions within the plume but decreases over power plants as the result of other reactions with the emissions. This plume was observed to stretch for a total distance of 240 km, but under conditions of an intense pollution source, steady large-scale surface

airflow and vertical atmospheric stability, pollution plumes may extend downwind for hundreds of kilometres. Plumes originating in the Chicago–Gary conurbation have been observed from high-flying aircraft to extend almost to Washington, DC, 950 km away.

2 Modification of the heat budget

The energy balance of the built surface is similar to those described earlier in this chapter, except for the heat production resulting from human energy consumption by combustion, which may even exceed R_n during the winter in some cities. Although R_n may not be greatly different from that in nearby rural areas (except during times of significant pollution) heat storage by surfaces is greater (20–30 per cent of R_n by day), leading to greater nocturnal values of H and, in particular, LE is much less in city centres. After long dry periods, evapotranspiration may be zero in city centres, except for certain industrial operations, and in the case of irrigated parks and gardens, where LE may exceed R_n. This lack of LE means that by day 70–80 per cent of R_n may be transferred to the atmosphere as sensible heat (H). Beneath the urban canopy, the microclimates of the streets and 'urban canyons' are dominated by the effects of elevation and aspect on

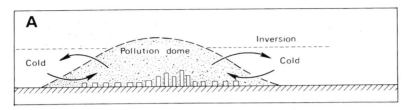

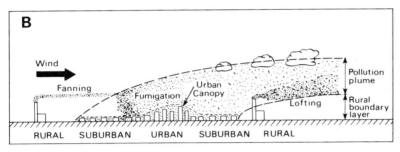

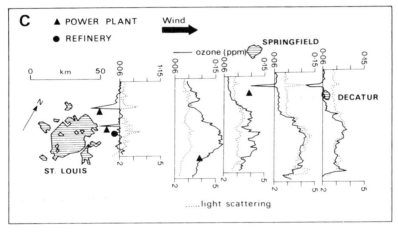

Figure 10.30 Configurations of urban pollution. A: urban pollution dome. B: urban pollution plume in a stable situation (i.e. early morning following a clear night). Fanning is indicative of vertical atmospheric stability. C: pollution plume north-east of St Louis, Missouri, on 18 July 1975.

Sources: B: After Oke 1978; C: after White *et al.* 1976 and Oke 1978.

the energy balance, which may vary strikingly even within one street.

The complex nature of the urban modification of the heat budget is well illustrated by observations made in and around the city of Vancouver. Figure 10.31 compares the summer diurnal energy balances for rural and suburban locations, with the former showing considerable consumption of net radiation (R_n) by evapotranspiration (LE) during the day, giving lower temperatures than in the suburbs, whereas the suburban gain of net radiation is greater by day but the loss greater during the evening and night due to release of turbulent sensible heat from the suburban fabric (i.e. ΔS negative). The diurnal energy balance for a regularly watered suburban lawn (Figure 10.32B) shows features similar to those of irrigated crops (see Figure 10.13) in that some three-quarters of the net radiation is consumed by evaporative processes (LE), which exceed R_n after about 1400 hours, the deficiency being made up by the conduction of sensible heat (H) from the air to the ground after that time. The symmetrical energy balance for the dry top of an urban canyon (Figures 10.32A and 10.33C) shows that two-thirds of the net radiation is transferred into atmospheric sensible heat and one-third into heat storage in the building material (ΔS). Figures 10.33A–C rationalize this canyon top energy balance symmetry in terms of the behaviour of its components (i.e. canyon floor, east-facing wall and west-facing wall), making up a white, windowless urban canyon in early September aligned north–south and with a canyon height equal to its

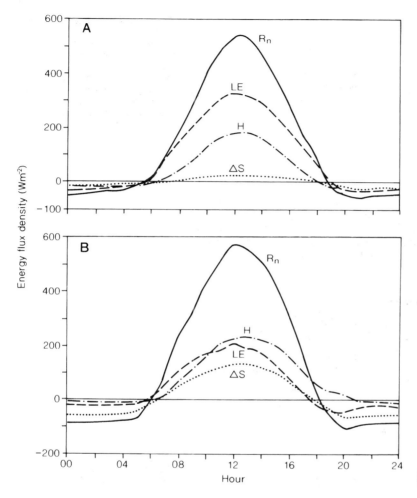

Energy flux density (Wm⁻²)

Figure 10.31 Average diurnal energy balances for (A) rural and (B) suburban locations in Greater Vancouver for 30 summer days.

width. The east-facing wall receives the first radiation in the early morning, reaching a maximum at 1000 hours, but being totally in shadow after 1200 hours. The low total R_n is because the east-facing wall is often in shadow. The street level (i.e. canyon floor) is sunlit only in the middle of the day and R_n and H dispositions are symmetrical. The third component of the urban canyon total energy balance is the west-facing wall, which is a mirror image (centred on noon) of that of the east-facing wall. Consequently, the symmetry of the street level energy balance and the mirror images of the east- and west-facing walls produce the symmetrical diurnal energy balance of R_n, H and ΔS observed at the canyon top.

The thermal characteristics of urban areas are in marked contrast to those of the surrounding countryside, and the generally higher urban temperatures are the result of the interaction of the following factors:

1 Changes in the radiation balance due to atmospheric composition.
2 Changes in the radiation balance due to the albedo and thermal capacity of urban surface materials, and to canyon geometry.
3 The production of heat by buildings, traffic and industry.
4 The reduction of heat diffusion due to changes in airflow patterns as the result of urban surface roughness.
5 The reduction in thermal energy required for evaporation and evapotranspiration due to the surface character, rapid drainage and generally lower wind speeds of urban areas.

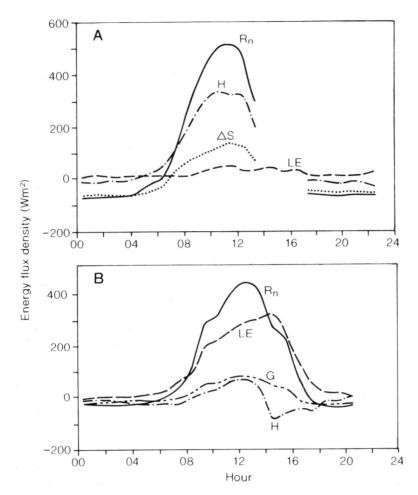

Figure 10.32 Diurnal energy balances for (A) the top of an urban canyon and (B) an irrigated suburban lawn in Vancouver, British Columbia, for 9–11 September 1973 and 6 August 1978, respectively.

Source: After Oke and Nunez, from Oke 1988.

Consideration of the last two factors will be left to D.3, this chapter.

a Atmospheric composition

Air pollution makes the transmissivity of urban atmospheres significantly lower than that of nearby rural areas. For example, during the period 1960–9 the atmospheric transmissivity over Detroit averaged 9 per cent less than that for nearby areas, and reached 25 per cent less under calm conditions. The increased absorption of solar radiation by aerosols plays a role in daytime heating of the boundary layer pollution dome (see Figure 10.30A) but is less important within the urban canopy layer, which extends to mean rooftop height (see Figure 10.30B). Table 10.2 gives comparable urban and rural energy budget figures for the Cincinnati region during the summer of 1968 under anticyclonic conditions with < 3/10 cloud and a wind speed of <2 m s^{-1}. The data show that pollution reduces the incoming short-wave radiation, but this is counterbalanced by a lower albedo and the greater surface area within urban canyons. The increased urban L_n at 1200 and 2000 LST is largely offset by anthropogenic heating (see below).

b Urban surfaces

Primary controls over a city's thermal climate are the character and density of urban surfaces, that is, the total *surface* area of buildings and roads, as well as the building geometry. Table 10.2 shows the relatively high heat absorption of the city surface.

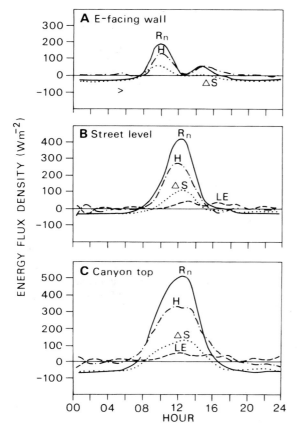

Figure 10.33 Diurnal variation of energy balance components for a N–S-oriented urban canyon in Vancouver, British Columbia, having white concrete walls, no windows, and a width/height ratio of 1:1, during the period 9–11 September 1973. A: the average for the E-facing wall. B: the average for the floor. C: averages of fluxes through the canyon top.

Source: After Nunez and Oke, from Oke 1978.

A problem of measurement is that the stronger the urban thermal influence, the weaker is the heat absorption *at street level*, and, consequently, observations made only in streets may lead to erroneous results. The geometry of urban canyons is particularly important, since it involves an increase in effective surface area and the trapping by multiple reflection of short-wave radiation, as well as a reduced 'sky view' (proportional to the areas of the hemisphere open to the sky), which decreases the loss of infrared radiation. From analyses by T. R. Oke, there appears to be an inverse linear relationship on calm, clear summer nights between the sky view factor (0–1.0) and the maximum urban–rural temperature difference. The difference is 10–12°C for a sky view factor of 0.3, but only 3°C for a sky view factor of 0.8–0.9.

c Human heat production

Numerous studies show that large urban conurbations now produce energy through combustion at rates comparable with incoming solar radiation in winter. Winter receipts from solar radiation average around 25 W m⁻² in Europe, compared with similar heat production from large cities. Figure 10.34 illustrates the magnitude and spatial scale of artificial and natural energy fluxes. In Cincinnati, a significant proportion of the energy budget is generated by human activity, even in summer (see Table 10.2). This heat production averages 26 W m⁻² or more, two-thirds of which was produced by industrial, commercial and domestic sources and one-third by cars. In terms of the future, it has been estimated that by AD 2000 the Boston–Washington megapolis may accommodate up to 56 million people in a continuous urban area of 30,000 km², and that this concentration of human activity would produce heat

Table 10.2 Energy budget figures (W m⁻²) for the Cincinnati region during the summer of 1968.

| Area | Central business district | | | Surrounding country | | |
Time	0800	1300	2000	0800	1300	2000
Short-wave, incoming $(Q + q)$	288*	763	–	306	813	–
Short-wave, reflected $[(Q + q)a]$	42†	120†	–	80	159	–
Net long-wave radiation (L_n)	–61	–100	–98	–61	–67	–67
Net radiation (R_n)	184	543	–98	165	587	–67
Heat produced by human activity	36	29	26‡	0	0	0

Source: From Bach and Patterson 1969.

Notes: *Pollution peak.

†An urban surface reflects less than agricultural land, and a rough skyscraper complex can absorb up to six times more incoming radiation.

‡Replaces more than 25 per cent of the long-wave radiation loss in the evening.

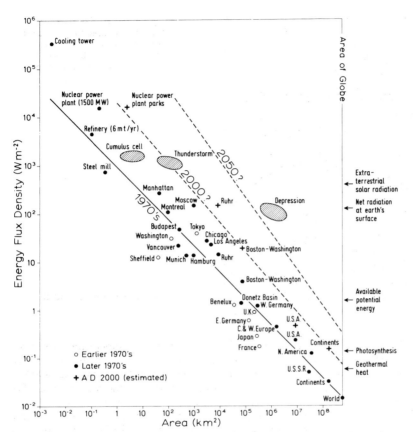

Figure 10.34 A comparison of natural and artificial heat sources in the global climate system on small, meso- and synoptic scales. Generalized regressions are given for artificial heat releases in the 1970s (early 1970s circles, late 1970s dots), together with predictions for the years 2000 (crosses) and 2050.

Sources: After Pankratti 1980 and Bach 1979.

equivalent to 50 per cent of the total winter solar radiation recorded on a horizontal surface and 15 per cent of the total summer solar radiation. In the extreme situation of Arctic settlements during polar darkness, the energy balance during calm conditions depends only on net long-wave radiation and heat production by anthropogenic activities.

d Heat islands

The net effect of the urban thermal processes is to make city temperatures generally higher than those in the surrounding rural areas. This is most pronounced after sunset (for mid-latitude cities during calm, clear weather), when cooling rates in the rural areas greatly exceed those in the urban areas. The energy balance differences that cause this effect depend on the radiation geometry and thermal properties of the surface. It is thought that the canyon geometry effect dominates in the urban canopy layer, whereas the sensible heat input from urban surfaces determines the boundary layer heating. By day the urban boundary layer is heated by increased absorption of short-wave radiation due to the pollution, as well as by sensible heat transferred from below and entrained by turbulence from above.

The *heat island* effect may result in minimum urban temperatures being 5–6°C greater than those of the surrounding countryside, and these differences may be as much as 6–8°C in the early hours of calm, clear nights in large cities, when the heat stored by urban surfaces during the day (augmented by combustion heating) is released. It should be noted that, because this is a *relative* phenomenon, the heat island effect also depends on the rate of rural cooling, which is influenced by the magnitude of the regional environmental lapse rate.

For the period 1931–60, the centre of London had a mean annual temperature of 11.0°C, comparing with 10.3°C for the suburbs, and 9.6°C

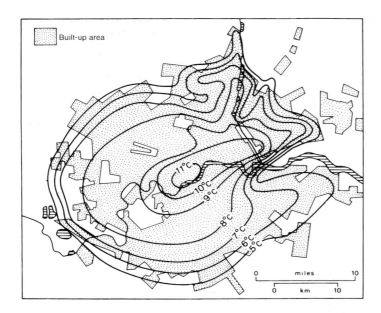

Figure 10.35 Distribution of minimum temperatures (°C) in London on 14 May 1959, showing the relationship between the 'urban heat island' and the built-up area.

Source: After Chandler 1965.

for the surrounding countryside. Calculations for London in the 1950s indicated that domestic fuel consumption gave rise to a 0.6°C warming in the city in winter and that accounted for one-third to one-half of the average city heat excess compared with adjacent rural areas. Differences are even more marked during still air conditions, especially at night under a regional inversion (Figure 10.35). For this heat island effect to operate effectively there must be wind speeds of less than 5–6 m s^{-1}, and it is especially in evidence on calm nights during summer and early autumn, when it has steep cliff-like margins on the upwind edge of the city and the highest temperatures associated with the highest density of urban dwellings. In the absence of regional winds, a well-developed heat island may even generate its own inward local wind circulation at the surface. Thus the thermal contrasts of a city, like those of many other of its climatic features, depend on its topographic situation, and are greatest for sheltered sites with light winds. The fact that for London urban–rural temperature differences are greatest in summer, when direct heat combustion and atmospheric pollution are at a minimum, indicates that heat loss from buildings by radiation is the most important single factor contributing to the heat island effect. Seasonal differences are not necessarily the same, however, in other macroclimatic zones.

The effects on minimum temperatures are especially significant. Cologne, for example, has an average of 34 per cent fewer days with minima below 0°C than its surrounding area, the corresponding figure for Basle being 25 per cent fewer. In London, Kew has an average of 72 more days with frost-free screen temperatures than rural Wisley. Precipitation characteristics are also affected: in pre-1917 Berlin, 21 per cent of the incidences of rural snowfall were associated with either sleet or rain in the city centre.

Although it is difficult to isolate changes in temperatures that are due to urban controls from those due to other influences (see Chapter 11), it has been suggested that city growth is often accompanied by an increase in mean annual temperature, that of Osaka, Japan, rising by 2.6°C in the last 100 years. Under calm conditions, the maximum difference in urban–rural temperatures is related statistically to population size, being nearly linear with the logarithm of the number of inhabitants. In North America, the maximum urban–rural temperature difference reaches 2.5°C for towns of 1,000, 8°C for cities of 100,000 and 12°C for cities of one million people. European cities show a smaller temperature difference for equivalent populations, perhaps as a result of the generally lower buildings and shallower urban canyons.

One of the most convincing examples of the possible relationship between urban growth and climate is that of the city of Tokyo, which expanded greatly after 1880 and particularly after 1946 (Figure 10.36A), when the population increased to 10.38 million in 1953 and to 11.67 million in 1975. During the period 1880–1975, there was a significant increase in mean January minimum temperatures and a decrease in the number of days within minimum temperatures below 0°C (Figures 10.36B and C). Although these graphs suggest a reversal of these trends during the Second World War (1942–5), when evacuation almost halved Tokyo's population, it is clear that the basis of correlations of urban climate with population is complex. Urban density, industrial activity and the production of anthropogenic heat are all involved. Leicester, for example, when it had a population of 270,000, exhibited warming comparable in intensity with that of central London over smaller sectors. This suggests that the thermal influence of city size is not as important as that of urban density. The vertical extent of the heat island is little known, but is thought to exceed 100–300 m, especially early in the night. In the case of cities with skyscrapers, the vertical and horizontal patterns of wind and temperature are very complex (see Figure 10.37 for wind conditions).

3 Modification of surface characteristics

a Airflow

On the average, city wind speeds are lower than those recorded in the surrounding open country owing to the sheltering effect of the buildings, and central average wind speeds are usually at least 5 per cent less than those of the suburbs. In 1935, for example, winds exceeding 10.5 m s^{-1} were recorded on the relatively open Croydon Airport (London suburbs) for a total of 371 hours, whereas the corresponding figure was only 13 hours for the closely built-up South Kensington. However, the urban effect on air motion varies greatly depending on the time of day and the season. During the day, city wind speeds are considerably lower than those of surrounding rural areas, but during the night the greater mechanical turbulence over the city means that the higher wind speeds aloft are transferred to the air at lower levels by turbulent mixing. During the day (1300 hours), the mean annual wind speed for London Airport (open country within the suburbs) was 2.9 m s^{-1}, compared with a 2.1 m s^{-1} in central London, for the period 1961–2. The

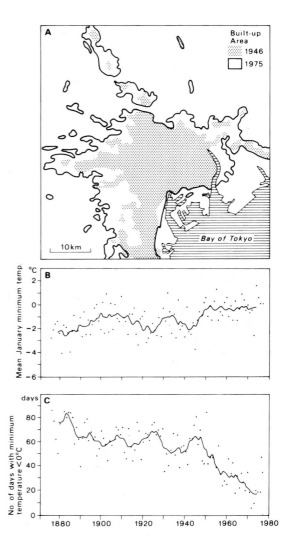

Figure 10.36 The built-up area of Tokyo in 1946 (A), the mean January minimum temperature (B), and number of days with sub-zero temperatures (C) between 1880 and 1975. During the Second World War, the population of the city fell from 10.36 million to 3.49 million and then increased to 10.38 million in 1953 and 11.67 million in 1975.
Source: After Maejima et al. 1982.

comparable figures for night (0100 hours) were 2.2 m s^{-1} and 2.5 m s^{-1}. Rural–urban wind speed differences are most marked with strong winds, and the effects are therefore more evident during winter than during summer, when a higher proportion of low speeds is recorded in temperate latitudes.

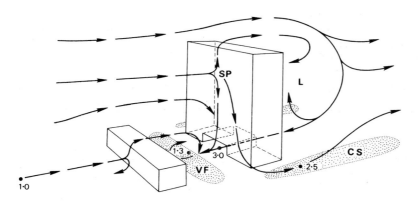

Figure 10.37 Details of urban airflow around two buildings of differing size and shape. Numbers give relative wind speeds; stippled areas are those of high wind velocity and turbulence at street level.

Sources: After Plate 1972 and Oke 1978.

Notes: SP = stagnation point; CS = corner stream; VF = vortex flow; L = lee eddy.

Urban structures have considerable effects on the movement of air both by producing turbulence as a result of their roughening the surface and by the channelling effects of the urban canyons. Some idea of the complexity of airflow around urban structures is given by Figure 10.37, illustrating the great differences in ground-level wind velocity and direction, the development of vortices and lee eddies, and the reverse flows that may occur. Structures play a major role in the diffusion of pollution within the urban canopy; for example, narrow streets often cannot be flushed by vortices. The formation of high-velocity streams and eddies in the usually dry and dusty urban atmosphere, where there is an ample debris supply, leads to general urban airflows of only 5 m s^{-1} being annoying, and those of more than 20 m s^{-1} being dangerous.

b Moisture

In urban areas, the absence of large bodies of standing water and the rapid removal of surface runoff through drains decreases local evaporation. In addition, the lack of an extensive vegetational cover eliminates much evapotranspiration, and this is an important source of augmenting urban heat. For these reasons, the air of mid-latitude cities has a tendency towards lower absolute humidities than that of their surroundings, especially under conditions of light wind and cloudy skies. On other occasions of calm, clear weather the streets trap warm air, which retains its moisture because lower dew is deposited on the warm surfaces of the city. Humidity contrasts between urban and rural areas are most noticeable in the case of relative humidity, which can be as much as 30 per cent less in the city by night as a result of the higher temperatures.

Urban influences on precipitation (excluding fog) are much more difficult to determine with any precision, partly because there are few rain gauges in cities and partly because air turbulence makes their 'catch' unreliable. It is now fairly certain, however, that urban areas in Europe and North America are responsible for local conditions that, in summer especially, can trigger off excesses of precipitation under marginal conditions. These triggers mainly involve the thermal effects and increased frictional convergence of built-up areas. These effects are well exemplified in the case of London for rainstorms on 21 August 1959 and 1 September 1960, respectively (Figure 10.38). Recordings for Munich showed 11 per cent more days of light rain (0.1–0.5 mm or 0.004–0.12 in) than in the surrounding countryside, and Nürnberg (Nuremberg) has recorded 14 per cent more thunderstorms than its rural environs. European and North American cities apparently record 6–7 per cent more days with rain per year than their surrounding regions, giving a 5–10 per cent increase in urban precipitation. The effect is generally more marked in the cold season in North America, although urban areas in the Mid-west of the United States significantly increase summer convective activity with more frequent thunderstorms and hail downwind of industrial areas of St Louis (for a distance of 30–40 km) compared with rural areas (Figure 10.39). The anomalies shown in Figure 10.39 are among the best documented of urban effects. Over south-east England during 1951–60, summer thunderstorm rain (which composed 5–15 per cent of the total precipitation) was especially concentrated in west, central and south London (Figure 10.40) and contrasted strikingly with the distribution of mean annual total rainfall. During this period,

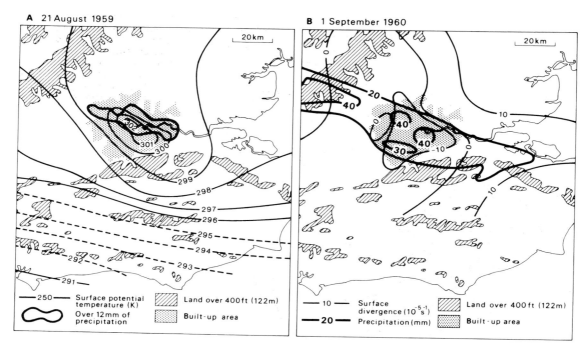

Figure 10.38 The climatic effect of London. A: surface potential temperature (°K) at noon on 21 August 1959 and the area receiving more than 12 mm total precipitation on that day. B: divergence of the surface wind field (10^{-5} s^{-1}) at 0900 on 1 September 1960 over south-east England and the total precipitation (mm) on that day. The high frictional effect of the city surface caused airflow to converge over the city (i.e. negative divergence to occur).

Source: From Atkinson 1987, in K. J. Gregory and D. E. Walling, (eds) *Human Activity and Environmental Processes* (Copyright © 1987; reprinted by permission of John Wiley & Sons, Inc.).

London's thunderstorm rain was of the order of 20–25 cm greater than that in rural south-east England.

Many of the results discussed in connection with urban influences are based on limited case studies, but a summary of average climatic differences between cities and their surrounding rural areas is presented in Table 10.3.

4 Tropical urban climates

A striking feature of recent and projected world population growth is the relative increase in the tropical and subtropical regions. At present, there are thirty-four world cities with more than five million people, twenty-one of which are in the less-developed countries, but by the year 2025 it is predicted that, of the thirteen world cities that will have populations in the 20–30 million range, eleven will be in less-developed countries (*i.e.* Mexico City, São Paulo, Lagos, Cairo, Karachi, Delhi, Bombay, Calcutta, Dhaka, Shanghai and Jakarta).

Table 10.3 Average urban climatic conditions compared with those of surrounding rural areas.

Atmospheric composition	carbon dioxide	×2
	sulphur dioxide	×200
	nitrogen oxides	×10
	carbon monoxide	×200(+)
	total hydrocarbons	×20
	particulate matter	×3 to 7
Radiation	global solar	−15 to 20%
	ultraviolet (winter)	−30%
	sunshine duration	−5 to 15%
Temperature	winter minimum (average)	+1 to 2°C
	heating degree days	−10%
Wind speed	annual mean	−20 to 30%
	number of calms	+5 to 20%
Fog	winter	+100%
	summer	+30%
Cloud		+5 to 10%
Precipitation	total	+5 to 10%
	days with <5 mm	+10%

Source: Partly after World Meteorological Organization 1970.

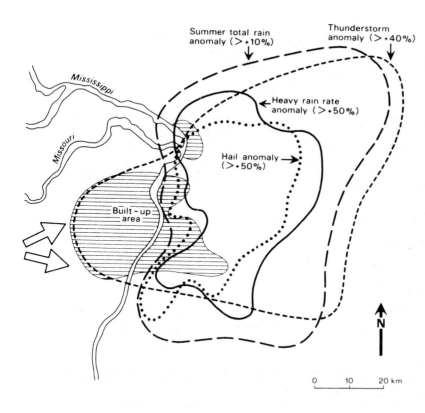

Summer total rain anomaly (>+10%)

Thunderstorm anomaly (>+40%)

Heavy rain rate anomaly (>+50%)

Hail anomaly (>+50%)

Built-up area

Mississippi

Missouri

N

0   10   20 km

Figure 10.39 Anomalies of summer rainfall, rate of heavy rains, hail frequency and thunderstorm frequency downwind of the St Louis metropolitan area. Large arrows indicate the prevailing direction of motion of summer rain systems.

Source: After Changnon 1979.

Despite the difficulties in extrapolating knowledge of urban climatic effects from one region to another, it is clear that the ubiquitous high-technology architecture of most modern city centres and multi-storey residential areas imposes strikingly similar tendencies on their differing background climates. Nevertheless, most tropical urban built land differs from that in higher latitudes in that it is commonly composed of high-density, single-storey buildings with few open spaces and poor drainage. In such a setting, as Oke points out, the composition of roofs is more important than that of walls in terms of thermal energy exchanges, and the production of anthropogenic heat is more uniformly distributed spatially and of lower intensity than in European and North American cities. In the dry tropics, buildings have a relatively high thermal mass to delay heat penetration and this, combined with the low soil moisture in the surrounding bare rural areas, makes the ratio of urban to rural thermal admittance greater than in temperate regions. However, it is not always possible to generalize regarding the thermal effects of cities in the dry tropics due to the differing 'oasis' effects of urban vegetation. Building construction in the humid tropics is characteristically lightweight to promote the much-needed ventilation. These cities differ strikingly from temperate ones in that the surrounding rural thermal admittance is greater than the urban because of the high rural soil moisture levels and the high urban albedos.

Tropical heat island tendencies are somewhat similar to those of temperate cities but are generally weaker, with different timings for temperature maxima, and with complications introduced by the effects of afternoon and evening convective rainstorms and by diurnal breezes. The thermal characteristics of tropical cities differ from those of the mid-latitudes in that their urban morphology is commonly dissimilar (e.g. building density, materials, geometry, green areas, etc.) and because they have fewer sources of anthropogenic heat. Urban areas in the tropics tend to have slower rates of cooling and warming than do the surrounding rural areas, and this causes the major nocturnal heat island effect to manifest itself later than in the mid-latitudes – i.e.

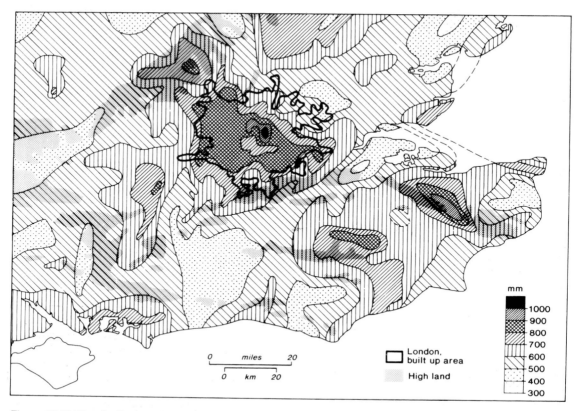

Figure 10.40 The distribution of total thunderstorm rain in south-east England during the period 1951–60.
Source: After Atkinson 1968.

at around sunrise (Figure 10.41A). Urban climates in the subtropics are well illustrated by four cities in Mexico (Table 10.4). The heat island effect is, expectably, greater for larger cities and best exemplified at night during the dry season (November–April), when anticyclonic conditions, clear skies and inversions are most common (Figure 10.41B). It is of note that in some tropical coastal cities (e.g. Veracruz; Figure 10.41A) the afternoon urban heating may produce such instability as to reinforce the sea-breeze effect to the point where there is a 'cool island' urban effect (Figure 10.41A). Elevation may play a significant thermal role (Table 10.4), as with Mexico City, where the urban heat island may be accentuated by rapid nocturnal cooling of the surrounding countryside. Quito, Ecuador (2,851 m), shows a maximum heat island effect by day (as much as 4°C) and weaker night-time effects, probably due to the nocturnal drainage of cold air from the nearby volcano Pichincha.

At 7°N in West Africa, Ibadan, Nigeria (population over one million; elevation 210 m), records higher rural temperatures than urban in the morning and higher urban temperatures in the afternoon, especially in the dry season (November to mid-March). In December, the harmattan dust haze tends to reduce city maximum temperatures. During this season, mean monthly minimum temperatures

Table 10.4 Population (1990) and elevation for four Mexican cities.

	Population (millions)	Elevation (metres)
Mexico City (19°25′N)	15.05	2,380
Guadalajara (20°40′N)	1.65	1,525
Monterey (25°49′N)	1.07	538
Veracruz (19°11′N)	0.33	SL

Source: Jauregui 1987.

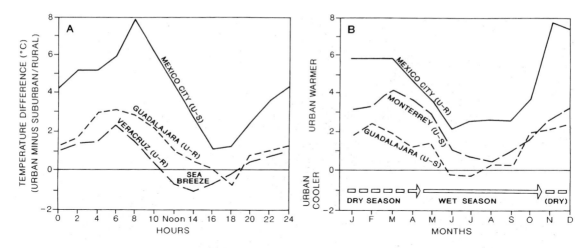

Figure 10.41 Diurnal (A) and seasonal (B) heat island intensity variations (i.e. urban minus rural or suburban temperature differences) for four Mexican cities.
Source: Jauregui 1987. Copyright © *Erdkunde*. Published by permission.

are significantly greater in the urban heat island than in rural areas (March +12°C, but December only +2°C due to the atmospheric dust effect). In general, urban–rural minimum temperature differences vary between –2 and +15°C. Two other tropical cities exhibiting urban heat islands are Nairobi, Kenya (+3.5°C for minimum temperatures and +1.6°C for maximum temperatures) and Delhi, India (+3 to 5°C for minimum temperatures and +2 to 4°C for maximum temperatures).

Despite a marked lack of data, there seems to be some urban precipitation enhancement in tropical regions, which is maintained for a larger part of the year than that associated with the shorter-term high-sun convective period in temperate latitudes.

SUMMARY

Small-scale climates are determined largely by the relative importance of the surface energy budget components, which vary in amount and sign depending on time of day and season. Bare land surfaces may have wide temperature variations controlled by H and G, whereas those of surface water bodies are strongly conditioned by LE and advective flows. Snow and ice surfaces have small energy transfers in winter with net outgoing radiation offset by transfers of H and G towards the surface. After snow melt, the net radiation is large and positive, balanced by turbulent energy losses. Vegetated surfaces have more complex exchanges, which are usually dominated by LE; this may account for more than 50 per cent of the incoming radiation, especially where there is an ample water supply (including irrigation). Forests have a lower albedo (<0.10 for conifers) than most other vegetated surfaces (0.20–0.25). Their vertical structure produces a number of distinct microclimatic layers, particularly in tropical rainforests. Wind speeds are characteristically low in forests and trees form important shelter belts. Unlike short vegetation, various types of tree exhibit a variety of rates of evapotranspiration and thereby differentially affect local temperatures and forest humidity. The effect of forests on precipitation has not yet been resolved but they may have a marginal topographic effect under convective conditions in temperate regions. The disposition of forest moisture is very much affected by canopy interception and evaporation, but forested catchments appear to have greater evapotranspiration losses than ones with a grass cover. Another major feature of forest micro-climates is their lower temperatures and smaller diurnal ranges compared with surrounding areas.

Urban climates are dominated by the geometry and composition of built-up surfaces and by the

continued

effects of human urban activities. The composition of the urban atmosphere is modified by the addition of aerosols, producing smoke pollution and fogs, by industrial gases such as sulphur dioxide, and by a chain of chemical reactions, initiated by automobile exhaust fumes, which causes smog and inhibits both incoming and outgoing radiation. Pollution domes and plumes are produced around cities under appropriate conditions of vertical temperature structure and wind velocity. The urban heat budget is dominated by H and G, except in city parks, and as much as 70–80 per cent of incoming radiation may become sensible heat, which is very variably distributed between the complex urban built forms. Urban influences combine to give generally higher temperatures than in the surrounding countryside, not least because of the growing importance of heat production by human activities. These factors give rise to the urban heat island, which may be 6–8°C warmer than surrounding areas in the early hours of calm, clear nights, when heat stored by urban surfaces is being released. The urban–rural temperature difference under calm conditions is statistically related to the city population size; the urban canyon geometry and sky view factor are major controlling factors. The heat island may be a few hundred metres deep, depending on the building configuration. Urban wind speeds are generally lower than in rural areas by day, but the wind flow is extremely complex, depending on the geometry of city built forms. Cities naturally tend to be less humid than rural areas, but their topography, roughness and thermal qualities tend to intensify the effects of summer convective activity over and downwind of the urban areas, giving more thunderstorms and heavier storm rainfall. Tropical cities have heat islands, but the diurnal phase tends to be delayed relative to mid-latitude ones. The temperature amplitude is largest during dry season conditions.

11

Climate change

A GENERAL CONSIDERATIONS

Realization that climate has changed radically with time came only during the 1840s, when indisputable evidence of former ice ages was obtained, yet in many parts of the world the climate has altered sufficiently, even within the last few thousand years, to affect the possibilities for agriculture and settlement. Reliable weather records have been kept only during the last hundred years or so, but *proxy* indicators of past conditions from tree rings, pollen in bog and lake sediments, ice core records of physical and chemical parameters, and ocean foraminifera in sediments provide a wealth of paleoclimatic data. The discussion in this chapter focuses on climate forcings and observed changes, as well as future projections, but first it is worth while to consider the nature of climate variations.

The standard interval adopted by the World Meteorological Organization for climatic statistics is thirty years: 1901–30, 1931–60, for example. However, for historical records and proxy indicators of climate, longer, arbitrary time intervals may be necessary to calculate average values. Tree rings and ice cores can give seasonal/annual records, while peat bog and ocean sediments may provide records with only 100- to 1,000-year time resolution. Hence, short-term changes and the true rates of change may, or may not, be identifiable.

The present climatic state is usually described in terms of an average value (arithmetic mean, or the median value in a frequency distribution), a measure of variation about the mean (the standard deviation, or interquartile range), the extreme values, and often the shape of the frequency distribution (see Note 1). A change in climate can occur in several different ways (Figure 11.1). For example, there may be a shift in the mean level (B), or a gradual trend in the mean (C). The variability may be

periodic (A), quasi-periodic (B) or non-periodic (C), or alternatively it may show a progressive trend (D). First, it is important to determine whether such changes are real or whether they are an artefact of changes in instrumentation, observational practices, station location, or the surroundings of the instrumental site, or due to errors in the transcribed data. Even when changes are real, it may be difficult to ascribe them to unique causes because of the complexity of the climate system. Natural variability operates over a wide range of time scales, and superimposed on these natural variations in climate are the effects of human activities.

B CLIMATE FORCINGS AND FEEDBACKS

The average state of the climate system is controlled by a combination of *forcings* external to the system (solar variability, astronomical effects, tectonic processes and volcanic eruptions), internal radiative forcings (atmospheric composition, cloud cover), anthropogenically induced changes (in atmospheric composition, surface land cover) and *feedback effects* (such as changes in atmospheric water vapour content or cloudiness caused by global temperature changes). Figure 11.2 illustrates schematically the complexity of the climate system. It is useful to try and assess the magnitude of such effects, globally and regionally, and the time scales over which they operate.

1 External forcing

Solar variability. The sun is a variable star, and it is known that early in the earth's history (during the Archean three billion years ago) solar irradiance was about 80 per cent of the modern value. Paradoxically, however, the effect of this 'faint early sun' was offset, most likely, by a concentration of

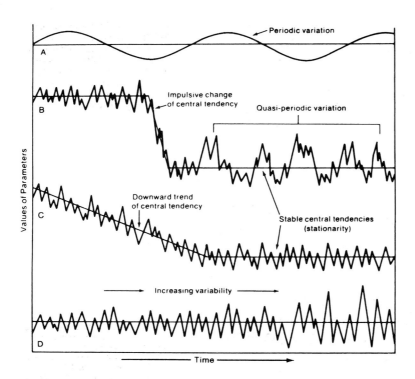

Figure 11.1 Different types of climatic variation. The scales are arbitrary.

Source: Hare 1979. Courtesy World Meteorological Organization.

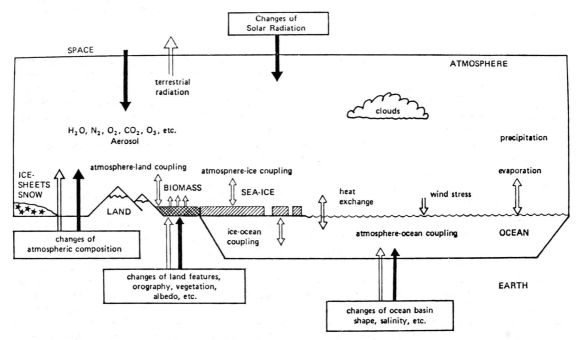

Figure 11.2 The 'climate system'. External processes are shown by solid arrows; internal processes by open arrows.

Source: After US GARP Committee.

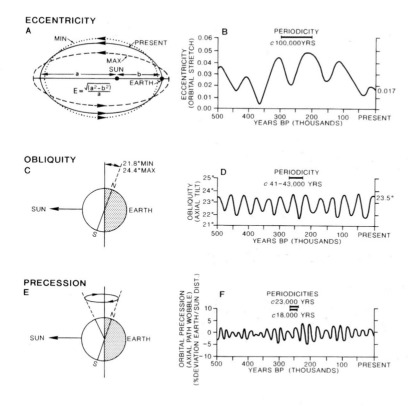

ECCENTRICITY
A

OBLIQUITY
C

PRECESSION
E

Figure 11.3 The astronomical (orbital) effects on the solar irradiance and their time scales over the past 500,000 years. A and B: Eccentricity or orbital stretch; C and D: Obliquity or axial tilt; E and F: Precession or axial path wobble.

Sources: Partly after Broecker and Van Donk 1970, and Henderson-Sellers and McGuffie 1984. B, D and F: from *Review of Geophysics and Space Physics* **8**, 1970. Reproduced by kind permission of the American Geophysical Union.

carbon dioxide that was perhaps 100 times higher than now, but also perhaps by the effects of a largely water-covered earth. The approximately 11-year solar cycle (and 22-year magnetic field cycle) is well known. As discussed in Chapter 2, this causes a ±1 W m⁻² fluctuation in solar irradiance and much larger effects on ultraviolet radiation. Intervals when sunspot and solar flare activity were much reduced (especially the Maunder minimum of AD 1650–1700) may have caused cumulative effects leading to temperature decreases of about 1°C.

Tectonic processes. On geological time scales, there have been great changes in continental positions and sizes and in the configuration of ocean basins as a result of crustal processes (known as plate tectonics). These movements have also altered the size and location of mountain ranges and plateaus. As a result, the global circulation of the atmosphere and the pattern of ocean circulation and surface currents have also been modified. Changes in continental location have contributed substantially to major ice age episodes (such as the Permo-

Carboniferous glaciation of Gondwanaland) as well as to intervals with extensive arid (Permo-Triassic) or humid (coal deposits) environments during other geological periods. Over approximately the last few million years, the uplift of the Tibetan Plateau and the Himalayan ranges has caused the onset, or intensification, of desert conditions in western China and Central Asia.

Astronomical periodicities. As discussed in Chapter 2A.2, the earth's orbit around the sun is subject to long-term variations. There are three principal effects on incoming solar radiation: the eccentricity (or stretch) of the orbit, with a period of approximately 95,000 years and 410,000 years; the tilt of the earth's axis (an approximately 41,000-year period); and a wobble in the earth's axis of rotation, which causes changes in the timing of perihelion (Figure 11.3). This precessional effect, with a period of about 21,000 years, is illustrated in Figure 2.3. The range of variation of these three components and their consequences are summarized in Table 11.1.

Table 11.1 Orbital forcings and characteristics.

Element	Index range	Present value	Average periodicity
Obliquity of Ecliptic (ε) (Tilt of axis of rotation)	22–24.5°	23.4°	41 ka
Effects equal in both hemispheres, effect intensifies polewards (for caloric seasons)			

Low ε	High ε
Weak seasonality, steep poleward radiation gradient	Strong seasonality, more summer radiation at poles, weaker radiation gradient

Element	Index range	Present value	Average periodicity
Precession of Equinox (ω) (Wobble of axis of rotation)	0.05 to −0.05	0.0164	19, 23 ka
Changing earth–sun distance alters seasonal cycle structure; complex effect, modulated by eccentricity of orbit			
Eccentricity of Orbit (e)	0.005 to 0.0607	0.0167	410, 95 ka
Gives 0.02% variation in annual incoming radiation; modifies amplitude of precession cycle changing seasonal duration and intensity; effects opposite in each hemisphere; greatest in low latitudes.			

Volcanic eruptions. Major explosive eruptions inject dust and sulphur dioxide aerosols into the stratosphere, where they may circle the earth for several years causing brilliant sunsets (see Figure 1.12). Equatorial eruption plumes spread into both hemispheres, whereas plumes from eruptions in mid to high latitudes are confined to that hemisphere. Records of such eruptions are preserved in the Antarctic and Greenland ice sheets for at least the last 150,000 years. Observational evidence from the last 100 years demonstrates that major eruptions cause a hemisphere/global cooling of 0.5–1.0°C in the year following the event.

Atmospheric composition. The greenhouse effect has been examined in Chapter 1, but it is appropriate to recall that there is a large 'natural' greenhouse, as a result of the atmospheric composition, distinct from human-induced changes over the last few centuries. Glacial–interglacial changes in terrestrial vegetation and in the oceanic uptake of trace gases, as a result of changes in the thermohaline circulation of the global ocean (see Chapter 6D),

have caused major fluctuations in atmospheric carbon dioxide (±50 ppm) and methane (±150 ppb). Negative (positive) excursions are associated with cold (warm) intervals, as illustrated in Figure 1.6. The changes in greenhouse gases (CO_2 and CH_4) and global temperatures are virtually coincident during both glacial and interglacial transitions, so that there is no clear causative agent. Both the long-term and rapid changes in atmospheric CO_2 seen in polar ice cores seem to result from the combined effects of ocean and land biological activity and ocean circulation shifts.

Rates of change. Obviously, changes in climate resulting from changes in the earth's geography through geological processes (e.g. position and size of ocean basins, continents and mountain ranges) are only perceptible on time scales of millions of years. Although geographical changes have had immense paleoclimatic significance, they are of less immediate concern to contemporary climatologists than the radiative forcing agents. Radiative forcing agents affect the supply and disposition of solar

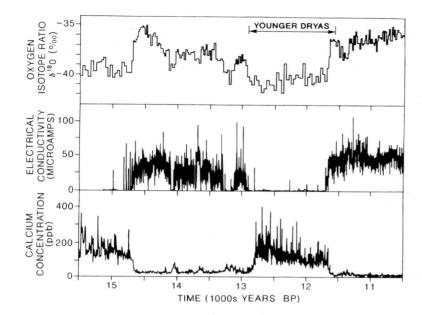

Figure 11.4 The late glacial to interglacial transition (14.7 to 11.6 ka) as indicated by δ18O (ppm), electrical conductivity (microamps) and calcium concentrations (ppb) in the Greenland Ice Sheet Program (GISP) ice core from central Greenland.

Source: Reprinted by permission from *Natural Climate Variability on Decade-to-Century Time Scales* by P. Grootes. Copyright © National Academy of Sciences. Courtesy of the National Academy of Sciences, Washington, DC.

radiation. Solar radiation changes, like the non-radiative forcing agents, are external inputs into the atmosphere–earth–ocean–ice system but, unlike them, occur at a range of time scales from tens to hundreds of thousands and, probably, millions of years. Thus solar radiation is both a long-term and a short-term external forcing agent. Astronomical forcings give rise to global temperature fluctuations of ±2–5°C per 10,000 years. The timing of orbital forcing is also clearly represented in glacial–interglacial fluctuations with major glacial cycles spanning about 100,000 years (or 100 Ka). However, the most striking fact to emerge from analysis of two recent deep ice cores in central Greenland is the great rapidity of large changes in atmospheric temperature, precipitation and aerosol levels, presumably as a result of major readjustments of atmospheric circulation. The onset and termination of the Younger Dryas cold episode 12,900–11,600 BP (before present) (see Figure 11.4), with a switch from glacial to interglacial conditions and back again, apparently occurred within a five-year time interval for both transitions!

2 Short-term forcing and feedback

Internal radiative forcing agents are mainly those that involve changes in atmospheric composition, cloud cover, aerosols and surface albedo. Although subject to long-term changes, it is their susceptibility to short-term anthropogenic changes that makes them of particular interest to contemporary climatologists. The interactive relations between short-term external solar radiative forcing and these internal radiative forcing agents lie at the heart of the understanding and prediction of short-term global climatic changes, through a complex set of *feedback mechanisms*, which can be either *positive* (i.e. self-enhancing) or *negative* (i.e. self-regulating or damping).

Positive feedback mechanisms affecting global climate appear to be widespread and to be particularly effective in response to temperature changes, which is a matter of especial current concern. Increases in global temperature produce increases in atmospheric water vapour, increases in plant respiration, decreases in CO_2 dissolved in the oceans, and an increase in methane emissions from wetlands. All of these, in turn, tend to increase the global concentration of greenhouse gases and, hence, to increase global temperature further. Ice and snow cover is involved in especially important positive feedback effects in that a more extensive cover creates higher albedo and lower temperatures, which, in turn, will further extend the ice and snow cover, producing additional cooling. Conversely, a warming effect, which melts ice and snow cover, decreases the surface albedo, allows the absorption of more incoming solar radiation and leads to increased surface heating and higher temperatures.

Unfortunately, negative feedback mechanisms appear to be much less important in the face of short-term radiative forcing and it is important to understand that, for example, they can only reduce the rate of warming but cannot, of themselves, cause global cooling. Cloud cover is a particularly complex global feedback mechanism, producing both positive and negative effects. For example, negative feedback may operate when increased global heating leads to greater evaporation and greater amounts of high-altitude cloud cover, which will reflect more incoming solar radiation and thus damp down the global temperature rise.

C THE CLIMATIC RECORD

1 The geological record

To understand the significance of climatic trends over the last hundred years, they need to be viewed against the background of our general knowledge about past conditions. On geological time scales, global climate periodically undergoes major shifts between a generally warm, ice-free state and an ice age state with continental ice sheets. There have been at least seven major ice ages through geological time. The first occurred 2,500 million years ago (Ma) in the Archean period, followed by three more between 900 and 600 Ma, in the Proterozoic. There were two ice ages in the Paleozoic era (the Ordovician, 500–430 Ma; and the Permo-Carboniferous, 345–225 Ma). The late Cenozoic glaciation began 10 Ma, or earlier, in Antarctica and extended into northern middle and high latitudes about three million years ago.

Major ice ages are determined by a combination of external and internal forcing (solar luminosity, continental location, tectonics and atmospheric CO_2 concentration). The ice sheets of the Ordovician and Permo-Carboniferous periods formed in high southern latitudes on the former mega-continent of Gondwanaland, whereas bipolar glaciation developed in the Pliocene–Pleistocene epoch. Alfred Wegener proposed continental drift as a major determinant of climates and biota in 1912, but this idea remained controversial until the motion of crustal plates was recognized in the 1960s. Uplift of the western cordilleras of North America and the Tibetan Plateau by plate movements during the Tertiary period (50–2 Ma), caused regional aridity to develop in the respective continental interiors. However, geographical factors are only part of the explanation of climate variations. The mid-Cretaceous period, about 100 Ma, was notably warm in high latitudes. This warmth may be attributable to atmospheric concentrations of carbon dioxide three to seven times those at present, augmented by the effects of alterations in land–sea distribution and ocean heat transport.

Within the major ice ages, there are recurrent oscillations between glacial and interglacial conditions. Eight such cycles of global ice volume are recorded by land and ocean sediments during the last 0.8–0.9 Ma, each averaging ~100 Ka, with only 10 per cent of each cycle as warm as the twentieth century (Figure 11.5, D and E). Prior to 0.9 Ma, ice volume records have a dominant 41 Ka periodicity, while ocean records of calcium carbonate indicate fluctuations of 400 Ka. These periodicities represent the forcing of global climate by changes in incoming solar radiation associated with the earth's orbital variations (see Chapter 2A). Building on the work of nineteenth-century astronomers and geologists, calculations published by Milutin Milankovich in 1920 established an astronomical theory of glacial cycles. This became widely accepted only in the 1970s; the inadequate dating of glacial records, as well as the complex nature of radiation-climate relationships and the time lags involved in ice sheet build-up and decay allowed considerable scope for alternative explanations. The precession signature (19 and 23 Ka) is most apparent in low-latitude records, whereas that of obliquity (41 Ka) is represented in high latitudes. However, the 100 Ka orbital eccentricity signal is generally dominant overall. This is surprising because the effect of eccentricity on incoming solar radiation is modest, implying some non-linear interaction in the climate system. Several models of ice sheet growth and decay, incorporating bedrock isostatic response with orbital radiative forcing, also generate ice oscillations with periods of about 100 Ka, implying that internal dynamics are important.

Only four or five glaciations are identified in the land evidence due to the absence of continuous sedimentary records, but it is likely that in each of these glacial intervals large ice sheets covered northern North America and northern Europe. Sea levels were also lowered by about 100–150 m due to the large volume of water locked up in the ice. Records from tropical lake basins show that these regions were generally arid at these times. The last such glacial maximum climaxed about 20,000 years ago, but 'modern' climatic conditions only became

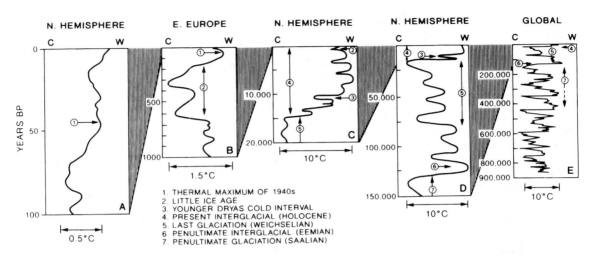

Figure 11.5 Main trends in global climate during the past million years or so. A: northern hemisphere, average land air temperatures. B: Eastern Europe, winter temperatures. C: northern hemisphere, average land–air temperatures. D: northern hemisphere, average air temperatures based partly on sea-surface temperatures. E: global average temperatures derived from deep-sea cores.

Source: Data taken from *Understanding Climatic Change: A Program for Action* (1975), published by National Academy Press, Washington, DC.

established during post-glacial (Holocene) time – conventionally dated to 10,000 years BP. This time scale is used whenever dates are based on radio-carbon (carbon-14, ^{14}C) dating or other radiometric methods involving isotopic decay processes, such as potassium-argon (K–Ar).

For the last glacial cycle and post-glacial time, information on climatic conditions is obtained indi-rectly from proxy records. For example, the advance and retreat of glaciers represents a response to winter snowfall and summer melt. The history of vegetation, which indicates temperature and mois-ture conditions, can be traced from pollen types preserved in lake sediments and peat bogs. Former lake shorelines indicate changes in moisture balance. Estimates of seasonal climatic elements can be made from studies of annual snow/ice layers in cores taken from polar ice sheets, where no melt occurs. These layers also record past volcanic events through the inclusion of micro-particles and chemical com-pounds in the ice. In forested areas, where trees form an annual growth layer, the ring width can be interpreted through dendroclimatological studies in terms of moisture availability (in semi-arid regions) and summer warmth (near the polar and alpine tree lines). Pollen sequences, lake level records and ice cores usually span the last 10,000–23,000 years (and, exceptionally, back to 150,000 BP), while tree

rings rarely extend more than 1,500 years ago. For more recent time, historical documents often record crop harvests or extreme weather events (droughts, river and lake freezing, etc.).

2 Late glacial and post-glacial conditions

The 100 Ka glacial cycles of the last 0.8 Ma show characteristic abrupt terminations. The last glacia-tion, for example, ended with abrupt warming about 14,700–13,000 BP, interrupted by a short cold regression (the Younger Dryas, 12,900–11,600 BP), followed by a renewed sharp warming trend (see Figure 11.4). Such shifts clearly involve more than the cyclical variations in solar radiation associated with astronomical effects, but the mechanisms involved are poorly understood.

Early Holocene warmth around 10,000 BP is attributed to July solar radiation being 30–40 W m^{-2} greater than now in northern mid-latitudes, due to the combined orbital effects. Following the final retreat of the continental ice sheets from Europe and North America between 10,000 and 7,000 years ago, the climate rapidly ameliorated in middle and higher latitudes. In the subtropics, this interval was also generally wetter, with high lake levels in Africa and the Middle East. A *thermal maximum* was reached in the mid-latitudes about 5,000 years ago, when summer temperatures are known to have been 1–2°C higher than today (see Figure 11.5B) and the Arctic tree line was several hundred kilometres further north in Eurasia and North America. By this time, subtropical desert regions were again very dry and were largely abandoned by primitive man. A temperature decline set in around 2,000 years ago with colder, wetter conditions in Europe and North America. Although temperatures have not since equalled those of the thermal maximum, there was certainly a warmer period in some areas of the world from the ninth to the mid-fifteenth centuries AD. In particular, summer temperatures in Scandinavia, China, the Sierra Nevada (California), Canadian Rocky Mountains and Tasmania were above those that prevailed over the next five centuries. The reconstruction shown for this period in central England (Figure 11.6) has subsequently been criticized. More rigorous analysis of documentary records for Western and Central Europe (excluding Britain) shows a warm phase around AD 1300. Icelandic records indicate mild conditions up to the late twelfth century, and this phase was marked by the Viking colonization of Greenland and the occupation of Ellesmere Island in the Canadian Arctic by the Inuit. A further deterioration followed, and severe winters between AD 1550 and 1700 gave a 'Little Ice Age' and extensive Arctic pack-ice and glacier advances in some areas to maximum positions since the end of the Ice Age. These advances occurred at dates ranging from the mid-seventeenth to the late nineteenth century in Europe, as a result of the lag in glacier response and minor climatic fluctuations. Figure 11.6 attempts to summarize seasonal trends in central England, but it must be stressed that at present only the gross features are represented, as we know little or nothing about short-term fluctuations before the medieval period, for example, and even the relative magnitudes of the changes before about AD 1700 can only be indicated in a very general way.

3 The last 500 years or so

Climate records over the last 500 years suggest fluctuations on three time intervals: 15–35 years with a peak-to-peak amplitude of 0.3°C; 50–100 years with a 1.0°C amplitude over the North Atlantic–Arctic; and 100–400-year global oscillations of about 0.75°C (warmer pre-1500 followed by a Little Ice Age). The shortest interval is apparently linked to ENSO/PNA circulation modes and the

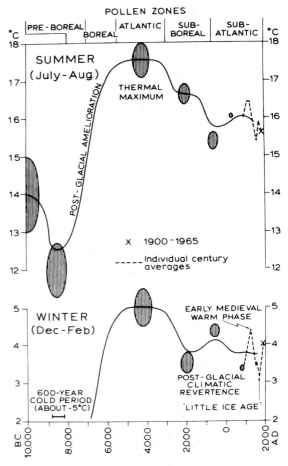

Figure 11.6 Air temperatures in the lowlands of central England. Trends of the supposed 1,000-year and 100-year averages since 10,000 BC (the latter calculated for the last millennium). Shaded ovals indicate the approximate ranges within which the temperature estimates lie and error margins of the radiocarbon dates.

Source: After Lamb 1966.

50–100-year scale to low-frequency changes in the thermohaline circulation. Interdecadal variability seems to be mostly induced by atmospheric dynamics. Much of the variance in winter temperatures is associated with the varying strength of land/ocean contrasts in the 1,000–500 mb thickness field.

Long instrumental records for stations in Europe and the eastern United States indicate that the warming trend that ended the 'Little Ice Age' began early

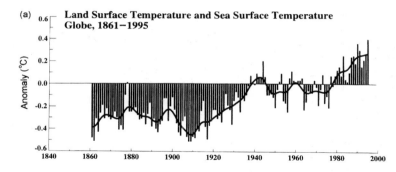

(a) **Land Surface Temperature and Sea Surface Temperature Globe, 1861–1995**

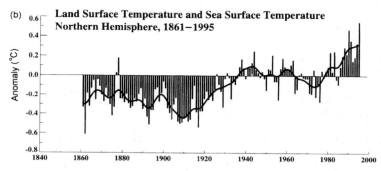

(b) **Land Surface Temperature and Sea Surface Temperature Northern Hemisphere, 1861–1995**

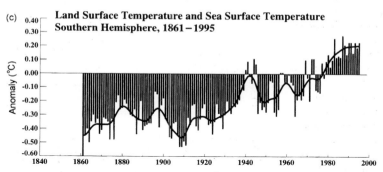

(c) **Land Surface Temperature and Sea Surface Temperature Southern Hemisphere, 1861–1995**

Figure 11.7 Annual combined land-air and sea-surface temperature anomalies between 1861 and 1995, relative to the 1961–90 average (designated zero). A: northern hemisphere; B: southern hemisphere; and C: globe.

Source: After Parker *et al.* 1996. Reproduced from *Weather* by permission of the Royal Meteorological Society. Copyright ©.

in the nineteenth century. Global records since 1861 (Figure 11.7) show a significant, but irregular, temperature rise of between 0.3 and 0.6°C, probably closer to the upper estimate. This trend was least in the tropics and greatest in cloudy, maritime regions of high latitude (Figure 11.8). Winter temperatures were most affected (Figure 11.9), and on Svalbard (77°N) the 1920–39 mean January temperature was 7.8°C greater than the 1900–19 mean.

The general temperature rise has not been continuous, however, and four phases can be identified:

1 1861–1920, during which there was mean annual oscillation within extreme limits of 0.4 °C but no consistent trend.

2 1920–mid-1940s, during which there was considerable warming averaging 0.4°C.

3 Mid-1940s–early 1970s, during which there were oscillations within extreme limits of less than 0.4°C, with the northern hemisphere cooling slightly on average and the southern hemisphere remaining fairly constant in temperature. Regionally, northern Siberia, the eastern Canadian Arctic and Alaska experienced a mean lowering of winter temperatures by 2–3°C between 1940-9 and 1950-9; this was partly compensated by a slight warming in the western United States, Eastern Europe and Japan.

4 Early 1970s–1989, during which there was a marked overall warming of about 0.2°C, except

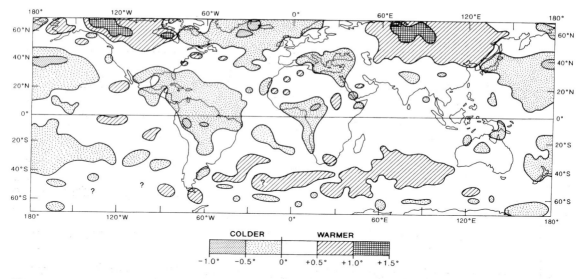

Figure 11.8 Global temperature trends (°C) observed between 1955–74 and 1975–94.
Source: After Houghton *et al.* 1996. Reproduced by permission of the IPCC.

for areas of the North and South Pacific, North Atlantic, Europe, Amazonia and Antarctica (see Figure 11.8). The tropospheric temperature increase in the northern hemisphere after 1976 began over the tropical oceans, especially in winter, when mid-latitudes were still cooling. In the tropics, the tropospheric warming (1976–90) exceeded that at the surface, while the opposite was found in high latitudes (except the Arctic).

Global temperatures reached their highest observed levels during the 1990s, which include *many* of the warmest years on record (see Figure 11.7). In the southern hemisphere, where the twentieth-century warming was delayed, there has been a more or less continuous warming of the order of 0.4°C. The twentieth-century global warming was approximately 0.5°C, which appears to exceed natural trends (estimated from statistical modelling) of 0.3°C/100 years. It is now widely considered likely that this global warming may be a result of increasing greenhouse gas concentrations, but the causal linkage is not yet definitively established. For example, experiments with general circulation models for doubled CO_2 concentrations (see Figure 11.21) indicate significant amplification of warming in high latitudes in winter. This signal, attributable to feedback effects between surface albedo and temperature associated with reduced snow cover and sea ice, is apparent in northern high latitudes

for the interval between 1955–74 and 1975–94 (see Figure 11.8). A further tendency of the last forty years or so is a decrease in the diurnal temperature range; night-time minimum temperatures increased by 0.8°C during 1951–90 over at least half of the northern land areas compared with only 0.3°C for daytime maximum temperatures. This appears to be mainly a result of increased cloudiness, which, in turn, *may* be a response to increased greenhouse gases and tropospheric aerosols. However, the linkages are not yet adequately determined.

Precipitation records are much more difficult to characterize. Since the mid-twentieth century, decreases dominate much of the tropics and subtropics from North Africa eastwards to Southeast Asia and Indonesia (Figure 11.10). Many of the dry episodes are associated with El Niño events. Equatorial South America and Australasia also show ENSO influences. The Indian monsoon area shows wetter and drier intervals; the drier periods are evident in the early twentieth century and during 1961–90.

The West African records for this century (Figure 11.11) show a tendency for both wet and dry years to occur in runs of up to 10–18 years. Precipitation minima were experienced in the 1910s, 1940s and post-1968, with intervening wet years, in all of sub-Saharan West Africa. Throughout the two northern zones outlined in Figure 11.11, means for 1970–84 were generally <50 per cent of those for 1950–9,

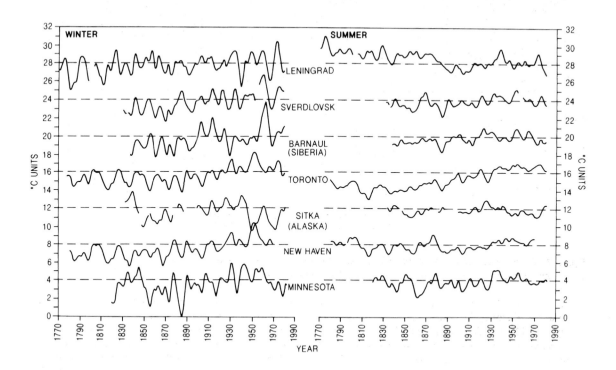

Figure 11.9 Long-term records of winter (December–February) and summer (June–August) temperatures over the past 200 years for a number of northern hemisphere stations, smoothed with a 10-year Gaussian filter.
Source: Jones and Bradley 1992. Reprinted from *Climate Since AD 1500*, R. S. Bradley and P. D. Jones (eds) 1992. Copyright © Routledge, London.

with deficits during 1981–4 equal to or exceeding those of the disastrous early 1970s' drought. The deficits continued in the 1990s. It has been suggested that the severe drought is related to weakening of the tropical easterly jet stream and limited north-ward penetration of the West Africa south-westerly monsoon flow. However, S. E. Nicholson attributes the precipitation fluctuations to contraction and expansion of the Saharan arid core, rather than to north–south shifts of the desert margin. In Australia, rainfall changes have been related to changes in the location and intensity of subtropical anticyclones and associated changes in atmospheric circulation. Winter rainfall decreased in the south-west while summer rainfall increased in south-eastern Australia, particularly after 1950. North-eastern

Australia shows decadal oscillations and large inter-annual variability.

In the middle latitudes, precipitation changes are usually less pronounced. Figure 11.12 illustrates long-term fluctuations for England and Wales and for individual stations. For the country as a whole, decadal departures are only about ±10 per cent. The individual station graphs show that even over relatively short distances there may be considerable differences in the magnitude of anomalies (e.g. Manchester and Oxford). This is further illustrated by the pattern of fluctuations in Central Europe (Figure 11.13). The rising trend in the Bauer series, representing primarily West Germany (the Federal Republic), is distinctly different from the other series. Poland in the early part of its record also differs from the regions to the south.

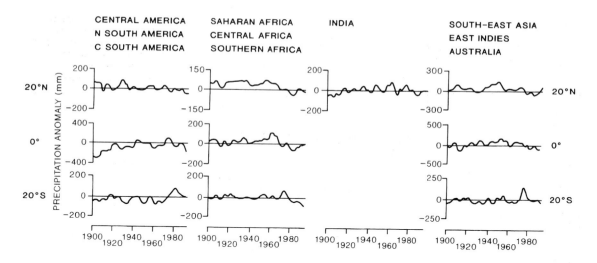

Figure 11.10 Variations of tropical and sub-tropical land area precipitation anomalies relative to 1961–90; 9-point binomially smoothed curves superimposed on the annual anomalies.
Source: Houghton *et al.* 1996.

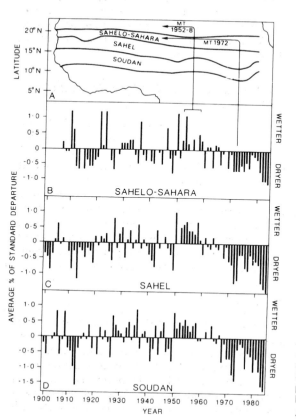

For these countries, the decadal range is approximately ±7 per cent about the longer-term means.

Growing concern for possible anthropogenic causes of the greenhouse effect (see Chapter 1A.4 and D.3, this chapter) has generated a strong body of opinion that the general temperature increase since the middle of the last century, and in particular the temperature increase and supposed unusual climatic fluctuations during the past twenty years are due to changes in atmospheric composition induced by human activities. For example, during the past three decades Britain has experienced several major droughts (1976, 1984, 1989–1992 and 1995); seven severe winter cold spells occurred between 1978 and 1987 (compared with only three in the preceding forty years); and several major windstorms (1987, 1989 and 1990) were recorded. However, the driest 28-month spell (1988–92) recorded in Britain since 1850 was followed by the wettest 32-month interval of the twentieth century. In the United States, the last twenty years has shown a marked increase in the interannual variability of mean winter temperatures

Figure 11.11 Rainfall variations (percentage of standard departure) since 1900 for the Sahelo–Sahara, Sahel and Soudan zones of West Africa. The average positions of the Monsoon Trough (MT) in northern Nigeria during 1952–8 and in 1972 are shown.
Source: From Nicholson 1985.

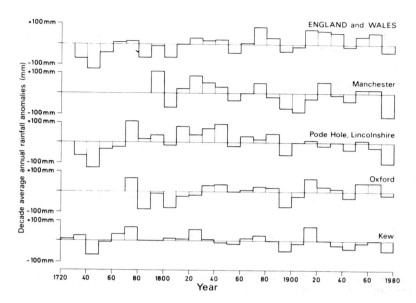

Figure 11.12 Decadal anomalies of average annual rainfall (m) for England and Wales and for four individual stations in England (the 'last decade' includes data up to September 1978).

Source: After Kelly 1980.

and total precipitation, particularly when compared with the preceding twenty years. Worldwide, 1990 and 1995 were the two warmest years on record, 1983 was the year of the most intense El Niño event for a century, and Hurricane Gilbert (1988) was the most severe storm on record.

In the face of these changes and speculations, however, it must be remembered that non-anthropogenic temperature changes of considerable magnitude have been common since the end of the last glaciation, and it is possible that much of the temperature increase during the past 100 years or so may be part of the general temperature upturn following the Little Ice Age (see Figure 11.5B). It is therefore to be expected that global or hemispheric temperatures may vary through several tenths of a degree Celsius within a few years. Neither is it easy to explain the existence of periods of relatively stable global temperatures (e.g. 1861–1920 and mid-1940s–early 1970s) at times of continuing increases of greenhouse gas concentration. Nevertheless, the general warming since the middle of the last century and the most recent climatic events seem to be very much in harmony with the changes in atmospheric composition noted in Chapter 1A.4 (rather than with other forcing mechanisms such as solar variations). In addition, it is recognized that the course of climatic change is undoubtedly complicated by little-known global mechanisms such as the inertial effects of the oceans and the possibly periodic effects of deep ocean currents. Overall, a strong and growing impression remains that natural climatic variabilities may from time to time be merely offsetting an inexorable and increasingly menacing anthropogenic global temperature enhancement.

D POSSIBLE CAUSES OF RECENT CLIMATIC CHANGE

Remarkably, the causes of the observed climatic changes during the last few centuries are less well understood than those of the last glaciation. A multiplicity of possible explanations exists and, indeed, more than one factor is likely to be involved. It is useful to try to distinguish between the natural causes that have existed throughout the earth's history and those that are attributable to human activities in modifying the global environment. However, in many instances – such as the production of tropospheric aerosols – the two operate together.

1 Circulation changes

The immediate cause of the recent climatic fluctuations appears to be the strength of the global wind circulation. The first thirty years of this century saw a pronounced increase in the vigour of the westerlies over the North Atlantic, the north-east trades, the summer monsoon of South Asia and the southern hemisphere westerlies (in summer). Over the North

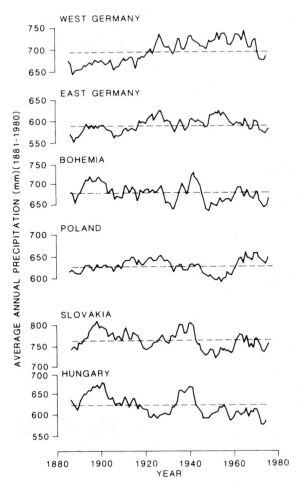

Figure 11.13 Area-averaged annual precipitation data, smoothed by 11-year running means, for central Europe, 1881–1980. The series for West Germany is an arithmetic mean of 12 stations in the Federal Republic and two in the German Democratic Republic (GDR); the GDR series is a mean of 18 stations, the Hungarian series a mean of 10 stations. The series for Poland and Slovakia made double use of weighted arithmetic means of station data, first, for station heights and, second, for the geographical area of height intervals. The Bohemian series was derived by planimetry of monthly isohyetal maps.

Source: From Brádzil *et al.* 1985 (by permission of the Royal Meteorological Society).

which spread westwards. These changes were accompanied by more northerly depression tracks, and this resulted in a significant increase in the frequency of mild south-westerly airflow over the British Isles between about 1900 and 1930, as reflected by the average annual frequency of Lamb's westerly airflow type (see Chapter 8A.3). For 1873–97, 1898–1937, 1938–61 and 1962–95 the figures are 27, 38, 30 and 21 per cent, respectively. Coinciding with the westerly decline, cyclonic and anticyclonic types increased substantially (Figure 11.14). The decrease in westerly airflow during the last thirty-year interval, especially in winter, is linked with greater continentality in Europe. These regional indicators reflect a general decline in the overall strength of the mid-latitude circumpolar westerlies, accompanying an apparent expansion of the polar vortex.

As has already been pointed out (see Chapter 6A.4), global climate is closely related to the position and strength of the subtropical high-pressure cells. It has been estimated that a warming of the Arctic tropopause (winter +10°C; summer +3°C; annual +7°C), without changing equatorial or Antarctic temperatures, would cause an annual shift of the subtropical high-pressure belt from its present average position of 37°N to 41–43°N (i.e. some 100–200 km in summer but as much as 800 km in winter). This would bring drought to the Mediterranean, California, the Middle East, Turkestan and the Punjab, as well as displacing the thermal equator from 6°N to 9–10°N, increasing the desertification in the belt 0–20°S.

2 Energy budgets

The key to these atmospheric variations must be linked to the heat balance of the earth–atmosphere system and this forces us to return to the fundamental energy considerations with which we began this book. The evidence for fluctuations greater than 0.1 per cent in the 'solar constant' is inconclusive, although significant variations apparently do occur in the emission of high-energy particles and ultraviolet radiation during brief solar flares. All solar activity follows the well-known cycle of approximately eleven years, which is usually measured with reference to the period between sunspot maximum and minimum (see Figure 2.2), but numerous attempts to establish secure correlations between sunspot activity and terrestrial climates have produced mostly negative results. Nevertheless, a

Atlantic, these changes consisted of an increased pressure gradient between the Azores high and the Icelandic low, as the latter deepened, and also between the Icelandic low and the Siberian high,

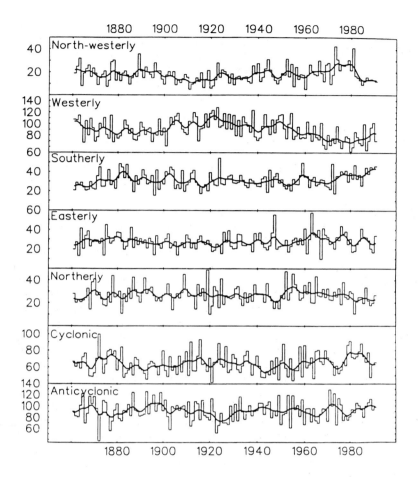

Figure 11.14 Annual totals and ten-year moving averages of the frequency (days) of Lamb's circulation types over the British Isles, 1961–91.

Source: Lamb 1994. Reprinted from the Climate Monitor with permission of the University of East Anglia.

statistical relationship has been found between the occurrence of drought in the western United States over the last 300 years and the approximately 22-year double (Hale) cycle of the reversal of the solar magnetic polarity. Drought areas are most extensive in the two to five years following a Hale sunspot minimum (i.e. alternate eleven-year sunspot minima).

Changes in atmospheric composition may also have modified the atmospheric heat budget. The presence of increased amounts of volcanic dust and sulphate aerosols in the stratosphere is one suggested cause of the 'Little Ice Age'. Major eruptions can result in a surface cooling of perhaps 0.2°C for a few years after the event. Hence, frequent volcanic activity would be required for persistently cooler conditions. Conversely, it is suggested that reduced volcanic activity after 1914 may have contributed in part to the early twentieth-century warming. New

interest in this question has been aroused by eruptions of El Chichón (March 1982) and Mount Pinatabo (June 1991) (see Chapter 1A.4). It has been estimated that huge volcanic eruptions such as these, can, during a given decade, produce a forcing effect on global temperature about one-third as great as that exerted by greenhouse gases – *but in the opposite direction* (i.e. to produce surface cooling). The role of low-level aerosols is also complex. These originate naturally, from wind-blown soil and silt for example, as well as from atmospheric pollution due to human activities (industry, domestic heating and modern transportation).

3 Anthropogenic factors

The growing influence of human activities on the environment is being increasingly recognized and concern over the potential for global warming

Table 11.2 The four categories of climatic variable subject to change.

Variable changed	Scale of effect	Sources of change
Atmospheric composition	Local–global	Release of aerosols and trace gases
Surface properties; energy budgets	Regional	Deforestation; desertification; urbanization
Wind regime	Local–regional	Deforestation; urbanization
Hydrological cycle components	Local–regional	Deforestation; desertification, irrigation; urbanization

caused by such anthropogenic effects is growing. Four categories of climatic variable are subject to change (Table 11.2) and will now be considered in turn.

Changes in atmospheric composition associated with the explosive growth of world population, industry and technology have been described in Chapter 1A.4, and it is clear that these have led to drastic increases in the concentration of greenhouse gases. The tendency of these increases is to increase radiative forcing and global temperatures; the percentage apportionment of radiative forcing of these greenhouse gas increases since the pre-industrial era is summarized in Table 11.3, together with the associated ranges of uncertainty and levels of confidence assigned to each factor. The radiative forcing effect of the minor trace gases is projected to increase steadily. Up to 1960, the cumulative CO_2 contribution since AD 1750 was about 67 per cent of the calculated 1.2 W m^{-2} forcing, whereas for 1980–90 the CO_2 contribution decreased to 56 per cent, with CFCs contributing 24 per cent and methane 11 per cent. For the entire period from AD 1765 to 2050, the CO_2 contribution is projected to range from 4.15 W m^{-2}, out of a 6.5 W m^{-2} total (65 per cent), for a 'business-as-usual' scenario to 2.6 W m^{-2}, out of a 4.0 W m^{-2} total (65 per cent), if emission control policies are implemented rapidly. Figure 11.15 provides a summary of the global annual mean forcing from 1850 to 1992 and the

level of confidence that can be ascribed to each factor.

The recent increase in global temperature forcing by the release of CFCs is particularly worrying. Ozone, which at high altitudes absorbs incoming short-wave radiation, is being dramatically destroyed above 25 km in the stratosphere (see Chapter 1A.4) by emissions of H_2O and NO_x by jet aircraft and by surface emissions of N_2O by combustion and, especially, of CFCs. It is estimated that CFCs are now accumulating in the atmosphere five times faster than they can be destroyed by ultraviolet radiation. Ozone circulates in the stratosphere from low to high latitudes and thus the occurrence of ozone in polar regions is particularly diagnostic of its global concentration. In October 1984, an area of marked ozone depletion (the so-called 'ozone hole') was observed in the lower stratosphere (i.e. 12–24 km) centred on, but extending far beyond, the Antarctic continent. Ozone depletion is always greatest in the Antarctic spring, but in this year the ozone concentration was more than 40 per cent lower than that of October 1977. By 1990, Antarctic ozone concentrations had fallen to about 200 Dobson units in September–October (see Figure 1.9), compared with 400 units in the 1970s. In the extreme years (1993–5), record minima of 116 D.U. have been recorded at South Pole. It has been estimated that, because of the slowness of the global circulation of CFCs and of its reaction with ozone, even a cut in CFC emissions to the level of that in 1970 would not eliminate the Antarctic ozone hole for at least fifty years. Predictions regarding the future climatic effects of changes in atmospheric composition will be treated in section F, this chapter.

The role of tropospheric aerosols in climate forcing and the magnitude of such effects are poorly known (see Figure 11.15). There are four key aerosol types (see Chapter 1A.1), and these have a variety of effects:

Table 11.3 Percentage apportionment of radiative forcing due to greenhouse gas increases.

	CO_2	CH_4	CFCs	Stratospheric H_2O*	N_2O
Since 1795	61	17	12	6	4
1980–90	56	11	24	4	6

Note: *From the breakdown of CH_4.

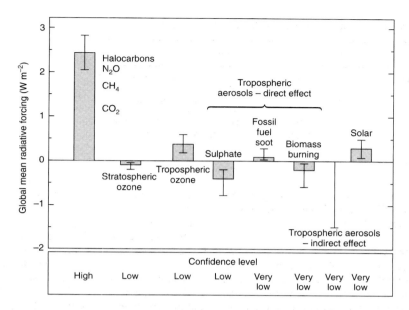

Figure 11.15 Estimates of global mean radiative forcing (W m⁻²) due to changes in concentrations of greenhouse gases and anthropogenic aerosols (pre-industrial to 1992) and to natural changes in solar output (1950–present). The error bars indicate the uncertainty ranges. Confidence levels are also given.

Source: Houghton *et al.* 1996. Reproduced by permission of the IPCC.

1 black carbon – absorbs solar radiation; changes the vertical temperature gradient.
2 water-soluble inorganic species (SO_2, NO_3, NH_4) – backscatter of direct beam solar radiation, indirect effect of CCN on cloud albedo and cloud droplet lifetime.
3 condensed organic species – as (2)
4 mineral dust – as (1), (2) and absorption/emission of infrared radiation.

The global mean forcing exerted by the principal aerosols is as follows:

• sulphate aerosols –0.6 W m⁻².
• biomass burning aerosols –0.8 W m⁻².
• mineral dust –1.0 W m⁻².

However, it should be emphasized that about 88 per cent of the total aerosols (see Table 1.2) input is of natural origin. The indirect effects of cloud condensation nuclei (CCN) from anthropogenic sources are undetermined. Nevertheless, a ±15 per cent change of CCN within marine stratus clouds, which cover about 25 per cent of the earth, could change the global energy balance by ±1 W m⁻².

Indirect anthropogenic factors, such as increasing population pressures leading to overgrazing and forest clearance, may increase desertification which also contributes to the increase of wind-blown soil. The 'dust-bowl' years of the 1930s in the United States and the African Sahel drought since 1972 illustrate this. Evidence from the Soviet Union shows a sharp rise in dust-fall on mountain snow-fields from 1930 to the 1960s, and atmospheric turbidity has increased by 57 per cent over Washington, DC, over the period 1905–64, and by 85 per cent over Davos, Switzerland (1920–58). The presence of particles in the atmosphere increases the backscatter of short-wave radiation, thereby increasing the planetary albedo and causing cooling, but the effect on infrared radiation is one of surface warming. The net result is complicated by the surface albedo. Man-made aerosols cause net warming over snow and ice and most land surfaces, but cooling over the oceans, which have a low albedo. Natural aerosols probably cause general cooling. The overall effect on global surface temperature remains uncertain.

Changes in surface albedo occur naturally with season, but climatic forcing is also caused by anthropogenic vegetation changes. Human effects on vegetation cover have a long history. Deliberate burning of vegetation by Aborigines in Australia has been practised for perhaps 40,000 years. However, significant deforestation began in Eurasia during Neolithic times (*c.* 5000 BP), as evidenced by the appearance of agricultural species and weeds. Deforestation expanded in these areas between about AD 700 and 1700 as populations slowly grew, but it did not take place in North America until the westward movement of settlement in the eighteenth and nineteenth centuries. During the last half-century extensive deforestation has occurred in the tropical rainforests of South-east Asia, Africa and South America. Estimates of current tropical deforestation suggest losses of 10^5 km^2/year, out of a total tropical forest area of 9×10^6 km^2. This annual figure is more than half the total land surface at present under irrigation and twice the annual loss of marginal land to desertification. Forest destruction causes an increase in albedo of perhaps 10 per cent locally, with consequences for surface energy and moisture budgets. However, the large-scale effect of deforestation in temperate and tropical latitudes on global surface albedo is estimated to be <0.001. It should also be noted that deforestation is difficult to define and monitor; it can refer to loss of forest cover with complete clearance and conversion to a different land use, or species' impoverishment without major changes in physical structure. The term desertification, applied in semi-arid regions, creates similar difficulties. The process of vegetation change and associated soil degradation is not solely attributable to human-induced changes but is triggered by natural rainfall fluctuations.

Deforestation affects world climate in two main ways – first, by altering the atmospheric composition and, second, by affecting the hydrological cycle and local soil conditions:

1 Forests store great amounts of carbon dioxide, so buffering the carbon dioxide cycle in the atmosphere. The carbon retained in the vegetation of the Amazon basin is equivalent to at least 20 per cent of the entire atmospheric CO_2. Destruction of the vegetation would release about four-fifths of this to the atmosphere, about one-half of which would dissolve in the oceans, but the other half would be added to the 16 per cent increase of atmospheric CO_2 already observed this century. The effect of this would be to accelerate the increase of world temperatures. A further effect of tropical forest destruction would be to reduce the natural production of nitrous oxide. Tropical forests and their soils produce up to one-half of the world's nitrous oxide, which helps to destroy stratospheric ozone. Any increase in ozone would warm the stratosphere, but lower global surface temperatures.

2 Dense tropical forests have a great effect on the hydrological cycle through their high evapotranspiration and their reduction of surface runoff (about one-third of the rain never reaches the ground, being intercepted and evaporating off the leaves). Forest destruction decreases evapotranspiration, atmospheric humidity, local rainfall amounts, interception, effective soil depth, the height of the water table and surface roughness (and thereby atmospheric turbulence and heat transfer). Conversely, deforestation increases the seasonality of rainfall, surface runoff, soil erosion, soil temperatures and surface albedo (and therefore near-surface air temperatures). All these tendencies operate to degrade existing primary and secondary tropical forests into savanna. Models designed to simulate the operation of Amazonian forests having a 27°C air temperature and a mean monthly rainfall of 220 mm (falling in four showers every third day, each lasting 30 minutes at an intensity of 0.003 mm s^{-1}) predict that their degradation to savanna conditions would lead to a decrease of evapotranspiration by up to 40 per cent, an increase of runoff from 14 per cent of rainfall to 43 per cent, and an average increase of soil temperature from 27 to 32°C.

E MODEL STRATEGIES FOR THE PREDICTION OF CLIMATE CHANGE

Long-range forecasting, which has been described earlier (see Chapter 7I), looks forward for only a month or season. However, concern regarding the longer-term results of possible anthropogenic effects has prompted much recent research directed towards global climatic change during the coming century. For this purpose, global mathematical models (see Chapter 6E) have been employed, assuming radiative forcing by changes in atmospheric composition and based on assumptions of feedback within the atmosphere–earth–ocean–ice system (see B.2, this chapter).

More generally, three modelling strategies can be identified, as described below.

1 *Black box modelling* involves the statistical extrapolation of an historical time series (e.g. of

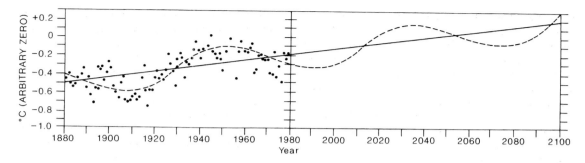

Figure 11.16 Plot of mean annual temperatures for the northern hemisphere for the period 1880–1980, to which have been fitted a linear trend and a sinusoidal oscillation. The latter have been extended to attempt a prediction to the year 2100.

past temperatures) into the future, without real concern for the mechanisms involved. Figure 11.16 for example, shows a crude attempt to describe the variation of mean annual temperature in the northern hemisphere between 1880 and 1980 in terms of a linear increase on which is superimposed a sinusoidal change. This has been projected to the end of the next century to predict a temperature increase of 0.6°C in 120 years. This prediction disregards the poorness of fit of the simple mathematical relationships to the complex scatter of points representing temperature and also makes the very questionable assumption that during the next 120 years the mix and weighting of the factors controlling temperature in the northern hemisphere will be unchanged over those that operated during the past century. It clearly represents a primitive and unsatisfactory attempt at prediction.

2 *Grey box modelling* is based on the assumption that the effects of the most important controlling variables can be identified, measured and superimposed to produce a satisfactory simulation of a past record, and that the resultant mathematical model is capable of being projected meaningfully into the future for predictive purposes. Figure 11.17 shows one such attempt involving the variation of atmospheric CO_2 concentration and sulphate aerosols in the atmosphere. The difficulties of this type of approach to prediction are, again, quite obvious and fall into two categories:

(a) The fitting of a composite curve to the past record. Have all the important variables been included? Have their actual and relative effects on hemispheric temperature been

accurately determined? Is it legitimate to assume that the effect of each variable operates independently of that of the others to allow the cumulative effect to be calculated?

(b) As well as the problems listed above, it is clear that the accurate prediction of future changes in CO_2 concentration is difficult and that of sulphate and other aerosols is even more uncertain.

Thus grey box predictions, although giving a greater impression of scientific exactitude than black box ones, are also far from satisfactory. It is clear that the complexity of the global climate system is such as to rule out the use of statistically or empirically based models, with their total reliance on past data, in favour of numerical ones.

3 *White box modelling* is based on a detailed understanding of the structure and operation of the earth–atmosphere–ocean system, such that its possible future states can be simulated by applying assumed forcing mechanisms, particularly anthropogenic ones. This numerical model building thus involves bringing together locational, temporal and attribute information into a database that allows hypotheses regarding climatic processes and interactions to be simulated. Mathematical simulation of the white box kind, unlike black or grey box varieties, is potentially very powerful but involves the need for a very complete understanding of the variables, states, feedbacks, transfers and forcings of the complex system (i.e. the parameters), together with the laws of physics of the atmosphere and oceans on which they are based. The most powerful of such models are the coupled atmosphere–ocean General Circulation Models

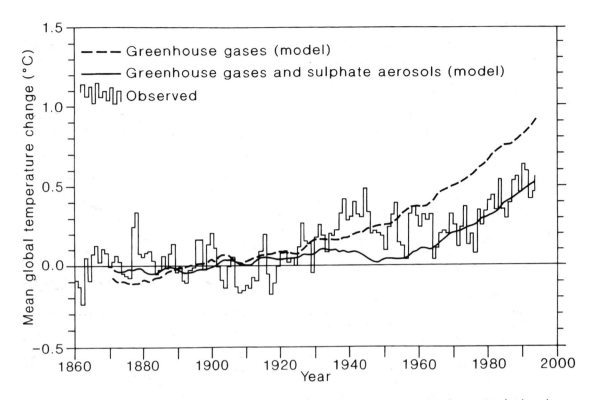

Figure 11.17 A grey box model of global mean warming from 1850 to 1990 using greenhouse gas forcing alone and greenhouse gas plus anthropogenic sulphate aerosol forcing. It is clear that the latter provides much the better fit to the observed temperature record.

Source: Houghton *et al.* 1996. Reproduced by permission of the IPCC.

(GCM), which treat three-dimensional parcels of air and water as they move horizontally and vertically (see Chapter 6E).

The equations used in the GCM to describe the physical and dynamic relationships that determine climate are commonly of two types – governing (1–4) and storage (5–6). These include:

1 The equation of motion (i.e. conservation of momentum).
2 The equation of continuity (i.e. conservation of mass or the hydrodynamic equation).
3 The equation of continuity for atmospheric water vapour (i.e. conservation of water vapour).
4 The equation of energy (i.e. the thermodynamic equation derived from the first law of thermodynamics).
5 The equation of state (i.e. the hydrostatic equation).
6 The equation of surface pressure tendency.

Such complex models require very detailed calibration and adjustment. They are thus run with actual empirical data relating to past decades, in the light of which adjustments are made to model parameters (such as those relating to fluxes, feedbacks and forcing) in order to improve the overall performance of the model. Some such adjustments may be quite large, as in the past has been the case in respect of sea ice extent.

The advantage of the GCM is that it attempts to take into account the total structure and dynamism of the earth–atmosphere–ocean system. Its disadvantages are its obvious failure to do this completely, the huge amount of data required to establish, test and run (i.e. force) this white box predictive model and also, extremely importantly, the lack of knowledge regarding the future forcing conditions.

F THE IPCC MODELS

The most sophisticated coupled atmosphere–ocean General Circulation Model in current use was developed under the aegis of the Intergovernmental Panel on Climate Change (IPCC). The first report by the panel in 1990 was naturally dominated by predictions of the increase in greenhouse gas concentrations during the twenty-first century (Figure 11.18). These assumed increases were classified under four possible scenarios:

BAU   business as usual. This envisages modest controls and efficiency improvements over industrial emissions; uncontrolled agricultural emissions; depletion of tropical forests at the present rate. This would result in the world CO_2 concentration being more than twice the pre-industrial era level by the year 2070.

B   A shift towards lower-carbon fuels and natural gas. Large increases in efficiency, stringent carbon monoxide controls and a reversal of deforestation envisaged.

C   Scenario B; plus immediate phasing out of CFCs and limitation of agricultural emissions; plus shift to renewable and nuclear energy sources in the second half of the next century.

D   Shift to renewable and nuclear energy sources immediately and CO_2 emissions reduced to 50 per cent of 1985 levels by 2050.

Figure 11.19 gives a 1990 prediction of radiative forcing mechanisms during the period 2000–2050 for the different scenarios, showing the heavy stress on the effects of CO_2.

It was suggested that to stabilize the concentration of greenhouse gases at 1990 levels would require the following *percentage reductions* in emissions resulting from human activities: CO_2 >60 per cent; CH_4 15–20 per cent; N_2O 70–80 per cent; CFCs 70–85 per cent. With 'business as usual' the concentration of CO_2 by the year 2100 was expected to be 236 per cent of that in 1980; of CH_4, 242 per cent; and of CFC-11, 350 per cent (see Figure 11.18). Assuming the most severe cutback scenario (D), the resulting percentage changes in the concentration of greenhouse gases by the year 2100 (compared with 1980) were expected to be: CO_2, 129 per cent; CH_4, 91 per cent; CFC-11, 57 per cent.

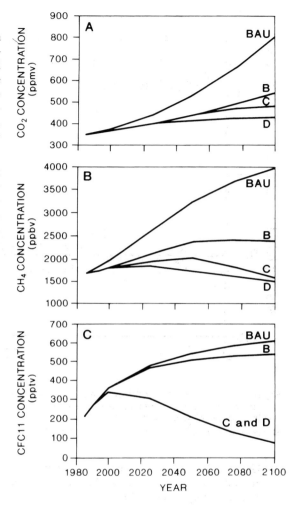

Figure 11.18 Predicted changes of CO_2, CH_4 and CFC-11 between 1980 and 2100 with 'business as usual' (BAU) and with three other scenarios (B–D).

Source: Houghton *et al.* 1990. Reproduced by permission of the IPCC.

Note: Units are in parts per million by volume (ppmv), parts per billion (ppbv) and parts per trillion (pptv), respectively.

The IPCC Scientific Assessment for 1990 predicted, with business as usual, an overall global temperature rise of 3.5°C by the end of the next century (Figure 11.20A), with a huge error band. The shortcomings of this model were highlighted when a test applied to the last 100 years of record 'predicted' a temperature rise of 1°C, whereas the actual temperature increase was only half that figure. Twenty numerical simulation models were run to equilibrium (reached at *c.* AD 2030) for the

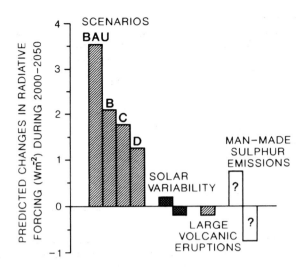

Figure 11.19 Comparison of different radiative forcing mechanisms for a 50-year period in the future. The greenhouse gas forcings are for the period 2000–2050, using the four policy scenarios. Forcings due to changes in solar variability and sulphur emissions could be either positive or negative over the two periods.
Source: Houghton *et al.* 1990. Reproduced by permission of the IPCC.

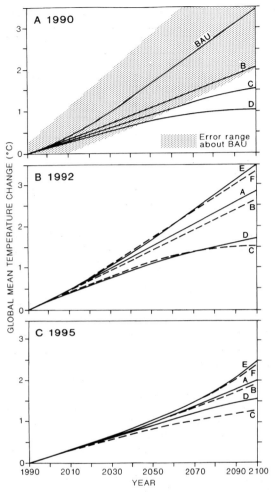

Figure 11.20 IPCC simulated increases in global mean temperature (above that for 1990) for the period 1990–2100. A: the 1990 simulation showing four scenarios: business as usual and three others, which are described in the text. B: the 1992 simulation showing six scenarios (see Table 11.5 in the text), which differ from those in 1990. C: the 1995 simulation with scenarios A–F being the same as those in the 1992 simulation.

Sources: Houghton *et al.* 1990, 1992 and 1996. Reproduced by permission of the IPCC.

period 1850–2050, assuming a doubling of CO_2 during that period but employing only a relatively simple model of the upper 50–100 m of the ocean. The major predictions were as follows (Table 11.4):

1 A general warming of the earth's surface and troposphere (by up to more than 12°C in some high southern latitudes), and a cooling of the stratosphere and upper atmosphere as a result of the necessity for equilibrium between the incoming solar radiation and the outgoing terrestrial radiation. The stratosphere and mesosphere would cool by about 10°C and the thermosphere by 50°C, for CO_2 doubling, perhaps causing more frequent noctilucent clouds in the polar mesosphere.

2 A consequent increase in the overall temperature lapse rate, making the atmosphere more turbulent, increasing the frequency and intensity of depressions, tropical storms and hurricanes, and making mid-latitude winter storm tracks more stable in their positions.

3 Stronger warming of the earth's surface and troposphere in higher latitudes (especially at about 60° latitude) in late autumn and winter (by 4–8°C) (Figure 11.21).

4 A tropical warming of less than the global mean (i.e. 2–3°C).

5 Amplified warming over northern mid-latitude continents in summer (by 4–6°C).

6 Increased precipitation in high latitudes and in the tropics throughout the year (by 10–20 per

Table 11.4 Predictions of seasonal changes at specific locations.

Location	Season	Temperature	Rainfall	Soil moisture
Central N. America	Winter	+2–4°C	+0–15%	–
	Summer	+2–3°C	–5–10%	–15–20%
India	Winter ⎫	+1–2°C	–	–
	Summer ⎭	annually	+5–15%	+5–10%
Sahel	Winter ⎫	+1–3°C ⎫	Changes marginal and	
	Summer ⎭	annually ⎭	regularly variable	
Southern Europe	Winter	+2°C	Some increase	–
	Summer	+2–3°C	–5–15%	–15–25%
Australia	Winter	+2°C	–	–
	Summer	+1–2°C	+10%	–

cent), and in mid-latitudes in winter. This increase is mostly associated with zones of lower-level convergence (i.e. mid-latitude storm tracks and the ITCZ).

7 Small changes only in the dry subtropics.

8 Soil moisture increases in winter and decreases in summer in northern mid-latitude continents.

9 A reduced extent of sea ice extent and thickness.

Two years later, the IPCC model was adjusted to take account of the potential cooling effect of anthropogenic sulphate aerosols; other changes in greenhouse forcing; a possible slowdown of the thermohaline deep ocean circulation expected to accompany global warming; an upper mixed ocean layer 90 m deep with vertical mixing processes operating at 1 cm² s⁻¹; revised estimates of the thermal differences between land and sea; and improved feedback assumptions, particularly relating to the effects of cloud cover. Its predictions were based on six forcing scenarios (A–F) (Figure 11.20B) operating over two time periods, 1990–2025 and 1990–2100 (Table 11.5). Scenarios C and D were based on low estimates of world economic growth, forest clearances and CO_2 emissions, whereas scenarios E and F were based on high estimates. The best estimate of global mean temperature rise was given by the moderate assumptions underlying scenarios A and B, giving an estimate of mean global warming by the year 2100 of around 2.5°C (range 1.5–4.5°C), assuming a doubling of CO_2.

These predictions were broadly similar to those associated with the 1990 model, but more moderate.

Table 11.5 Six scenarios (A–F) of assumptions regarding economic growth, forest clearance and CO_2 emissions used in making the 1992 IPCC forecasts.

	Economic growth (%)	Total forest cleared (million ha)	Annual CO_2 emissions (GtC)	
A	2.9	678	12.2	1990–2025
	2.3	1,447	20.3	1990–2100
B	2.9	678	11.8	1990–2025
	2.3	1,447	19.1	1990–2100
C	2.0	675	8.8	1990–2025
	1.2	1,343	4.6	1990–2100
D	2.7	402	9.3	1990–2025
	2.0	651	10.3	1990–2100
E	3.5	678	15.1	1990–2025
	3.0	1,447	35.8	1990–2100
F	2.9	725	14.4	1990–2025
	2.3	1,686	26.6	1990–2100

Sources: Houghton et al. 1992.

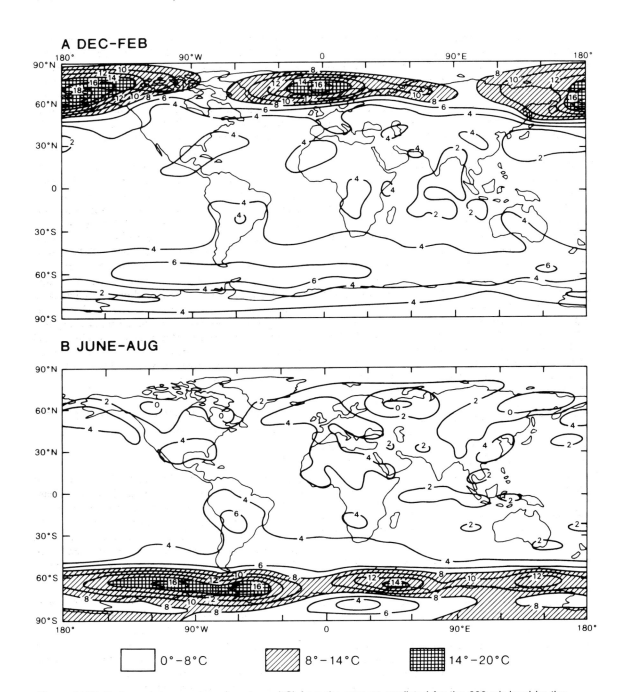

Figure 11.21 Surface air temperature departures (°C) from the present predicted for the 990-mb level by the NCAR Community Climate Model coupled with a mixed-layer ocean model, assuming a doubled CO_2 concentration.

Source: From Meehl and Washington 1990.

They included a generally greater warming of the land than of the oceans in winter; a minimum warming, or even cooling, around Antarctica and in the northern North Atlantic associated with deep ocean mixing; somewhat less warming in high northern latitudes in late autumn and winter associated with reduced sea ice and snow cover than was predicted in 1990; little warming over the Arctic in summer; and a reduction in diurnal temperature range over land in most seasons and most regions. Precipitation predictions associated with the enhanced global hydrological cycle included increased rainfall over southern Europe, increased precipitation and soil moisture in high latitudes in winter, and, most important, some modification of view regarding Asian summer monsoonal rainfall, which, far from being enhanced, might even be decreased due to the anthropogenic aerosol effect. More consideration was given to possible extreme climatic events than was possible in 1990. Predictions included an increase in extremely high temperatures and a decrease in winter days with extremely low temperatures; a decrease in diurnal temperature variability in certain regions; increased diurnal precipitation variability over a few areas, such as north-west North America in winter; an increase in precipitation intensities and extreme rainfall events; and possibly more frequent or severe drought periods in warmer climates.

Since 1992, there have been more minor changes to the model, including those to anthropogenic aerosols and non-CO_2 greenhouse gases. Overall, anthropogenic aerosols are believed to reduce the effects of greenhouse gas forcing by an average of one-third (range 20–40 per cent). The incorporation of an improved knowledge of ocean dynamics into the models has, however, been limited by the fact that, although the effects of mesoscale ocean eddies can now be well modelled, their computational complexity limits their full integration into coupled atmosphere–ocean models. Figure 11.15 summarizes the anthropogenic and natural forcing agents for 1850 to 1992. Model forcings for the period from the present to 2100 have been based on projections of these factors into the future. Anthropogenic forcing scenarios are constantly being refined to take account of assumptions regarding population growth, economic growth, land use changes, technological developments, energy availability and fuel mix.

The current estimate for global warming by the year 2100, assuming a doubling of CO_2, now stands at about 2°C (range 1–3.5°C) under scenario A (see Figure 11.20C). However, it must be held in mind that natural lags in the system response mean that in all probability only 50–90 per cent of the final equilibrium climate change will have occurred by 2100. Several different spatial simulation variants of the model are in use and Figure 11.22 shows one model estimate of seasonal temperature changes from 1880–9 to 2040–9 and Figure 11.23 the projected annual temperature change predicted by another for 1995–2100.

Tests are demonstrating considerable improvements in model replications of lumped global temperature changes that result from using assumptions that include both greenhouse gas and anthropogenic aerosol forcings (see Figure 11.17). Large-scale global patterns of past temperatures and precipitation rates can also be replicated quite well. However, due to the four-dimensional nature of the climatic responses (i.e. including latitude, longitude, elevation and time), it is clear that it is difficult to achieve a good match between modelled and predicted (i.e. subsequently observed) climatic patterns using forcings that cannot be reliably anticipated. Thus model predictions of global patterns of surface temperature change by, for example, the year 2100 are still quite speculative (Figure 11.23).

G OTHER ENVIRONMENTAL IMPACTS OF CLIMATE CHANGE

1 Sea level

The mechanisms influencing sea level over the globe are extremely complex. Present sea level is not easy to define, estimates of sea levels over the past 100 years are difficult to make, and predictions over the next 100 years are highly speculative. Sea level changes are influenced by the following mechanisms (those of short time scales – i.e. tens of years – italicized):

1 *Changes in ocean water mass – e.g. exchanges with glaciers; changes in the atmosphere–earth–ocean–ice–water distribution.*
2 *Changes in ocean water volume – e.g. thermal expansion and contraction; salinity changes; changes in atmospheric pressure.*
3 Changes in earth crustal levels.
    (a) Tectonic – e.g. rise of ocean ridges; sea-floor subsidence; plate movements.
    (b) Isostatic – e.g. tectonic loading; *ice and water loading.*
4 Changes in the global distribution of water.

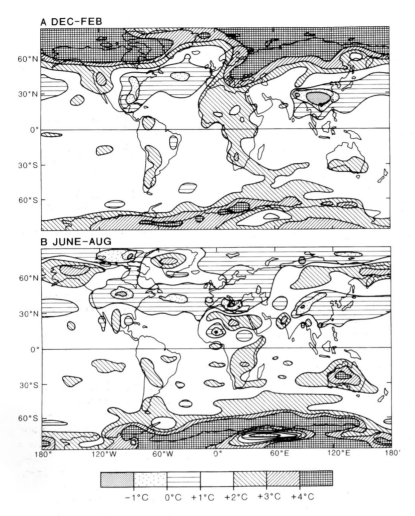

Figure 11.22 Simulations of changes in surface temperatures (°C) from 1880–89 to 2040–49 employing greenhouse gas and anthropogenic aerosol forcing.

Source: Houghton et al. 1996. Reproduced by permission of the IPCC.

(a) Terrestrial rotation effects.
(b) Global axis changes.
(c) Terrestrial gravity variations.
(d) Changes in the attraction of sun and moon.
(e) *Changes in the velocity of ocean currents.*

Over the past 100 years, the general global sea level has risen by 10–25 cm or more, accelerating during the century. This rise has been attributed proportionately to the following causes:

1 Thermal expansion of ocean waters (2–7 cm; i.e. possibly 50 per cent or more). This is difficult to estimate due to lack of knowledge regarding oceanic circulations, such that estimates vary from 30 to 60 per cent.

2 Glacier and small ice cap melting (2–5 cm; i.e. possibly 30 per cent). Estimates of this contribution to sea level rise go as high as 48 per cent.

3 Greenland ice cap melting (very indeterminate). This could be as great as 25 per cent and as little as 5 per cent.

4 Antarctic ice sheet melting. This is very uncertain; the Antarctic ice sheet is a large and complex system with its own internal mechanisms and a mass balance that changes slowly. Some workers believe the balance is positive, which would offset sea level rise. This source has probably not yet contributed greatly to the global sea level rise but may do so in the future.

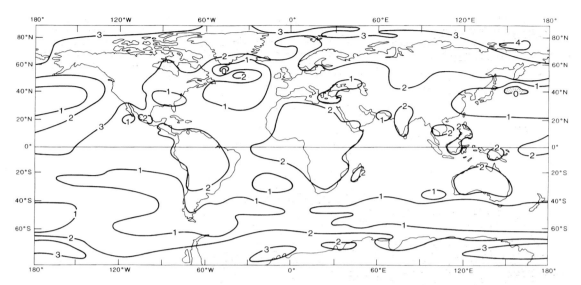

Figure 11.23 Predicted pattern of increases in annual surface temperature (°C) produced by an atmosphere–ocean coupled model for the period from the present to the year 2100, based on the forcing assumptions of a doubling of CO_2 plus increases in anthropogenic aerosols.
Source: Houghton *et al.* 1996. Reproduced by permission of the IPCC.

Figure 11.24 shows the 1990 (business as usual) and the 1992 (scenario A) IPCC model predictions, together with their error and uncertainty ranges. The 1992 revision of predicted forcing has lowered the predicted sea level rise by the year 2100 from about 66 to 49 cm. The choice between the emission scenarios A–F would seem to make little difference to the estimated rise in sea level, particularly for the period until the year 2050. This is because of the long lags in the responses of the huge oceans and ice sheet masses, but, by the same token, sea level will continue to rise long after atmospheric forcing mechanisms have stabilized. Uncertainties regarding sea level rise are still considerable, mainly because of our lack of knowledge concerning the behaviour of the large ice sheets, especially Antarctica. There is even the possibility that increased global warming may introduce a tendency for sea level to *fall* because of increased snow accumulation rates in high latitudes. Another outside possibility is that a rise in sea level might cause the West Antarctic ice sheet to be buoyed up and melt bodily (not just around the edges, as in the past) and cause a further catastrophic sea level rise but spread over several hundred years. Final considerations are the possible effects of extreme sea level events (such as tides, waves and storm surges), but these are extremely difficult to predict.

2 Snow and ice

The effects of twentieth-century climate change on global snow and ice cover are apparent in many ways, but the responses differ widely as a result of the different factors and time scales involved. Snow cover is essentially seasonal, related to storm system precipitation and temperature levels. Sea ice is also seasonal around much of the Antarctic continent and in the marginal seas of the Arctic Ocean (see Figure 8.39), but the central Arctic has thick multi-year ice. Seasonal (or first-year) ice grows and decays in response to ocean surface temperature, radiation balance, snowfall and ice motion due to winds and currents. The loss of multi-year ice from the Arctic is mainly through ice export. Glacier ice builds up from the net balance of snow accumulation and summer melt (ablation), but glacier flow transports ice towards the terminus, where it may melt or calve into water. In small glaciers, the ice may have a residence time of 10s–100s of years, but in ice caps and ice sheets this increases to 10^3–10^6 years.

In the twentieth century, there has been a rapid retreat of most of the world's glaciers. Glaciers in the North Atlantic area retreated during the 1920s to the mid-1960s and since 1980, due largely to temperature increases, which have the effect of

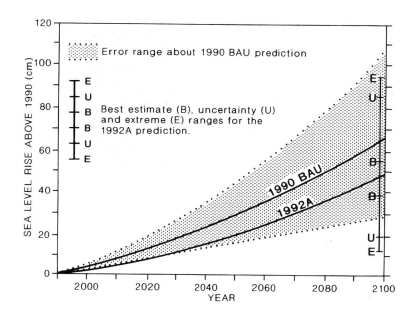

Figure 11.24 Predictions of sea level rise to the year 2100, together with error, best estimate, uncertainty and extreme ranges.

Sources: Houghton et al. 1990 and 1996. Reproduced by permission of the IPCC.

lengthening the ablation season with a corresponding raising of the snowline. In the last 10–15 years, the freezing level in the troposphere has risen in the inner tropics by 100–150 m, contributing to rapid ice loss on equatorial glaciers in East Africa and the northern Andes. Also in the last decade or so, some glaciers in maritime climates (western North America and Scandinavia) have shown advances, due to heavier snowfalls during warmer winters. Major alpine glaciers in many areas of the world have lost mass and shrunk since the late nineteenth century, whereas smaller ones show short-term fluctuations in response to climatic variability (Figure 11.25). Projections for AD 2050 suggest that one quarter of the present glacier mass may disappear.

Another tendency illustrating world warming is the retreat of Arctic sea ice. Ports in the Arctic remained free of ice for longer periods during the 1920s–50s for example. This trend was reversed in the 1960s–70s, but since the late 1980s the summer extent of Arctic ice has decreased, with large reductions, particularly in the Eurasian Arctic, in 1990, 1993 and 1995. There appears to be no general trend in Antarctic ice extent, although comprehensive records began only with all-weather satellite coverage in 1973. Sea ice in both polar regions is expected to shrink and thin with continued warming, but modelling of these processes remains rudimentary.

Major iceberg calving events have occurred along the Ross Ice Shelf and on the Larsen Ice Shelf of the Antarctic Peninsula, but the causes of such calving are more related to the long history of the ice shelves and ice dynamics than to recent climatic trends.

Snow cover extent shows perhaps the clearest indication of a response to recent temperature trends. Northern hemisphere snow cover has been mapped by visible satellite images since 1966. Compared with the 1970s–mid-1980s, annual snow cover since 1988 has shrunk by about 10 per cent. The decrease is most pronounced in spring and is well-correlated with springtime warming (Figure 11.26). Winter snow extent shows little or no change. Nevertheless, annual snowfall in North America north of 55°N increased during 1950–90. Much work remains to be done to analyse station snowfall and snow depth records for other countries, particularly since these variables are difficult to measure and the design of gauges and wind shields has changed through time. Scenarios for AD 2050 suggest a shorter snow cover period in North America, with a decrease of 70 per cent over the Great Plains. In alpine areas, snow lines will rise by 100–400 m, depending on precipitation.

3 Hydrology

The difficulties of deducing the possible effects of climate change on hydrological regimes stem from

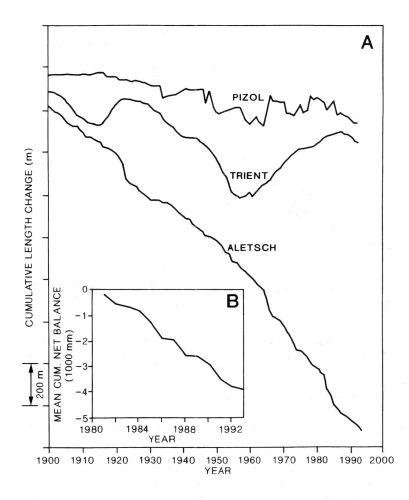

Figure 11.25 Trends in cumulative net mass balance for 35 glaciers in 11 mountain ranges in North America, Europe and Africa for 1980–93 (B) and cumulative length changes since 1900 of characteristic glacier types in the Swiss Alps (A). Pizol is a cirque glacier; Trient is a medium-size mountain glacier; and the Aletsch is a large valley glacier.

Source: Haeberli 1995. Reproduced by permission of *Geografia Fisica e Dinamica Quaternaria*.

attempts to adapt the essentially large-scale climatic predictions derived from General Circulation Models to the smaller catchment scales appropriate to hydrological modelling; from errors in the climatic and hydrological data; and from converting climatic inputs into hydrological responses.

The climatic changes predicted by current modelling may be expected to lead to:

1 A more vigorous world hydrological cycle.
2 More severe droughts and/or floods in some places and less severe ones in others.
3 An increase in precipitation intensities with possibly more extreme rainfall events.
4 Greater hydrological effects of climate change in drier areas than in wetter ones.
5 An increase in overall potential evapotranspiration.

6 An increase in the variability of river discharges along with that of rainfall.
7 A shift of runoff peak times from spring to winter in continental and mountain areas if snowfall decreases.
8 The greatest falls in lake water levels in dry regions with high evaporation.

The implication that the hydrological impacts of climate change will be greatest in currently arid or semi-arid regions may well mean that the more severe runoff events there will be particularly destructive in terms of soil erosion.

4 Vegetation

An increase in CO_2 may be expected to enhance global plant growth up to a saturation value of

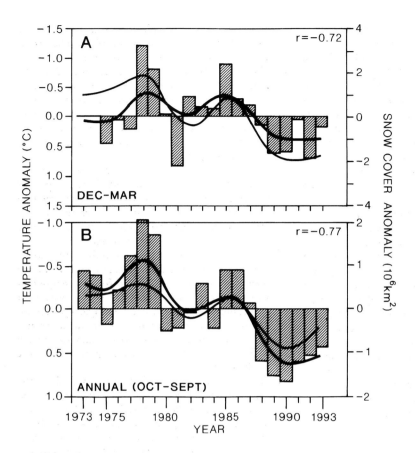

Figure 11.26 Recent changes in (A) spring (Dec.–March) and B) annual (Oct.–Sept.) snow cover extent in the northern hemisphere and associated changes in air temperature (scale inverted). Yearly anomalies in snow cover extent (Greenland excluded) are shown as bars. The smooth curves show smoothed snow cover anomalies (thick) or temperature anomalies (thin). The correlations between the two variables are –0.72 and –0.77, respectively, for spring and annual values.

Source: Houghton *et al.* 1996. Reproduced by permission of the IPCC.

possibly around 1,000 ppmv, when a saturation limit may be reached. However, deforestation could decrease the biosphere's capacity to act as a carbon sink. A sustained increase of only 1°C can cause considerable change in tree growth, regeneration and species extent. Species migrate only slowly but, eventually, extensive forested areas may change to new vegetation types, and it has been estimated that 33 per cent of the present forest area could be affected, with as high as 65 per cent of the boreal zone being subject to change. Alpine tree lines appear to be quite resistant to climatic fluctuations. However, surveys of plant species on peaks in the European Alps indicate an upward migration of alpine plants by 1–4 m per decade during this century.

Tropical forests are likely to be affected more by human deforestation than by climate change. However, decreases of soil moisture are particularly destructive in hydrologically marginal areas. In the Amazon, climatic predictions support the idea of increased convection, and therefore of rainfall, in its western equatorial portion, where present rainfall is most abundant. Because of the particularly high temperature rises predicted for the high northern latitudes, boreal forests are expected to be strongly affected by their advance northwards into tundra regions. This may produce the positive feedback effect of further regional warming because of the lower albedo of forests during the snow season. Climate change over the next 100 years may be expected to exert the least changes on temperate forests.

Wetlands at present cover 4–6 per cent of the land surface, having been reduced by human activities by more than half during the past century. Climate change will affect wetlands mainly by altering their hydrological regimes. Although general predictions are difficult to make, it is believed that eastern China, the USA and southern Europe will suffer a natural decline in the area of wetlands during the next century, decreasing the methane flux to the atmosphere.

Drier regions may be expected to be more profoundly affected than wet ones. Rangelands (including grasslands, shrublands, savannas, hot and cold deserts, and tundra) occupy 51 per cent of the terrestrial land surface, contain 36 per cent of the world's total carbon in their biomass and support half the world's livestock. The lower-latitude rangelands are most at risk both because an increase in CO_2 (increasing the carbon/nitrogen ratio) will decrease the nutrient value of forage and because the increasing frequency of extreme events will cause environmental degradation. Most deserts are likely to become hotter and not significantly wetter, increases in rainfall being generally associated with increased storm intensity. Greater wind speeds and evaporation may be expected to increase wind erosion, capillary rise and salinization of soils. Central Australia is one of the few places where desert conditions may improve.

It is obvious from the foregoing that a major effect of climate change involving global warming is that desiccation and soil erosion will increase in currently semi-arid regions, rangelands and savannas adjacent to the world's deserts. This will increase the current rate of desertification, which is proceeding at six million hectares per year partly due to high rainfall variability and partly to unsuitable human agricultural activities such as overgrazing and over-intensive cultivation.

H POSTSCRIPT

Although we can view our ability to anticipate climate change with greater confidence than when the first IPCC Report appeared in 1990, many problems, uncertainties and research needs continue to face the long-term forecaster. These include (not in order of importance):

- The development of more refined *forcing* scenarios through a better understanding of forcing mechanisms (past, present and future) such as those relating to economic growth, forest clearances, land use changes, sulphate aerosols, carbonaceous aerosols generated by biomass burning, and radiative trace gases other than CO_2 (e.g. methane and ozone). The incorporation of past anthropogenic forcings will generate more realistic warming in model simulations – so avoiding the present 'cold start' problem.
- The acquiring of a better understanding of *feedback* processes, notably those involving clouds and the surface radiation budget, interactions between the land biosphere and the carbon cycle and between climate and atmospheric chemistry, and those involving sea ice and vegetation.
- The need for more information regarding the distribution of *clouds and their radiative effects*. It is clear that high and low clouds may have very different radiative effects and that under certain conditions cloud albedo may have the capacity to counterbalance much of the potential CO_2 warming effect.
- The need to increase the resolution of global climate models so that *small-scale physical processes* can be represented (e.g. those relating to clouds). This is part of the move to improve scale-coupling between global climate models and regional and smaller-scale models.
- The gaining of more understanding of certain *oceanic processes*, including the heat flux at the ocean surface, the upwelling diffusion energy balance in the oceans, the role of the oceans in absorbing CO_2, especially by biological processes, and the role of the oceans in heat transport and its role in causing delays to the climate system achieving equilibrium with respect to forcing mechanisms.
- The need for more information regarding the *processes involved in the coupling between the atmosphere and the oceans*. A particular problem concerns the coupling of sea-surface temperatures (a vital part of the atmospheric model) with ocean surface energy flux, fresh water supplies and momentum or wind stress conditions (important parts of the ocean model). Attempts to link large parts of complex systems generally leads to 'drift' – in this case a tendency of the model's climate system to drift into a new, unrealistic, mean state.
- The *links between land-surface and atmospheric processes*, including the hydrological cycle and interactions between ice sheets and glaciers, on the one hand, and climate, on the other, need to be examined.
- The imperative to make a clear distinction between climate change related to anthropogenic causes and the *natural variability* of the space–time climate structure, which has nothing to do with anthropogenic forcing. Information on natural climatic variability can be gained from instrumental data, paleoclimatic reconstructions and numerical models. An important feature of natural variability is rapid changes, which are at present little understood. These changes are

characteristic of complex non-linear systems that are rapidly forced. Examples of such changes include rapid circulation changes in the North Atlantic and feedbacks associated with terrestrial ecosystem changes.

- The need for *the systematic collection of long-term instrumental and proxy observations* to do with solar output, atmospheric energy balance components, hydrological cycles, ocean characteristics, atmosphere and ocean coupling, and ecosystem changes.

- Lastly, because numerical models of the world climate system are becoming ever more complex, there is a growing requirement for *massive computing facilities*; sometimes, weeks of expensive computer time are needed to run a single climatic simulation to achieve an equilibrium state.

SUMMARY

Changes in climate involve factors both external to and within the climate system. External ones include solar variability, astronomical effects on the earth's orbit, and volcanic activity. Internal factors include natural variability within the climate system, and feedbacks between the atmosphere, ocean and land surface. During the last century, human-induced climatic change on local and global scales has become a reality, primarily through changes in atmospheric composition and surface properties.

Climatic changes on geological time scales involve continental drift, volcanic activity and possible changes in solar output. Over the last few million years, glacial–interglacial cycles appear to have been strongly controlled by astronomical variations in the earth's orbit, although atmosphere–ocean–cryosphere feedbacks must also be involved in amplifying the initial changes in solar radiation.

During the twentieth century there has been a significant average global temperature increase of 0.5°C, greatest in the higher middle latitudes with warming, particularly during the periods 1920–40s and since the early 1970s; the early 1990s have included several of the warmest years on record. Diurnal temperature ranges show a decrease over the last few decades and some regions are experiencing more frequent extreme conditions for temperature levels and precipitation amounts. Precipitation trends are less clear, particularly in the mid-latitudes, but the precipitation of dry subtropical regions has tended to oscillate widely. Climatic behaviour during the past twenty years has tended to support a growing impression that the anthropogenically induced increase of greenhouse gases is permanently affecting global climate.

Possible causes of climatic change are examined from the point of view of the global atmosphere–earth–ocean–ice system and with respect to forcing and feedback mechanisms. Whereas longer-term changes are probably due to astronomical forcing mechanisms, short-term changes (i.e. the last 100 years) appear to be more obviously linked to anthropogenic factors. These are mainly changes in atmospheric composition, including aerosol loading, depletion of ozone and destruction of world vegetation. Natural and anthropogenic aerosol effects appear to be particularly important, but their net effect remains uncertain.

Climate predictions are being made using a variety of modelling strategies, of which coupled atmosphere–ocean GCMs are the most sophisticated. Predictions covering the next 100 years assuming a doubling of CO_2 with 'business as usual' indicate a global temperature increase of about 2°C by the year 2100, together with sea level rises of about 50 cm. The magnitude of such predictions, based on computer modelling, are still uncertain, however, and are subject to large error bands due to our restricted knowledge of the operations of the global atmosphere–earth–ocean–ice system.

Alpine glaciers show a general twentieth-century shrinkage, but even the sign of the mass balance of Greenland and Antarctica is uncertain. Continental snow cover has decreased over the last decade, especially in spring, and Arctic sea ice was also less extensive in recent summers. Hydrological models suggest that spring snow melt runoff would occur earlier and the variability in rainfall amounts would intensify floods and droughts. Vegetation cover and croplands will also be affected in the long term, but human-induced changes will predominate. Subtropical semi-arid areas are most likely to be affected by climatic trends.

continued

Critical research needs include better data on cloud cover and radiation, ocean processes and their atmospheric coupling, feedback processes in general, and the relationships between large-scale and local/regional-scale processes and phenomena.

APPENDIX 1
Climate classification

The purpose of any classification system is to obtain an efficient arrangement of information in a simplified and generalized form. Thus, climate statistics can be organized in order to describe and delimit the major types of climate in quantitative terms. Obviously, no single classification can serve more than a limited number of purposes satisfactorily and many different schemes have therefore been developed. Some schemes merely provide a convenient nomenclature system, whereas others are an essential preliminary to further study. Many climatic classifications, for instance, are concerned with the relationships between climate and vegetation or soils, but surprisingly few attempts have been made to base a classification on the direct effects of climate on man.

Only the basic principles of the four groups of the most widely known classification systems are summarized here. Further information may be found in the listed references.

A GENERIC CLASSIFICATIONS RELATED TO PLANT GROWTH OR VEGETATION

The numerous schemes that have been suggested for relating climate limits to plant growth or vegetation groups rely on two basic criteria – the degree of aridity and of warmth.

Aridity is not simply a matter of low precipitation, but of the 'effective precipitation' (i.e. precipitation minus evaporation). The ratio of rainfall/temperature has been used as such an index of precipitation effectiveness, on the grounds that higher temperatures increase evaporation.

The work of W. Köppen is the prime example of this type of classification. Between 1900 and 1936, he published several classification schemes involving considerable complexity in their full detail. Nevertheless, the system has been used extensively in geographical teaching. The key features of Köppen's final classification are temperature criteria and aridity criteria.

Temperature criteria

Five of the six major climate types are recognized on the basis of monthly mean temperature.

A Tropical rainy climate: coldest month >18°C.
B Dry climates.
C Warm temperate rainy climates: coldest month between −3° and +18°C, warmest month >10°C.
D Cold boreal forest climates: coldest month <−3°, warmest month >10°C. Note that many American workers use a modified version with 0°C as the C/D boundary.
E Tundra climate: warmest month 0–10°C.
F Perpetual frost climate: warmest month <0°C.

The arbitrary temperature limits stem from a variety of criteria, the supposed significance of the selected values being as follows: the 10°C summer isotherm correlates with the poleward limit of tree growth; the 18°C winter isotherm is critical for certain tropical plants; and the −3°C isotherm indicates a few weeks of snow cover. However, these correlations are far from precise! The criteria were determined from the study of vegetation groups defined on a physiological basis (i.e. according to the internal functions of plant organs) by De Candolle in 1874.

Aridity criteria

Precipitation	Steppe (BS)/ desert (BW) boundary	Forest/steppe boundary
Winter precipitation maximum	$r/t = 1$	$r/t = 2$
Precipitation evenly distributed	$r/(t + 7) = 1$	$r/(t + 7) = 2$
Summer precipitation maximum	$r/(t + 14) = 1$	$r/(t + 14) = 2$

where: r = annual precipitation (cm)
t = mean annual temperature (°C)

The criteria imply that, with winter precipitation, arid (desert) conditions occur where $r/t < 1$, semi-arid conditions where $1 < r/t < 2$. If the rain falls in summer, a larger amount is required to offset evaporation and maintain an equivalent total of effective precipitation.

Subdivisions of each major category are made with reference, first, to the seasonal distribution of precipitation (the most common of which are: f = no dry season; m = monsoonal, with a short dry season and heavy rains during the rest of the year; s = summer dry season; w = winter dry season) and, second, to additional temperature characteristics. (For the B climates: h = mean annual temperature >18°C; k = mean annual temperature <18°C (warmest month >18°C); k' = mean annual temperature (and warmest month) <18°C.)

Figure A1.1A illustrates the distribution of the major Köppen climate types on a hypothetical continent of low and uniform elevation. Experiments using GCMs with and without orography show that, in fact, the poleward orientation of BS/BW climatic zones inland from the west coast is largely determined by the western cordilleras. It would not be found on a low, uniform-elevation continent.

The Köppen climatic classification has recently proved useful in evaluating the accuracy of GCMs

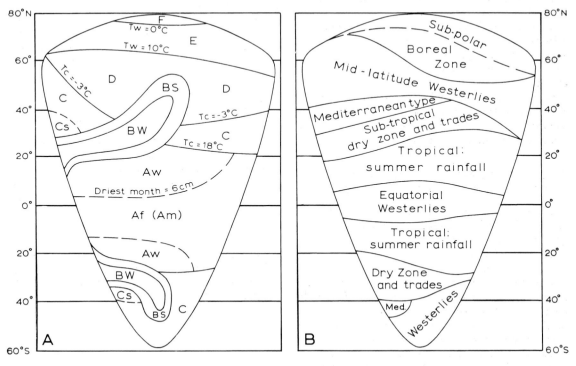

Figure A1.1 A: the distribution of the major Köppen climatic types on a hypothetical continent of low and uniform elevation. Tw = mean temperature of warmest month; Tc = mean temperature of coldest month. B: the distribution of Flohn's climatic types on a hypothetical continent of low and uniform elevation (see Note 1).
Source: From Flohn 1950.

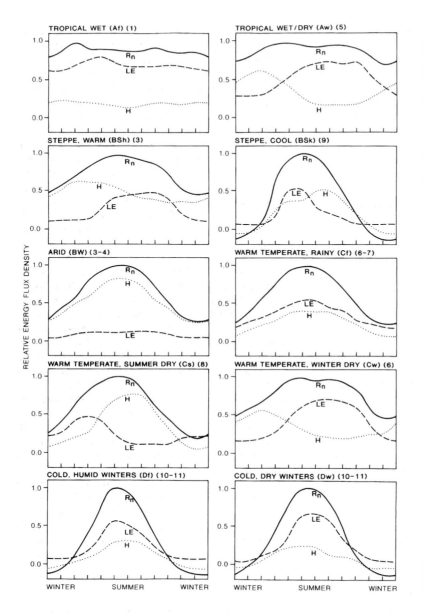

Figure A1.2 Characteristic annual energy balances for ten different climatic types (Köppen symbols and Strahler classification numbers shown). The ordinate shows energy flux density normalized with the maximum monthly net all-wavelength radiation (R_n) normalized with the maximum monthly value as unity. The abscissa intervals indicate the months of the year with summer in the centre. H = turbulent flux of sensible heat and LE = turbulent latent heat flux to the atmosphere.

Source: From Kraus and Alkhalaf 1995. From International Journal of Climatology, copyright © John Wiley & Sons Ltd. Reproduced with permission.

in simulating present climatic patterns and as a convenient index of change for climate scenarios projected for CO_2 doubling.

The ten main Köppen climate categories each have distinct annual energy budget regimes. Figure A1.2 illustrates these.

C. W. Thornthwaite introduced a complex, empirical classification in 1931. An expression for *precipitation efficiency* was obtained by relating measurements of pan evaporation to temperature and precipitation. The second element of the classification is an index of *thermal efficiency* $(T - E)$, expressed by the positive departure of monthly mean temperatures from freezing point. The index is thus the annual sum of $(t - 32)/4$ for each month. On this scale, zero is 'frost climate' and over 127 is 'tropical'. Maps of the distribution of these climatic provinces in North America and over the world have been published, but the classification is now largely of historical interest.

A new classification and map of world climate types has been prepared by W. Lauer and colleagues based on thermal and hygric thresholds for both natural vegetation and agricultural crops. The limits of the four primary zones (tropical, subtropical, mid-latitude and polar regions) are determined from a radiation index (duration of daily sunshine hours). Climate types are then based on a thermal index (temperature sums) and a moisture index, which takes account of the difference between monthly precipitation and potential evaporation.

B ENERGY AND MOISTURE BUDGET CLASSIFICATIONS

Thornthwaite's most important contribution was his second (1948) classification. It is based on the concept of potential evapotranspiration and the moisture budget (see Chapters 3C and 8B.3c). Potential evapotranspiration (PE) is calculated from the mean monthly temperature (in °C), with corrections for day length. For a 30-day month (12-hour days):

$$PE \text{ (in cm)} = 1.6(10t/I)^a$$

where: I = the sum for 12 months of $(t/5)^{1.514}$
a = a further complex function of I.

Tables have been prepared for the easy computation of these factors.

The monthly water surplus (S) or deficit (D) is determined from a moisture budget assessment, taking into account stored soil moisture (Thornthwaite and Mather 1955; Mather 1985). A moisture index (Im) is given by:

$$Im = 100(S - D)/PE$$

This allows for a variable soil moisture storage according to vegetation cover and soil type, and it permits the evaporation rate to vary with the actual soil moisture content. The average water balance is calculated through a book-keeping procedure. For each month, the mean values of the following variables are determined in turn: PE, potential evapotranspiration (from the relationship given above); precipitation minus PE; Ws, soil water storage, a value assumed appropriate for that soil type at field capacity. Ws is reduced as the soil dries (ΔWs). AE is actual evapotranspiration. There are two cases: $AE = PE$, when Ws is at field capacity, or $(P - PE) > 0$; otherwise $AE = P + \Delta Ws$. The monthly moisture deficit, D, or surplus, S, is determined from $D = (PE - AE)$, or $S = (P - PE) > 0$, when $Ws \leq$ field capacity. Monthly deficits or surpluses are carried forward to the subsequent month.

A novel feature of the system is that the thermal efficiency is derived from the PE value, because this itself is a function of temperature. The climate types defined by these two factors are shown in Table A1.1, both elements being subdivided according to the season of moisture deficit or surplus and the seasonal concentration of thermal efficiency.

The system has been applied to many regions, although no world map has yet been published. In tropical and semi-arid areas the method is not very satisfactory, but in eastern North America, for example, vegetation boundaries have been shown to coincide reasonably closely with particular PE values. This classification, unlike that of 1931, Köppen's and many others, does not use vegetation boundaries to determine climatic ones.

M. I. Budyko in the Soviet Union developed a similar, but more fundamental approach using net radiation rather than temperature (see Chapter 3A). He related the net radiation available for evaporation from a wet surface (R_o) to the heat required to evaporate the mean annual precipitation (Lr).

Table A1.1 Thornthwaite's climatic classification.

Im (1955 system)*		PE		Climatic type
		cm	in	
>100	Perhumid (A)	>114	>44.9	Megathermal (A′)
20 to 100	Humid (B_1 to B_4)	57 to 114	22.4 to 44.9	Mesothermal (B'_1 to B'_4)
0 to 20	Moist subhumid (C_2)	28.5 to 57	11.2 to 22.4	Microthermal (C'_1 to C'_2)
−33 to 0	Dry subhumid (C_1)	14.2 to 28.5	5.6 to 11.2	Tundra (D′)
−67 to −33	Semi–arid (D)	<14.2	<5.6	Frost (E′)
−100 to −67	Arid (E)			

Note: *Im = 100(S − D)/PE is equivalent to 100(r/PE − 1), where r = annual precipitation.

This ratio R_0/Lr (where L = latent heat of evaporation) is called the *radiational index of dryness*. It has a value of less than unity in humid areas and greater than unity in dry areas. Boundary values, with R_0/Lr in parentheses, are: Desert (>3.0); Semi-desert (2.0–3.0); Steppe (1.0–2.0); Forest (0.33–1.0); Tundra (<0.33). By way of comparison with the revised Thornthwaite index ($Im = 100(r/PE - 1)$), it may be noted that $Im = 100(Lr/R_0 - 1)$ if all the net radiation is used for evaporation from the wet surface (i.e. none is transferred into the ground by conduction or into the air as sensible heat). A general world map of R_0/Lr has appeared, but over large parts of the earth there are few measurements of net radiation.

Energy fluxes have also been used by Terjung and Louie (1972) to categorize the magnitude of energy input (net radiation and advection) and outputs (sensible heat and latent heat), and their seasonal range. On this basis, sixty-two climatic types are distinguished (in six broad groups), and a world map is presented.

C GENETIC CLASSIFICATIONS

The genetic basis of large-scale (or macro-) climates is the atmospheric circulation, and this can be related to regional climatology in terms of wind regimes or air masses.

One system was proposed in 1950 by H. Flohn. His major categories, which are based on the global wind belts and the precipitation characteristics, are as follows:

1 Equatorial westerly zone: constantly wet.
2 Tropical zone, winter trades: summer rainfall.
3 Subtropical dry zone (trades or subtropical high pressure): dry conditions prevail.
4 Subtropical winter-rain zone (mediterranean type): winter rainfall.
5 Extra-tropical westerly zone: precipitation throughout the year.
6 Subpolar zone: limited precipitation throughout the year.
6a Boreal, continental subtype: summer rainfall; limited winter snowfall.
7 High polar zone: meagre precipitation; summer rainfall, early winter snowfall.

It will be noted that temperature does not appear explicitly in the scheme. Figure A1.1B shows the distribution of these types on a hypothetical continent. Rough general agreement between these types

and those of Köppen's scheme is apparent. Note that the boreal subtype is restricted to the northern hemisphere and that the subtropical zones do not occur on the eastern side of a land mass. Flohn's approach has much to commend it as an introductory teaching outline. Although no world map of the distribution of these zones has been published, two maps prepared along similar lines by E. Neef and E. Kupfer were presented by Flohn in 1957.

Another simple, but extremely effective, genetic classification of world climates has been proposed by Strahler. He makes a major tripartite diversion into:

1 Low-latitude climates, controlled by equatorial and tropical air masses.
2 Mid-latitude climates, controlled by both tropical and polar air masses.
3 High-latitude climates, controlled by polar and arctic air masses.

These are subdivided into fourteen climatic regions, to which is added that of highland climates (Table A1.2). Figure A1.3 shows the world distribution of these fifteen regions, and Figure A1.4 gives mean monthly climatic data for representatives of thirteen of them.

D CLASSIFICATIONS OF CLIMATIC COMFORT

The body's thermal equilibrium is determined by metabolic rate, heat storage in body tissues, radiative and convective exchanges with the surroundings, and evaporative heat loss by sweating. In indoor conditions, about 60 per cent of body heat is lost by radiation and 25 per cent by evaporation from the lungs and skin. Outdoors, additional heat is lost by convective transfer due to the wind. Human comfort depends primarily on air temperature, relative humidity and wind speed (Buettner 1962)(Figure A1.5). Comfort indices have been developed by physiological experiments in test chambers. They include measures of heat stress and windchill.

Windchill describes the cooling effect of low temperature and wind on bare skin. It is commonly expressed via a windchill equivalent temperature. For example, a 15 m s^{-1} wind with an air temperature of $-10°C$ has a windchill equivalent of $-25°C$. A windchill of $-30°C$ denotes a high risk of frostbite and corresponds to a heat loss of approximately 1,600 W m^{-2}. Nomograms to determine windchill have been proposed as well as other formulae that include the protective effect of clothing.

Table A1.2 Strahler's climatic classification.

Climate name	Köppen symbol		Air mass source regions and frontal zones, general characteristics

Group 1: Low-latitude climates (controlled by equatorial and tropical air masses)

1 Wet equatorial climate 10°–10°S lat. (Asia 10°–20°N)	Af Am	Tropical rain forest climate, and Tropical rain forest climate, monsoon type	Equatorial trough (convergence zone) climates are dominated by warm, moist tropical maritime (mT) and equatorial (mE) air masses yielding heavy rainfall through convectional storms. Remarkably uniform temperatures prevail throughout the year.
2 Trade wind littoral climate 10°–25°N and S lat	Af-Am	Included in climate 1	Tropical easterlies (trades) bring maritime tropical (mT) air masses from moist western sides of oceanic subtropical high-pressure cells to give narrow east-coast zones of heavy rainfall and uniformly high temperatures. Rainfall shows strong seasonal variation.
3 Tropical desert and steppe climates 15°–35°N and S lat.	BWh BSh	Desert climate, hot, and Steppe climate, hot	Source regions of continental-tropical (cT$_s$) air masses in high-pressure cells at high level over lands astride the Tropics of Cancer and Capricorn give arid to semi-arid climate with very high maximum temperatures and moderate annual range.
4 West-coast desert climate 15°–30°N and S lat.	BWk BWh	Desert climate, cool, and Desert climate, hot (*BWn* in earlier versions, *n* meaning frequent fog)	On west coasts bordering the oceanic subtropical high-pressure cells, subsiding maritime tropical (mT$_s$) air masses are stable and dry. Extremely dry, but relatively cool, foggy desert climates prevail in narrow coastal belts. Annual temperature range is small.
5 Tropical wet-dry climate 5°–25°N and S lat.	Aw	Tropical rainy climate, savanna	Seasonal alternation of moist mT or mE air masses with dry cT air masses gives climate with wet season at time of high sun, dry season at time of low sun.

Group 2: Middle-latitude climates (controlled by both tropical and polar air masses)

6 Humid subtropical climate 20°–35°N and S lat.	Cfa Cwa	Temperate rainy (humid mesothermal) climate, hot summers Temperate rainy (humid mesothermal) climate, dry winter, hot summer	Subtropical, eastern continental margins dominated by moist maritime (mT) air masses flowing from the western sides of oceanic high-pressure cells. In high-sun season, rainfall is copious and temperatures high. Winters are cool with frequent continental polar (cP) air mass invasions. Frequent cyclonic storms
7 Marine west-coast climate 40°–60°N and S lat.	Cfb Cfc	Temperate rainy (humid mesothermal) climate, warm summers, and same but cool, short summers	Windward, middle-latitude west coasts receive frequent cyclonic storms with cool, moist maritime polar (mP) air masses. These bring much cloudiness and well-distributed precipitation, but with winter maximum. Annual temperature range is small for middle latitudes.
8 Mediterranean climate 30°–45°N and S lat.	Csa Csb	Temperate rainy (humid mesothermal) climate, dry, hot summer, and same, but dry, warm summer	This wet-winter, dry-summer climate results from seasonal alternation of conditions causing climates 4 and 7; mP air masses dominate in winter with cyclonic storms and ample rainfall, mT air masses dominate in summer and extreme drought. Moderate annual temperature range.

Table A1.2 Strahler

Climate name	Köppen symbol		Air mass source regions and frontal zones, general characteristics
9 Middle-latitude desert and steppe-climates 35°–50°N and S lat.	BWk BWk′ BSk′ BSk′	Desert climate, cool same, but cold; and Steppe climate, cool same, but cold	Interior, middle-latitude deserts and steppes of regions shut off by mountains from invasions of maritime air masses (mT or mP), but dominated by continental tropical (cT) air masses in summer and continental polar (cP) air masses in winter. Great annual temperature range; hot summers, cold winters.
10 Humid continental climate 35°–60°N lat.	Dfa Dfb Dwa Dwb	Cold, snowy forest (humid microthermal) climate, moist all year, hot summers; and same, but warm summers; also Cold, snowy forest (humid microthermal) climate, dry winters, hot summers; and same, but warm summers	Located in central and eastern parts of continents of middle latitudes, these climates are in the polar front zone, the 'battle ground' of polar and tropical air masses. Seasonal contrasts are strong and weather highly variable. Ample precipitation throughout the year is increased in summer by invading maritime tropical (mT) air masses. Cold winters are dominated by continental polar (cP) air masses invading frequently from northern source regions.

Group 3: High-latitude climates (controlled by polar and arctic air masses)

Climate name	Köppen symbol		Air mass source regions and frontal zones, general characteristics
11 Continental sub-arctic climate 50°–70°N lat.	Dfc Dfd Dwc Dwd	Cold, snowy forest (humid microthermal) climate, moist all year, cool summers; and same, but very cold winters; also Cold, snowy forest (humid microthermal) climate, dry winter, cool summer; and same, but very cold winter	This climate lies in source region of continental polar (cP) air masses, which in winter are stable and very cold. Summers are short and cool. Annual temperature range is enormous. Cyclonic storms, into which maritime polar (mP) air is drawn, supply light precipitation, but evaporation is small and climate is therefore effectively moist.
12 Marine sub-arctic climate 50°–60°N and 45°–60°S	ET	Polar, tundra climate	Located in the arctic frontal zones of the winter season, these windward coasts and islands of subarctic latitudes are dominated by cool mP air masses. Precipitation is relatively large and annual temperature range small for so high a latitude.
13 Tundra climate north of 55°N south of 50°S		Polar, tundra climate	The arctic coastal fringes lie along a frontal zone, in which polar (mP, cP) air masses interact with arctic (A) air masses in cyclonic storms. Climate is humid and severely cold with no warm season or summer. Moderating influence of ocean water prevents extreme winter severity as in climate 11.
14 Ice-cap climate (Greenland, Antarctica)	EF	Polar climate, perpetual frost	Source regions of arctic (A) and antarctic (AA) air masses situated upon the great continental ice-caps have climate with annual temperature average far below all other climates and no above-freezing monthly average. High altitudes of ice plateaus intensify air mass cold.
Highland climates			Cool to cold moist climates, occupying high-altitude zones of the world's mountain ranges, are localized in extent and not included in classification system.

Source: From A. N. Strahler, *Physical Geography* (3rd edn), 1969. Copyright © John Wiley and Sons, Inc. Reproduced by permission.

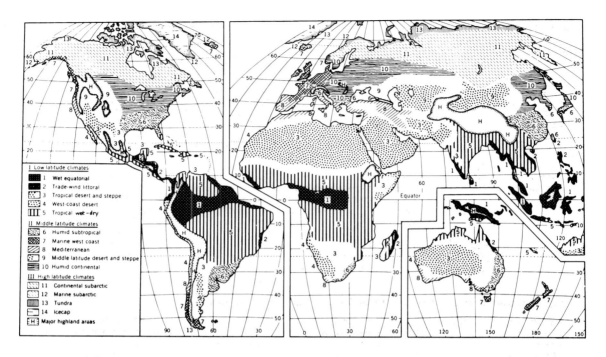

Figure A1.3 Simplified world map showing the distribution of Strahler's genetic climatic regions.
Source: From A. N. Strahler, *Physical Geography* (3rd edn), 1969. Copyright © John Wiley and Sons, Inc. Reproduced by permission.

Heat discomfort is assessed from measurements of air temperature and relative humidity. The US National Weather Service uses a Heat Index based on a measure of *apparent temperature* developed by R. G. Steadman for normally clothed individuals. The value of apparent temperature in the shade (TAPP) is approximately:

$$TAPP = -2.7 + 1.04\,T_A + 2.0e - 0.65\,V_{10}$$

where T_A = midday temperature (°C), e = vapour pressure (mb), V_{10} = 10 m wind speed (m s^{-1}). Warnings are issued in the United States when the apparent temperature reaches 40.5°C for more than three hours/day on two consecutive days.

Another approach measures the thermal insulation provided by clothing. One 'clo' unit maintains a seated/resting person comfortable in surroundings of 21°C, relative humidity below 50 per cent and air movement of 10 cm s^{-1}. For example, the clo values of representative clothing are: tropical wear ≤0.25, light summer clothes 0.5, typical male/female day wear ~1.0, winter wear with hat and overcoat 2.0–2.5, woollen winter sportswear ~3.0, and polar clothing 3.6–4.5. The clo units correlate closely with windchill and inversely with the heat index.

Figure A1.5A illustrates the ranges of human comfort, discomfort and danger (heat stroke, frostbite) for one type of bioclimatic chart developed in the United States by V. Olgyay. The frequency of mean climatic conditions outside the *comfort range* in New York and Phoenix is shown in Figure A1.5B, indicating the need for supplementary heat or clothing and air conditioning or evaporative cooling systems. A bioclimatic classification that incorporates estimates of comfort using temperature, relative humidity, sunshine and wind speed data has been proposed for the United States by W. H. Terjung (1966).

BIBLIOGRAPHY

Bailey, H. P. (1960) A method for determining the temperateness of climate, *Geografiska Annaler* **42**, 1–16.

Basile, R. M. and Corbin, S. W. (1969) A graphical method for determining Thornthwaite climatic classification, *Ann. Assn. Amer. Geog.* **59**, 561–72.

Budyko, M. I. (1956) *The Heat Balance of the Earth's Surface* (trans. by N. I. Stepanova), US Weather Bureau, Washington, DC.

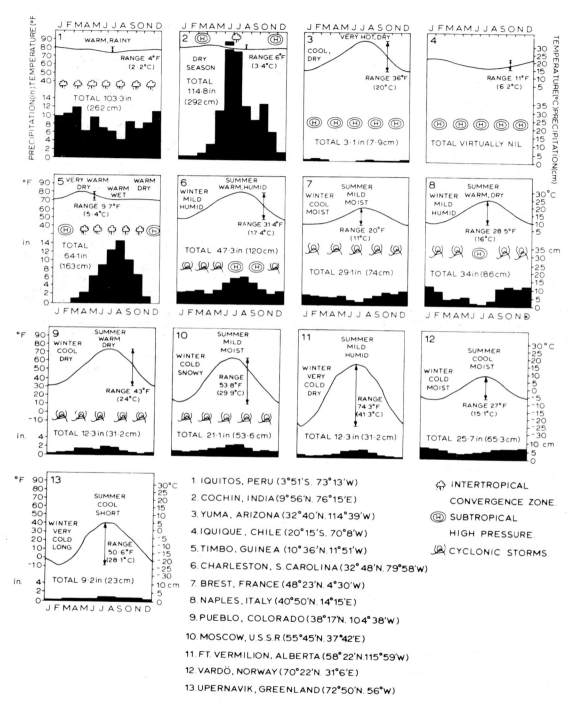

Figure A1.4 Climatic data for representative stations in thirteen of Strahler's climatic regions.
Source: Mostly after Strahler 1969.

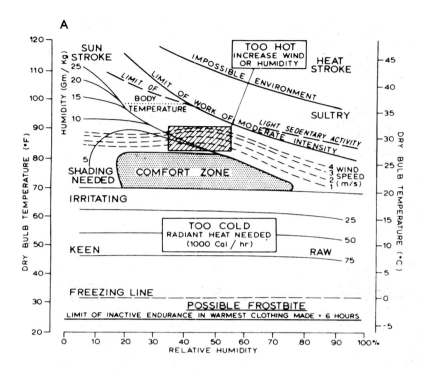

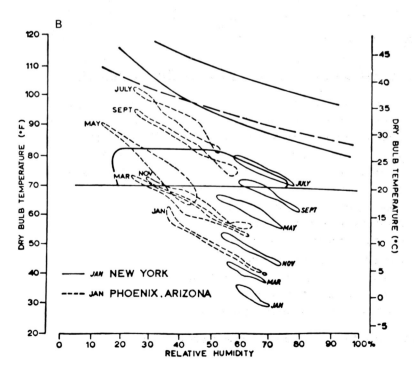

Figure A1.5 The human relevance of climatic ranges. A: atmospheric comfort, discomfort and danger for inhabitants of temperate climatic zones. Within the stippled limits of dry-bulb temperature and relative humidity the body feels comfortable at elevations of less than 300 m with customary indoor clothing, and performing sedentary or light work. Outside these limits corrective measures are necessary to restore the feeling of comfort. Below the comfort zone radiant heat is needed, above it the skin temperature can be lowered by increasing the evaporation and heat convection either by raising the wind speed (for humid air) or by an increase of atmospheric moisture (for dry air). B: mean daily dry-bulb temperatures and relative humidities during alternate months for New York City and Phoenix, Arizona, plotted on the 'comfort chart', indicating the need for both central heating and air conditioning.

Source: After Olgyay 1993.

Budyko, M. I. (1974) *Climate and Life* (trans. by D. H. Miller), Academic Press, New York (508 pp.).

Buettner, K. J. (1962) Human aspects of bioclimatological classification; in Tromp, S. W. and Weihe, W. H. (eds) *Biometeorology*, Pergamon, Oxford and London, pp. 128–40.

Carter, D. B. (1954) Climates of Africa and India according to Thornthwaite's 1948 classification, *Publications in Climatology* 7(4), Laboratory of Climatology, Centerton, NJ.

Chang, J.-H. (1959) An evaluation of the 1948 Thornthwaite classification, *Ann. Assn. Amer. Geog.* 49, 24–30.

Dixon, J. C. and Prior, M. J. (1987) Wind-chill indices – a review, *Met. Mag.* 116, 1–17.

Flohn, H. (1950) Neue Anschauungen über die allgemeine Zirkulation der Atmosphäre und ihre klimatische Bedeutung, *Erdkunde* 4, 141–62.

Flohn, H. (1957) Zur Frage der Einteilung der Klimazonen, *Erdkunde* 11, 161–75.

Gentilli, J. (1958) *A Geography of Climate*, University of Western Australia Press, pp. 120–66.

Gregory, S. (1954) Climatic classification and climatic change, *Erdkunde* 8, 246–52.

Kraus, H. and Alkhalaf, A. (1995) Characteristic surface energy budgets for different climate types. *Internat. J. Climatol.* 15, 275–84.

Lauer, W., Rafiqpoor, M. D. and Frankenberg, P. (1996) Die Klimate der Erde. Eine Klassifikation auf ökophysiologicher Grundlage auf der realen Vegetation. *Erdkunde*, 50(4), 275–300.

Lohmann, U. *et al.* (1993) The Köppen climate classification as a diagnostic tool for general circulation models. *Climate Res.* 3, 277–94.

Mather, J. R. (1985) The water budget and the distribution of climates, vegetation and soils, *Publications in Climatology* 38(2), Center for Climatic Research, University of Delaware, Newark (36 pp.).

Olgyay, V. (1963) *Design with Climate: Bioclimatic Approach to Architectural Regionalism*, Princeton University Press, Princeton, NJ (190 pp.).

Oliver, J. E. (1970) A genetic approach to climatic classification, *Ann. Assn. Amer. Geog.* 60, 615–37. (Commentary, see 61, 815–20.)

Oliver, J. E. and Wilson, L. (1987) Climatic classification. In J. E. Oliver and R. W. Fairbridge (eds) *The Encyclopedia of Climatology*, Van Nostrand Reinhold, New York, 231–6.

Papadakis, J. (1975) *Climates of the World and their Potentialities*, Buenos Aires (200 pp.).

Rees, W. G. (1993) New wind-chill nomogram. *Polar Rec.* 29(170), 229–34.

Steadman, R. G. (1984) A universal scale of apparent temperature. *J. Clim. Appl. Met.* 23(12), 1,674–87.

Steadman, R. G., Osczewski, R. J. and Schwerdt, R. W. (1995) Comments on 'Wind chill errors'. *Bull. Amer. Met. Soc.* 76(9), 1,628–37.

Strahler, A. N. (1965) *Introduction to Physical Geography*, Wiley, New York (455 pp.).

Strahler, A. N. (1969) *Physical Geography* (3rd edn), Wiley, New York (733 pp.).

Terjung, W. H. (1966) Physiologic climates of the conterminous United States: a bioclimatological classification based on man, *Ann. Assn. Amer. Geog.* 56, 141–79.

Terjung, W. H. and Louie, S. S.-F. (1972) Energy input–output climates of the world. *Archiv. Met. Geophys. Biokl. B*, 20, 127–66.

Thornthwaite, C. W. (1933) The climates of the earth, *Geog. Rev.* 23, 433–40.

Thornthwaite, C. W. (1948) An approach towards a rational classification of climate, *Geog. Rev.* 38, 55–94.

Thornthwaite, C. W. and Mather, J. R. (1955) The water balance, *Publications in Climatology* 8(1), Laboratory of Climatology, Centerton, NJ (104 pp.).

Thornthwaite, C. W. and Mather, J. R. (1957) Instructions and tables for computing potential evapotranspiration and the water balance, *Publications in Climatology* 10(3), Laboratory of Climatology, Centerton, NJ (129 pp.).

Troll, C. (1958) Climatic seasons and climatic classification, *Oriental Geographer* 2, 141–65.

Yan, Y. Y. and Oliver, J. E. (1996) The clo: A utilitarian unit of measure weather/climate comfort. *Int. J. Climatol.* 16(9), 1,045–56.

APPENDIX 2
Système International (SI) units

Quantity	Dimensions	SI	cgs metric	British
length	L	m	10^2 cm	3.2808 ft
area	L^2	m^2	10^4 cm^2	10.7640 ft^2
volume	L^3	m^3	10^6 cm^3	35.3140 ft^3
mass	M	kg	10^3 g	2.2046 lb
density	ML^{-3}	kg m^{-3}	10^{-3} g cm^{-3}	
time	T	s	s	
velocity	LT^{-1}	m s^{-1}	10^2 cm s^{-1}	2.24 mi hr^{-1}
acceleration	LT^{-2}	m s^{-2}	10^2 cm s^{-2}	
force	MLT^{-2}	newton (kg m s^{-2})	10^5 dynes (10^5 g cm^{-1} s^{-2})	
pressure	$ML^{-1}T^{-2}$	N m^{-2} (pascal)	10^{-2} mb	
energy, work	ML^2T^{-2}	joule (kg m^2 s^{-2})	10^7 ergs (10^7 g cm^2 s^{-2})	
power	ML^2T^{-3}	watt (kg m^2 s^{-1})	10^7 ergs s^{-1}	1.340 × 10^{-3}hp
temperature	θ	kelvin (K)	°C	1.8°F
heat energy	ML^2T^{-2} (or H)	joule (J)	0.2388 cal	9.470 × 10^{-4} BTU
heat/radiation flux	HT^{-1}	watt (W) or J s^{-1}	0.2388 cal s^{-1}	3.412 BTU hr^{-1}
heat flux density	$HL^{-2}T^{-1}$	W m^{-2}	2.388 × 10^{-5} cal cm^{-2} s^{-1}	

The *basic SI units* are metre, kilogram, second (m, kg, s):

1 m	= 3.2808 feet	1 ft	= 0.3048 m
1 km	= 0.6214 miles	1 mi	= 1.6090 km
1 kg	= 2.2046 lb	1 lb	= 0.4536 kg
1 m s^{-1}	= 2.2400 mi hr^{-1}	1 mi hr^{-1}	= 0.4460 m s^{-1}
1 m^2	= 10.7640 ft^2	1 ft^2	= 0.0929 m^2
1 km^2	= 0.3861 mi^2	1 mi^2	= 0.5900 km^2
1°C	= 1.8°F	1°F	= 0.555°C

Temperature conversions can be determined by noting that:

$$\frac{T(°C)}{5} = \frac{T(°F) - 32}{9}$$

Radius of the sun = 7×10^8 m
Radius of the earth = 6.37×10^6 m
Mean earth–sun distance = 1.495×10^{11} m

Energy conversion factors:

4.1868 J	= 1 calorie
J cm^{-2}	= 0.2388 cal cm^{-2}
Watt	= J s^{-1}
W m^{-2}	= 1.433 x 10^{-8} cal^{-2} min^{-1}
697.8 W m^{-2}	= 1 cal cm^{-2} min^{-1}

For time sums:

Day:	1 W m^{-2}	= 8.64 J cm^{-2} dy^{-1}
		= 2.064 cal cm^{-2} dy^{-1}
Day:	1 W m^{-2}	= 8.64 x 10^4 J m^{-2} dy^{-1}
Month:	1 W m^{-2}	= 2.592 M J m^{-2} (30 dy)$^{-1}$
		= 61.91 cal cm^{-2} (30 dy)$^{-1}$
Year:	1 W m^{-2}	= 31.536 M J m^{-2} yr^{-1}
		= 753.4 cal cm^{-2} yr^{-1}

Gravitational acceleration (g) = 9.81 m s^{-2}
Latent heat of vaporization (288 K) = 2.47×10^6 J kg^{-1}
Latent heat of fusion (273 K) = 3.33×10^5 J kg^{-1}

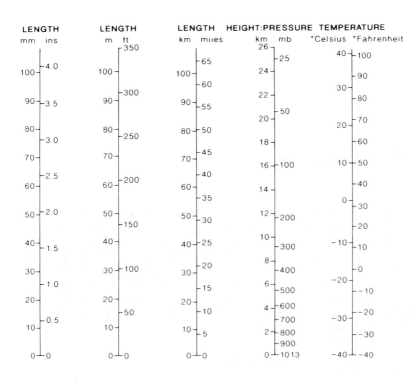

Figure A2.1 Nomograms of height, pressure, length and temperature.

APPENDIX 3
Synoptic weather maps

The synoptic weather map provides a generalized view of weather conditions over a large area at a given time. The map analysis smooths out local pressure and wind departures from the broad pattern. Such maps are usually prepared at 6- or 12-hourly intervals. Maps are generally prepared for mean sea-level pressure (or of height contours for the 1000 mb pressure surface) and at standard isobaric surfaces – 850, 700, 500, 300 mb, etc. The MSL pressure map typically shows isobars at 4 or 5 mb intervals, surface fronts and weather information.

Weather phenomena shown on the map are as follows:

temperature	type and height of cloud base
dew point	present weather
wind direction	past weather (last 6 hours)
wind speed	pressure tendency
pressure	pressure change (last 3 hours)
cloud amount	visibility

These data are presented in coded or symbolic form for each weather station. The plotting convention ('station model') is illustrated in Figure A3.1. The basic weather symbols are illustrated in Figure A3.2, and the synoptic code is given in Table A3.1.

Note: Meteorological Office Leaflet No. 12. HMSO, London, gives titles and prices of meteorological reports, maps, reporting forms and diagrams.

REFERENCE

Stubbs, M. W. (1981) New code for reporting surface observations – an introduction, *Weather* **36**, 357–66.

MODEL (enlarged)

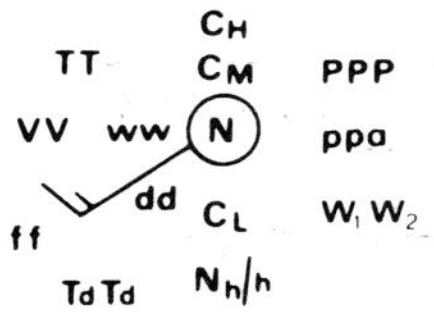

CH
TT
CM
PPP
VV
ww
N
ppa
dd
CL
W1 W2
ff
TdTd
Nh/h

EXAMPLE

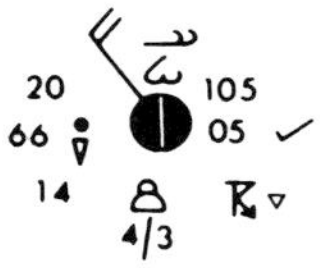

20
105
66
05
14
4/3

	KEY	EXAMPLE
N	Total cloud (oktas)[1]	7
dd	Wind direction (tens of degrees)	32
ff	Wind speed (knots)	20
VV	Visibility (code)	66
ww	Present weather (coded symbol)	80
W_1	Past weather (coded symbol)	9
W_2	" " " "	8
PPP	Sea-level pressure (mb)[2]	105
TT	Temperature (°C)[4]	20
N_h	Low cloud (oktas)	4
C_L	Low cloud type (coded symbol)	2
h	Height of C_L (code)	3
C_M	Medium cloud type (coded symbol)	5
C_H	High cloud type (coded symbol)	2
T_dT_d	Dew-point temperature (°C)[4]	14
a	Barograph trace (coded symbol)	3
pp	3-hour pressure change (mb)[3]	05

[1] okta = eighth

[2] Pressure in tens units and tenths mb: omitting initial 9 or 10 i.e. 105 = 1010·5

[3] Pressure change in units and tenths mb

[4] Rounded to nearest °C

Figure A3.1 Basic station model for plotting weather data. The key and example are tabulated in the internationally agreed sequence for teletype messages. These data would be preceded by an identifying station number, date and time.

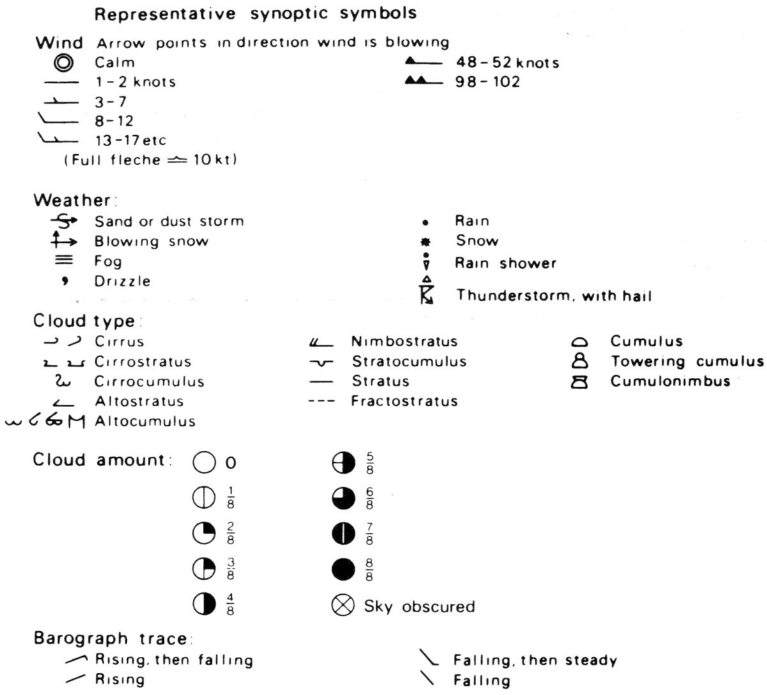

Representative synoptic symbols
Wind Arrow points in direction wind is blowing
Calm
1-2 knots
3-7
8-12
13-17 etc
(Full fleche = 10 kt)
48-52 knots
98-102
Weather:
Sand or dust storm
Blowing snow
Fog
Drizzle
Rain
Snow
Rain shower
Thunderstorm, with hail
Cloud type:
Cirrus
Cirrostratus
Cirrocumulus
Altostratus
Altocumulus
Nimbostratus
Stratocumulus
Stratus
Fractostratus
Cumulus
Towering cumulus
Cumulonimbus
Cloud amount:
0
1/8
2/8
3/8
4/8
5/8
6/8
7/8
8/8
Sky obscured
Barograph trace:
Rising, then falling
Rising
Falling, then steady
Falling

Figure A3.2 Representative synoptic symbols.

Table A3.1 Synoptic code (World Meteorological Organization, January 1982).

Symbol	Key	Example	Comments
yy	Day of the month (GMT)	05	⎧ All groups are in blocks
GG	Time (GMT) to nearest hour	06	⎨ of 5 digits
i_w	Indicator for type of wind speed observation and units	4	Measured by anemometer (knots)
IIiii	International index number of station		
i_R	Indicator: precipitation data included/omitted (code)	3	Data omitted
i_x	Indicator: station type + ww W_1 W_2 included/omitted (code)	1	Manned station with ww W_1 W_2 included
h	Height of lowest cloud (code)	3	
vv	Visibility (code)	66	
N	Total cloud amount (oktas)	7	
dd	Wind direction (tens of degrees)	32	
ff	Wind speed (knots, or m s^{-1})	20	Knots
1	Header	1	
S_n	Sign of temperature (code)	0	Positive value
TTT	Temperature (0.1°C), plotted rounded to nearest 1°C	203	(1 = negative value)
2	Header	2	
S_n	Sign of temperature (code)	0	
$T_dT_dT_d$	Dewpoint temperature (as TTT)	138	
4	Header	4	
PPPP	Mean sea-level pressure (tenths of mb, omitting thousands)	0105	
5	Header	5	
a	Characteristic of pressure tendency (coded symbol)	3	
ppp	3-hour pressure tendency (tenths of mb)	005	
7	Header	7	
ww	Present weather (coded symbol)	80	
W_1	Past weather (coded symbol)	9 ⎫	(W_1 must be greater
W_2	Past weather (coded symbol)	8 ⎭	than W_2)
8	Header	8	
N_h	Amount of low cloud (oktas)	4	
C_L	Low cloud type (coded symbol)	2	
C_M	Medium cloud type (coded symbol)	5	
C_H	High cloud type (coded symbol)	2	

Note: Group 3 is for a report of surface pressure and group 6 for precipitation data.

APPENDIX 4
Data sources

A DAILY WEATHER MAPS AND DATA

Western Europe/North Atlantic: *Daily Weather Summary* (synoptic chart, data for the UK). London Weather Centre, 284 High Holborn, London WC1V 7HX, England.

Western Europe/North Atlantic: *Monthly Weather Report* (published about 15 months in arrears; tables for approximately 600 stations in the UK). London Weather Centre.

Europe – eastern/North Atlantic: *European Daily Weather Report* (synopic chart). Deutsche Wetterdienst, Zentralamt D6050, Offenbach, Germany.

Europe – eastern/North Atlantic: *Weather Log* (daily synopic chart, supplement to *Weather* magazine). Royal Meteorological Society, Bracknell, Berkshire, England.

North America: *Daily Weather Reports* (weekly publication). National Environmental Satellite Data and Information Service, NOAA, US Government Printing Office, Washington, DC 20402, USA.

B SATELLITE DATA

NOAA operational satellites (imagery, digital data). Satellite Data Services NCDC, NOAA/NESDIS, Asheville, NC 28801–5001, USA.

Defence Meteorological Satellite Program (data). National Geophysical Data Center, NOAA/NESDIS, 325 Broadway, Boulder, CO 80303, USA.

Metsat (imagery, digital data). ESOC, Robert-Bosil Str. 5, D-6100 Darmstadt, Federal Republic of Germany.

NASA research satellites (digital data). National Space Science Data Center, Goddard Space Flight Center, Greenbelt, MD 20771, USA.

UK Direct readout data from NOAA and Meteosat satellites are received at Dundee, Scotland. Dr P. E. Baylis, Department of Electrical Engineering, University of Dundee, Scotland, UK.

C CLIMATIC DATA

Canadian Climate Center: *Climatic Perspectives* (197 weekly and monthly summary charts). Atmospheric Environment Service, 4905 Dufferin Street, Downsview, Ontario, Canada M3H 5T4.

Carbon Dioxide Information Center (CDIC): Data holdings and publications on climate – related variables and indices. Carbon Dioxide Information Center, Oak Ridge National Laboratory, Oak Ridge, TN 37931, USA.

National Center for Atmospheric Research, Boulder, CO 80307–3000, USA. Archives most global analyses and many global climate records.

National Climatic Center, NOAA/NESDIS: *Local Climatological Data* (1948–) (monthly tabulations, charts); *Monthly Climatic Data for the World* (May 1948–). National Climatic Data Center, Federal Building, Asheville, NC 28801, USA.

Climate Analysis Center, NOAA/NESDIS: *Climatic Diagnostics Bulletin* (1983–) (monthly summaries of selected diagnostic product from NMC analyses). Climate Analysis Center, NMC NOAA/NWS, World Weather Building, Washington, DC 20233, USA.

Climatic Research Unit, University of East Anglia: *Climate Monitor* (1976–) (monthly summaries, global and UK). Climatic Research Unit, University of East Anglia, Norwich NR4 7TJ, England.

World Climate Data Programme: *Climate System Monitoring Bulletin* (1984–) (monthly). World Climate Data Programme, WMO Secretariat, CP5, Geneva 20 CH-1211, Switzerland.

World Meteorological Centre, Melbourne: *Climate Monitoring Bulletin for the Southern Hemisphere* (1986–). Bureau of Meteorology, GPO Box 1289 K, Melbourne, Victoria 3001, Australia.

D SELECTED SOURCES OF INFORMATION ON THE WORLD WIDE WEB

World Meteorological Organization, Geneva, Switzerland
http://www.wmo.ch

National Oceanic and Atmospheric Administration, Washington, DC, USA
http://www.noaa.gov

National Climate Data Center, Asheville, NC, USA
hhtp://www.ncdc.noaa.gov/ncdc.html

National Snow and Ice Data Center, Boulder, Colo., USA
http://www-nsidc.colorado.edu

Environment Canada
http://www.on.doe.ca

European Centre for Medium-range Weather Forecasting
http://www.ecmwf.int

Climate Diagnostics Bulletin (US)
http://nic.fb4.noaa.gov

UK Meteorological Office
http://www.meto.gov.uk

US National Severe Storms Laboratory
http://www.nssl.uoknor.edu

BIBLIOGRAPHY

Ahlquist, J. (1993) Free software and information via the computer network. *Bull. Amer. Met. Soc.* **74** (3), 377–86.

Brugge, R. (1994) Computer networks and meteorological information. *Weather*, **49**(9), 298–306.

Carleton, A. M. (1991) *Satellite Remote Sensing in Climatology*, Belhaven Press, London and CRC Press, Boca Raton, Fla. (291 pp.).

European Space Agency (1978) *Introduction to the Metsat System*, European Space Operations Centre, Darmstadt (54 pp.).

European Space Agency (1978) *Atlas of Meteosat Imagery, Atlas Meteosat*, ESA-SP-1030, ESTEC, Nordwijk, Netherlands (494 pp.).

Finger, F. G., Laver, J. D., Bergman, K. H. and Patterson, V. L. (1958) The Climate Analysis Center's user information service, *Bull. Amer. Met. Soc.* **66**, 413–20.

Hastings, D. A., Emery, W. J., Weaver, R. L., Fisher, W. J. and Ramsey, J. W. (1987) *Proceedings North American NOAA Polar Orbiter User Group First Meeting*, NOAA/NESDIS, US Dept of Commerce, Boulder, CO, National Geophysical Data Center (273 pp.).

Hattemer-Frey, H. A., Karl, T. R. and Quinlan, F. T. (1986) *An Annotated Inventory of Climatic Indices and Data Sets*, DOE/NBB-0080, Office of Energy Research, US Dept of Energy, Washington, DC (195 pp.).

Jenne, R. L. and McKee, T. B. (1985) Data; in Houghton, D. D. (ed.) *Handbook of Applied Meteorology*, Wiley, New York, pp. 1175–281.

Meteorological Office (1958) *Tables of Temperature, Relative Humidity and Precipitation for the World*, HMSO, London.

Singleton, F. (1985) Weather data for schools, *Weather* **40**, 310–13.

Stull, A. and Griffin, D. (1996) *Life on the Internet – Geosciences*. Prentice Hall, Upper Saddle River, NJ.

US Department of Commerce (1983) *NOAA Satellite Programs Briefing*, National Oceanic and Atmospheric Administration, Washington, DC (203 pp.).

US Department of Commerce (1984) *North American Climate Data Catalog. Part 1*, National Environmental Data Referral Service, Publication NEDRES-1, National Oceanic and Atmospheric Administration, Washington, DC (614 pp.).

World Meteorological Organization (1965) *Catalogue of Meteorological Data for Research*, WMO No. 174. TP-86, World Meteorological Organization, Geneva.

Notes

1 ATMOSPHERIC COMPOSITION, MASS AND STRUCTURE

1   Mixing ratio = ratio of number of molecules of ozone to molecules of air (parts per million by volume, ppm(v)). Concentration = mass per unit volume of air (molecules cubic metre).

2   K = degrees Kelvin (or Absolute). The degree symbol is omitted.
    °C = degrees Celsius
    °C = K − 273
    Conversions for °C and °F are given in Appendix 2.

3   Joule = 0.2388 cal. The units of the International Metric System are given in Appendix 2. At present the data in many references are still in calories; a calorie is the heat required to raise the temperature of 1 g of water from 14.5°C to 15.5°C. In the United States, another unit in common use is the Langley (ly) (ly min^{-1} = 1 cal cm^{-2} min^{-1}).

4   The equation for the so-called 'reduction' (actually the adjusted value is normally greater!) of station pressure (p_h) to sea level pressure (p_0) is written:

$$p_0 = p_h \exp \left[\frac{g_0}{R_d \bar{T}_v} \right] Z_p$$

where R_d = gas content for dry air; g_0 = global average of gravitational acceleration (9.8 ms^{-2}); Z_h = geopotential height of the station ($\simeq$ geometric height in the lowest kilometre or so); $\bar{T}_v$ = mean virtual temperature. This is a fictitious temperature used in the ideal gas equation to compensate for the fact that the gas constant of moist air exceeds that of dry air. Even for hot moist air, $\bar{T}_v$ is only a few degrees greater than the air temperature.

5   The official definition is the lowest level at which the lapse rate decreases to less than, or equal to, 2°C/km (provided that the average lapse rate of the 2-km layer does not exceed 2°C/km).

2 SOLAR RADIATION AND GLOBAL ENERGY BUDGET

1   The radiation flux (per unit area) received normal to the beam at the top of the earth's atmosphere is calculated from the total solar output weighted by $1/(4\pi D^2)$, where the solar distance D = 1.5 × 10^{11} m, since the surface area of a sphere of radius r (here equivalent to D) is 4πr^2 – i.e. the radiation flux is (6.24 × 10^7 W m^{-2}) (61.58 × 10^{23} m^2)/4π (2.235 × 10^{22}) = 1367 W m^{-2}.

2   The albedos refer to the solar radiation received on each given surface: thus the incident radiation is different for planet earth, the global surface and global cloud cover, as well as between any of these and the individual cloud types or surfaces.

4 ATMOSPHERIC INSTABILITY, CLOUD FORMATION AND PRECIPITATION PROCESSES

1   It is significant that some storms also occur downwind of other and plateau regions in Mexico, the Iberian peninsula and West Africa.

5 ATMOSPHERIC MOTION: PRINCIPLES

1   The centrifugal force is equal in magnitude and opposite in sign to the centripetal acceleration.

2   Apparent gravity, g = 9.78 m s^{-2} at the equator, 9.83 m s^{-2} at the poles.

3   The vorticity, or circulation, about a rotating circular fluid disc is given by the product of the rotation on its boundary (ωR) and the circumference (2πR) where R = radius of the disc. The vorticity is then 2ωπR^2, or 2ω per unit area.

6 PLANETARY-SCALE MOTION IN THE ATMOSPHERE AND OCEAN

1   The geostrophic wind concept is equally applicable to contour charts. Heights on these charts are given in geopotential metres (g.p.m.) or dekametres (g.p.dkm).

2   The World Meteorological Organization recommends an arbitrary lower limit of 30 m s^{-1}.

3   Equatorial speed or rotation is 465 m s^{-1}.

4   Note that, at the equator, an east/west wind of 5 m s^{-1} represents an absolute motion of 460/470 m s^{-1} towards the east.

7 MID-LATITUDE SYNOPTIC SYSTEMS

1   Resultant wind is the vector average of all wind directions and speeds.
2   This latter term is tending to be restricted to the tropical (hurricane) variety.

8 WEATHER AND CLIMATE IN TEMPERATE AND HIGH LATITUDES

Standard indices of continentality developed by Gorcynski (see p. 89), Conrad and others are based on the annual range of temperature, scaled by the sine of the latitude angle as a reciprocal in the expression. This index is unsatisfactory for several reasons. The small amplitude of annual temperature range in humid tropical climates renders it unworkable for low latitudes. The latitude weighting is intended to compensate for summer–winter differences in solar radiation and thus temperatues, which were thought to increase uniformly with latitude. For North America, the differences peak about 55°N. The index used in Figure 8.20 is based on *departures* from the regression line of annual temperature range versus latitude. The specific regression constants will differ between continents. It should be noted that indices of the Gorcynski type are appropriate for regions of limited latitudinal extent as shown in Figure 8.2.

11 CLIMATE CHANGE

1   Statistics commonly reported for climatic data are: the *arithmetic mean*,

$$\bar{x} = \sum \frac{x_i}{n}$$

where $\sum$ = sum of all values for i = 1 to n
            x_i = an individual value

n = number of cases

and the *standard deviation*, σ (pronounced sigma).

$$\sigma = \sqrt{\frac{\sum (x_i - \bar{x})^2}{n}}$$

which expresses the variability of observations.

For precipitation data, the *coefficient of variation*, CV is often used:

$$CV = \frac{\sigma}{\bar{x}} \times 100 \ (\%)$$

For a *normal* (or Gaussian) bell-shaped symmetrical frequency distribution, the arithmetic mean is the central value; 68.3 per cent of the distribution of values are within $\pm 1 \ \sigma$ of the mean and 94.5 per cent within $\pm 2 \ \sigma$ of the mean.

The frequency distribution of mean daily temperatures is usually approximately normal. However, the frequency distribution of annual (or monthly) totals of rainfall over a period of years may be 'skewed' with some years (months) having very large totals whereas most years (months) have low amounts. For such distributions the *median* is a more representative average statistic; the median is the middle value of a set of data ranked according to magnitude. Fifty per cent of the frequency distribution is above the median and fifty per cent below it. The variability may be represented by the 25 and 75 percentile values in the ranked distribution.

A third measure of central tendency is the *mode* – the value which occurs with greatest frequency. In a normal distribution the mean, median and mode are identical.

Frequency distributions for cloud amounts are commonly bimodal with more observations having small or large amounts of cloud cover than are in the middle range.

Bibliography

GENERAL

Anthes, R. A., Panofsky, H. A., Cahir, J. J. and Rango, A. (1981) *The Atmosphere* (3rd edn), C. E. Merrill, Columbus, Ohio (531 pp.).

Anthes, R. A. (1997) *Meteorology* (7th edn), Prentice-Hall, Upper Saddle River, NJ (240 pp.).

Atkinson, B. W. (1981a) *Meso-Scale Atmospheric Circulations*, Academic Press, London (496 pp.).

*Atkinson, B. W. (ed.) (1981b) *Dynamical Meteorology*, Methuen, London (250 pp.).

Barry, R. G. (1992) *Mountain Weather and Climate* (2nd edn), Routledge, London and New York (402 pp.).

Barry, R. G. and Perry, A. H. (1973) *Synoptic Climatology: Methods and Applications*, Methuen, London (555 pp.).

Battan, L. J. (1984) *Fundamentals of Meteorology* (2nd edn), Prentice-Hall, Englewood Cliffs, NJ (304 pp.).

*Berry, F. A., Bollay, E. and Beers, N. R. (eds) (1945) *Handbook of Meteorology*, McGraw-Hill, New York (1,068 pp.).

Bigg, G. R. (1996) *The Oceans and Climate*, Cambridge University Press, Cambridge (266 pp.).

Blüthgen, J. (1966) *Allgemeine Klimageographie* (2nd edn), W. de Gruyter, Berlin (720 pp.).

Bradley, R. S., Ahern, L. G. and Keimig, F. T. (1994) A computer-based atlas of global instrumental climate data, *Bull. Amer. Met. Soc.* 75(1), 35–41.

Browning, K.A. (1996) Current research in atmospheric sciences, *Weather* 51, 167–72.

Bruce, J. P. and Clark, R. H. (1966) *Introduction to Hydrometeorology*, Pergamon, Oxford (319 pp.).

*Byers, H. R. (1974) *General Meteorology* (4th edn), McGraw-Hill, New York (480 pp.).

Carleton, A. M. (1991) *Satellite Remote Sensing in Climatology*, Belhaven Press, London (291 pp.).

Chang, J. H. (1972) *Atmospheric Circulation Systems and Climates*, Oriental Publishing Co., Honolulu (326 pp.).

Cole, F. W. (1980) *Introduction to Meteorology* (3rd edn), Wiley, New York (505 pp.).

Crowe, P. R. (1971) *Concepts in Climatology*, Longman, London (589 pp.).

Crowley, T. J. and North, G. R. (1991) *Paleoclimatology*, New York, Oxford University Press (339 pp.).

Fleagle, R. G. and Businger, J. A. (1980) *Introduction to Atmospheric Physics* (2nd edn), Academic Press, New York (432 pp.).

*Flohn, H. (ed.) (1969) *General Climatology* 2, World Survey of Climatology 2, Elsevier, Amsterdam (266 pp.).

*Flohn, H. (ed.) (1981) *General Climatology* 3, World Survey of Climatology 3, Elsevier, Amsterdam (408 pp.).

*Garratt, J. R. (1992) *The Atmospheric Boundary Layer*, Cambridge University Press, Cambridge (316 pp.).

Garratt, J. R. (1994) Review: The atmospheric boundary layer, *Earth-Science Reviews*, 37, 89–134.

Gedzelman, S. D. (1980) *The Science and Wonders of the Atmosphere*, Wiley, New York (535 pp.).

Geiger, R. (1965) *The Climate Near the Ground* (2nd edn), Harvard University Press, Cambridge, Mass. (611 pp.).

Graedel, T. E. and Crutzen, P. J. (1993) *Atmospheric Change: An Earth System Perspective*, Freeman, New York (446 pp.).

*Haltiner, G. J. and Martin, F. L. (1957) *Dynamical and Physical Meteorology*, McGraw-Hill, New York (470 pp.).

Hartmann, D. L. (1994) *Global Physical Climatology*, Academic Press, New York (408 pp.).

Hecht, A. D. (ed.) (1985) *Paleoclimate Analysis and Modeling*, Wiley, New York (445 pp.).

Henderson-Sellers, A. and Robinson, P. J. (1986) *Contemporary Climatology*, Longman, London (439 pp.).

Herman, J. R. and Goldberg, R. A. (1985) *Sun, Weather and Climate*, Dover, New York (360 pp.).

*Hess, S. L. (1959) *Introduction to Theoretical Meteorology*, Henry Holt, New York (362 pp.).

Hewson, E. W. and Longley, R. W. (1944) *Meteorology, Theoretical and Applied*, Wiley, New York (468 pp.).

Hobbs, J. (1980) *Applied Climatology*, Westview Press, Boulder, Colo. (250 pp.).

Houghton, D. D. (1985) *Handbook of Applied Meteorology*, Wiley, New York (1,461 pp.).

Houghton, H. G. (1985) *Physical Meteorology*, MIT Press, Cambridge, Mass. (442 pp.).

Houghton, J. T. (ed.) (1984) *The Global Climate*, Cambridge University Press, Cambridge (233 pp.).

*Humphreys, W. J. (1929) *Physics of the Air*, McGraw-Hill, New York (654 pp.).

Huschke, R. E. (ed.) (1959) *Glossary of Meteorology*, American Meteorological Society, Boston (638 pp.).

*Indicates more advanced texts with reference particularly to the physical bases of meteorology, often including mathematical material.

Hutchinson, G. E. (1957) *A Treatise on Limnology* **1**, Wiley, New York (1,015 pp.).

Kendrew, W. G. (1961) *The Climates of the Continents* (5th edn), Oxford University Press, London (608 pp.).

Lamb, H. H. (1972) *Climate: Present, Past and Future* **1**: *Fundamentals and Climate Now*, Methuen, London (613 pp.).

Liljequist, G. H. (1970) *Klimatologi*, Generalstabens Litografiska Anstalt, Stockholm.

List, R. J. (1951) *Smithsonian Meteorological Tables* (6th edn), Smithsonian Institution, Washington (527 pp.).

Lockwood, J. G. (1974) *World Climatology: An Environmental Approach*, Arnold, London (330 pp.).

Lockwood, J. G. (1979) *Causes of Climate*, Arnold, London (260 pp.).

Lutgens, F. K. and Tarbuck, E. J. (1986) *The Atmosphere: An Introduction to Meteorology* (2nd edn), Prentice-Hall, Englewood Cliffs, NJ (492 pp.).

McBoyle, G. (ed.) (1973) *Climate in Review*, Houghton Mifflin, Boston (314 pp.).

McIlveen, R. (1986) *Basic Meteorology: A Physical Outline*, Van Nostrand Reinhold, UK (457 pp.).

McIntosh, D. H. and Thom, A. S. (1972) *Essentials of Meteorology*, Wykeham Publications, London (239 pp.).

*Malone, T. F. (ed.) (1951) *Compendium of Meteorology*, American Meteorological Society, Boston (1,334 pp.).

Manley, G. (1952) *Climate and the British Scene*, Collins, London (314 pp.).

Martyn, D. (1992) *Climates of the World*, Elsevier, Amsterdam (435 pp.).

Miller, A. and Anthes, R. A. (1980) *Meteorology* (4th edn), C. E. Merrill, Columbus, Ohio (170 pp.).

Monteith, J. L. (1973) *Principles of Environmental Physics*, Arnold, London (241 pp.).

Monteith, J. L. (ed.) (1975) *Vegetation and the Atmosphere*, Vol. 1: *Principles*, Academic Press, London (278 pp.).

Moran, J. M. and Morgan, M. D. (1997) *Meteorology: The Atmosphere and Science of Weather* (5th edn), Prentice-Hall, Upper Saddle River, NJ (550 pp.).

Munn, R. E. (1966) *Descriptive Micrometeorology*, Academic Press, New York (245 pp.).

Musk, L. F. (1988) *Weather Systems*, Cambridge University Press, Cambridge (160 pp.).

Neiburger, M., Edinger, J. G. and Bonner, W. D. (1982) *Understanding Our Atmospheric Environment* (2nd edn), W. H. Freeman, San Francisco (453 pp.).

Oliver, J. E. and Fairbridge, R. W. (eds) (1987) *The Encyclopedia of Climatology*, Van Nostrand Reinhold, New York (986 pp.).

*Palmén, E. and Newton, C. W. (1969) *Atmosphere Circulation Systems: Their Structure and Physical Interpretation*, Academic Press, New York (603 pp.).

Pedgley, D. E. (1962) *A Course of Elementary Meteorology*, HMSO, London (189 pp.).

*Peixoto, J. P. and Oort, A. H. (1992) *Physics of Climate*, American Institute of Physics, New York (520 pp.).

Perry, A. H. and Walker, J. M. (1977) *The Ocean–Atmosphere System*, Longman, London (160 pp.).

*Petterssen, S. (1956) *Weather Analysis and Forecasting* (2 vols), McGraw-Hill, New York (428 and 266 pp.).

Petterssen, S. (1969) *Introduction to Meteorology* (3rd edn), McGraw-Hill, New York (333 pp.).

Pickard, G. L. and Emery, W. J. (1990) *Descriptive Physical Oceanography* (5th edn), Pergamon Press, Oxford.

*Reiter, E. R. (1963) *Jet Stream Meteorology*, University of Chicago Press, Chicago, Ill. (515 pp.).

Rex, D. F. (ed.) (1969) *Climate of the Free Atmosphere*, World Survey of Climatology **4**, Elsevier, Amsterdam (450 pp.).

Schaefer, V. J. and Day, J. A. (1981) *A Field Guide to the Atmosphere*, Houghton Mifflin, Boston, Mass. (359 pp.).

Schwerdtfeger, W. (1984) *Weather and Climate of the Antarctic*, Elsevier, Amsterdam (261 pp.).

*Sellers, W. D. (1965) *Physical Climatology*, University of Chicago Press, Chicago, Ill. (272 pp.).

Singh, H. B. (ed.) (1992) *Composition, Chemistry, and Climate of the Atmosphere*, Van Nostrand Reinhold, New York (527 pp.).

Smith, W. L., Bishop, W. P., Dvorak, V. F., Hayden, C. M., McElroy, J. H., Mosher, F. R., Oliver, V. J., Purdom, J. F. and Wark, D. Q. (1986) The meteorological satellite: overview of 25 years of operations, *Science* **231**, 455–62.

Strahler, A. N. (1965) *Introduction to Physical Geography*, Wiley, New York (455 pp.).

Strahler, A. N. and Strahler, A. H. (1987) *Modern Physical Geography* (3rd edn), John Wiley & Sons, New York (332 pp.).

Stringer, E. T. (1972a) *Foundations of Climatology*, W. H. Freeman, San Francisco (586 pp.).

Stringer, E. T. (1972b) *Techniques of Climatology*, W. H. Freeman, San Francisco (539 pp.).

Sverdrup, H. V. (1945) *Oceanography for Meteorologists*, Allen & Unwin, London (235 pp.).

Sverdrup, H. V., Johnson, M. W. and Fleming, R. H. (1942) *The Oceans: Their Physics, Chemistry and General Biology*, Prentice-Hall, New York (1,087 pp.).

Taljaard, J. J., van Loon, H., Crutcher, H. L. and Jenne, R. L. (1969) *Climate of the Upper Air, Part 1. Southern Hemisphere* **1**, Naval Weather Service Command, Washington, DC, NAVAIR 50–1C–55.

Taylor, J. A. and Yates, R. A. (1967) *British Weather in Maps* (2nd edn), Macmillan, London (315 pp.).

Trenberth, K. E. (ed.) (1992) *Climate System Modeling*, Cambridge University Press, Cambridge (788 pp.).

Trewartha, G. T. (1981) *The Earth's Problem Climates* (2nd edn), University of Wisconsin Press, Madison (371 pp.).

Trewartha, G. T. and Horne, L. H. (1980) *An Introduction to Climate* (5th edn), McGraw-Hill, New York (416 pp.).

van Loon, H. (ed.) (1984) *Climates of the Oceans: World Survey of Climatology* **15**, Elsevier, Amsterdam (716 pp.).

*Wallace, J. M. and Hobbs, P. V. (1977) *Atmospheric Science: An Introductory Survey*, Academic Press, New York (467 pp.).

Willett, H. C. and Sanders, F. (1959) *Descriptive Meteorology* (2nd edn), Academic Press, New York (355 pp.).

World Meteorological Organization (1962) *Climatological Normals (CLINO) for CLIMAT and CLIMAT SHIP stations for the period 1931–60*, World Meteorological Organization, Geneva.

World Meteorological Organization (1973) The use of satellite pictures in weather analysis and forecasting, *WMO Technical Note* 124, Geneva (275 pp.).

CHAPTER 1: ATMOSPHERIC COMPOSITION, MASS AND STRUCTURE

Ahmad, S. A. and Lockwood, J. G. (1979) Albedo, *Prog. Phys. Geog.* 3, 520–43.

Andreae, M. O. and Schimel, D. S. (1989) *Exchange of Trace Gases Between Terrestrial Ecosystems and the Atmosphere*, J. Wiley & Sons, Chicester (347 pp.).

Bach, W. (1976) Global air pollution and climatic change, *Rev. Geophys. Space Phys.* 14, 429–74.

Bojkov, R. D. and Fioletov, V. E. (1995) Estimating the global ozone characteristics during the last 30 years, *J. Geophys. Res.* 100(D8), 16,537–51.

Bolin, B., Degens, E. T., Kempe, S. and Ketner, P. (eds) (1979) *The Global Carbon Cycle* (SCOPE 13), J. Wiley & Sons, Chichester (528 pp.).

Bolin, B., Döös, B. R., Jäger, J. and Warrick, R. A. (eds) (1986) *The Greenhouse Effect, Climatic Change, and Ecosystems* (SCOPE 29), J. Wiley & Sons, Chichester (541 pp.).

Bolle, H.-J., Seiler, W. and Bolin, B. (1986) Other greenhouse gases and aerosols; in Bolin, B. *et al.* (eds) *The Greenhouse Effect, Climatic Change, and Ecosystems*, J. Wiley & Sons, Chichester, 157–203.

Bridgman, H. A. (1990) *Global Air Pollution: Problems for the 1990s*, Belhaven Press, London (201 pp.).

Brimblecombe, P. (1986) *Air: Composition and Chemistry*, Cambridge University Press, Cambridge (224 pp.).

Brugge, R. (1996) Back to basics: Atmospheric stability. Part I – Basic concepts, *Weather* 51(4), 134–40.

Craig, R. A. (1965) *The Upper Atmosphere: Meteorology and Physics*, Academic Press, New York (509 pp.).

Crowley, T. J. and North, G. R. (1991) *Paleoclimatology*, Oxford University Press, New York and Oxford (339 pp.).

Defant, F. R. and Taba, H. (1957) The threefold structure of the atmosphere and the characteristics of the tropopause, *Tellus* 9, 259–74.

Gardiner, B. G. (1989) The Antarctic ozone hole. *Weather* 44(7), 291–8.

Hales, J. (1996) Scientific background for AMS policy statement on atmospheric ozone, *Bull. Amer. Met. Soc.* 77(6), 1,249–53.

Hare, F. K. (1962) The stratosphere, *Geog. Rev.* 52, 525–47.

Hastenrath, S. L. (1968) Der regionale und jahrzeithliche Wandel des vertikalen Temperaturgradienten und seine Behandlung als Wärmhaushaltsproblem, *Meteorologische Rundschau* 1, 46–51.

Houghton, J. T., Jenkins, G. J. and Ephraums, J. J. (eds) (1990) *Climate Change: The IPCC Scientific Assessment*, Cambridge University Press, Cambridge (365 pp.).

Jiang, Y. B., Yung, Y. L. and Zurek, R. W. (1996) Decadal evolution of the Antarctic ozone hole, *J. Geophys. Res.* 101(D4), 8,985–9,000.

Kellogg, W. W. and Schware, R. (1981) *Climate Change and Society*, Westview Press, Boulder, Colo. (178 pp.).

Kiehl, J. T. and Trenberth, K. E. (1997) Earth's annual global mean energy budget, *Bull. Amer. Met. Soc.* 78, 197–208.

Kondratyev, K. Ya. and Moskalenko, N. I. (1984) The role of carbon dioxide and other minor gaseous components and aerosols in the radiation budget; in Houghton, J. T. (ed.) *The Global Climate*, Cambridge University Press, Cambridge, 225–33.

Kyle, H. L., *et al.* (1993) The Nimbus Earth Radiation Budget (ERB) experiment: 1975–1992, *Bull. Amer. Met. Soc.* 74, 815–30.

LaMarche, V. C., Jr and Hirschboeck, K. K. (1984) Frost rings in trees as records of major volcanic eruptions, *Nature* 307, 121–6.

Lashof, D. A. and Ahnja, D. R. (1990) Relative contributions of greenhouse gas emissions to global warming, *Nature* 344, 529–31.

Lautensach, H. and Bogel, R. (1956) Der Jahrsgang des mittleren geographischen Hohengradienten der Lufttemperatur in den verschiedenen Klimagebieten der Erde, *Erdkunde* 10, 270–82.

Lean, J. (1991) Variations in the sun's radiative output, *Rev. Geophys.* 29, 505–35.

Lean, J. and Rind, D. (1994) Solar variability: implications for global change, *EOS* 75(1), 1 and 5–7.

London, J. (1985) The observed distribution of atmospheric ozone and its variations; in Whitten, R. C. and Prasad, T. S. (eds) *Ozone in the Free Atmosphere*, Van Nostrand Reinhold, New York, 11–80.

McElroy, M. B. and Salawitch, R. J. (1989) Changing composition of the global stratosphere, *Science* 243, 763–70.

Machta, L. (1972) The role of the oceans and biosphere in the carbon dioxide cycle; in Dyrssen, D. and Jagner, D. (eds) *The Changing Chemistry of the Oceans*, Nobel Symposium 20, Wiley, New York, 121–45.

NERC (1989) *Our Future World: Global Environmental Research*, NERC, London (28 pp.).

Neuendorffer, A. C. (1996) Ozone monitoring with the TIROS-N operational vertical sounders, *J. Geophys. Res.* 101(D13), 8,807–28.

Newell, R. E. (1964) The circulation of the upper atmosphere, *Sci. American* 210, 62–74.

Paffen, K. (1967) Das Verhältniss der Tages – zur Jahreszeitlichen Temperaturschwankung, *Erdkunde* 21, 94–111.

Pearce, F. (1989) Methane: the hidden greenhouse gas, *New Scientist* 122, 37–41.

Plass, G. M. (1959) Carbon dioxide and climate, *Sci. American* 201, 41–7.

Ramanathan, V., Cicerone, R. J., Singh, H. B. and Kiehl, J. T. (1985) Trace gas trends and their potential role in climatic change, *J. Geophys. Res.* 90(D3), 5,547–66.

Rampino, M. R. and Self, S. (1984) The atmospheric effects of El Chichón, *Sci. American* 250(1), 34–43.

Raval, A. and Ramanathan, V. (1989) Observational determination of the greenhouse effect, *Nature* 342, 758–61.

Robock, A. and Matson, M. (1983) Circumglobal transport of the El Chichón volcanic dust cloud, *Science* 221, 195–7.

Rodhe, H. (1990) A comparison of the contribution of various gases to the greenhouse effect, *Science* 244, 763–70.

Roland, F. S. and Isaksen, I. S. A. (eds) (1988) *The Changing Atmosphere*, J. Wiley & Sons, Chichester.

Schimel, D. and 26 others (1996) Radiative forcing of climate change; in Houghton, J. T. *et al.* (eds) *Climate Change 1995. The Science of Climate Change*, Cambridge University Press, Cambridge, 65–131.

Sellers, W. D. (1980) A comment on the cause of the diurnal and annual temperature cycles, *Bull. Amer. Met. Soc.* **61**, 741–55.

Shine, K. (1990) Effects of CFC substitutes, *Nature* **344**, 492–3.

Slinn, W. G. N. (1983) Air-to-sea transfer of particles; in Liss, P. S. and Slinn, W. G. N. (eds) *Air–Sea Exchange of Gases and Particles*, D. Reidel, Dordrecht, 299–407.

Solomon, S. (1988) The mystery of the Antarctic ozone hole, *Rev. Geophys.* **26**, 131–48.

Tetlow-Smith, A. (1995) Environmental factors affecting global atmospheric methane concentrations, *Prog. Phys. Geog.* **19**, 322–35.

Thompson, R. D. (1995) The impact of atmospheric aerosols on global climate, *Prog. Phys. Geog.* **19**, 336–50.

Trenberth, K. E., Houghton, J. T. and Meira Filho, L. G. (1996) The climate system: An overview; in Houghton, J.T. *et al.* (eds), *Climate Change 1995. The Science of Climate Change*, Cambridge University Press, Cambridge, 51–64.

Valley, S. L. (ed.) (1965) *Handbook of Geophysics and Space Environments*, McGraw-Hill, New York.

Webb, A. R. (1995) To burn or not to burn, *Weather* **50**(5), 150–4.

World Meteorological Organization (1964) Regional basic networks, *WMO Bulletin* **13**, 146–7.

CHAPTER 2: SOLAR RADIATION AND GLOBAL ENERGY BUDGET

Ardanuy, P. E., Kyle, H. L. and Hoyt, D. (1992) Global relationships among the earth's radiation budget, cloudiness, volcanic aerosols and surface temperature, *J. Climate* **5**(10), 1,120–39.

Barry, R. G. (1985) The cryosphere and climatic change; in MacCracken, M. C. and Luther, F. M. (eds) *Detecting the Climatic Effects of Increasing Carbon Dioxide*, DOE/ER-0235, US Department of Energy, Washington, DC, 109–48.

Barry, R. G. and Chambers, R. E. (1966) A preliminary map of summer albedo over England and Wales, *Quart. J. Roy. Met. Soc.* **92**, 543–8.

Beckinsale, R. P. (1945) The altitude of the zenithal sun: a geographical approach, *Geog. Rev.* **35**, 596–600.

Berger, A. (1996) Orbital parameters and equations; in Schneider, S. H. (ed.) *Encyclopedia of Climate and Weather*, vol. 2, New York, Oxford University Press, 552–7.

Budyko, M. I., Nayefimova, N. A., Aubenok, L. I. and Strokhina, L. A. (1962) The heat balance of the surface of the earth, *Soviet Geography* **3**(5), 3–16.

Campbell, I. M. (1986) *Energy and the Atmosphere. A Physical–Chemical Approach* (2nd edn), John Wiley & Sons, Chichester (337 pp.).

Currie, R. G. (1993) Luni–solar 18.6 and solar cycle 10–11 year signals in U.S.A. air temperature records, *Int. J. Climatology* **13**, 31–50.

Foukal, P. V. (1990) The variable sun, *Sci. American* **262**, 26–33.

Fröhlich, C. and London, J. (1985) *Radiation Manual*, World Meteorological Organization, Geneva.

Garnett, A. (1937) Insolation and relief, *Trans. Inst. Brit. Geog.* **5** (71 pp.).

Henderson-Sellers, A. and Wilson, M. F. (1983) Surface albedo data for climate modeling, *Rev. Geophys. Space Phys.* **21**, 1,743–78.

Kiehl, J. T. and Trenbreth, K. E. (1997) Earth's annual global mean energy budget, *Bull. Amer. Met. Soc.* **78**, 197–208.

Kraus, H. and Alkhalaf, A. (1995) Characteristic surface energy balances for different climate types, *Int. J. Climatology* **15**, 275–84.

Kung, E. C., Bryson, R. A. and Lenschow, D. H. (1964) Study of a continental surface albedo on the basis of flight measurements and structure of the earth's surface cover over North America, *Monthly Weather Review* **92**, 543–64.

Kyle, H. L. *et al.* (1993) The Nimbus Earth Radiation Budget (ERB) experiment: 1975–1992, *Bull. Amer. Met. Soc.* **74**, 815–30.

Lean, J. and Rind, D. (1994) Solar variability: implications for global change, *EOS* **75**(1), 1 and 5–7.

London, J. and Sasamori, T. (1971) Radiative energy budget of the atmosphere, *Space Research* **11**, 639–49.

London, J., Warren, S. G. and Hahn, C. J. (1989) The global distribution of observed cloudiness – a contribution to the ISCCP, *Adv. Space Res.* **9**, 161–5.

Lumb, F. E. (1961) *Seasonal variation of the sea surface temperature in coastal waters of the British Isles*, Sci. Paper No. 6, Meteorological Office, HMSO, London (21 pp.).

McFadden, J. D. and Ragotzkie, R. A. (1967) Climatological significance of albedo in central Canada, *J. Geophys. Res.* **72**, 1,135–43.

Miller, D. H. (1968) *A survey course: the energy and mass budget at the surface of the earth*, Assn. Amer. Geog., Pub. No. 7, Washington, DC (142 pp.).

Minami, K. and Neue, H.-U. (1994) Rice paddies as a methane source, *Climatic Change* **27**, 13–26.

NASA (n.d.) *From Pattern to Process: The Strategy of the Earth Observing System* (Vol. III), EOS Science Steering Committee Report, NASA, Houston, Texas.

Newton, H. W. (1958) *The Face of the Sun*, Pelican, London (208 pp.).

Ramanathan, V., Cess, R. D., Harrison, E. F., Minnis, P., Barkstrom, B. R., Ahmad, E. and Hartmann, D. (1989) Cloud-radiative forcing and climate: results from the Earth Radiation Budget Experiment, *Science* **243**, 57–63.

Ramanathan, V., Barkstrom, B. R. and Harrison, E. F. (1990) Climate and the earth's radiation budget, *Physics Today* **42**, 22–32.

Ransom, W. H. (1963) Solar radiation and temperature, *Weather* **8**, 18–23.

Simpkin, T. and Fiske, R. S. (1983) *Krakatau 1883*, Smithsonian Institution Press, Washington, DC (464 pp.).

Stephens, G. L., Campbell, G. G. and Vonder Haar, T.

H. (1981) Earth radiation budgets, *J. Geophys. Res.* **86**(C10), 9,739–60.

Stone, R. (1955) Solar heating of land and sea, *J. Geography* **40**, 288.

Tully, J. P. and Giovando, L. F. (1963) Seasonal temperature structure in the eastern subarctic Pacific Ocean, *Roy. Soc. Canada, Sp. Pub.* **5**, Dunbar, J. J. (ed.), 10–36.

Weller, G. and Wendler G. (1990) Energy budgets over various types of terrain in polar regions, *Ann. Glac.* **14**, 311–14.

Wilson, R. C. and Hudson, H. S. (1991) The sun's luminosity over a complete solar cycle, *Nature* **351**, 42–3.

CHAPTER 3: ATMOSPHERIC MOISTURE BUDGET

Acreman, M. (1989) Extreme rainfall in Calderdale, 19 May 1989, *Weather* **44**, 438–46.

Armstrong, C. F. and Stidd, C. K. (1967) A moisture-balance profile in the Sierra Nevada, *J. of Hydrology* **5**, 258–68.

Atlas, D., Chou, S.-H. and Byerly, W. P. (1983) The influence of coastal shape on winter mesoscale air-sea interactions, *Monthly Weather Review* **111**, 245–52.

Ayoade, J. A. (1976) A preliminary study of the magnitude, frequency and distribution of intense rainfall in Nigeria, *Hydro. Sci. Bull.* **21**(3), 419–29.

Bannon, J. K. and Steele, L. P. (1960) Average water-vapour content of the air, *Geophysical Memoirs* **102**, Meteorological Office (38 pp.).

Baumgartner, A. and Reichel, E. (1975) *The World Water Balance: Mean Annual Global, Continental and Maritime Precipitation, Evaporation and Runoff*, Elsevier, Amsterdam (179 pp.).

Biswas, M. R. and Biswas, A. K. (eds) (1980) *Desertification*, Pergamon, Oxford (523 pp.).

Borchert, J. R. (1971) The dust bowl in the 1970s, *Ann. Assn Amer. Geogr.* **61**, 1–22.

Bridgman, H. A. (1990) *Global Air Pollution*, Belhaven Press, London (261 pp.).

Browning, K. (1993) The global energy and water cycle, *NERC News* **July**, 21–3.

Bryson, R. A. (1973) Drought in the Sahel: who or what is to blame? *The Ecologist* **3**(10), 366–71.

Chacon, R. E. and Fernandez, W. (1985) Temporal and spatial rainfall variability in the mountainous region of the Reventazon River Basin, Costa Rica, *J. Climatology* **5**, 175–88.

Dalby, D. and Harrison Church, R. J. (eds) (1973) *Drought in Africa*, Centre for African Studies, School of Oriental and African Studies, University of London (124 pp.).

Deacon, E. L. (1969) Physical processes near the surface of the earth; in Flohn, H. (ed.) *General Climatology*, World Survey of Climatology **2**, Elsevier, Amsterdam, 39–104.

Dobbie, C. H. and Wolf, P. O. (1953) The Lynmouth flood of August 1952, *Pro. Inst. Civ. Eng.* Part III, 522–88.

Doornkamp, J. C. and Gregory, K. J. (eds) (1980) *Atlas of Drought in Britain 1975–6*, Institute of British Geographers, London (82 pp.).

Dorman, C. E. and Bourke, R. H. (1981) Precipitation over the Atlantic Ocean, 30°S to 70°N, *Monthly Weather Review* **109**, 554–63.

Garcia-Prieto, P. R., Ludlam, F. H. and Saunders, P. M. (1960) The possibility of artificially increasing rainfall on Tenerife in the Canary Islands, *Weather* **15**, 39–51.

Gilman, C. S. (1964) Rainfall; in Chow, V. T. (ed.) *Handbook of Applied Hydrology*, McGraw-Hill, New York, section 9.

Harrold, T. W. (1966) The measurement of rainfall using radar, *Weather* **21**, 247–9 and 256–8.

Hastenrath, S. L. (1967) Rainfall distribution and regime in Central America, *Archiv. Met. Geophys. Biokl. B.* **15**(3), 201–41.

Hershfield, D. M. (1961) Rainfall frequency atlas of the United States for durations from 30 minutes to 24 hours and return periods of 1 to 100 years, *US Weather Bureau, Tech. Rept.* 40.

Howarth, D. A. and Rayner, J. N. (1993) An analysis of the atmospheric water balance over the southern hemisphere, *Phys. Geogr.* **14**, 513–35.

Howe, G. M. (1956) The moisture balance in England and Wales, *Weather* **11**, 74–82.

Howells, G. (1990) *Acid Rain and Acid Waters*, Ellis Horwood, New York (215 pp.).

Ilesanmi, O. O. (1971) An empirical formulation of an ITD rainfall model for the tropics: a case study for Nigeria, *J. App. Met.* **10**(5), 882–91.

Jaeger, L. (1976) Monatskarten des Niederschlags für die ganze Erde, *Berichte des Deutsches Wetterdienstes* **18**(139), Offenbach am Main (38 pp. + plates).

Jiusto, J. E. and Weickmann, H. K. (1973) Types of snowfall, *Bull. Amer. Met. Soc.* **54**, 148–62.

Kelly, P. M. and Wright, P. B. (1978) The European drought of 1975–6 and its climatic context, *Prog. Phys. Geog.* **2**, 237–63.

Landsberg, H. E. (1974) Drought, a recurring element of climate, Graduate Program in Meteorology, University of Maryland, Contribution No. 100 (47 pp.).

Legates, D. R. (1995) Global and terrestrial precipitation: a comparative assessment of existing climatologies, *Int. J. Climatol.* **15**, 237–58.

Legates, D. R. (1996) Precipitation; in Schneider, S. H. (ed.) *Encyclopedia of Climate and Weather,* Oxford University Press, New York, 608–12.

Likens, G. E., Wright, R. F., Galloway, J. N. and Butler, T. J. (1979) Acid rain, *Sci. American* **241**, 39–47.

Linsley, R. K. and Franzini, J. B. (1964) *Water-Resources Engineering*, McGraw-Hill, New York, (654 pp.).

Lott, J. N. (1994) The U.S. summer of 1993: A sharp contrast in weather extremes, *Weather* **49**, 370–83.

MacDonald, J. E. (1962) The evaporation–precipitation fallacy, *Weather* **17**, 168–77.

Markham, C. G. and McLain, D. R. (1977) Sea-surface temperature related to rain in Ceará, north-eastern Brazil, *Nature* **265**, 320–3.

Marsh, T. J. and Turton, P. S. (1996) The 1995 drought – a water resources perspective, *Weather* **51**(2), 46–53.

Mather, J. R. (1985) The water budget and the distribution of climates, vegetation and soils, *Publications in Climatology* **38**(2), University of Delaware, Center for Climatic Research, Newark, Del. (36 pp.).

McCallum, E. and Waters, A. J. (1993) Severe thunderstorms over southeast England, 20/21 July 1992, *Weather* 48, 198–208.

Miller, D. H. (1977) *Water at the Surface of the Earth*, Academic Press, New York (557 pp.).

Möller, F. (1951) Vierteljahrkarten des Niederschlags für die ganze Erde, *Petermanns Geographische Mitteilungen*, 95 Jahrgang, 1–7.

More, R. J. (1967) Hydrological models and geography; in Chorley, R. J. and Haggett, P. (eds) *Models in Geography*, Methuen, London, 145–85.

Palmer, W. C. (1965) Meteorological drought, Research Paper No. 45, US Weather Bureau, Washington, DC.

Parrett, C., Melcher, N. B. and James, R. W., Jr (1993) Flood discharges in the upper Mississippi River basin, *U.S. Geol. Sur. Circular* 1120–A (14 pp.).

Paulhus, J. L. H. (1965) Indian Ocean and Taiwan rainfall set new records, *Monthly Weather Review* 93, 331–5.

Pearl, R. T. *et al.* (1954) *The Calculation of Irrigation Need*, Tech. Bull. No. 4, Min. Agric., Fish and Food, HMSO, London (35 pp.).

Peixoto, J. P. and Oort, A. H. (1983) The atmospheric branch of the hydrological cycle and climate; in Street-Perrott, A., Beran, M. and Ratcliffe, R. (eds) *Variations in the Global Water Budget*, D. Reidel, Dordrecht, 5–65.

Peixoto, J. P. and Oort, A. H. (1992) *Physics of Climate*, American Institute of Physics, New York, Ch. 12.

Penman, H. L. (1963) *Vegetation and Hydrology*, Tech. Comm. No. 53, Commonwealth Bureau of Soils, Harpenden (124 pp.).

Pike, W. S. (1993) The heavy rainfalls of 22–23 September 1992, *Met. Mag.* 122, 201–9.

Ratcliffe, R. A. S. (1978) Meteorological aspects of the 1975–6 drought, *Proc. Roy. Soc. Lond. Sect. A* 363, 3–20.

Reitan, C. H. (1960) Mean monthly values of precipitable water over the United States, 1946–56, *Monthly Weather Review* 88, 25–35.

Roach, W. T. (1994) Back to basics: Fog. Part 1–Definitions and basic physics, *Weather* 49(12), 411–15.

Rodda, J. C. (1970) Rainfall excesses in the United Kingdom, *Trans. Inst. Brit. Geog.* 49, 49–60.

Rodhe, H. (1989) Acidification in a global perspective, *Ambio* 18, 155–60.

Schwartz, S. E. (1989) Acid deposition: unravelling a regional phenomenon, *Science* 243, 753–63.

Sevruk, B. (ed.) (1985) Correction of precipitation measurements, *Zürcher Geogr. Schriften* No. 23 (also appears as WMO Rep. No. 24, Instruments and Observing Methods, WMO, Geneva) (288 pp.).

Smith, F. B. (1991) An overview of the acid rain problem, *Met. Mag.* 120, 77–91.

So, C. L. (1971) Mass movements associated with the rainstorm of June 1966 in Hong Kong, *Trans. Inst. Brit. Geog.* 53, 55–65.

Sutcliffe, R. C. (1956) Water balance and the general circulation of the atmosphere, *Quart. J. Roy. Met. Soc.* 82, 385–95.

Twort, A. C., Hoather, R. C. and Law, F. M. (1985) *Water Supply*, Arnold, London.

Ward, R. C. (1963) Measuring potential evapotranspiration, *J. Geography* 47, 49–55.

Weischet, W. (1965) Der tropische-konvektive und der ausser tropischeadvektive Typ der vertikalen Niederschlagsverteilung, *Erdkunde* 19, 6–14.

Wilhite, D. A. and Glantz, M. H. (1982) Understanding the drought phenomenon: the role of definitions, *Water Internat.* 10, 111–30.

World Meteorological Organization (1972) *Distribution of Precipitation in Mountainous Areas* (2 vols), WMO No. 326, Geneva (228 and 587 pp.).

Yarnell, D. L. (1935) Rainfall intensity–frequency data, US Dept. Agr., Misc. Pub. no. 204.

CHAPTER 4: ATMOSPHERIC INSTABILITY, CLOUD FORMATION AND PRECIPITATION PROCESSES

Andersson, T. (1980) Bergeron and the oreigenic (orographic) maxima of precipitation, *Pure Appl. Geophys.* 119, 558–76.

Barry, R. G., Scharfen, G. R., Knowles, K. W. and Goodman, S. J. (1994) Global distribution of lighting mapped from night-time visible band DMSP satellite data, *Révue generale d'Electricité* 6, 13–16.

Bennetts, D. A., McCallum, E. and Grant, J. R. (1986) Cumulonimbus clouds: an introductory review, *Met. Mag.* 115, 242–56.

Bergeron, T. (1960) Problems and methods of rainfall investigation; in *The Physics of Precipitation*, Geophysical Monograph 5, Amer. Geophys. Union, Washington, DC, 5–30.

Braham, R. R. (1959) How does a raindrop grow? *Science* 129, 123–9.

Browning, K. A. (1980) Local weather forecasting, *Proc. Roy. Soc. Lond. Sect. A* 371, 179–211.

Browning, K. A. (1985) Conceptual models of precipitation systems, *Met. Mag.* 114, 293–319.

Browning, K. A. and Hill, F. F. (1981) Orographic rain, *Weather* 36, 326–9.

Brugge, R. (1996) Back to basics. Atmospheric stability: Part 1. Basic concepts. *Weather* 51(4), 134–40.

Byers, H. R. and Braham, R. R. (1949) *The Thunderstorm*, US Weather Bureau.

Chinn, T. J. (1979) How wet is the wettest of the wet West Coast? *New Zealand Alpine Journal* 32, 85–7.

Dudhia, J. (1996) Back to basics: Thunderstorms. Part 1, *Weather* 51(11), 371–6.

Dudhia, J. (1997) Back to basics: Thunderstorms. Part 2 – Storm types and associated weather, *Weather* 52(1), 2–7.

Dunne, T. and Leopold, L. B. (1978) *Water in Environmental Planning*, W. H. Freeman and Co., San Francisco (818 pp.).

Durbin, W. G. (1961) An introduction to cloud physics, *Weather* 16, 71–82 and 113–25.

East, T. W. R. and Marshall, J. S. (1954) Turbulence in clouds as a factor in precipitation, *Quart. J. Roy. Met. Soc.* 80, 26–47.

Eyre, L. A. (1992) How severe can a 'severe thunderstorm' be? *Weather* 47, 374–83.

Griffiths, D. J., Colquhoun, J. R., Batt, K. L. and Casinader, T. R. (1993) Severe thunderstorms in New South Wales: Climatology and means of assessing the impact of climate change, *Climatic Change* 25, 369–88.

Henderson, R. (1993) Extreme storm rainfalls in the Southern Alps, New Zealand; in *Extreme Hydrological Events: Precipitation, Floods and Droughts (Proceedings of the Yokohama Symposium)*, IAHS Pub. No. 213, 113–20.

Hirschboeck, K. K. (1987) Catastrophic flooding and atmospheric circulation anomalies; in Mayer, L. and Nash, D. (eds) *Catastrophic Flooding*, Allen & Unwin, Boston, 23–56.

Hopkins, M. M., Jr (1967) An approach to the classification of meteorological satellite data, *J. Appl. Met.* **6**, 164–78.

Houze, R. A., Jr and Hobbs, P. V. (1982) Organization and structure of precipitating cloud systems, *Adv. Geophys.* **24**, 225–315.

Jonas, P. R. (1994) Back to basics: Why do clouds form? *Weather* **49**(5), 176–80.

Jonas, P. R. (1994) Back to basics: Why does it rain? *Weather* **49**(7), 258–60.

Kyle, H. L. *et al.* (1993) The Nimbus Earth Radiation Budget (ERB) experiment: 1975–1992, *Bull. Amer. Met. Soc.* **74**, 815–30.

Latham, J. (1966) Some electrical processes in the atmosphere, *Weather* **21**, 120–7.

London, J., Warren, S. G. and Hahn, C. J. (1989) The global distribution of observed cloudiness – a contribution to the ISCPP, *Adv. Space Res.* **9**(7), 161–5.

Ludlam, F. H. (1980) *Clouds and Storms. The Behavior and Effect of Water in the Atmosphere*, Pennsylvania State University, University Park and London (405 pp.).

Mason, B. J. (1959) Recent developments in the physics of rain and rainmaking, *Weather* **14**, 81–97.

Mason, B. J. (1962) Charge generation in thunderstorms, *Endeavour* **21**, 156–63.

Mason, B. J. (1975) *Clouds, Rain and Rainmaking* (2nd edn), Cambridge University Press, Cambridge and New York (189 pp.).

Pearce, F. (1994) Not warming, but cooling, *New Scientist* **143**, 37–41.

Pike, W. S. (1993) The heavy rainfalls of 22–23 September 1992, *Met. Mag.* **122**, 201–9.

Sawyer, J. S. (1956) The physical and dynamical problems of orographic rain, *Weather* **11**, 375–81.

Schermerhorn, V. P. (1967) Relations between topography and annual precipitation in western Oregon and Washington, *Water Resources Research* **3**, 707–11.

Smith, R. B. (1989) Mechanisms of orographic precipitation, *Met. Mag.* **118**, 85–8.

Sumner, G. (1996) Precipitation weather, *J. Geography* **81**(4), 327–45

Weston, K. J. (1977) Cellular cloud patterns, *Weather* **32**, 446–50.

World Meteorological Organization (1956) *International Cloud Atlas*, Geneva.

Wratt, D. S. *et al.* (1996) The New Zealand Southern Alps Experiment, *Bull. Amer. Met. Soc.* **77**(4), 683–92.

CHAPTER 5: ATMOSPHERIC MOTION: PRINCIPLES

Barry, R. G. (1967) Models in meteorology and climatology; in Chorley, R. J. and Haggett, P. (eds) *Models in Geography*, Methuen, London, 97–144.

Beran, W. D. (1967) Large amplitude lee waves and chinook winds, *J. Appl. Met.* **6**, 865–77.

Brinkmann, W. A. R. (1971) What is a foehn? *Weather* **26**, 230–9.

Brinkmann, W. A. R. (1974) Strong downslope winds at Boulder, Colorado, *Monthly Weather Review* **102**, 592–602.

Buettner, K. J. and Thyer, N. (1965) Valley winds in the Mount Rainer area, *Archiv. Met. Geophys. Biokl.* B **14**, 125–47.

Eddy, A. (1966) The Texas coast sea-breeze: a pilot study, *Weather* **21**, 162–70.

Ernst, J. A. (1976) SMS-1 Night-time infrared imagery of low-level mountain waves, *Monthly Weather Review* **104**, 207–9.

Flohn, H. (1969) Local wind systems; in Flohn, H. (ed.) *General Climatology*, World Survey of Climatology **2**, Elsevier, Amsterdam, 139–71.

Geiger, R. (1969) Topoclimates; in Flohn, H. (ed) *General Climatology*, World Survey of Climatology **2**, Elsevier, Amsterdam, 105–38.

Glenn, C. L. (1961) The chinook, *Weatherwise* **14**, 175–82.

Johnson, A. and O'Brien, J. J. (1973) A study of an Oregon sea breeze event, *J. Appl. Met.* **12**, 1,267–83.

Lockwood, J. G. (1962) Occurrence of föhn winds in the British Isles, *Met. Mag.* **91**, 57–65.

McDonald, J. E. (1952) The Coriolis effect, *Sci. American* **186**, 72–8.

Oke, T. R. (1978) *Boundary Layer Climates*, Methuen, London (372 pp.).

Scorer, R. S. (1958) *Natural Aerodynamics*, Pergamon Press, Oxford (312 pp.).

Scorer, R. S. (1961) Lee waves in the atmosphere, *Sci. American* **204**, 124–34.

Steinacker, R. (1984) Area–height distribution of a valley and its relation to the valley wind, *Contrib. Atmos. Phys.* **57**, 64–74.

Thompson, B. W. (1986) Small-scale katabatics and cold hollows, *Weather* **41**, 146–53.

Waco, D. E. (1968) Frost pockets in the Santa Monica Mountains of southern California, *Weather* **23**, 456–61.

Wallington, C. E. (1960) An introduction to lee waves in the atmosphere, *Weather* **15**, 269–76.

Wells, N. (1986) *The Atmosphere and Ocean. A Physical Introduction*, Taylor and Francis, London (345 pp.).

Wickham, P. G. (1966) Weather for gliding over Britain, *Weather* **21**, 154–61.

CHAPTER 6: PLANETARY-SCALE MOTIONS IN THE ATMOSPHERE AND OCEAN

Barry, R. G. (1967) Models in meteorology and climatology; in Chorley, R. J. and Haggett, P. (eds) *Models in Geography*, Methuen, London, 97–144.

Barry, R. G. (1979) Recent advances in climate theory based on simple climate models, *Prog. Phys. Geog.* **3**, 259–86.

Bearman, G. (ed.) (1989) *Ocean Circulation*, The Open University, Pergamon Press, Oxford (238 pp.).

Borchert, J. R. (1953) Regional differences in world atmospheric circulation, *Ann. Assn Amer. Geog.* **43**, 14–26.

Bottomley, M. *et al.* (1990) *Global Ocean Surface Temperature Atlas*, Meteorological Office, London.

Boville, B. A. and Randel, W. J. (1986) Observations and simulation of the variability of the stratosphere and troposphere in January, *J. Atmos. Sci.* **43**, 3,015–34.

Bowditch, N. (1966) American practical navigator, US Navy Hydrographic Office, Pub. No. 9, US Naval Oceanographic Office, Washington, DC.

Broecker, W. S. and Denton, G. H. (1990) What drives glacial cycles?, *Sci. American* **262**(1), 43–50.

Broecker, W. S., Peteet, D. M. and Rind, D. (1985) Does the ocean–atmosphere system have more than one stable mode of operation? *Nature* **315**, 21–6.

Corby, G. A. (ed.) (1970) *The Global Circulation of the Atmosphere*, Roy. Met. Soc., London (257 pp.).

Crowe, P. R. (1949) The trade wind circulation of the world, *Trans. Inst. Brit. Geog.* **15**, 38–56.

Crowe, P. R. (1950) The seasonal variation in the strength of the trades, *Trans. Inst. Brit. Geog.* **16**, 23–47.

Crowley, T. J. and North, G. R. (1991) *Paleoclimatology*, Oxford University Press, New York and Oxford (339 pp.).

Defant, F. and Taba, H. (1957) The threefold structure of the atmosphere and the characteristics of the tropopause, *Tellus* **9**, 259–74.

Dietrich, G. (1963) *General Oceanography: An Introduction*, Wiley, New York (588 pp.).

Druyan, L. M., Somerville, R. J. C. and Quirk, W. J. (1975) Extended-range forecasts with the GISS model of the global atmosphere, *Monthly Weather Review* **103**, 779–95.

Dunn, C. R. (1957) The weather and circulation of December 1957: High index and abnormal warmth, *Monthly Weather Review* **85**, 490–516.

Flohn, H. and Fantechi, R. (eds) (1984) *The Climate of Europe: Past, Present and Future*, D. Reidel, Dordrecht (356 pp.).

Foreman, S. J. (1992) The role of ocean models in FOAM, *Met. Mag.* **121**, 113–22.

Gates, W. L. (1992) The Atmospheric Model Inter-comparison Project, *Bull. Amer. Met. Soc.* **73**(120), 1,962–70.

Hare, F. K. (1965) Energy exchanges and the general circulation, *J. Geography* **50**, 229–41.

Hastenrath, S. (1980) Heat budget of the tropical ocean and atmosphere, *J. Phys. Oceanography* **10**, 159–70.

Hoskins, B. J. (1996) On the existence and strength of the summer subtropical anticyclones, *Bull. Amer. Met. Soc.* **77**(6), 1,287–91.

Houghton, J. (ed.) (1984) *The Global Climate*, Cambridge University Press, Cambridge (233 pp.).

Indian Meteorological Department (1960) *Monsoons of the World*, Delhi (270 pp.).

Kerr, R. A. (1988) Linking earth, ocean and air at the AGU, *Science* **239**, 259–60.

Kiehl, J. T. (1992) Atmospheric general circulation modelling; in Trenberth, K. E. (ed.) *Climate System Modelling*, Cambridge University Press, Cambridge, 319–69.

Klein, W. H. (1958) The weather and circulation of February 1958: A month with an expanded circumpolar vortex of record intensity, *Monthly Weather Review* **86**, 60–70.

Kuenen, Ph. H. (1955) *Realms of Water*, Cleaver-Hulme Press, London (327 pp.).

Lamb, H. H. (1960) Representation of the general atmospheric circulation, *Met. Mag.* **89**, 319–30.

LeMarshall, J. F., Kelly, G. A. M. and Karoly, D. J. (1985) An atmospheric climatology of the southern hemisphere based on 10 years of daily numerical analyses (1972–1982): I. Overview, *Austral. Met. Mag.* **33**, 65–86.

Levitus, S. (1982) *Climatological Atlas of the World Ocean*, NOAA Professional Paper No. 13, Rockville, Md. (173 pp.).

Lorenz, E. N. (1967) *The Nature and Theory of the General Circulation of the Atmosphere*, World Meteorological Organization, Geneva (161 pp.).

Martyn, D. (1992) *Climates of the World*, Elsevier, Amsterdam (435 pp.).

Meehl, G. A. (1984) Modelling the earth's climate, *Climatic Change* **6**, 259–86.

Meehl, G. A. (1987a) The annual cycle and interannual variability in the tropical Pacific and Indian Ocean regions, *Monthly Weather Review* **115**, 51–74.

Meehl, G. A. (1987b) The tropics and their role in the global climate system, *Geographical Journal* **153**, 21–36.

Meehl, G. A. (1992) Global coupled models: Atmosphere, ocean, sea ice; in Trenberth, K. E. (ed.) *Climate System Modelling*, Cambridge University Press, Cambridge, 555–81.

Monin, A. S. (1975) Role of oceans in climate models; in *Physical Basis of Climate: Climate Modelling*, GARP Publications Series, Rept. no. 16, WMO, Geneva, 201–5.

Namias, J. (1972) Large-scale and long-term fluctuations in some atmospheric and ocean variables; in Dyrssen, D. and Jagner, D. (eds) *The Changing Chemistry of the Oceans*, Nobel Symposium 20, Wiley, New York, 27–48.

NASA (n.d.) *From Pattern to Process: The Strategy of the Earth Observing System* (Vol. II), EOS Science Steering Committee Report, NASA, Houston.

Niiler, P.P. (1992) The ocean circulation; in Trenberth, K. E. (ed.) *Climate System Modelling*, Cambridge University Press, Cambridge, 117–48.

O'Connor, J. F. (1961) Mean circulation patterns based on 12 years of recent northern hemispheric data, *Monthly Weather Review* **89**, 211–28.

Palmén, E. (1951) The role of atmospheric disturbances in the general circulation, *Quart. J. Roy. Met. Soc.* **77**, 337–54.

Pfeffer, R. L. (1964) The global atmospheric circulation, *Trans. New York Acad. Sci., Ser. 11*, **26**, 984–97.

Phillips, T. J. (1996) Documentation of the AMIP models on the World Wide Web, *Bull. Amer. Met. Soc.* **77**(6), 1,191–6.

Riehl, H. (1962a) General atmospheric circulation of the tropics, *Science* **135**, 13–22.

Riehl, H. (1962b) *Jet streams of the atmosphere*, Tech. Paper No. 32, Colorado State University (117 pp.).

Riehl, H. (1969) On the role of the tropics in the general circulation of the atmosphere, *Weather* **24**, 288–308.

Riehl, H. *et al.* (1954) The jet stream, *Met. Monogr.* **2**(7), American Meteorological Society, Boston, Mass. (100 pp.).

Riley, D. and Spalton, L. (1981) *World Weather and*

Climate (2nd edn), Cambridge University Press, London (128 pp.).

Rodwell, M. J. and Hoskins, B. J. (1996) Monsoons and the dynamics of deserts, *Quart. J. Roy. Met. Soc.* **122**, 1,385–404.

Rossby, C.-G. (1941) The scientific basis of modern meteorology, US Dept of Agriculture Yearbook *Climate and Man*, 599–655.

Rossby, C.-G. (1949) On the nature of the general circulation of the lower atmosphere; in Kulper, G. P. (ed.) *The Atmosphere of the Earth and Planets*, University of Chicago Press, Chicago, Ill., 16–48.

Saltzman, B. (1983) Climatic systems analysis, *Adv. Geophys.* **25**, 173–233.

Sawyer, J. S. (1957) Jet stream features of the earth's atmosphere, *Weather* **12**, 333–4.

Shapiro, M. A. *et al.* (1987) The Arctic tropopause fold, *Monthly Weather Review* **115**, 444–54.

Shapiro, M. A. and Keyser, D. A. (1990) Fronts, jet streams, and the tropopause; in Newton, C. W. and Holopainen, E. D. (eds) *Extratropical Cyclones*, American Meteorological Society, Boston, 167–91.

Starr, V. P. (1956) The general circulation of the atmosphere, *Sci. American* **195**, 40–5.

Stowe, K. (1983) *Ocean Science* (2nd edn), John Wiley & Sons, New York (673 pp.).

Strahler, A. N. and Strahler, A. H. (1992) *Modern Physical Geography* (4th edn), John Wiley & Sons, New York (638 pp.).

Streten, N. A. (1980) Some synoptic indices of the southern hemisphere mean sea level circulation 1972–77, *Monthly Weather Review* **108**, 18–36.

Taljaard, J. J., van Loon, H., Crutcher, H. L. and Jenne, R. L. (1969) *Climate of the Upper Air, Part 1, Southern Hemisphere* 1, Naval Weather Service Command, Washington, DC, NAVAIR 50–1C–55.

Tolmazin, D. (1994) *Elements of Dynamic Oceanography*, Chapman & Hall, London.

Trenberth, K. E. (ed.) (1992) *Climate System Modelling*, Cambridge University Press, Cambridge (788 pp.).

Troen, I. and Petersen, E. L. (1989) *European Wind Atlas*, Commission of the Economic Community, Risø National Laboratory, Roskilde, Denmark (656 pp.).

Tucker, G. B. (1962) The general circulation of the atmosphere, *Weather* **17**, 320–40.

Tyson, P. D. (1986) *Climatic Change and Variability in Southern Africa*, Oxford University Press, Cape Town (220 pp.).

van Arx, W. S. (1962) *Introduction to Physical Oceanography*, Addison-Wesley, Reading, Mass. (422 pp.).

van Loon, H. (1964) Mid-season average zonal winds at sea level and at 500 mb south of 25°S and a brief comparison with the northern hemisphere, *J. Appl. Met.* **3**, 554–63.

van Loon, H. (ed.) (1984) *Climates of the Oceans*; in Landsberg, H. E. (ed.) *World Survey of Climatology* 15, Elsevier, Amsterdam (716 pp.).

Walker, J. M. (1972) Monsoons and the global circulation, *Met. Mag.* **101**, 349–55.

Wallington, C. E. (1969) Depressions as moving vortices, *Weather* **24**, 42–51.

Washington, W. M. (1992) Climate-model responses to increased CO_2 and other greenhouse gases; in

Trenberth, K. E. (ed.) *Climate System Modelling*, Cambridge University Press, Cambridge, 643–68.

CHAPTER 7: MID-LATITUDE SYNOPTIC SYSTEMS

Bader, M. J. *et al.* (1995) *Images in Weather Forecasting*, Cambridge University Press, Cambridge (499 pp.).

Bates, F. C. (1962) Tornadoes in the central United States, *Trans. Kansas Acad. Sci.* **65**, 215–46.

Belasco, J. E. (1952) Characteristics of air masses over the British Isles, Meteorological Office, *Geophysical Memoirs* **11**(87) (34 pp.).

Bennetts, D. A., Grant, J. R. and McCallum, E. (1988) An introductory review of fronts: Part I Theory and observations, *Met. Mag.* **117**, 357–70.

Bosart, L. (1985) Weather forecasting; in Houghton, D. D. (ed.) *Handbook of Applied Meteorology*, Wiley, New York, 205–79.

Boucher, R. J. and Newcomb, R. J. (1962) Synoptic interpretation of some TIROS vortex patterns: a preliminary cyclone model, *J. Appl. Met.* **1**, 122–36.

Boyden, C. J. (1963) Development of the jet stream and cut-off circulations, *Met. Mag.* **92**, 287–99.

Browning, K. A. (1968) The organization of severe local storms, *Weather* **23**, 429–34.

Browning, K. A. (1980) Local weather forecasting, *Proc. Roy. Soc. Lond. Sect. A* **371**, 179–211.

Browning, K. A. (ed.) (1983) *Nowcasting*, Academic Press, New York (256 pp.).

Browning, K. A. (1985) Conceptual models of precipitation systems, *Met. Mag.* **114**, 293–319.

Browning, K. A. (1986) Weather radar and FRONTIERS, *Weather* **41**, 9–16.

Browning, K. A. (1990) Organization of clouds and precipitation in extratropical cyclones; in Newton, C. W. and Holopainen, E. D. (eds) *Extratropical Cyclones. The Erik Palmén Memorial Volume*, American Meteorological Society, Boston, 129–53.

Browning, K. A., Bader, M. J., Waters, A. J., Young, M. V. and Monk, G. A. (1987) Application of satellite imagery to nowcasting and very short range forecasting, *Met. Mag.* **116**, 161–79.

Browning, K. A. and Hill, F. F. (1981) Orographic rain, *Weather* **36**, 326–9.

Browning, K. A. and Roberts, N. M. (1994) Structure of a frontal cyclone. *Quart. J. Roy. Met. Soc.* **120**, 1,535–57.

Businger, S. (1985) The synoptic climatology of polar low outbreaks, *Tellus* **37A**, 419–32.

Carleton, A. M. (1985) Satellite climatological aspects of the 'polar low' and 'instant occlusion', *Tellus* **37A**, 433–50.

Carleton, A. M. (1996) Satellite climatological aspects of cold air mesocyclones in the Arctic and Antarctic, *Global Atmos. Ocean System* **5**, 1–42.

Crowe, P. R. (1949) The trade wind circulation of the world, *Trans. Inst. Brit. Geog.* **15**, 38–56.

Crowe, P. R. (1965) The geographer and the atmosphere, *Trans. Inst. Brit. Geog.* **36**, 1–19.

Cullen, M. J. P. (1993) The Unified Forecast/Climate model, *Met. Mag.* **122**, 81–94.

Freeman, M. H. (1961) Fronts investigated by the Meteorological Research Flight, *Met. Mag.* **90**, 189–203.

Galloway, J. L. (1958a) The three-front model: its philosophy, nature, construction and use, *Weather* **13**, 3–10.

Galloway, J. L. (1958b) The three-front model, the tropopause and the jet stream, *Weather* **13**, 395–403.

Galloway, J. L. (1960) The three-front model, the developing depression and the occluding process, *Weather* **15**, 293–309.

Godson, W. L. (1950) The structure of North American weather systems, *Cent. Proc. Roy. Met. Soc.* London, 89–106.

Gyakum, J. R. (1983) On the evolution of the *QE II* storm, I: Synoptic aspects, *Monthly Weather Review* **111**, 1,137–55.

Hare, F. K. (1960) The westerlies, *Geog. Rev.* **50**, 345–67.

Harman, J. R. (1971) Tropical waves, jet streams, and the United States weather patterns, Association of American Geographers, Commission on College Geography, Resource Paper No. 11 (37 pp.).

Harrison, M. J. S. (1995) Long-range forecasting since 1980 – empirical and numerical prediction out to one month for the United Kingdom, *Weather* **50**(12), 440–9.

Harrold, T. W. (1973) Mechanisms influencing the distribution of precipitation within baroclinic disturbances, *Quart. J. Roy. Met. Soc.* **99**, 232–51.

Hindley, K. (1977) Learning to live with twisters, *New Scientist* **70**, 280–2.

Hobbs, P. V. (1978) Organization and structure of clouds and precipitation on the meso-scale and micro-scale of cyclonic storms, *Rev. Geophys. and Space Phys.* **16**, 741–55.

Hobbs, P. W., Locatelli, J. D. and Martin, J. E. (1996) A new conceptual model for cyclones generated in the lee of the Rocky Mountains, *Bull. Amer. Met. Soc.* **77**(6), 1,169–78.

Houze, R. A. and Hobbs, P. V. (1982) Organization and structure of precipitating cloud systems, *Adv. Geophys.* **24**, 225–315.

Hughes, P. and Gedzelman, S. D. (1995) Superstorm success, *Weatherwise* **48**(3), 18–24.

Hunt, J. C. R. (1994) Developments in forecasting the atmospheric environment, *Weather* **49**(9), 312–18.

Jackson, M. C. (1977) Meso-scale and small-scale motions as revealed by hourly rainfall maps of an outstanding rainfall event: 14–16 September 1968, *Weather* **32**, 2–16.

Kalnay, E., Kanamitsu, M. and Baker, W. E. (1990) Global numerical weather prediction at the National Meteorological Center, *Bull. Amer. Met. Soc.* **71**, 1,410–28.

Kalnay, E. *et al.* (1996) The NCEP/NCAR 40-year reanalysis project, *Bull. Amer. Met. Soc.* **77**(3), 437–71.

Kessler, E. (ed.) (1983) *The Thunderstorm in Human Affairs* (2nd edn), University of Oklahoma Press, Norman, Okla. (250 pp.).

Klein, W. H. (1948) Winter precipitation as related to the 700 mb circulation, *Bull. Amer. Met. Soc.* **29**, 439–53.

Klein, W. H. (1957) *Principal tracks and mean frequencies of cyclones and anticyclones in the Northern Hemisphere*, Research Paper No. 40, Weather Bureau, Washington, DC (60 pp.).

Klein, W. H. (1982) Statistical weather forecasting on different time scales, *Bull. Amer. Met. Soc.* **63**, 170–7.

Kocin, P. J., Schumacher, P. N., Morales, R. F., Jr and Uccellini, L. W. (1995) Overview of the 12–14 March 1993 Superstorm, *Bull. Amer. Met. Soc.* **76**(2), 165–82.

Liljequist, G. H. (1970) *Klimatologi*, Generalstabens Litografiska Anstalt, Stockholm.

Locatelli, J. D., Martin, J. E., Castle, J. A. and Hobbs, P. V. (1995) Structure and evolution of winter cyclones in the central United States and their effects on the distribution of precipitation: Part III. The development of a squall line associated with weak cold frontogenesis aloft, *Monthly Weather Review* **123**, 2,641–62.

Ludlam, F. H. (1961) The hailstorm, *Weather* **16**, 152–62.

Lyall, I. T. (1972) The polar low over Britain, *Weather* **27**, 378–90.

Maddox, R. A. (1980) Mesoscale convective complexes, *Bull. Amer. Met. Soc.* **61**, 1,374–87.

McCallum, E. and Mansfield, D. (1996) Weather forecasts in 1996, *Weather* **51**(5), 181–8.

McPherson, R. D. (1994) The National Centers for Environmental Prediction: Operational climate, ocean and weather prediction for the 21st century, *Bull. Amer. Met. Soc.* **75**(3), 363–73.

Miles, M. K. (1962) Wind, temperature and humidity distribution at some cold front over SE England, *Quart. J. Roy. Met. Soc.* **88**, 286–300.

Miller, R. C. (1959) Tornado-producing synoptic patterns, *Bull. Amer. Met. Soc.* **40**, 465–72.

Miller, R. C. and Starrett, L. G. (1962) Thunderstorms in Great Britain, *Met. Mag.* **91**, 247–55.

Monk, G. A. (1992) Synoptic and mesoscale analysis of intense mid-latitude cyclones, *Met. Mag.* **121**, 269–83.

National Weather Service (1991) *Experimental Long-Lead Forecast Bulletin*, NOAA, Climate Prediction Center, Washington, DC.

Newton, C. W. (1966) Severe convective storms, *Adv. Geophys.* **12**, 257–308.

Newton, C. W. (ed.) (1972) *Meteorology of the Southern Hemisphere*, Met. Monogr. No. 13 (35), American Meteorological Society, Boston, Mass. (263 pp.).

Newton, C. W. and Holopainen, E. D. (eds) (1990) *Extratropical Cyclones: Palmén Memorial Symposium*, American Meteorological Society, Boston (262 pp.).

Palmer, T. N. and Anderson, D. L. T. (1994) The prospects for seasonal forecasting – a review paper, *Quart. J. Roy. Met. Soc.* **120**, 755–93.

Pedgley, D. E. (1962) A meso-synoptic analysis of the thunderstorms on 28 August 1958, *Geophys. Memo. Meteorolog. Office* **14**(1) (30 pp.).

Penner, C. M. (1955) A three-front model for synoptic analyses, *Quart. J. Roy. Met. Soc.* **81**, 89–91.

Petterssen, S. (1950) *Some aspects of the general circulation of the atmosphere*, *Cent. Proc. Roy. Met. Soc.*, London, 120–55.

Portelo, A. and Castro, M. (1996) Summer thermal lows in the Iberian Peninsula: a three-dimensional simulation, *Quart. J. Roy. Met. Soc.* **122**, 1–22.

Reed, D. N. (1995) Developments in weather forecasting for the medium range, *Weather* **50**(12), 431–40.

Reed, R. J. (1960) Principal frontal zones of the northern hemisphere in winter and summer, *Bull. Amer. Met. Soc.* **41**, 591–8.

Richter, D. A. and Dahl, R. A. (1958) Relationship of heavy precipitation to the jet maximum in the eastern United States, *Monthly Weather Review* 86, 368–76.

Riley, D. and Spalton, L. (1974) *World Weather and Climate*, Cambridge University Press, Cambridge (120 pp.).

Roebber, P. J. (1989) On the statistical analysis of cyclone deepening rates, *Monthly Weather Review* 117, 2, 293–8.

Sanders, F. and Gyakum, J. R. (1980) Synoptic-dynamic climatology of the 'bomb', *Monthly Weather Review* 108, 1,589–606.

Shapiro, M. A. and Keyser, D. A. (1990) Fronts, jet streams and the tropopause; in Newton, C. W. and Holopainen, E. O. (eds) *Extratropical Cyclones. The Erik Palmén Memorial Volume*, Amer. Met. Soc., Boston, Mass., 167–91.

Showalter, A. K. (1939) Further studies of American air mass properties, *Monthly Weather Review* 67, 204–18.

Slater, P. M. and Richards, C. J. (1974) A memorable rainfall event over southern England, *Met. Mag.* 103, 255–68 and 288–300.

Smagorinsky, J. (1974) Global atmospheric modeling and the numerical simulation of climate; in Hess, W. N. (ed.) *Weather and Climate Modification*, Wiley, New York, 633–86.

Smith, W. L. (1985) Satellites; in Houghton, D. D. (ed.) *Handbook of Applied Meteorology*, Wiley, New York, 380–472.

Snow, J. T. (1984) The tornado, *Sci. American* 250(4), 56–66.

Sumner, G. (1996) Precipitation weather, *Geography* 81, 327–45.

Sutcliffe, R. C. and Forsdyke, A. G. (1950) The theory and use of upper air thickness patterns in forecasting, *Quart. J. Roy. Met. Soc.* 76, 189–217.

Vederman, J. (1954) The life cycles of jet streams and extratropical cyclones, *Bull. Amer. Met. Soc.* 35, 239–44.

Wagner, A. J. (1989) Medium- and long-range weather forecasting, *Weather and Forecasting* 4, 413–26.

Wallington, C. E. (1963) Meso-scale patterns of frontal rainfall and cloud, *Weather* 18, 171–81.

Wendland, W. M. and Bryson, R. A. (1981) Northern hemisphere airstream regions, *Monthly Weather Review* 109, 255–70.

Wendland, W. M. and McDonald, N. S. (1986) Southern hemisphere airstream climatology, *Monthly Weather Review* 114, 88–94.

Wick, G. (1973) Where Poseidon courts Aeolus, *New Scientist*, 18 January, 123–6.

Yoshino, M. M. (1967) Maps of the occurrence frequencies of fronts in the rainy season in early summer over east Asia, Science Reports of the Tokyo University of Education 89, 211–45.

CHAPTER 8: WEATHER AND CLIMATE IN TEMPERATE AND HIGH LATITUDES

Axelrod, D. I. (1992) What is an equable climate? *Palaeogeogr., Palaeoclim., Palaeoecol.* 91, 1–12.

Balling, R. C., Jr (1985) Warm seasonal nocturnal precipitation in the Great Plains of the United States, *J. Climate Appl. Met.* 24, 1,383–7.

Barry, R. G. (1963) Aspects of the synoptic climatology of central south England, *Met. Mag.* 92, 300–8.

Barry, R. G. (1967a) Seasonal locational of the arctic front over North America, *Geog. Bull.* 9, 79–95.

Barry, R. G. (1967b) The prospects for synoptic climatology: a case study; in Steel, R. W. and Lawton, R. (eds) *Liverpool Essays in Geography*, Longman, London, 85–106.

Barry, R. G. (1973) A climatological transect on the east slope of the Front Range, Colorado, *Arct. Alp. Res.* 5, 89–110.

Barry, R. G. (1983) Arctic Ocean ice and climate: Perspectives on a century of polar research, *Ann. Assn Amer. Geog.* 73(4), 485–501.

Barry, R. G. (1986) Aspects of the meteorology of the seasonal sea ice zone; in Untersteiner, N. (ed.) *The Geophysics of Sea Ice*, Plenum Press, New York 993–1,020.

Barry, R. G. (1996) Arctic; in Schneider, S. H. (ed.) *Encyclopedia of Climate and Weather*, Oxford University Press, New York, 43–7.

Barry, R. G. and Hare, F. K. (1974) Arctic climate; in Ives, J. D. and Barry, R. G. (eds) *Arctic and Alpine Environments*, Methuen, London, 17–54.

Belasco, J. E. (1952) Characteristics of air masses over the British Isles, Meteorological Office, *Geophysical Memoirs* 11(87) (34 pp.).

Blackall, R. M. and Taylor, P. L. (1993) The thunderstorms of 19/20 August 1992 – a view from the United Kingdom, *Met. Mag.* 122, 189.

Boast, R. and McQuingle, J. B. (1972) Extreme weather conditions over Cyprus during April 1971, *Met. Mag.* 101, 137–53.

Borchert, J. (1950) The climate of the central North American grassland, *Ann. Assn Amer. Geog.* 40, 1–39.

Browning, K. A. and Hill, F. F. (1981) Orographic rain, *Weather* 36, 326–9.

Bryson, R. A. (1966) Air masses, streamlines and the boreal forest, *Geog. Bull.* 8, 228–69.

Bryson, R. A. and Hare, F. K. (eds) (1974) *Climates of North America*, World Survey of Climatology 11, Elsevier, Amsterdam (420 pp.).

Bryson, R. A. and Lahey, J. F. (1958) *The March of the Seasons*, Meteorological Department, University of Wisconsin (41 pp.).

Bryson, R. A. and Lowry, W. P. (1955) Synoptic climatology of the Arizona summer precipitation singularity, *Bull. Amer. Met. Soc.* 36, 329–39.

Burbridge, F. E. (1951) The modification of continental polar air over Hudson Bay, *Quart. J. Met. Soc.* 77, 365–74.

Butzer, K. W. (1960) Dynamic climatology of large-scale circulation patterns in the Mediterranean area, *Meteorologische Rundschau* 13, 97–105.

Carleton, A. M. (1986) Synoptic-dynamic character of 'bursts' and 'breaks' in the southwest US summer precipitation singularity, *J. Climatol.* 6, 605–23.

Carleton, A. M. (1987) Antarctic climates; in Oliver, J. E. and Fairbridge, R. W. (eds) *The Encyclopedia of Climatology*, Van Nostrand Reinhold, New York, 44–64.

Chandler, T. J. and Gregory, S. (eds) (1976) *The Climate of the British Isles*, Longman, London (390 pp.).

Chinn, T. J. (1979) How wet is the wettest of the West Coast? *New Zealand Alp. J.* 32, 84–7.

Climate Prediction Center (1996) Jet streams, pressure distribution and climate for the USA during the winters of 1995–6 and 1994–5, *The Climate Bull.* 96(3), US Dept of Commerce.

Cooter, E. J. and Leduc, S. K. (1995) Recent frost date trends in the north-eastern USA, *Int. J. Climatology* 15, 65–75.

Derecki, J. A. (1976) Heat storage and advection in Lake Erie, *Water Resources Research* 12(6), 1,144–50.

Driscoll, D. M. and Yee Fong, J. M. (1992) Continentality: a basic climatic parameter re-examined, *Int. J. Climatol.* 12, 185–92.

Durrenberger, R. W. and Ingram, R. S. (1978) *Major storms and floods in Arizona 1862–1977*, State of Arizona, Office of the State Climatologist, Climatological Publications, Precipitation Series No. 4 (44 pp.).

Easterling, D. R. and Robinson, P. J. (1985) The diurnal variation of thunderstorm activity in the United States, *J. Climate Appl. Met.* 24, 1,048–58.

Elsom, D. M. and Meaden, G. T. (1984) Spatial and temporal distribution of tornadoes in the United Kingdom 1960–1982, *Weather* 39, 317–23.

Environmental Science Services Administration (1968) *Climatic Atlas of the United States*, US Department of Commerce, Washington, DC (80 pp.).

Evenari, M., Shanan, L. and Tadmor, N. (1971) *The Negev*, Harvard University Press, Cambridge, Mass. (345 pp.).

Ferguson, E. W., Ostby, F. P., Leftwich, P. W., Jr, and Hales, J. E., Jr (1986) The tornado season of 1984, *Monthly Weather Review* 114, 624–35.

Flohn, H. (1954) *Witterung und Klima in Mitteleuropa*, Zurich (218 pp.).

Forrest, B. and Nishenko, S. (1996) Losses due to natural hazards, *Natural Hazards Observer* 21(1), University of Colorado, Boulder, 16–17.

Gentilli, J. (ed.) (1971) *Climates of Australia and New Zealand*, World Survey of Climatology 13, Elsevier, Amsterdam (405 pp.).

Gorcynski, W. (1920) Sur le calcul du degré du continentalisme et son application dans la climatologie, *Geografiska Annaler* 2, 324–31.

Green, C. R. and Sellers, W. D. (1964) *Arizona Climate*, University of Arizona Press, Tucson (503 pp.).

Hales, J. E., Jr (1974) South-western United States summer monsoon source – Gulf of Mexico or Pacific Ocean, *J. Appl. Met.* 13, 331–42.

Hare, F. K. (1968) The Arctic, *Quart. J. Roy. Met. Soc.* 74, 439–59.

Hare, F. K. and Thomas, M. K. (1979) *Climate Canada* (2nd edn), Wiley, Canada (230 pp.).

Hawke, E. L. (1933) Extreme diurnal range of air temperature in the British Isles, *Quart. J. Roy. Met. Soc.* 59, 261–5.

Hill, F. F., Browning, K. A. and Bader, M. J. (1981) Radar and rain gauge observations of orographic rain over South Wales, *Quart. J. Roy. Met. Soc.* 107, 643–70.

Horn, L. H. and Bryson, R. A. (1960) Harmonic analysis of the annual march of precipitation over the United States, *Ann. Assn Amer. Geog.* 50, 157–71.

Hulme, M. *et al.* (1995) Construction of a 1961–1990 European climatology for climate change modelling and impact applications, *Int. J. Climatol.* 15, 1,333–63.

Huttary, J. (1950) Die Verteilung der Niederschläge auf die Jahreszeiten im Mittelmeergebiet, *Meteorologische Rundschau* 3, 111–19.

Klein, W. H. (1963) Specification of precipitation from the 700 mb circulation, *Monthly Weather Review* 91, 527–36.

Knappenberger, P. C. and Michaels, P. J. (1993) Cyclone tracks and wintertime climate in the mid-Atlantic region of the USA, *Int. J. Climatol.* 13, 509–31.

Knox, J. L. and Hay, J. E. (1985) Blocking signatures in the northern hemisphere: frequency distribution and interpretation, *J. Climatology* 5, 1–16.

Lamb, H. H. (1950) Types and spells of weather around the year in the British Isles: annual trends, seasonal structure of the year, singularities, *Quart. J. Roy. Met. Soc.* 76, 393–438.

Linacre, W. and Hobbs, J. (1977) *The Australian Climatic Environment*, Wiley, Brisbane (354 pp.).

Longley, R. W. (1967) The frequency of Chinooks in Alberta, *The Albertan Geographer* 3, 20–2.

Lott, J. N. (1994) The US summer of 1993: a sharp contrast in weather extremes, *Weather* 49, 370–83.

Lumb, F. E. (1961) *Seasonal variations of the sea surface temperature in coastal waters of the British Isles*, Met. Office Sci. Paper No. 6, MO 685 (21 pp.).

Lydolph, P. E. (1977) *Climates of the Soviet Union*, World Survey of Climatology 7, Elsevier, Amsterdam (435 pp.).

Manley, G. (1944) Topographical features and the climate of Britain, *Geog. J.* 103, 241–58.

Manley, G. (1945) The effective rate of altitude change in temperate Atlantic climates, *Geog. Rev.* 35, 408–17.

Martyn, D. (1992) *Climates of the World*, Elsevier, Amsterdam (435 pp.).

Mather, J. R. (1985) The water budget and the distribution of climates, vegetation and soils, *Publications in Climatology* 38(2), Center for Climatic Research, University of Delaware, Newark (36 pp.).

Maytham, A. P. (1993) Sea ice – a view from the Ice Bench, *Met. Mag.* 122, 190–5.

Meteorological Office (1952) *Climatological Atlas of the British Isles*, MO 488, HMSO, London (139 pp.).

Meteorological Office (1962) Weather in the Mediterranean I, *General Meteorology* (2nd edn), MO 391, HMSO, London (362 pp.).

Meteorological Office (1964a) *Weather in the Mediterranean II* (2nd edn), MO 391b, HMSO, London (372 pp.).

Meteorological Office (1964b) *Weather in Home Fleet Waters I, The Northern Seas*, Part 1, MO 732a, HMSO, London (265 pp.).

Namias, J. (1964) Seasonal persistence and recurrence of European blocking during 1958–60, *Tellus* 16, 394–407.

Nickling, W. G. and Brazel, A.J. (1984) Temporal and spatial characteristics of Arizona dust storms (1965–1980), *Climatology* 4, 645–60.

Nkemdirim, L. C. (1996) Canada's chinook belt, *Int. J. Climatol.* 16(4), 427–39.

O'Hare, G. and Sweeney, J. (1993) Lamb's circulation types and British weather: an evaluation, *Geography* 78, 43–60.

Palz, W. (ed.) (1984) *European Solar Radiation Atlas*, 2 vols (2nd edn), Verlag Tüv Rheinland, Cologne (297 and 327 pp.).

Parrett, C., Melcher, N. B. and James, R. W., Jr (1993) Flood discharges in the upper Mississippi River basin, *U.S. Geol. Sur. Circular* **1120-A** (14 pp.).

Pickard, G. L. and Emery, W. J. (1982) *Descriptive Physical Oceanography. An Introduction* (4th edn), Pergamon Press, Oxford.

Poltaraus, B. V. and Staviskiy, D. B. (1986) The changing continentality of climate in central Russia, *Soviet Geography* **27**, 51–8.

Rayner, J. N. (1961) *Atlas of Surface Temperature Frequencies for North America and Greenland*, Arctic Meteorological Research Group, McGill University, Montreal.

Rex, D. F. (1950–1) The effect of Atlantic blocking action upon European climate, *Tellus* **2**, 196–211 and 275–301; **3**, 100–11.

Salinger, M. J., Basher, R. E., Fitzharris, B. B., Hay, J. E., Jones, P. D., McVeigh, J. P. and Schmidely-Leleu, I. (1995) Climate trends in the southwest Pacific, *Int. J. Climatol.* **15**, 285–302.

Schick, A. P. (1971) A desert flood, *Jerusalem Studies in Geography* **2**, 91–155.

Schwartz, M. D. (1995) Detecting structural climate change: an air mass-based approach in the north-central United States, 1958–92, *Ann. Assn Amer. Geog.* **76**, 553–68.

Schwerdtfeger, W. (1984) *Weather and Climate of the Antarctic*, Elsevier, Amsterdam (261 pp.).

Sellers, P. *et al.* (1995) The boreal ecosystem–atmosphere study (BOREAS): an overview and early results from the 1994 field year, *Bull. Am. Met. Soc.* **76**, 1,549–77.

Serreze, M. C. *et al.* (1993) Characteristics of arctic synoptic activity, 1952–1989, *Met. Atmos. Phys.* **51**, 147–64.

Shaw, E. M. (1962) An analysis of the origins of precipitation in Northern England, 1956–60, *Quart. J. Roy. Met. Soc.* **88**, 539–47.

Sivall, T. (1957) Sirocco in the Levant, *Geografiska Annaler* **39**, 114–42.

Stearns, C. R. *et al.* (1993) Mean cluster data for Antarctic weather studies; in Bromwich, D. H. and Stearns, C. R. (eds) *Antarctic Meteorology and Climatology: Studies Based on Automatic Weather Stations*, Antarctic Research Series, Am. Geophys. Union **61**, 1–21.

Stone, J. (1983) Circulation type and the spatial distribution of precipitation over central, eastern and southern England, *Weather* **38**, 173–7, 200–5.

Storey, A. M. (1982) A study of the relationship between isobaric patterns over the UK and central England temperature and England–Wales rainfall, *Weather* **37**, 2–11, 46, 88–9, 122, 151, 170, 208, 244, 260, 294, 327, 360.

Sturman, A. P. and Tapper, N. J. (1996) *The Weather and Climate of Australia and New Zealand*, Oxford University Press (496 pp.).

Sumner, E. J. (1959) Blocking anticyclones in the Atlantic–European sector of the northern hemisphere, *Met. Mag.* **88**, 300–11.

Sweeney, J. C. and O'Hare, G. P. (1992) Geographical variations in precipitation yields and circulation types in Britain and Ireland, *Trans. Inst. Brit. Geog.* (n.s.) **17**, 448–63.

Thomas, M. K. (1964) *A Survey of Great Lakes Snowfall*, Great Lakes Research Division, University of Michigan, Publication No. 11, 294–310.

Thorn, P. (1965) *The Agro-Climatic Atlas of Europe*, Elsevier, Amsterdam.

Thornthwaite, C. W. and Mather, J. R. (1955) The moisture balance, *Publications in Climatology* **8**(1), Laboratory of Climatology, Centerton, NJ (104 pp.).

Tout, D. G. and Kemp, V. (1985) The named winds of Spain, *Weather* **40**, 322–9.

Trenberth, K. E. and Guillemot, C. J. (1996) Physical processes involved in the 1988 drought and 1993 floods in North America, *J. Climate* **9**(6), 1,288–98.

Troen, I. and Petersen, E. L. (1989) *European Wind Atlas*, Commission of the Economic Community, Risø National Laboratory, Roskilde, Denmark (656 pp.).

United States Weather Bureau (1947) *Thunderstorm Rainfall*, Vicksburg, Miss. (331 pp.).

Villmow, J. R. (1956) The nature and origin of the Canadian dry belt, *Ann. Assn Amer. Geog.* **46**, 221–32.

Visher, S. S. (1954) *Climatic Atlas of the United States*, Harvard University Press, Cambridge, Mass. (403 pp.).

Wallace, J. M. (1975) Diurnal variations in precipitation and thunderstorm frequency over the coterminous United States, *Monthly Weather Review* **103**, 406–19.

Wallén, C. C. (1960) Climate; in Somme, A. (ed.) *The Geography of Norden*, Cappelens Forlag, Oslo, 41–53.

Wallén, C. C. (ed.) (1970) *Climates of Northern and Western Europe*, World Survey of Climatology 5, Elsevier, Amsterdam (253 pp.).

Weller, G. and Holmgren, B. (1974) The microclimates of the arctic tundra, *J. App. Met.* **13**(8), 854–62.

Woodroffe, A. (1988) Summary of the weather pattern developments of the storm of 15/16 October 1987, *Met. Mag.* **117**, 99–103.

Wratt, D. S. *et al.* (1996) The New Zealand Southern Alps Experiment, *Bull. Amer. Met. Soc.* **77**(4), 683–92.

CHAPTER 9: TROPICAL WEATHER AND CLIMATE

Academica Sinica (1957–8) On the general circulation over eastern Asia, *Tellus* **9**, 432–46; **10**, 58–75 and 299–312.

Anthes, R. A. (1982) Tropical cyclones: their evolution, structure, and effects, *Met. Monogr.* **19**(41), Amer. Met. Soc., Boston, Mass. (208 pp.).

Arakawa, H. (ed.) (1969) *Climates of Northern and Eastern Asia*, World Survey of Climatology 8, Elsevier, Amsterdam (248 pp.).

Atkinson, G. D. (1971) *Forecasters Guide to Tropical Meteorology*, Headquarters Air Weather Service, US Air Force Tech. Rep. 240 (360 pp.).

Avila, L. A. (1990) Atlantic tropical systems of 1989, *Monthly Weather Review* **118**, 1,178–85.

Barnston, A. G. (1995) Our improving capability in ENSO forecasting, *Weather* **50**(12), 419–30.

Barry, R. G. (1978) Aspects of the precipitation characteristics of the New Guinea mountains, *J. Trop. Geog.* **47**, 13–30.

Beckinsale, R. P. (1957) The nature of tropical rainfall, *Tropical Agriculture* 34, 76–98.

Bhalme, H. N. and Mooley, D. A. (1980) Large-scale droughts/floods and monsoon circulation, *Monthly Weather Review* 108, 1,197–211.

Blumenstock, D. I. (1958) Distribution and characteristics of tropical climates, *Proc. 9th Pacific Sci. Congr.* 20, 3–23.

Breed, C. S. *et al.* (1979) Regional studies of sand seas, using Landsat (ERTS) imagery, US Geological Survey Professional Paper No. 1052, 305–97.

Browning, K. A. (1996) Current research in atmospheric science, *Weather* 51(5), 167–72.

Chang, J.-H. (1962) Comparative climatology of the tropical western margins of the northern oceans, *Ann. Assn Amer. Geog.* 52, 221–7.

Chang, J.-H. (1967) The Indian summer monsoon, *Geog. Rev.* 57, 373–96.

Chang, J.-H. (1971) The Chinese monsoon, *Geog. Rev.* 61, 370–95.

Chopra, K. P. (1973) Atmospheric and oceanic flow problems introduced by islands, *Adv. Geophys.* 16, 297–421.

Clackson, J. R. (1957) The seasonal movement of the boundary of northern air, Nigerian Meteorological Service, Technical Note 5 (see Addendum 1958).

Crowe, P. R. (1949) The trade wind circulation of the world, *Trans. Inst. Brit. Geog.* 15, 37–56.

Crowe, P. R. (1951) Wind and weather in the equatorial zone, *Trans. Inst. Brit. Geog.* 17, 23–76.

Cry, G. W. (1965) Tropical cyclones of the North Atlantic Ocean, Tech. Paper No. 55, Weather Bureau, Washington, DC (148 pp.).

Curry, L. and Armstrong, R. W. (1959) Atmospheric circulation of the tropical Pacific Ocean, *Geografiska Annaler* 41, 245–55.

Das, P. K. (1987) Short- and long-range monsoon prediction in India; in Fein, J. S. and Stephens, P. L. (eds) *Monsoons*, John Wiley and Sons, New York, 549–78.

Dhar, O. N. and Nandargi, S. (1993) Zones of extreme rainstorm activity over India, *Int. J. Climatology* 13, 301–11.

Dickinson, R. E. (ed.) (1987) *The Geophysiology of Amazonia: Vegetation and Climate Interactions*, John Wiley & Sons, New York (526 pp.).

Domrös, M. and Gongbing, P. (1988) *The Climate of China*, Springer-Verlag, Berlin (361 pp.).

Dubief, J. (1963) Le climat du Sahara. Memoire de l'Institut de Recherches Sahariennes, Université d'Alger, Algiers (275 pp.).

Dunn, G. E. and Miller, B. I. (1960) *Atlantic Hurricanes*, Louisiana State University Press, Baton Rouge, La. (326 pp.).

Eldridge, R. H. (1957) A synoptic study of West African disturbance lines, *Quart. J. Roy. Met. Soc.* 83, 303–14.

Fett, R. W. (1964) Aspects of hurricane structure: new model considerations suggested by TIROS and Project Mercury observations, *Monthly Weather Review* 92, 43–59.

Findlater, J. (1971) Mean monthly airflow at low levels over the Western Indian Ocean, *Geophysical Memoirs* 115, Meteorological Office (53 pp.).

Findlater, J. (1974) An extreme wind speed in the low-level jet-stream system of the western Indian Ocean, *Met. Mag.* 103, 201–5.

Flohn, H. (1968) *Contributions to a Meteorology of the Tibetan Highlands*. Atmos. Sci. Paper No. 130, Colorado State University, Fort Collins (120 pp.).

Flohn, H. (1971) Tropical circulation patterns, *Bonn. Geogr. Abhandl.* 15 (55 pp.).

Fosberg, F. R., Garnier, B. J. and Küchler, A. W. (1961) Delimitation of the humid tropics, *Geog. Rev.* 51, 333–47.

Fraedrich, K. (1990) European Grosswetter during the warm and cold extremes of the El Niño/Southern Oscillation, *Int. J. Climatology* 10, 21–31.

Frank, N. L. and Hubert, P. J. (1974) Atlantic tropical systems of 1973, *Monthly Weather Review* 102, 290–5.

Frost, R. and Stephenson, P. H. (1965) Mean streamlines and isotachs at standard pressure levels over the Indian and west Pacific oceans and adjacent land areas, *Geophys. Mem.* 14(109), HMSO, London (24 pp.).

Gao, Y.-X. and Li, C. (1981) Influence of Qinghai-Xizang plateau on seasonal variation of general atmospheric circulation; in *Geoecological and Ecological Studies of Qinghai-Xizang Plateau*, Vol. 2, Science Press, Beijing, 1,477–84.

Garnier, B. J. (1967) *Weather conditions in Nigeria*, Climatological Research Series No. 2, McGill University Press, Montreal (163 pp.).

Gentilli, J. (ed.) (1971) *Climates of Australia and New Zealand*, World Survey of Climatology 13, Elsevier, Amsterdam (405 pp.).

Glantz, M. H., Katz, R. W. and Nicholls, N. (eds) (1990) *Teleconnections Linking Worldwide Climate Anomalies*, Cambridge University Press, Cambridge (535 pp.).

Goudie, A. and Wilkinson, J. (1977) *The Warm Desert Environment*, Cambridge University Press, Cambridge (88 pp.).

Gray, W. M. (1968) Global view of the origin of tropical disturbances and hurricanes, *Monthly Weather Review* 96, 669–700.

Gray, W. M. (1979) Hurricanes: their formation, structure and likely role in the tropical circulation; in Shaw, D. B. (ed.) *Meteorology over the Tropical Oceans*, Royal Meteorological Society, Bracknell, 155–218.

Gray, W. M. (1984) Atlantic seasonal hurricane frequency, *Monthly Weather Review* 112, 1,649–68; 1,669–83.

Gray, W. M. and Jacobson, R. W. (1977) Diurnal variation of deep cumulus convection, *Monthly Weather Review* 105, 1,171–88.

Gray, W. M., Mielke, P. W. and Berry, K. J. (1992) Predicting Atlantic season hurricane activity 6–11 months in advance, *Wea. Forecasting* 7, 440–55.

Gregory, S. (1965) *Rainfall over Sierra Leone*, Geography Department, University of Liverpool, Research Paper No. 2 (58 pp.).

Gregory, S. (1988) El Niño years and the spatial pattern of drought over India, 1901–70; in Gregory, S. (ed.) *Recent Climatic Change*, Belhaven Press, London, 226–36.

Griffiths, J. F. (ed.) (1972) *Climates of Africa*, World Survey of Climatology 10, Elsevier, Amsterdam (604 pp.).

Halpert, M. S. and Ropelewski, C. F. (1992) Surface temperature patterns associated with the Southern Oscillation, *J. Climate* 5, 577–93.

Hamilton, M. G. (1979) *The South Asian Summer Monsoon*, Arnold, Australia (72 pp.).

Hastenrath, S. (1985) *Climate and Circulation of the Tropics*, D. Reidel, Dordrecht (455 pp.).

Hastenrath, S. (1995) Recent advances in tropical climate prediction, *J. Climate* 8(6), 1,519–32.

Hayward, D. F. and Oguntoyinbo, J. S. (1987) *The Climatology of West Africa*, Hutchinson, London (271 pp.).

Houze, R. A., Goetis, S. G., Marks, F. D. and West, A. K. (1981) Winter monsoon convection in the vicinity of North Borneo, *Monthly Weather Review* 109, 591–614.

Houze, R. A. and Hobbs, P. V. (1982) Organizational and structure of precipitating cloud systems, *Adv. Geophys.* 24, 225–315.

Hutchings, J. W. (ed.) (1964) *Proceedings of the Symposium on Tropical Meteorology*, New Zealand Meteorological Service, Wellington (737 pp.).

Indian Meteorological Department (1960) *Monsoons of the World*, Delhi (270 pp.).

Jackson, I. J. (1977) *Climate, Water and Agriculture in the Tropics*, Longman, London (248 pp.).

Jalu, R. (1960) Etude de la situation météorologique au Sahara en Janvier 1958, *Ann. de Géog.* 69(371), 288–96.

Jordan, C. L. (1955) Some features of the rainfall at Guam, *Bull. Amer. Met. Soc.* 36, 446–55.

Kamara, S. I. (1986) The origins and types of rainfall in West Africa, *Weather* 41, 48–56.

Kiladis, G. N. and Diaz, H. F. (1989) Global climatic anomalies associated with extremes of the Southern Oscillation, *J. Climate* 2, 1,069–90.

Knox, R. A. (1987) The Indian Ocean: interaction with the monsoon; in Fein, J. S. and Stephens, P. L. (eds) *Monsoons*, John Wiley & Sons, New York, 365–97.

Koteswaram, P. (1958) The easterly jet stream in the tropics, *Tellus* 10, 45–57.

Kousky, V. E. (1980) Diurnal rainfall variation in northeast Brazil, *Monthly Weather Review* 108, 488–98.

Kreuels, R., Fraedrich, K. and Ruprecht, E. (1975) An aerological climatology of South America, *Met. Rundsch.* 28, 17–24.

Krishna Kumar, K., Soman, M. K. and Rupa Kumar, K. (1996) Seasonal forecasting of Indian summer monsoon rainfall: a review, *Weather* 50(12), 449–67.

Krishnamurti, T. N. (ed.) (1977) Monsoon meteorology, *Pure Appl. Geophys.* 115, 1,087–529.

Kurashima, A. (1968) Studies on the winter and summer monsoons in east Asia based on dynamic concept, *Geophys. Mag.* (Tokyo) 34, 145–236.

Kurihara, Y. (1985) Numerical modeling of tropical cyclones; in Manabe, S. (ed.) *Issues in Atmospheric and Oceanic Modeling. Part B, Weather Dynamics, Advances in Geophysics*, Academic Press, New York, 255–87.

Lander, M. A. (1990) Evolution of the cloud pattern during the formation of tropical cyclone twins symmetrical with respect to the Equator, *Monthly Weather Review* 118, 1,194–202.

Landsea, C. W., Gray, W. M., Mielke, P. W., Jr and Berry, K. J. (1994) Seasonal forecasting of Atlantic hurricane activity, *Weather* 49, 273–84.

Lau, K.-M. and Li, M.-T. (1984) The monsoon of East Asia and its global associations – a survey, *Bull. Amer. Met. Soc.* 65, 114–25.

Le Borgue, J. (1979) Polar invasion into Mauretania and Senegal, *Ann. de Géog.* 88(485), 521–48.

Lighthill, J. and Pearce, R. P. (eds) (1979) *Monsoon Dynamics*, Cambridge University Press, Cambridge (735 pp.).

Lin, Chenyu (1982) The establishment of the summer monsoon over the middle and lower reaches of the Yangtze River and the seasonal transition of circulation over East Asia in early summer. *Proc. Symp. Summer Monsoon South East Asia*, People's Press of Yunnan Province, Kunming, 21–8 (in Chinese).

Lockwood, J. G. (1965) The Indian monsoon – a review, *Weather* 20, 2–8.

Logan, R. F. (1960) *The Central Namib Desert, Southwest Africa*, National Academy of Sciences, National Research Council, Publication 758, Washington, DC (162 pp.).

Lowell, W. E. (1954) Local weather of the Chicama Valley, Peru, *Archiv. Met. Geophys. Biokl.* B 5, 41–51.

Lydolph, P. E. (1957) A comparative analysis of the dry western littorals, *Ann. Assn Amer. Geog.* 47, 213–30.

Maejima, I. (1967) Natural seasons and weather singularities in Japan, *Geog. Report No. 2. Tokyo Metropolitan University*, 77–103.

Maley, J. (1982) Dust, clouds, rain types, and climatic variations in tropical North Africa, *Quaternary Res.* 18, 1–16.

Malkus, J. S. (1955–6) The effects of a large island upon the trade-wind air stream, *Quart. J. Roy. Met. Soc.* 81, 538–50; 82, 235–8.

Malkus, J. S. (1958) Tropical weather disturbances: why do so few become hurricanes? *Weather* 13, 75–89.

Malkus, J. S. and Riehl, H. (1964) *Cloud Structure and Distributions over the Tropical Pacific Ocean*, University of California Press, Berkeley and Los Angeles (229 pp.).

Mason, B. J. (1970) Future developments in meteorology: an outlook to the year 2000, *Quart. J. Roy. Met. Soc.* 96, 349–68.

Matsumoto, J. (1985) Precipitation distribution and frontal zones over East Asia in the summer of 1979, *Bull. Dept Geog., Univ. Tokyo*, 17, 45–61.

Meehl, G. A. (1987) The tropics and their role in the global climate system, *Geog. J.* 153, 21–36.

Mink, J. F. (1960) Distribution pattern of rainfall in the leeward Koolau Mountains, Oahu, Hawaii, *J. Geophys. Res.* 65, 2,869–76.

Mohr, K. and Zipser, E. (1996) Mesoscale convective systems defined by their 85–GHz ice scattering signature: size and intensity comparison over tropical oceans and continents, *Monthly Weather Review*, 124, 2,417–37.

Molion, L. C. B. (1987) On the dynamic climatology of the Amazon Basin and associated rain-producing mechanisms; in Dickinson, R. E. (ed.) *The Geophysiology of Amazonia*, John Wiley & Sons, New York, 391–405.

Musk, L. (1983) Outlook – changeable, *Geog. Mag.* 55, 532–3.

Neal, A. B., Butterworth, L. J. and Murphy, K. M. (1977) The morning glory, *Weather* 32, 176–83.

Nicholson, S. E. (1989) Long-term changes in African rainfall, *Weather* 44, 46–56.

Nicholson, S. E. and Flohn, H. (1980) African environmental and climatic changes and the general atmospheric circulation in late Pleistocene and Holocene, *Climatic Change* 2, 313–48.

Nieuwolt, S. (1977) *Tropical Climatology*, Wiley, London (207 pp.).

NOAA (1992) *Experimental Long-Lead Forecast Bulletin*, NOAA, Washington, DC.

Omotosho, J. B. (1985) The separate contributions of line squalls, thunderstorms and the monsoon to the total rainfall in Nigeria, *J. Climatology* 5, 543–52.

Palmén, E. (1948) On the formation and structure of tropical hurricanes, *Geophysica* 3, 26–38.

Palmer, C. E. (1951) Tropical meteorology; in Malone, T. F. (ed.) *Compendium of Meteorology*, American Meteorological Society, Boston, Mass., 859–80.

Philander, S. G. (1990) *El Niño, La Niña, and the Southern Oscillation*, Academic Press, New York (289 pp.).

Physik, W. L. and Smith, R. K. (1985) Observations and dynamics of sea breezes in northern Australia, *Austral. Met. Mag.* 33, 51–63.

Pogosyan, K. P. and Ugarova, K. F. (1959) The influence of the Central Asian mountain massif on jet streams, *Meteorol. Gidrol.* 11, 16–25 (in Russian).

Quinn, W. H. and Neal, V. T. (1992) The historical record of El Niño; in Bradley, R. S. and Jones, P. D. (eds) *Climate Since A.D. 1500*, Routledge, London, 623–48.

Raghavan, K. (1967) Influence of tropical storms on monsoon rainfall in India, *Weather* 22, 250–5.

Ramage, C. S. (1952) Relationships of general circulation to normal weather over southern Asia and the western Pacific during the cool season, *J. Met.* 9, 403–8.

Ramage, C. S. (1964) Diurnal variation of summer rainfall in Malaya, *J. Trop. Geog.* 19, 62–8.

Ramage, C. S. (1968) Problems of a monsoon ocean, *Weather* 23, 28–36.

Ramage, C. S. (1971) *Monsoon Meteorology*, Academic Press, New York and London (296 pp.).

Ramage, C. S. (1986) El Niño, *Sci. American* 254, 76–83.

Ramage, C. S. (1995) *Forecaster's Guide to Tropical Meteorology*, AWS/TR–95/001, Air Weather Service, Scott Air Force Base, Ill. (392 pp.).

Ramage, C. S., Khalsa, S. J. S. and Meisner, B. N. (1980) The central Pacific near-equatorial convergence zone, *J. Geophys. Res.* 86(7), 6,580–98.

Ramaswamy, C. (1956) On the sub-tropical jet stream and its role in the department of large-scale convection, *Tellus* 8, 26–60.

Ramaswamy, C. (1962) Breaks in the Indian summer monsoon as a phenomenon of interaction between the easterly and the sub-tropical westerly jet streams, *Tellus* 14, 337–49.

Rasmusson, E. M. (1985) El Niño and variations in climate, *Amer. Sci.* 73, 168–77.

Ratisbona, L. R. (1976) The climate of Brazil; in Schwerdtfeger, W. (ed.) *Climates of Central and South America*, World Survey of Climatology 12, Elsevier, Amsterdam, 219–93.

Reynolds, R. (1985) Tropical meteorology, *Prog. Phys. Geog.* 9, 157–86.

Riehl, H. (1954) *Tropical Meteorology*, McGraw-Hill, New York (392 pp.).

Riehl, H. (1963) On the origin and possible modification of hurricanes, *Science* 141, 1,001–10.

Riehl, H. (1979) *Climate and Weather in the Tropics*, Academic Press, New York (611 pp.).

Rodwell, M. J. and Hoskins, B. J. (1996) Monsoons and the dynamics of deserts, *Quart. J. Roy. Met. Soc.* 122, 1,385–404.

Ropelewski, C. F. and Halpert, M. S. (1987) Global and regional scale precipitation patterns associated with the El Niño/Southern Oscillation, *Monthly Weather Review* 115, 1606–25.

Rossignol-Strick, M. (1985) Mediterranean Quaternary sapropels, an immediate response of the African monsoon to variation in isolation, *Palaeogeog., Palaeoclim., Palaeoecol.* 49, 237–63.

Sadler, J. C. (1975a) The monsoon circulation and cloudiness over the GATE area, *Monthly Weather Review* 103, 369–87.

Sadler, J. C. (1975b) *The Upper Tropospheric Circulation over the Global Tropics*, UHMET–75–05, Department of Meteorology, University of Hawaii (35 pp.).

Saha, R. R. (1973) Global distribution of double cloud bands over the tropical oceans, *Quart. J. Roy. Met. Soc.* 99, 551–5.

Saito, R. (1959) The climate of Japan and her meteorological disasters, *Proceedings of the International Geophysical Union*, Regional Conference in Tokyo, Japan, 173–83.

Sawyer, J. S. (1970) Large-scale disturbance of the equatorial atmosphere, *Met. Mag.* 99, 1–9.

Schwerdtfeger, W. (ed.) (1976) *Climates of Central and South America*, World Survey of Climatology 12, Elsevier, Amsterdam (532 pp.).

Shaw, D. B. (ed.) (1978) *Meteorology over the Tropical Oceans*, Royal Meteorological Society, Bracknell (278 pp.).

Sheng, Cheng-yu *et al.* (1986) *General Comments on the Climate of China*, Science Press, Beijing (538 pp.) (in Chinese).

Sikka, D. R. (1977) Some aspects of the life history, structure and movement of monsoon depressions, *Pure and Applied Geophysics* 15, 1,501–29.

Suppiah, R. (1992) The Australian summer monsoon: a review, *Prog. Phys. Geog.* 16(3), 283–318.

Tao, Shi-yan (ed. for the Chinese Geographical Society) (1984) *Physical Geography of China*, Science Press, Beijing (161 pp.) (in Chinese).

Thompson, B. W. (1951) An essay on the general circulation over South-East Asia and the West Pacific, *Quart. J. Roy. Met. Soc.* 569–97.

Trenberth, K. E. (1976) Spatial and temporal oscillations in the Southern Oscillation, *Quart. J. Roy. Met. Soc.* 102, 639–53.

Trenberth, K. E. (1990) General characteristics of El Niño–Southern Oscillation; in Glantz, M. H., Katz, R. W. and Nicholls, N. (eds) *Teleconnections Linking Worldwide Climate Anomalies*, Cambridge University Press, Cambridge.

Trewartha, G. T. (1958) Climate as related to the jet stream in the Orient, *Erdkunde* 12, 205–14.

Trewartha, G. T. (1981) *The Earth's Problem Climates* (2nd edn), University of Wisconsin Press, Madison (371 pp.).

Tyson, P. D. (1986) *Climatic Change and Variability in Southern Africa*, Oxford University Press, Cape Town (220 pp.).

Vincent, D. G. (1994) The South Pacific Convergence Zone (SPCZ): a review, *Monthly Weather Review* **122**(9), 1,949–70.

Watts, I. E. M. (1955) *Equatorial Weather, with Particular Reference to South-east Asia*, Oxford University Press, London (86 pp.).

Webster, P. J. (1987a) The elementary monsoon; in Fein, J. S. and Stephens, P. L. (eds) *Monsoons*, John Wiley & Sons, New York, 3–32.

Webster, P. J. (1987b) The variable and interactive monsoon; in Fein, J. S. and Stephens, P. L. (eds) *Monsoons*, John Wiley & Sons, New York, 269–330.

World Meteorological Organization (1972) Synoptic analysis and forecasting in the tropics of Asia and the south-west Pacific, WMO No. 321, Geneva (524 pp.).

World Meteorological Organization (n.d.) *The Global Climate System. A Critical Review of the Climate System during 1982–1984*, World Climate Data Programme, WMO, Geneva (52 pp.).

Wyrtki, K. (1982) The Southern Oscillation, ocean–atmosphere interaction and El Niño, *Marine Tech. Soc. J.* **16**, 3–10.

Yarnal, B. (1985) Extratropical teleconnections with El Niño/Southern Oscillation (ENSO) events, *Prog. Phys. Geog.* **9**, 315–52.

Ye, D. (1981) Some characteristics of the summer circulation over the Qinghai–Xizang (Tibet) plateau and its neighbourhood, *Bull. Amer. Met. Soc.* **62**, 14–19.

Ye, D. and Gao, Y.-X. (1981) The seasonal variation of the heat source and sink over Qinghai–Xizang plateau and its role in the general circulation; in *Geoecological and Ecological Studies of Qinghai–Xizang Plateau*, Vol. 2, Science Press, Beijing, 1,453–61.

Yoshino, M. M. (1969) Climatological studies on the polar frontal zones and the intertropical convergence zones over South, South-east and East Asia, *Climatol. Notes* **1**, Hosei University (71 pp.).

Yoshino, M. M. (ed.) (1971) *Water Balance of Monsoon Asia*, University of Tokyo Press (308 pp.).

Young, J. A. (co-ordinator) (1972) *Dynamics of the Tropical Atmosphere* (Notes from a Colloquium), National Center for Atmospheric Research, Boulder, Colo. (587 pp.).

Zhang, Jiacheng and Lin, Zhi-quang (1985) *Climate of China*, Science of Technology Press, Shanghai (603 pp.) (in Chinese).

CHAPTER 10: BOUNDARY LAYER CLIMATES

Adebayo, Y. R. (1991) 'Heat island' in a humid tropical city and its relationship with potential evaporation, *Theoret. and App. Climatology* **43**, 137–47.

Anderson, G. E. (1971) Mesoscale influences on wind fields, *J. App. Met.* **10**, 377–86.

Atkinson, B. W. (1968) A preliminary examination of the possible effect of London's urban area on the distrib-ution of thunder rainfall 1951–60, *Trans. Inst. Brit. Geog.* **44**, 97–118.

Atkinson, B. W. (1977) *Urban Effects on Precipitation: An Investigation of London's Influence on the Severe Storm of August 1975*, Dept Geog. Queen Mary Coll. London, Occasional Paper 8 (31 pp.).

Atkinson, B. W. (1987) Precipitation; in Gregory, K. J. and Walling, D. E. (eds) *Human Activity and Environmental Processes*, John Wiley & Sons, Chichester, 31–50.

Bach, W. (1971) Atmospheric turbidity and air pollution in Greater Cincinnati, *Geog. Rev.* **61**, 573–94.

Bach, W. (1979) Short-term climatic alterations caused by human activities, *Prog. Phys. Geog.* **3**(1), 55–83.

Bach, W. and Patterson, W. (1966) Heat budget studies in Greater Cincinnati, *Proc. Assn Amer. Geog.* **1**, 7–16.

Brimblecombe, P. (1986) *Air: Composition and Chemistry*, Cambridge University Press, Cambridge (224 pp.).

Bryson, R. A. and Kutzbach, J. E. (1968) *Air Pollution*, Association of American Geographers, Commission on College Geography, Resource Paper 2 (42 pp.).

Caborn, J. M. (1955) The influence of shelter-belts on microclimate, *Quart. J. Roy. Met. Soc.* **81**, 112–15.

Chandler, T. J. (1965) *The Climate of London*, Hutchinson, London (292 pp.).

Chandler, T. J. (1967) Absolute and relative humidities in towns, *Bull. Amer. Met. Soc.* **48**, 394–9.

Changnon, S. A. (1969) Recent studies of urban effects on precipitation in the United States, *Bull. Amer. Met. Soc.* **50**, 411–21.

Changnon, S. A. (1979) What to do about urban-generated weather and climate changes, *J. Amer. Plan. Assn.* **45**(1), 36–48.

Committee on Air Pollution (1955) *Report*, Cmnd 9322, HMSO, London.

Cotton, W. R. and Pielke, R. A. (1995) *Human Impacts on Weather and Climate*, Cambridge University Press, Cambridge (288 pp.).

Coutts, J. R. H. (1955) Soil temperatures in an afforested area in Aberdeenshire, *Quart. J. Roy. Met. Soc.* **81**, 72–9.

Dickinson, R. E. and Henderson-Sellers, A. (1988) Modelling tropical deforestation: a study of GCM land-surface parameterizations, *Quart. J. Roy. Met. Soc.* **114**, 439–62.

Duckworth, F. S. and Sandberg, J. S. (1954) The effect of cities upon horizontal and vertical temperature gradients, *Bull. Amer. Met. Soc.* **35**, 198–207.

Food and Agriculture Organization of the United Nations (1962) *Forest Influences*, Forestry and Forest Products Studies No. 15, Rome (307 pp.).

Garnett, A. (1967) Some climatological problems in urban geography with special reference to air pollution, *Trans. Inst. Brit. Geog.* **42**, 21–43.

Gay, L. W. and Stewart, J. B. (1974) *Energy balance studies in coniferous forests*, Report No. 23, Inst. Hydrol., Nat. Env. Res. Coun., Wallingford.

Goldreich, Y. (1984) Urban topo-climatology, *Prog. Phys. Geog.* **8**, 336–64.

Harriss, R. C. *et al.* (1990) The Amazon boundary layer experiment: wet season 1987, *J. Geophys. Res.* **95**(D10), 16,721–736.

Heintzenberg, J. (1989) Arctic haze: air pollution in polar regions, *Ambio* **18**, 50–5.

Hewson, E. W. (1951) Atmospheric pollution; in Malone, T. F. (ed.) *Compendium of Meteorology*, American Meteorological Society, Boston, Mass., 1,139–57.

Jäger, J. (1983) *Climate and Energy Systems. A Review of their Interactions*, Wiley, New York (231 pp.).

Jauregui, E. (1987) Urban heat island development in medium and large urban areas in Mexico, *Erdkunde* **41**, 48–51.

Jenkins, I. (1969) Increases in averages of sunshine in Greater London, *Weather* **24**, 52–4.

Kessler, A. (1985) Heat balance climatology; in Essenwanger, B. M. (ed.) *General Climatology*, World Survey of Climatology **1A**, Elsevier, Amsterdam (224 pp.).

Kittredge, J. (1948) *Forest Influences*, McGraw-Hill, New York (394 pp.).

Koppány, Gy. (1975) Estimation of the life span of atmospheric motion systems by means of atmospheric energetics, *Met. Mag.* **104**, 302–6.

Landsberg, H. E. (1981) City climate; in Landsberg, H. E. (ed.) *General Climatology 3*, World Survey of Climatology **3**, Elsevier, Amsterdam, 299–334.

Long, I. F., Monteith, J. L., Penman, H. L. and Szeicz, G. (1964) The plant and its environment, *Meteorolog. Rundschau* **17**(4), 97–101.

Lowry, W. P. (1969) *Weather and Life*, Academic Press, New York (305 pp.).

McNaughton, K. and Black, T. A. (1973) A study of evapotranspiration from a Douglas fir forest using the energy balance approach, *Water Resources Research* **9**, 1,579–90.

Maejima, I. *et al.* (1982) Recent climatic change and urban growth in Tokyo and its environs, *Japanese Prog. Climatology*, March 1983, 1–22.

Marshall, W. A. L. (1952) *A Century of London Weather*, Met. Office, Air Ministry, Rept MO 508, HMSO, London (103 pp.).

Meetham, A. R. *et al.* (1980) *Atmospheric Pollution* (4th edn, 1st edn 1952), Pergamon Press, Oxford and London.

Miess, M. (1979) The climate of cities; in Laurie, I. C. (ed.) *Nature in Cities*, Wiley, Chichester, 91–104.

Miller, D. H. (1965) The heat and water budget of the earth's surface, *Adv. Geophys.* **11**, 175–302.

Nicholas, F. W. and Lewis, J. E. (1980) Relationships between aerodynamic roughness and land use and land cover in Baltimore, Maryland, US Geol. Surv. Prof. Paper 1099–C (36 pp.).

Oke, T. R. (1978) *Boundary Layer Climates*, Methuen, London (372 pp.) (2nd edn 1987, 435 pp.).

Oke, T. R. (1979) *Review of Urban Climatology 1973–76*, WMO Technical Note No. 169, Geneva, World Meteorological Organization (100 pp.).

Oke, T. R. (1980) Climatic impacts of urbanization; in Bach, W., Pankrath, J. and Williams, J. (eds) *Interactions of Energy and Climate*, D. Reidel, Dordrecht, 339–56.

Oke, T. R. (1982) The energetic basis of the heat island, *Quart. J. Roy. Met. Soc.* **108**, 1–24.

Oke, T. R. (1986) *Urban Climatology and its Applications with Special Regard to Tropical Areas*, World Meteorological Organization Publication No. 652, Geneva (534 pp.).

Oke, T. R. (1988) The urban energy balance, *Prog. Phys. Geog.* **12**(4), 471–508.

Oke, T. R. and East, C. (1971) The urban boundary layer in Montreal, *Boundary-Layer Met.* **1**, 411–37.

Pankrath, J. (1980) Impact of heat emissions in the Upper-Rhine region; in Bach, W., Pankrath, J. and Williams, J. (eds) *Interactions of Energy and Climate*, D. Reidel, Dordrecht, 363–81.

Parry, M. (1966) The urban 'heat island'; in Tromp, S. W. and Weite, W. H. (eds) *Biometeorology* 2, Pergamon, Oxford and London, 616–24.

Pease, R. W., Jenner, C. B. and Lewis, J. E. (1980) The influences of land use and land cover on climate analysis: an analysis of the Washington–Baltimore area, US Geol. Surv. Prof. Paper 1099–A (39 pp.).

Peel, R. F. (1974) Insolation and weathering: some measures of diurnal temperature changes in exposed rocks in the Tibesti region, central Sahara, *Zeit. für Geomorph. Supp.* **21**, 19–28.

Peterson, J. T. (1971) Climate of the city; in Detwyler, T. R. (ed.) *Man's Impact on Environment*, McGraw-Hill, New York, 131–54.

Plate, E. (1972) Berücksichtigung von Windströmungen in der Bauleitplanung; in *Seminarberichte Rahmenthema Unweltschutz*, Institut für Stätebau und Landesplanung, Selbstverlag, Karlsruhe, 201–29.

Reynolds, E. R. C. and Leyton, L. (1963) Measurement and significance of throughfall in forest stands; in Whitehead, F. M. and Rutter, A. J. (eds) *The Water Relations of Plants*, Blackwell Scientific Publications, Oxford, 127–41.

Richards, P. W. (1952) *The Tropical Rain Forest*, Cambridge University Press, Cambridge (450 pp.).

Rutter, A. J. (1967) Evaporation in forests, *Endeavour* **97**, 39–43.

Scorer, R. (1968) *Air Pollution*, Pergamon, Oxford and London (151 pp.).

Seinfeld, J. H. (1989) Urban air pollution: state of the science, *Science* **243**, 745–52.

Sellers, W. D. (1965) *Physical Climatology*, University of Chicago Press, Chicago (272 pp.).

Shuttleworth, W. J. *et al.* (1985) Daily variation of temperature and humidity within and above the Amazonian forest, *Weather* **40**, 102–8.

Shuttleworth, W. J. (1989) Micrometeorology of temperate and tropical forest, *Phil. Trans. Roy. Soc. London* **B324**, 299–334.

Sopper, W. E. and Lull, H. W. (eds) (1967) *International Symposium on Forest Hydrology*, Pergamon, Oxford and London (813 pp.).

Stearn, A. C. (ed.) (1968) *Atmospheric Pollution* (3 vols), Academic, New York.

Sukachev, V. and Dylis, N. (1968) *Fundamentals of Forest Biogeocoenology*, Oliver and Boyd, Edinburgh (672 pp.).

Terjung, W. H. (1970) Urban energy balance climatology, *Geog. Rev.* **60**, 31–53.

Terjung, W. H. and Louis, S. S.-F. (1973) Solar radiation and urban heat islands, *Ann. Assn Amer. Geog.* **63**, 181–207.

Terjung, W. H. and O'Rourke, P. A. (1980) Simulating the causal elements of urban heat islands, *Boundary-Layer Met.* **19**, 93–118.

Terjung, W. H. and O'Rourke, P. A. (1981) Energy input and resultant surface temperatures for individual urban interfaces, *Archiv. Met. Geophys. Biokl.* B, **29**, 1–22.

Turner, W. C. (1955) Atmospheric pollution, *Weather* **10**, 10–19.

Tyson, P. D., Garstang, M. and Emmitt, G. D. (1973) *The Structure of Heat Islands*, Occasional Paper No. 12, Dept. of Geography and Environmental Studies, University of the Witwatersrand, Johannesburg (71 pp.).

US Department of Health, Education and Welfare (1970) *Air Quality Criteria for Photochemical Oxidants*, National Air Pollution Control Administration, US Public Health Service, Publication No. AP–63, Washington, DC.

Vehrencamp, J. E. (1953) Experimental investigation of heat transfer at an air–earth interface, *Trans. Amer. Geophys. Union* **34**, 22–30.

Weller, G. and Wendler, G. (1990) Energy budgets over various types of terrain in polar regions, *Ann. Glac.* **14**, 311–14.

White, W. H., Anderson, J. A., Blumenthal, D. L., Husar, R. B., Gillani, N. V., Husar, J. D. and Wilson, W. E. (1976) Formation and transport of secondary air pollutants: ozone and aerosols in the St Louis urban plume, *Science* **194**, 187–9.

World Meteorological Organization (1970) *Urban Climates*, WMO Technical Note No. 108 (390 pp.).

Zon, R. (1941) Climate and the nation's forests; in US Dept of Agriculture Yearbook, *Climate and Man*, 477–98.

CHAPTER 11: CLIMATIC CHANGE

Adger, W. N. and Brown, K. (1994) *Land Use and the Causes of Global Warming*, John Wiley & Sons, Chichester (271 pp.).

Bayce, A. (ed.) (1979) *Man's Influence on Climate*, D. Reidel, Dordrecht (113 pp.).

Bolin, B., Döös, Bo R., Jäger, J. and Warrick, R. A. (eds) (1986) *The Greenhouse Effect, Climatic Change, and Ecosystems*, John Wiley & Sons, Chichester (541 pp.).

Bradley, R. S. (1985) *Quaternary Paleoclimatology*, Allen & Unwin, Boston (472 pp.).

Bradley, R. S. (ed.) (1990) *Global Changes of the Past*, University Corporation for Atmospheric Research, Boulder, Colo. (514 pp.).

Bradley, R. S. and Jones, P. D. (eds) (1992) *Climate Since A.D. 1500*, Routledge, London (679 pp.).

Brádzil, R., Samaj, F. and Valovic, S. (1985) Variation of spatial annual precipitation sums in central Europe in the period 1881–1980, *J. Climatology* **5**, 617–31.

Broecker, W. S. and Denton, G. S. (1990) What drives glacial cycles? *Sci. American* **262**, 48–56.

Broecker, W. S. and Van Donk, J. (1970) Insolation changes, ice volumes and the O^{18} record in deep sea cores, *Rev. Geophys.* **8**, 169–96.

Bruce, J. P., Lee, H. and Haites, E. F. (eds) (1996) *Climate Change 1995: Economic and Social Dimensions of Climate Change*, Intergovernmental Panel on Climate Change (IPCC), Cambridge University Press, Cambridge (448 pp.).

Chu, P.-S., Yu, Z.-P. and Hastenrath, S. (1994) Detecting climate change concurrent with deforestation in the Amazon Basin, *Bull. Amer. Met. Soc.* **75**(4), 579–83.

Crowley, T. J. and North, G. R. (1991) *Paleoclimatology*, Oxford University Press, Oxford (339 pp.).

Davidson, G. (1992) Icy prospects for a warmer world, *New Scientist* **135**(1,833), 23–6.

Diaz, H. F. and Kiladis, G. N. (1995) Climatic variability on decadal to century time-scales; in Henderson-Sellers, A. (ed.) *Future Climates of the World: A Modelling Perspective*, World Surveys of Climatology, **16**, Elsevier, Amsterdam, 191–244.

Frakes, L. A. (1979) *Climates throughout Geologic Time*, Elsevier, Amsterdam (310 pp.).

French, J. R., Spencer, T. and Reed, D. J. (1995) Editorial – Geomorphic response to sea-level rise: Existing evidence and future impacts, *Earth Surface Processes and Landforms* **20**, 1–6.

Gilliland, R. L. (1982) Solar, volcanic and CO_2 forcing of recent climatic change, *Climatic Change* **4**, 111–31.

Goodess, C. M., Palutikof, J. P. and Davies, T. D. (eds) (1992) *The Nature and Causes of Climate Change: Assessing the Long-term Future*, Belhaven Press, London (248 pp.).

Goudie, A. (1981) *The Human Impact: Man's Role in Environmental Change*, Blackwell, Oxford (316 pp.).

Gregory, S. (1969) Rainfall reliability; in Thomas, M. F. and Whittington, G. W. (eds) *Environment and Land Use in Africa*, Methuen, London, 57–82.

Gribbin, J. (ed.) (1978) *Climatic Change*, Cambridge University Press, Cambridge (280 pp.).

Grootes, P. (1995) Ice cores as archives of decade-to-century scale climate variability; in *Natural Climate Variability on Decade-to-Century Time Scales*, National Academy Press, Washington, DC, 544–54.

Grove, J. M. (1988) *The Little Ice Age*, Methuen, London (498 pp.).

Haeberli, W. (1995) Glacier fluctuations and climate change detection, *Geogr. Fis. Dinam. Quat.* **18**, 191–9.

Hansen, J., Fung, I., Lacis, A., Rind, D., Lebedeff, S., Ruedy, R. and Russell, G. (1988) Global climatic changes as forecast by Goddard Institute for Space Studies three-dimensional model, *J. Geophys. Res.* **93**(D8), 9,341–64.

Hansen, J. E. and Lacis, A. A. (1990) Sun and dust versus greenhouse gases: an assessment of their relative roles in global climate change, *Nature* **346**, 713–19.

Hare, F. K. (1979) Climatic variation and variability: empirical evidence from meteorological and other sources; in *Proceedings of the World Climate Conference*, WMO Publication No. 537, WMO, Geneva, 51–87.

Henderson-Sellers, A. and McGuffie, K. (1987) *A Climate Modelling Primer*, John Wiley & Sons, Chichester (217 pp.).

Henderson-Sellers, A. and Wilson, M. F. (1983) Surface albedo data for climate modelling, *Rev. Geophys. Space Phys.* **21**, 1,743–8.

Houghton, J. T., Callander, B. A. and Varney, S. K. (eds) (1992) *Climate Change 1992: The Supplementary Report to the IPCC Scientific Assessment*, Cambridge University Press, Cambridge (200 pp.).

Houghton, J. T., Jenkins, G. J. and Ephraums, J. J. (eds) (1990) *Climate Change: The IPCC Scientific Assessment*, Cambridge University Press, Cambridge (365 pp.).

Houghton, J. T., Meira Filho, L. G., Callander, B. A., Harris, N., Kattenberg A. and Maskell, K. (eds) (1996) *Climate Change 1995: The Science of Climate Change*, Cambridge University Press, Cambridge (572 pp.).

Huggett, R. J. (1993) *Modelling the Human Impact on Nature*, Oxford University Press, Oxford (202 pp.).

Hughes, M. K. and Diaz, H. F. (eds) (1994) *The Medieval Warm Period*, Kluwer Academic Publishers, Dordrecht (342 pp.).

Hughes, M. K., Kelly, P. M., Pilcher, J. R. and La Marche, V. (eds) (1981) *Climate from Tree Rings*, Cambridge University Press, Cambridge (400 pp.).

Hulme, M. (1992) Rainfall changes in Africa: 1931–60 to 1961–90, *Int. J. Climatol.* **12**, 685–99.

Hulme, M. (1994) Global warming, *Prog. Phys. Geog.* **18**, 401–10.

Imbrie, J. and Imbrie, K. P. (1979) *Ice Ages: Solving the Mystery*, Macmillan, London (224 pp.).

Jäger, J. and Barry, R. G. (1991) Climate; in Turner, B. L. II (ed.) *The Earth as Transformed by Human Actions*, Cambridge University Press, Cambridge, 335–51.

Jones, P. D. and Bradley, R. S. (1992) Climatic variations in the longest instrumental records; in Bradley, R. S. and Jones, P. D. (eds) *Climate Since A.D. 1500*, Routledge, London, 246–68.

Jones, P. D., Wigley, T. M. L. and Farmer, G. (1991) Marine and land temperature data sets: a comparison and a look at recent trends; in Schlesinger, M. E. (ed.) *Greenhouse-Gas-Induced Climatic Change*, Elsevier, Amsterdam, 153–72.

Kelly, P. M. (1980) Climate: historical perspective and climatic trends; in Doornkamp, J. C. and Gregory, K. J. (eds) *Atlas of Drought in Britain, 1975–6*, Institute of British Geographers, London, 9–11.

Kutzbach, J. E. and Street-Perrott, A. (1985) Milankovitch forcings of fluctuations in the level of tropical lakes from 18 to 0 kyr BP, *Nature* **317**, 130–9.

Lamb, H. H. (1966) *The Changing Climate: Selected Papers*, Methuen, London (236 pp.).

Lamb, H. H. (1970) Volcanic dust in the atmosphere; with a chronology and an assessment of its meteorological significance, *Phil. Trans. Roy. Soc.* A **266**, 425–533.

Lamb, H. H. (1977) *Climate: Present, Past and Future, 2: Climatic History and the Future*, Methuen, London (835 pp.).

Lamb, H. H. (1994) British Isles daily wind and weather patterns 1588, 1781–86, 1972–1991 and shorter early sequences, *Climate Monitor* **20**, 47–71.

McCormac, B. M. and Seliga, T. A. (1979) *Solar-Terrestrial Influences on Weather and Climate*, D. Reidel, Dordrecht (340 pp.).

Manley, G. (1958) Temperature trends in England, 1698–957, *Archiv. Met. Geophy. Biokl.* (Vienna) B **9**, 413–33.

Mann, M. E., Park, J., and Bradley, R. S. (1995) Global interdecadal and century-scale climate oscillations during the past five centuries, *Nature* **378**(6,554), 266–70.

Mather, J. R. and Sdasyuk, G. V. (eds) (1991) *Global Change: Geographical Approaches* (Sections 3.2.2, 3.2.3), University of Arizona Press, Tucson.

Meehl, G. A. and Washington, W. M. (1990) CO_2 climate sensitivity and snow–sea-ice albedo parameterization in an atmospheric GCM coupled to a mixed-layer ocean model, *Climatic Change* **16**, 283–306.

Mitchell, J. F. B., Johns, T. C., Gregory, J. M. and Tett, S. F. B. (1995) Climate response to increasing levels of greenhouse gases and sulphate aerosols, *Nature* **376**, 501–4.

Mitchell, J. M., Jr (ed.) (1968) Causes of climatic change, *Amer. Met. Soc. Monogr.* **8**(30) (159 pp.).

Mitchell, J. M., Jr (1972) The natural breakdown of the present interglacial and its possible intervention by human activities, *Quat. Res.* **2**, 436–45.

Nicholson, S. E. (1980) The nature of rainfall fluctuations in subtropical West Africa, *Monthly Weather Review* **108**, 473–87.

Nicholson, S. E. (1985) Sub-Saharan rainfall 1981–84, *J. Applied Met.* **24**, 1,388–91.

Parker, D. E., Horton, E. B., Cullum, D. P. N. and Folland, C. K. (1996) Global and regional climate in 1995, *Weather* **51**(6), 202–10.

Penner, J. E. *et al.* (1994) Quantifying and minimizing uncertainty of climate forcing by anthropogenic aerosols, *Bull. Amer. Met. Soc.* **75**(3), 375–400.

Pfister, C. (1985) Snow cover, snow lines and glaciers in central Europe since the 16th century; in Tooley, M. J. and Sheail, G. M. (eds) *The Climatic Scene*, Allen & Unwin, London, 154–74.

Pittock, A. B., Frakes, L. A., Jenssen, D., Peterson, J. A. and Zillman, J. W. (eds) (1978) *Climatic Change and Variability: A Southern Perspective*, Cambridge University Press, Cambridge (455 pp.).

Schimel, D. and 26 lead authors (1996) Radiative forcing of climate change; in Houghton, J., Meira Filho, L. G., Callander, B. A., Harris, N., Kattenberg, A. and Maskell, K. (eds) *Climate Change 1995. The Science of Climate Change*, Cambridge University Press, Cambridge, 65–131.

Schuurmans, C. J. E. (1984) Climate variability and its time changes in European countries, based on instrumental observations; in Flohn, H. and Fantechi, R. (eds) *The Climate of Europe: Past, Present and Future*, D. Reidel, Dordrecht, 65–101.

Sewell, W. R. D. (ed.) (1966) *Human Dimensions of Weather Modification*, University of Chicago, Dept of Geography, Research Paper No. 105 (423 pp.).

Sioli, H. (1985) The effects of deforestation in Amazonia, *Geog. J.* **151**, 197–203.

Sokolik, I. N. and Toon, B. (1996) Direct radiative forcing by anthropogenic airborne mineral aerosols, *Nature* **381**, 501–4.

Stark, P. (1994) Climatic warming in the central Antarctic Peninsula area, *Weather* **49**(6), 215–20.

Stolarski, R. S. (1988) Changes in ozone over the Antarctic; in Roland, F. S. and Isaksen, I. S. A. (eds) *The Changing Atmosphere*, John Wiley & Sons, Chichester, 105–99.

Street, F. A. (1981) Tropical palaeoenvironments, *Prog. Phys. Geog.* **5**, 157–85.

Study of Man's Impact on Climate (SMIC) (1971) *Inadvertent Climate Modification*, Massachusetts Institute of Technology Press, Cambridge, Mass. (308 pp.).

Thompson, R. D. (1989) Short-term climatic change: evidence, causes, environmental consequences and strategies for action, *Prog. Phys. Geog.* **13**(3), 315–47.

Tooley, M. J. and Shennan, I. (eds) (1987) Sea-level changes, *Trans. Inst. Brit. Geog. Sp. Pub.* 20 (397 pp.).

Trenberth, K. E. (ed.) (1992) *Climate System Modeling*, Cambridge University Press, Cambridge (788 pp.).

Weber, G.R. (1995) Seasonal and regional variations of tropospheric temperatures in the Northern Hemisphere, 1976–1990, *Int. J. Climatol.* **15**, 259–74.

Wigley, M. L. and Raper, S. C. B. (1992) Implications for climate and sea level of revised IPCC emissions scenarios, *Nature* **357**, 293–300.

Wigley, T. M. L., Ingram, M. J. and Farmer, G. (eds) (1981) *Climate and History*, Cambridge University Press, Cambridge (530 pp.).

Williams, J. (ed.) (1978) *Carbon Dioxide, Climate and Society*, Pergamon, Oxford (332 pp.).

Zinyowera, M. C., Watson, R. T. and Moss, R. H. (eds) (1996) *Climate Change 1995: Impacts, Adaptations and Mitigation of Climate Change: Scientific–Technical Analysis*, Intergovernmental Panel on Climate Change (IPCC), Cambridge University Press, Cambridge (878 pp.).

Index

References to figures and tables in *italics*